International Series in Operations Research & Management Science

Volume 321

The book series **International Series in Operations Research and Management Science** encompasses the various areas of operations research and management science. Both theoretical and applied books are included. It describes current advances anywhere in the world that are at the cutting edge of the field. The series is aimed especially at researchers, advanced graduate students, and sophisticated practitioners.

The series features three types of books:

- Advanced expository books that extend and unify our understanding of particular areas.
- Research monographs that make substantial contributions to knowledge.
- Handbooks that define the new state of the art in particular areas. Each handbook will be edited by a leading authority in the area who will organize a team of experts on various aspects of the topic to write individual chapters. A handbook may emphasize expository surveys or completely new advances (either research or applications) or a combination of both.

The series emphasizes the following four areas:

Mathematical Programming: Including linear programming, integer programming, nonlinear programming, interior point methods, game theory, network optimization models, combinatorics, equilibrium programming, complementarity theory, multiobjective optimization, dynamic programming, stochastic programming, complexity theory, etc.

Applied Probability: Including queuing theory, simulation, renewal theory, Brownian motion and diffusion processes, decision analysis, Markov decision processes, reliability theory, forecasting, other stochastic processes motivated by applications, etc.

Production and Operations Management: Including inventory theory, production scheduling, capacity planning, facility location, supply chain management, distribution systems, materials requirements planning, just-in-time systems, flexible manufacturing systems, design of production lines, logistical planning, strategic issues, etc.

Applications of Operations Research and Management Science: Including telecommunications, health care, capital budgeting and finance, economics, marketing, public policy, military operations research, humanitarian relief and disaster mitigation, service operations, transportation systems, etc.

This book series is indexed in Scopus.

Adiel Teixeira de Almeida • Love Ekenberg •
Philip Scarf • Enrico Zio • Ming J. Zuo
Editors

Multicriteria and Optimization Models for Risk, Reliability, and Maintenance Decision Analysis

Recent Advances

Springer

Editors
Adiel Teixeira de Almeida
Department of Production Engineering
Federal University of Pernambuco
Recife, Pernambuco, Brazil

Love Ekenberg
Department of Computer and Systems Sciences
Stockholm University
Stockholm, Sweden

International Institute for Applied Systems Analysis
IIASA, Laxenburg, Austria

Philip Scarf
Cardiff Business School
Cardiff University
Cardiff, United Kingdom

Enrico Zio
Dipartimento di Energia
Polytechnic University of Milan
Milano, Italy

Ming J. Zuo
Department of Mechanical Engineering
University of Alberta
Edmonton, AB, Canada

ISSN 0884-8289 ISSN 2214-7934 (electronic)
International Series in Operations Research & Management Science
ISBN 978-3-030-89646-1 ISBN 978-3-030-89647-8 (eBook)
https://doi.org/10.1007/978-3-030-89647-8

This Springer imprint is published by the registered company Springer Nature Switzerland AG
The registered company address is: Gewerbestrasse 11, 6330 Cham, Switzerland

Contents

Part I
New Developments on Building MCDM/A Models

Multicriteria Decision Methods for RRM Models

Eduarda Asfora Frej and Adiel Teixeira de Almeida

1 Introduction

Complex decision-making situations involving several conflicting objectives are always present in practical real-life situations related to a very wide range of subjects, including RRM. Multiple criteria decision-making/aiding (MCDM/A) is a useful approach for tackling these problems. Nevertheless, building decision models for dealing with such situations while taking into account the complex variables involved is definitely not an easy job. Moreover, choosing a structured method for aiding such processes is also a challenging task. When characteristics of the available methodologies are known, however, it becomes easier to choose the most suitable approach for dealing with each specific situation.

In this context, this chapter presents an overview of various approaches for dealing with multiple criteria decision-making. In Sect. 2, a general view of issues involved in MCDM/A modeling is given. Then, Section 3 introduces and discusses a broad range of various MCDM/A methods, each of which takes a different perspective and has its own specific characteristics and peculiarities.

This section is divided into three main topics: first, additive aggregation methods within the scope of Multi-Attribute Utility/Value Theory (MAVT/MAUT) are presented. These deal with situations in which DMs have a compensatory rationality and are willing to perform trade-off analyses; secondly, outranking methods – i.e., methods that handle situations in which a non-compensatory rationality is more suitable for representing DMs' aspirations – are presented; and the third subsection gives an overview of other MCDM based on mathematical programming. Finally, in

E. A. Frej (✉) · A. T. de Almeida
CDSID – Center for Decision Systems and Information Development, Universidade Federal de Pernambuco, Recife, PE, Brazil
e-mail: eafrej@cdsid.org.br; almeida@cdsid.org.br

A. T. de Almeida et al. (eds.), *Multicriteria and Optimization Models for Risk, Reliability, and Maintenance Decision Analysis*, International Series in Operations Research & Management Science 321, https://doi.org/10.1007/978-3-030-89647-8_1

Sect. 4, new approaches and challenges for dealing with MCDM/A are presented. These address concepts related to decision-making for which there is only partial information and new trends for decision-making under uncertainty.

2 MCDM/A Models

Multiple criteria decision-making/aiding (MCDM/A) research addresses situations for modeling and solving decision problems in which multiple criteria (attributes) are involved (Roy 1996; Polmerol and Barba-Romero 2000; Belton and Stewart 2002; Figueira et al. 2005). With roots in Operational Research (OR), MCDM/A embraces different kinds of support within decision situations: decision-making (MCDM), decision aiding, or decision analysis (MCDA) (De Almeida et al. 2015). Several MCDM/A methods for modeling a DM's preferences have been developed. These methods generally seek to analyze the context in which the DM is involved and therefore evaluate promising decision alternatives by considering the conflicting objectives involved (Roy 1996; Belton and Stewart 2002). The construction of rational and efficient mathematical models for structuring such decision situations helps a DM to have a better understanding of the problem as a whole, and therefore of the various complex factors involved (Vincke 1992).

MCDM/A differs from classical OR methods mainly due to there being a DM who has preferences and aspirations with respect to the problematic situation. DMs' preference structure may be modeled in different ways in accordance with the method chosen for doing so. Some methods are based on rigorous axioms for characterizing an individual's behavior (Keeney and Raiffa 1976; Roy 1996; Belton and Stewart 2002; Figueira et al. 2005). Also, when preferences are imprecise, problem situations can be handled by fuzzy MCDM (Pedrycz et al. 2011).

There are different ways for classifying multicriteria methods. De Almeida et al. (2015) divide them according to the rationality that a DM uses to evaluate different criteria. This rationality may be compensatory or non-compensatory. DMs who have a compensatory rationality are willing to let a lower performance in some criterion be compensated for by a higher performance in another criterion, i.e., he/she performs tradeoffs among different criteria. In such models, criteria weights act as scaling constants, since an additive aggregation is performed (Keeney and Raiffa 1976; Belton and Stewart 2002). On the other hand, DMs whose rationality is non-compensatory do not allow this compensation between criteria, and therefore weights are related only to the relative importance of the criteria (Roy 1996; Vincke 1992). This classification based on rationality is important for choosing how DM's preference modeling will be performed, since there are methods that should be applied that are appropriate for a given type of rationality.

Another way into which MCDM/A methods are classified (Roy 1996; Vincke 1992; Belton and Stewart 2002; Pardalos et al., 2013) considers three kinds of methods: interactive methods, which are based on mathematical programming; outranking methods and unique criterion of synthesis methods, which usually aggregate criteria additively. Additive aggregation models are used when a compensatory rationality is applied. The two best-known and widely used axiomatically founded theories that underpin this approach are: MAUT – Multi-Attribute Utility Theory, for probabilistic situations and MAVT – Multi-Attribute Value Theory, when consequences are considered to be deterministic (Keeney and Raiffa 1976; Belton and Stewart 2002).

Examples of MCDM methods based on additive aggregation include MACBETH (Measuring Attractiveness by a Categorical Based Evaluation Technique), UTA (in French, UTilités Additives, rendered in English as Additive Utilities) methods, and SMARTS (Simple Multi-Attribute Rating Technique with Swing). As to dealing with non-compensatory rationality situations, however, outranking methods such as ELECTRE (Elimination and Choice Reflecting Reality, in translation from French) and PROMETHEE (Preference Ranking Organization Methods for Enrichment Evaluations) are more suitable. These approaches work based on an outranking preference relation between alternatives, in such a way that incomparability may arise.

Interactive methods are suitable for both continuous and discrete decision variables. Multi-objective linear programming (MOLP) approaches (Karasakal and Köksalan 2009; Allmendinger et al. 2017) are widely used. In interactive methods, DMs can revise their preferences during the process, since, in these methods, preference information steps are followed by computational steps throughout many interactions. Multi-Objective Combinatorial Optimization (MOCO) methods can also be applied when integer variables are involved in the model (Ehrgott and Gandibleux 2002).

Other possible ways to deal with MCDM/A problems are to use Decision Rules (Figueira et al. 2005) or a partial information approach (Weber 1987). Decision Rules are applied with the assumption that the individual gives preferences in the form of decision examples, and therefore simple rules that justify this decision are looked for. Partial information approaches are suitable when a DM does not have a well-defined preference structure and therefore a recommendation is built based on the incomplete preferential information provided. This approach is discussed in more detail in Sect. 4.

As can be seen from the above, a wide range of methodologies is available for aiding multicriteria decision situations. It is emphasized, however, that there is neither a best nor a worst method, but rather that there are methods that are more suitable than others for specific situations. In the following sections, these MCDM/A approaches are presented and discussed.

3 Multicriteria Decision Methods

In this section, several MCDM/A methods are described. This is done in order to give a brief overview of tools for aiding DMs in practical decision problems. First of all, additive aggregation methods – for both deterministic and probabilistic situations – are presented. These deal with compensatory rationality. Then, outranking methods for dealing with non-compensatory situations are described. Finally, an overview of some other approaches is also presented.

3.1 Additive Aggregation Methods for Deterministic Situations

Multi-attribute value theory (MAVT) embraces situations of deterministic consequences which characterize the decision alternatives in situations with multiple objectives (Belton and Stewart 2002). Also, a compensatory rationality is assumed from the DM, i.e., the assumption is made that he/she is willing to perform tradeoffs among criteria. In these models, alternatives are scored based on an additive aggregation function as shown in Eq. (1):

$$v\left(a_i\right)=\sum_{j=1}^{n} k_j v_j\left(x_{ij}\right) \quad (1)$$

In Eq. (1), $v(a_i)$ is the global value assigned to alternative a_i, which is calculated as a weighted sum of the scaling constants of the criteria k_j and the intracriteria value function $v_j(x_{ij})$, which represents the evaluation of consequence x_{ij} (performance of alternative a_i in criterion j). Criteria scaling constants have to be normalized in order to sum to 1 – see Eq. (2)

$$\sum_{j=1}^{n} k_j = 1 \quad (2)$$

In order to apply the additive model in Eq. (1), an important condition has to be verified: preferential independence between criteria (De Almeida et al. 2015), i.e., the direction of preference in a certain criterion should not change if the values of consequences in other criteria are modified. This means that preferences regarding the criteria of the problem do not depend on the consequences of other criteria.

Another important issue related to applying additive models concerns the meaning of criteria scaling constants k_j. This is to do not only with the level of importance of the criteria, but a scaling issue is also involved. This means that the range of consequences has to be considered in order to evaluate these values, since they act as substitution rates in additive models (Keeney and Raiffa 1976). This is why the determination of weights in additive models is not an easy task, and why there are specified procedures for doing so, which take account of a DM's preferences. The best-known procedures for eliciting criteria scaling constants are the tradeoff procedure (Keeney and Raiffa 1976) and the swing procedure (Von Winterfeldt and Edwards 1986), although there are some others (Weber and Borcherding 1993).

The traditional tradeoff procedure is structured based on a sequence of questions with regard to comparing consequences, in which the DM has to consider tradeoffs between different criteria. Based on these tradeoffs established by the DM, indifference relations between consequences are obtained, so that equations involving the scaling constants of the criteria can be built. With $n - 1$ (number of criteria – 1) equations obtained from tradeoff questions, and including also Eq. (2), it is possible to solve an equation system and therefore to find the values of criteria weights. Then, additive aggregation is performed based on Eq. (1) and alternatives are ranked according to their global value. This procedure has a strong axiomatic foundation based on MAVT concepts; nevertheless, it is criticized due to DMs needing to make a high amount of cognitive effort during the elicitation process in order to obtain indifference relations. This leads to a rate of inconsistencies of around 67% when MAUT is applied, according to behavioral studies (Weber and Borcherding 1993). The tradeoff elicitation procedure is used by the Flexible and Interactive Tradeoff method (FITradeoff) (De Almeida et al. 2016), but using only partial information about DM's preferences, and in such a way that the cognitive effort that the DM needs to make is much reduced. More details about partial information methods are given and further discussed in Sect. 4.

The swing weights procedure is also based on a sequence of questions to the DM, who first has to rank criteria based on the following reasoning: "imagine a hypothetical alternative with all criteria in the worst outcome, and it can be improved by turning the performance of one unique criterion into the best possible value; which criterion would you choose?" The criterion selected will be the one with the highest scaling constant value. Then, the same question is asked again, and so the DM chooses the criterion with the second-highest scaling constant value, and so on. Then, the DM considers the first criterion has a score of 100, and scores for the subsequent criteria are established based on a comparative analysis. Finally, these scores are normalized in order to find the values of the scaling constants. This procedure is used in SMARTS (Simple Multi-Attribute Rating Technique with Swing) method (Edwards and Barron 1994). The SMARTER (Simple Multi-Attribute Rating Technique Exploiting Ranks) method (Edwards and Barron 1994) also incorporates the swing weights approach but uses only its first part (ranking criteria weights), and then surrogate weights are calculated. The SMARTER method is another example of a method that works based only on partial information from DMs. Even though the swing procedure is considered easier to implement when compared to the tradeoff procedure, there is nevertheless a 50% rate of inconsistencies rate associated with this procedure (Weber and Borcherding 1993).

There are other MCDM/A methods that use additive aggregation models. The AHP (Analytic Hierarchy Process) method considers preferential information obtained from pairwise comparison between alternatives in each criterion, and the criteria of the problem are considered by following a hierarchical structure (Saaty 1980). The MACBETH (Measuring Attractiveness by a Categorical Based Evaluation Technique) method evaluated DM's preferences in accordance with the difference in attractiveness between alternatives. A qualitative scale is used for this, but it is then converted into a quantitative one that is based on linear programming models (Bana e Costa and Vansnick 1994). Finally, the Additive Veto Model approach considers limits for consequences under which they become unacceptable, and therefore some alternatives may be vetoed according to this rule (de Almeida 2013).

Additive aggregation models for MCDM/A are also applied when consequences are determined based on a probabilistic scenario, i.e., the probabilities of different states of nature occurring are considered. These cases are dealt with based on concepts of multi-attribute utility theory (MAUT), which is detailed in the following section.

3.2 *Additive Aggregation Methods for Probabilistic Situations*

This topic also leads to discussion of multicriteria additive models, but for probabilistic situations. First, a brief overview of Expected Utility Theory and its axioms is given. Then, Multi-Attribute Utility Theory is presented.

3.2.1 Expected Utility Theory

Expected Utility (EU) theory (von Neumann and Morgenstern 1944) is widely applied in order to aid decision-making situations under uncertainty. This approach is considered as a normative rational model for decision-making (Keeney and Raiffa 1976). The Expected Utility Theory model is characterized by:

- E: a set of n possible states of nature, which are exhaustive and mutually excluding events. $E = (E_1, E_2, \ldots, E_n)$.
- p: Subjective probability distribution $p = (p_1, p_2, \ldots, p_n)$, in which p_i is the probability of the occurrence of the state of nature E_i.
- X: consequences vector $X = (x_1, x_2, \ldots, x_n)$, in which x_i is the consequence related to the state of nature E_i.
- u: utility function, which represents a measure of benefit of a certain consequence for the DM.

Given these elements, the expected utility of a prospect X is calculated as follows Eq. (3):

$$EU(X) = \sum_{i=1}^{n} p_i u(x_i) \tag{3}$$

Therefore, Expected Utility Theory model incorporates not only preference modeling translated from utility functions, but there is also a probabilistic modeling task, which may require another actor to be involved in the decision process: an expert with knowledge about the probabilistic behavior of the state of nature, in order to determine the subjective probabilities.

Expected Utility Theory has a strong axiomatic structure, which contains the following axioms (Savage 1954):

A1: Preferences between alternatives are weakly ordered, i.e., they are complete and transitive.

A2: Preferences satisfy the independence condition, which means that choices between two prospects should not be affected by the consequence values corresponding to states for which both prospects have the same value. This leads to a natural definition of conditional preferences: if X is preferred to Y given an event E, then X is preferred to Y for a not-E event.

A3: Preferences among consequences are state independent, which means that conditional preferences between constant prospects (i.e., prospects that have the same consequence for all states) do not depend on the conditioning event.

A4: Events can be ordered by probability, in the following manner: let x_1 and x_2 be two consequences such that x_1 is preferred to x_2, if the prospect that leads to x_1 with probability p', and x_2 with probability $1 - p'$ is preferred to the prospect that leads to x_1 with probability p'' and x_2 with probability $1 - p''$, then $p' > p''$.

The utility function $u(x)$ is defined in such a way that $u(x^0) = 0$ and $u(x^*) = 1$, in which x^0 is the least desired outcome and x^* is the most desired outcome, and a wide set of procedures is available for eliciting preferences in order to obtain $u(x_i)$ (Raiffa 1968; Keeney and Raiffa 1976; Berger 1985). Most of them use the idea of a certain equivalent of lotteries, which is the indifference point between a lottery and a certain value, according to the DM's preferences. For instance, let us assess the utility function for money considering the range from \$0 to \$100. From the definition of the utility function, that $u(0) = 0$ and $u(100) = 1$. Now, intermediate points have to be found in order to obtain a good estimate of the form of $u(x)$. If the DM states, for instance, that he/she is indifferent between a lottery in which he/she may gain \$0 with 0.5 probability or \$100 with 0.5 probability and winning a certain value of \$45 for sure, then it can be inferred that, for this DM, $u(45) = 0.5$. By obtaining more indifference points like this, an estimate of the DM's utility function can be made. More detailed procedures can be found in Keeney and Raiffa (1976).

3.2.2 Multi-Attribute Utility Theory

The fundamental insight for Multi-Attribute Utility Theory (MAUT) was presented by Howard Raiffa in 1968, with the idea of extending concepts of Utility Theory and keeping its axiomatic structure for the multi-attribute case, i.e., when two or more criteria are involved in the decision problem. Similar to what happens in MAVT, criteria are also aggregated in MAUT, based on an additive function; but the main difference is that this is now done by considering probabilistic scenarios.

To apply MAUT, it is necessary to obtain a multi-attribute utility function. For instance, for a case of two attributes, x and z, consequences are evaluated based on the multi-attribute value function $u(x, z)$. This is a function of the marginal utility functions of the attributes; $u(x, z) = f[u(x), u(z)]$. The elicitation of the multi-attribute utility functions depends, therefore, on the values of the marginal utility functions of each attribute being known.

Keeney and Raiffa (1976) also present procedures for eliciting the multi-attribute utility (MAU) function. Usually the MAU function is the additive form $u(x, z) = k_x u(x) + k_z\, u(z)$, in which k_x and k_z are the scaling constants of the attributes x and z, normalized as in Eq. (2). This model can be generalized for cases with c criteria. Therefore, analogously to the preferential independence condition in MAVT – the necessary condition for using the additive model, in MAUT there are also two key independence conditions to be satisfied so as to make it possible to apply the additive model: additive independence and utility independence (for details, see Keeney and Raiffa 1976). If both the mutual independence condition and the additive independence condition between attributes are satisfied according to the DM's preferences, then the additive model can be applied. If these conditions do not hold, then other types of MAU function can be applied.

Given the analytical form of the MAU function, then it is necessary to define the scaling constants of the attributes. As to the additive model, scaling constants for a two-attribute problem are given by $k_x = u(x^*, z^0)$ and $k_z = u(x^0, z^*)$ (Keeney and Raiffa 1976). Thus, in order to obtain the value of k_x, the probability p which makes (x^*, z^0) the certain equivalent of the lottery $[(x^*, z^*), p; (x^0, z^0), 1 - p]$ must be found. The value of k_z can be obtained analogously.

3.3 *Outranking Methods*

Outranking methods are suitable for dealing with situations in which the DM has a non-compensatory rationality. These methods are based on an outranking preference relation. An alternative *a* is said to outrank another alternative *b* (*aSb*) if *a* is considered to be at least as good as *b*. These outranking relations allow incomparability between alternatives to arise (Roy 1996). Therefore, partial preorders of alternatives may be obtained when applying these methods, while in MAUT/MAVT methods a complete preorder is built. Another important issue that differs outranking methods from additive models in MAUT/MAVT is that, in outranking methods, criteria weights are considered based strictly on their level of importance compared to other criteria, without considering ranges of consequences or scaling factors.

Pairwise comparisons between all alternatives of the MCDM problem are performed in outranking methods. These methods work based on two main steps (Roy 1996; Vincke 1992): first, outranking relations are built considering these pairwise comparisons between alternatives; then, these outranking relations are explored in order to compute a recommendation for the DM, according to some algorithm (specified depending on the method applied).

There are two main families of outranking methods: ELECTRE (*Elimination Et Choix Traduisant la Réalité*) methods (Roy 1996; Vincke 1992) and PROMETHEE (Preference Ranking Organization Method for Enrichment Evaluation) methods (Brans and Vincke 1985; Vincke 1992). These are described in detail in the following subsections.

3.3.1 ELECTRE Methods

The family of ELECTRE methods embraces six different approaches, which differ one from another according to the decision problematic – choice, ranking, sorting, or portfolio (Roy 1996) – and to the type of criteria – true criteria (without thresholds) or pseudocriteria (with concordance/discordance thresholds):

- ELECTRE I: considers true criteria, and is suitable for the choice problematic;
- ELECTRE IS: considers pseudocriteria, and is suitable for the choice problematic;
- ELECTRE II: considers true criteria, and is suitable for the ranking problematic;
- ELECTRE III: considers pseudocriteria and is suitable for the ranking problematic;
- ELECTRE IV: considers pseudocriteria and is suitable for the ranking problematic;
- ELECTRE TRI: considers pseudocriteria and is suitable for the sorting problematic.

In order to illustrate how these methods work, let us consider the ELECTRE I method, which is suitable for the choice problematic with true criteria. Other ELECTRE methods differ from ELECTRE I in the construction phase of the outranking relation and differ even more in the exploitation phase, since the kind of recommendation is different.

The ELECTRE I method constructs outranking relations based on concordance and discordance notions. The concept of concordance is related to the fact that there is a meaningful subset of criteria that indicates the favorability for an outranking relation between two alternatives. The discordance concept is related to a notion of veto of this concordance. To evaluate the possibility of there being an outranking relation, it is necessary to define both the concordance and the discordance indexes. The concordance index between two alternatives a and b, $C(a, b)$, is calculated as follows:

$$C(a, b) = \sum_{j:g_j(a)\geq g_j(b)} w_j \tag{4}$$

In Eq. (4), w_j indicates the weight of criterion j, and the sum of these weights is equal to 1. The performance of alternative a in criterion j is denoted here by $g_j(a)$, and $g_j(b)$ indicates the performance of alternative b in criterion j. Therefore, the concordance index of an outranking relation between a and b is given by the sum of the weights of the criteria for which the performance of a is greater than the performance of b. To calculate the discordance index, different authors propose different ways to do so (Roy 1996; Vincke 1992; Belton and Stewart 2002). Equation 5 presents a possible formulation for calculating the discordance index between two alternatives a and b, $D(a, b)$:

$$D\left(a,b\right)=\max\left(\frac{g_j(b)-g_j(a)}{\max\left(g_j(c)-g_j(d)\right)}\right),\forall j \text{ such that } g_j(b)>(a);\quad \forall j,c,d \tag{5}$$

In order to define the outranking relations, two thresholds are defined: the concordance threshold $c^{'}$ and the discordance threshold $d^{'}$. These thresholds should be defined by the DM. An outranking relation between a and b is defined as follows:

$$aSb \text{ if and only if } \begin{cases} C\left(a,b\right)\geq c^{'}, \\ D\left(a,b\right)\leq d^{'} \end{cases} \tag{6}$$

This condition is tested for all pairs of alternatives of the decision problem. Two alternatives may be considered indifferent if the situation aSb and bSa happens. After these outranking relations are defined, it is time to perform the second step of the method, namely, to exploit these outranking relations in order to build a recommendation. This is done in ELECTRE I by searching for the kernel of the problem. Alternatives belong to the kernel if they are not outranked by any other alternative from the kernel. It may happen that the kernel is formed by more than one alternative; in this case, the recommendation for the DM is two (or more) incomparable alternatives that can be solutions for the MCDM choice problem.

Further details about the ELECTRE family of methods can be found in Roy (1996), Vincke (1992), Belton and Stewart (2002), and Figueira et al. (2005).

3.3.2 PROMETHEE Methods

*The family of PROMETHEE (*Preference Ranking Organization Method for Enrichment Evaluation) *methods work based on a valued outranking relation (*Brans and Vincke 1985; Vincke 1992)*, unlike the ELECTRE methods. Outranking relations are obtained based on the so-called outranking degrees between two alternatives, which are calculated as follows:*

$$\pi\left(a,b\right)=\sum_{j=1}^{n}w_jF_j\left(a,b\right) \tag{7}$$

In Eq. (7), w_j indicates the weight of criterion j, and $F_j(a,b)$ is the difference function whose value depends on the difference $g_j(a)-g_j(b)$. There are six different forms for $F_j(a,b)$. In its most basic form, a no thresholds definition is necessary, and $F_j(a,b)$ is equal to 1 if $g_j(a)>g_j(b)$, and 0 otherwise. Thus, the outranking degree is given by the sum of weights for criteria in which a is better than b. Other ways for obtaining $F_j(a,b)$ depend on how the DM defines the preference and indifference thresholds (Brans and Vincke 1985).

After calculating the outranking degrees, positive and negative flows for each alternative ($\phi^{+} e\ \phi^{-}$) are calculated for each alternative a. The positive flow of an alternative a indicates a measure of the advantage of a with respect to the other alternatives, while the negative outflow indicates a measure of the disadvantage of a compared to the other alternatives. The positive flow of an alternative a is calculated as shown in Eq. (8), and the negative flow is calculated as shown in Eq. (9).

$$\phi^{+}(a) = \sum_{b \in A} \pi\ (a, b) \tag{8}$$

$$\phi^{-}(a) = \sum_{b \in A} \pi\ (b, a) \tag{9}$$

The PROMETHEE I method uses these two flows to define outranking relations between all pairs of alternatives. It may happen that two alternatives are indifferent or even incomparable to each other, which leads to a partial preorder as a result of this method.

The PROMETHEE II method, on the other hand, works based on a net flow measure, which is also calculated for each alternative a, according to Eq. (10).

$$\phi(a) = \phi^{+}(a) - \phi^{-}(a) \tag{10}$$

This net flow leads to a score for each alternative, which enables one to build a complete preorder of the alternatives. There is no incomparability in the PROMETHEE II method.

There are other methods within the PROMETHEE family. PROMETHEE III and IV are suitable for stochastic situations, while the PROMETHEE V method has been developed for dealing with the portfolio problematic. Finally, the PROMETHEE VI method can be used when criteria weights are given from DMs in the form of ranges instead of exact values.

3.4 Other MCDM/A Methods

Besides the approaches presented in previous subsections, there are still other methods for aiding multicriteria decision situations, which can be applied in diverse contexts.

When DMs' preferences are imprecise, the concepts of fuzzy sets are suitable for application (Pedrycz et al. 2011). The fuzzy approach is not an MCDM method itself, but it is a tool for dealing with imprecise preferences that can be jointly applied with any method (Belton and Stewart 2002).

Disaggregation methods belong to another approach within MCDM, which is based on holistic judgments for evaluating alternatives (Pardalos et al. 2013;

Jacquet-Lagréze and Siskos 1982). A similar approach is used by Slowinski et al. (2012) in their preference learning approach.

Mathematical programming tools are also widely used for aiding multicriteria situations. Although nonlinear programming techniques are also applied, the use of linear programming is quite common, within the Multi-objective Linear Optimization (MOLP) approach (Korhonen 2009; Korhonen 2005; Korhonen and Wallenius 2010; Steuer 1986; Ehrgott 2006; Miettinen 1999; Coello et al. 2007). MOLP methods can be classified according to the way in which preferences are given by DM (a posteriori, a priori, or based on an interactive procedure).

Finally, another approach for tackling problems that have multiple objectives is to use Multi-objective Evolutionary Algorithms (MOEAs), such as: Multi-Objective Generic Algorithm – MOGA (Fonseca and Fleming 1993), Nondominated Sorting Genetic Algorithm – NSGA (Srinivas and Deb 1994), Strength Pareto Evolutionary Algorithm – SPEA (Zitzler and Thiele 1999), and NSGA II (Deb et al. 2002).

4 Challenges and Future Developments in MCDM/A for RRM

This section presents new approaches for dealing with decision-making situations, which have been put forward in response to the drawbacks of some classical decision methods that have been described earlier in this chapter. First, possibilities for dealing with incomplete preference information from the DM are discussed. Then, with regard to RRM stochastic situations, two models are presented: Rank-Dependent Utility (RDU) and Prospect Theory (PT).

4.1 Partial Information Methods in MCDM/A

In deterministic additive aggregation models within the scope of Multi-attribute Value Theory (MAVT), alternatives are scored straightforwardly with a value being given by Eq. 1. The global value of an alternative is given by a weighted sum of criteria scaling constants k_j and the respective value functions of each criterion. However, there is a great challenge related to this model, namely how to define criteria scaling constants, since these parameters do not represent only a degree of importance, but there is also a scaling factor involved.

Traditional approaches for eliciting criteria scaling constants such as the tradeoff and the swing weights procedures require DMs to provide information that they find demands a high cognitive effort to produce, such as indifference points between consequences, as previously explained in Sect. 3.1. DMs may not be able to provide the detailed information required, in a consistent way (Belton and Stewart 2002).

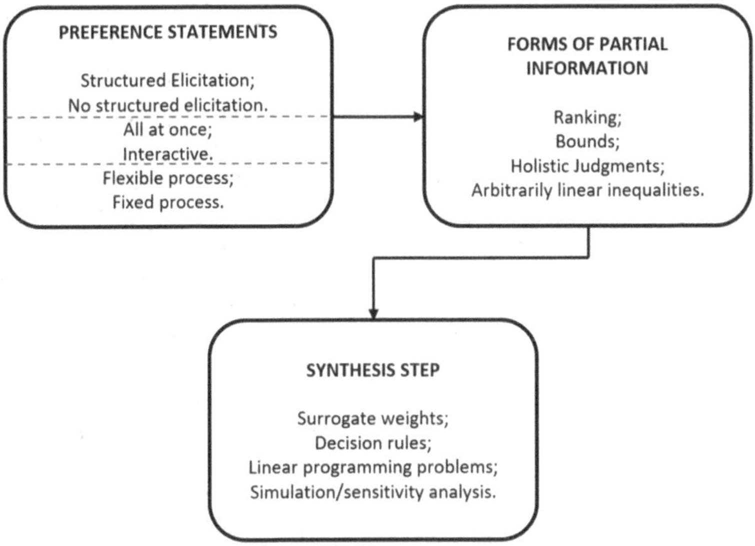

Fig. 1 Framework for classifying partial information methods (Adapted from De Almeida et al. 2016)

These issues may discourage people from engaging on such processes, which are tedious and time-consuming (Salo and Hämäläinen 1992).

In this context, a new trend has emerged in multicriteria decision-making: methods that consider only partial/incomplete information about the DMs' preferences. These approaches aim to facilitate the decision process, by making it easier for DMs to provide the information required. The main goal is to reduce the gap between methodological tools for decision-making and decision situations in practice, by having a process that DMs find cognitively easier to engage on.

Several methods have been developed within this line. De Almeida et al. (2016) present a framework for classifying such methods, based on the following characteristics: preference statements; forms of partial information, and synthesis step. Figure 1 illustrates this framework.

The first box in Fig. 1 deals with how DMs provide preference statements during the elicitation process. First, the analysis examines whether or not there is a structured elicitation process for eliciting preferences. An MCDM method can be considered to have a structured elicitation process if the way in which preferences are gathered from DMs follows a structured elicitation procedure, such as *swing weights* procedure or *tradeoff* procedure. It is simpler to apply the swing weights procedure, and that is why several partial information methods use this procedure to structure elicitation (e.g., Edwards and Barron 1994; Malakooti 2000; Salo and Hämäläinen 2001; Mustajóki et al. 2005). The strongest limitation of this procedure is that it is suitable only for linear intracriteria value functions, and, in practice, preferences can vary in a nonlinear way with the performance of the criteria – this is what Edwards and Barron (1994) call modeling error. The tradeoff procedure, on the other hand, is suitable also for nonlinear cases and has a strong axiomatic structure; however, it is more difficult for DMs because performing tradeoffs among criteria is definitely not an easy task. But the FITradeoff method presented by De Almeida et al. (2016) keeps the whole axiomatic structure of the traditional tradeoff, but with easier questions for the DMs, requiring only partial information, and is given in the form of preference relations between consequences, instead of indifference relations. Finally, there are also methods that do not explicitly assume that there is a formal elicitation procedure for gathering preferences, which are classified as having a non-structured elicitation process (see Kirkwood and Sarin 1985; Park et al. 1997; Dias and Clímaco 2000; Mármol et al. 2002; Punkka and Salo 2013; Mateos et al. 2014).

Still in the first box of Fig. 1, partial information can be classified as interactive or not interactive. In interactive processes, the elicitation is conducted interactively with DMs, and as he/she provides preference statements, partial results are computed. In interactive processes, people learn more about their own preferences during the elicitation. Examples of partial information approaches that are conducted based on interactive processes can be found in Salo and Hämäläinen (1992, 1995); Park and Kim (1997); Malakooti (2000); Dias and Clímaco (2000); Salo and Punkka (2005); De Almeida et al. (2016). On the other hand, methods in which the information of the DM is given all at once, i.e., when there is not an interactive process, can be found in Kirkwood and Sarin 1985; Park et al. 1997; Ahn et al. (2000); Mustajóki et al. (2005); Punkka and Salo (2013); and Danielson et al. (2014).

The last classification in the first box of Fig. 1 is that of determining whether or not the process is flexible. A method can be considered flexible when a wide set of possibilities is given to the DMs during the process. Visualization of partial results and different possibilities of providing information and graphical tools are examples of flexibility features that may characterize a flexible process. In general, methods operated by decision support systems are more likely to have flexibility features that DMs find useful. Examples of flexible methods can be found in Park and Kim (1997); Malakooti (2000); Dias and Clímaco (2000); Salo and Hämäläinen (2001); Sarabando and Dias (2010); Punkka and Salo (2013); Montiel and Bickel (2014), and De Almeida et al. (2016). Processes which do not have such flexibility features are considered to be fixed processes, in which the DM has a unique sequence of steps to follow in order to obtain a recommendation (e.g., Kirkwood and Sarin 1985; Salo and Hämäläinen 1992; Edwards and Barron 1994; Kim and Ahn 1999; Ahn et al. 2000; Mármol et al. 2002; Mustajóki et al. 2005; Jiménez et al. 2013; Mateos et al. 2014).

The second box of Fig. 1 concerns the form of partial information given by DMs, which may be: a ranking of criteria scaling constants; the boundaries of criteria scaling constants; holistic judgments, or even arbitrary linear inequalities involving criteria scaling constants and intracriteria value functions. Most partial information methods use more than a unique type of partial information from DMs. For instance, Malakooti (2000), Ahn and Park (2008), and Montiel and Bickel (2014) consider these four types of information in their approaches. Several methods work with rankings, boundaries, and arbitrary linear inequalities, without considering holistic judgments (e.g., Athanassopoulos and Podinovski 1997; Park and Kim 1997; Park et al. 1997; Kim and Han 2000; Dias and Clímaco 2000; Park 2004; Punkka and Salo 2013). Some other methods work based on rankings and boundaries only, without using either holistic judgments or arbitrary inequalities for intracriteria value functions (Salo and Punkka 2005; Danielson et al. 2014; de Almeida et al. 2016). Finally, there are also approaches that set out to ask the DM for the least possible information, and the only information provided concerns the ranking of criteria scaling constants (Stillwell et al. 1981; Kirkwood and Sarin 1985; Edwards and Barron 1994; Danielson and Ekenberg 2017).

Based on the partial information provided in the previous step, a recommendation must now be built for the DM. Therefore, a synthesis step to somehow compile all information obtained is conducted. This step may be performed in different ways, such as calculating surrogate weights, using decision rules to directly provide a recommendation, computing linear programming (LP) models, or even running simulations and/or sensitivity analysis. All these approaches can also be conjointly applied. Methods using surrogate weights can be found in Stillwell et al. (1981); Edwards and Barron (1994); Kim and Ahn (1999); Sarabando and Dias (2010); Danielson et al. (2014); and Danielson and Ekenberg (2017). Decision rules approaches are adopted by Park and Kim (1997); Park et al. (1997); Salo and Hämäläinen (2001); Salo and Punkka (2005); and Sarabando and Dias (2010). Simulation and sensitivity analysis are found in the studies by Salo and Hämäläinen (1992, 2001); and Montiel and Bickel (2014). However, the step approach to synthesis that is most applied is linear programming, since partial information is generally given in the form of linear inequalities (Kirkwood and Sarin 1985; Salo and Hämäläinen 1992, 1995; Park and Kim 1997; Malakooti 2000; Dias and Clímaco 2000; Salo and Punkka 2005; Ahn and Park 2008; Punkka and Salo 2013; De Almeida et al. 2016; Frej et al. 2019).

All these approaches mentioned above have been developed with the aim of addressing the challenge of how to make the decision-making process an easier task for DMs. Reducing cognitive effort and time spent on decision processes is necessary for real world cases, since DMs do not have much time and are not willing to engage on such cognitively demanding processes. Therefore, partial information approaches are intended to narrow greatly this huge gap between methodologically complex MCDM/A models developed in theory and real situations in practice.

4.2 *Decision Under Uncertainty: Rank-Dependent Utility and Prospect Theory*

Regarding decision situations under uncertainty, the assumption that a rational human being should behave according to the Expected Utility Axioms presented in Sect. 3.2.1 was adopted for many years; however, various behavioral studies show that people often violate these axioms when expressing their preferences in decision-making situations. Therefore, when, in the 1970s, violation of these axioms became a frequent finding of research and behavioral studies within this field, new theories started to be developed in order to relax these axioms and reflect DMs' preference structure in a more realistic way.

One of the strongest objections against the expected utility model emerged from a study by Allais (Allais 1953), who pointed to the following paradox. Let us assume three states of nature (E_1, E_2, E_3), whose probabilities are, approximately, $p_1 = 0.89$ (very likely), $p_2 = 0.1$ (rather unlikely), and $p_3 = 0.01$ (very unlikely), respectively. Now, a person has to choose between two possible prospects, X or Y; prospect X leads to a gain of 1 million dollars for sure, no matter which state happens, and prospect Y leads to a gain of 1 million dollars if E_1 happens, 5 million dollars if E_2 happens and no gain if E_3 happens. These prospects are represented in Table 1.

Experimental studies show that most individuals prefer X over Y. The logic behind this pattern is clear: most people prefer to win $1 million for sure rather than accepting a very low probability (1%) of winning nothing and a 10% chance of winning $5 million. Now, let us assume the consequence value for E_1 changes from $1 M to $0 M, which leads to two new prospects, X′ and Y′, as shown in Table 2.

Experimental studies show that most individuals now prefer Y′ over X′. The logic behind this pattern is also intuitive: since it is very likely that the person will not win anything ($0 M with probability 89%), it is better for the person to risk winning $5 M with 10% probability (Y′) rather than $1 M with 11% probability (X′).

This preference pattern (X P Y and Y′ P X′) is followed by many rational people; however, Expected Utility Theory is not able to rationalize this preference pattern, because it leads to a violation to the independence axiom (***A2***). In this case, X and Y have the same consequence value for E_1 ($1 M). Therefore, for any common value of consequence in E_1, prospect X should still be preferred to prospect Y. This is not what happens in this case; when state E_1 leads to a common consequence of $0 M, the preference changes (Y′ becomes preferred to X′). In this situation, it was possible to observe the "common consequence effect", which happens when preference direction is reversed due to a change in the common consequence value.

The so-called *Allais Paradox* has strongly motivated the development of new approaches that relax the axioms of Expected Utility Theory, such as Rank-

Table 1 Prospects X and Y

	E_1 ($p_1 = 0.89$)	E_2 ($p_2 = 0.1$)	E_3 ($p_3 = 0.01$)
X	$1 M	$1 M	$1 M
Y	$1 M	$5 M	$0 M

Table 2 Prospects X′ and Y′

	E_1 ($p_1 = 0.89$)	E_2 ($p_2 = 0.1$)	E_3 ($p_3 = 0.01$)
X′	$0 M	$1 M	$1 M
Y′	$0 M	$5 M	$0 M

Dependent Utility (RDU) and Prospect Theory (PT), which are described in detail in the sub-sections that follow below.

4.2.1 Rank-Dependent Utility (RDU)

The Rank-Dependent Utility is an extension of the Expected Utility model that has emerged as a widely studied alternative given the Allais paradox and other findings regarding the violation of axioms of EU theory.

RDU was mainly motivated by two key observations considering the violation of the independence axiom (Edwards et al. 2007). First, it was observed that the independence axiom may be violated when comparing two prospects in which the risk involved is radically changed when the value of the common consequence changes. The Allais paradox, for instance, illustrates a case in which \$1 M is replaced by \$0 M in state E_1; therefore, a safe alternative (winning \$1 M for sure) is replaced by a risky alternative (89% chance of winning \$0), and that is why there is a change in preferences. It would be easier to satisfy the independence condition if it was applied only to prospects with similar risk profiles. The second observation is that DMs perception of risk comes from their beliefs, which are represented by probabilities in the expected utility model. However, the independence axiom allows the expected utility to vary with probabilities only in a linear way; therefore, this axiom could be relaxed in such a way that the DM's evaluation of a prospect could be nonlinear with probabilities, and, analogously, nonlinear with the payoffs as well.

In RDU model, the utility of a prospect X is given by Eq. (11):

$$RDU(X) = \sum_{i=1}^{n} \pi_i u(x_i) \tag{11}$$

The coefficients π_i are called decision weights. These values are non-negative and sum 1, exactly in the same way as subjective probabilities in the expected utility model. The main difference between the EU and RDU models is that decision weights are not necessarily subjective probabilities, and the decision weight associated with some state is not necessarily the same for all prospects. The decision weight π_i associated with state E_i when prospect X is evaluated depends on how state E_i is ranked relative to other states in terms of payoffs, according to the DM's beliefs.

In the most general version of RDU, a probability-weighted function $w(p)$ is considered. This is a monotonically increasing function which satisfies the following conditions: $w(0) = 0$ and $w(1) = 1$. For a prospect X, the payoffs corresponding to each state should be ordered in a descending way, such that $x_1 \geq x_2 \geq \ldots \geq x_n$, and $p_1, p_2, \ldots p_n$ are the respective probabilities. Thus, the decision weights in the expression of RDU are given by Eq. (12):

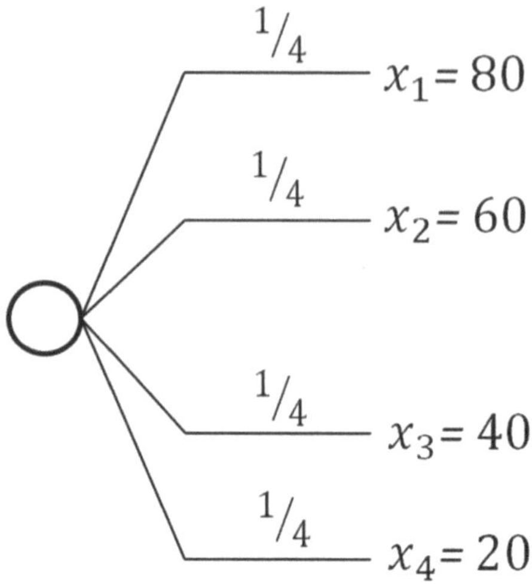

Fig. 2 Prospect X

$$\pi_i = \begin{cases} w\left(p_1\right), \text{if i} = 1 \\ w\left(p_1 + \cdots + p_i\right) - w\left(p_1 + \cdots + p_{i-1}\right), \text{if i} > 1 \end{cases} \tag{12}$$

From the expression above, it can be said that $\pi_1 + \ldots + \pi_i = w(p_1 + \ldots + p_i)$, i.e., the cumulative decision weight associated with the *i* best payoffs is equal to the cumulative probability transformed by the probability-weighted function of those states. If the probability weight function is linear, then the RDU model is reduced to the expected utility model with $\pi_i = p_i$. Nevertheless, if $w(p)$ is nonlinear, then the DM's behavior under risk situations is either optimistic or pessimistic. If $w(p)$ is a convex function, then the DM's behavior under risk is pessimistic, so much so that better payoffs are underestimated, i.e., their decision weights are greater than the probabilities of the respective states. On the other hand, a concave $w(p)$ function characterizes an optimistic DM, who overvalues better payoffs and underestimates worse payoffs.

In order to illustrate the ideas described above, let us consider a prospect X illustrated below in Fig. 2 (Wakker 2010). X presents the same probability (25%) for all possible payoffs: $x_1 = 80$; $x_2 = 60$; $x_3 = 40$; $x_4 = 20$. These payoffs can be considered to be in monetary values, and thus represent someone's monetary gain.

The first step now is to rank these payoffs in a descending order of preference. Since we are considering these values as monetary amounts, then these consequences are ranked as follows:

$$x_1 > x_2 > x_3 > x_4$$

Based on this ranking, decision weights π_i are now calculated as follows:

$$\boldsymbol{\pi_1} = w\left(p_1\right) = \boldsymbol{w}\left({}^{\mathbf{1}}/_{\mathbf{4}}\right)$$

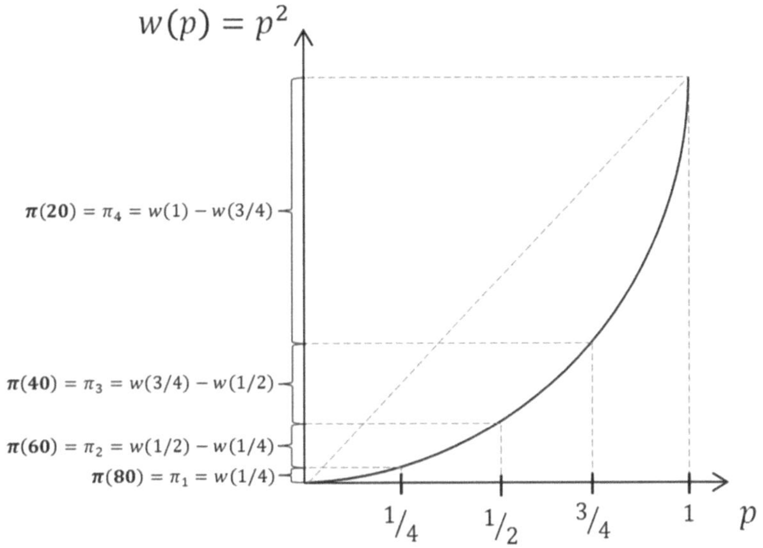

Fig. 3 Pessimistic DM (convex function). (Adapted from Wakker 2010)

$$\pi_2 = w(p_1 + p_2) - w(p_1) = w\left({}^1/_4 + {}^1/_4\right) - w\left({}^1/_4\right) = \mathbf{w}\left({}^{\mathbf{1}}/_{\mathbf{2}}\right) - \mathbf{w}\left({}^{\mathbf{1}}/_{\mathbf{4}}\right)$$

$$\begin{aligned}\pi_3 = w(p_1 + p_2 + p_3) - w(p_1 + p_2) &= w\left({}^1/_4 + {}^1/_4 + {}^1/_4\right)\\ &-w\left({}^1/_4 + {}^1/_4\right) = \mathbf{w}\left({}^{\mathbf{3}}/_{\mathbf{4}}\right) - \mathbf{w}\left({}^{\mathbf{1}}/_{\mathbf{2}}\right)\end{aligned}$$

$$\begin{aligned}\pi_4 = w(p_1 + p_2 + p_3 + p_4) - w(p_1 + p_2 + p_3) &= w\left({}^1/_4 + {}^1/_4 + {}^1/_4 + {}^1/_4\right)\\ &-w\left({}^1/_4 + {}^1/_4 + {}^1/_4\right) = \mathbf{w}(\mathbf{1}) - \mathbf{w}\left({}^{\mathbf{3}}/_{\mathbf{4}}\right)\end{aligned}$$

Now, let us consider two different profiles for characterizing DMs. The first profile is characterized by a person who is pessimistic, and has a convex probability weighting function $w(p) = p^2$, illustrated by Fig. 3. The second profile is characterized by a person who is optimistic person, and has a concave probability weighting function $w(p) = \sqrt{p}$, illustrated by Fig. 4.

Based on the graphics shown above, it can be seen that, for the pessimistic profile, the decision weights of better payoffs are underestimated, and the decision weights of worse payoffs are overestimated. For instance, the decision weight for payoff $x_1 = 80$ is much lower than the decision weight for $x_4 = 20$. For the optimistic

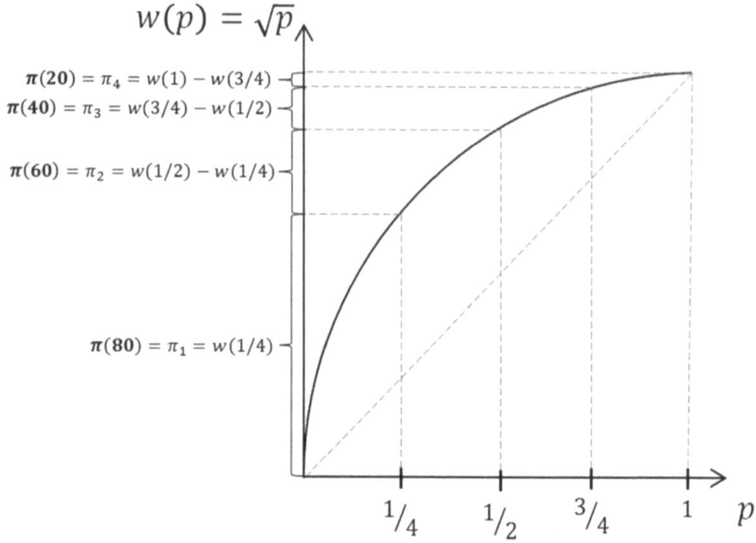

Fig. 4 Optimistic DM (concave function). (Adapted from Wakker 2010)

profile, the opposite situation happens: it can be seen that the decision weight for $x_1 = 80$ is much higher than the decision weight for $x_4 = 20$.

Now, let us go back to the situation of the Allais paradox – the main motivation for the development of RDU. A pessimistic DM (with a convex function $w(p)$) would overestimate the weight of state E_3 when comparing X and Y, since it leads to a payoff of \$0 M with 1% probability; on the other hand, the weighting effect would be much less expressive when comparing X′ and Y′, since the probability of winning \$0 M is much higher.

The key assumption that distinguishes the RDU model from the expected utility model is that the independence axiom is replaced with a weaker version of this condition: the *co-monotonic independence axiom*. Two prospects X and Y are considered co-monotonic if their states are ordered in the same way, such that the same state leads to the best payoff in both prospects, the same state leads to the second-best payoff in both prospects, and so on. The co-monotonic independence axiom states that when two co-monotonic prospects have the same consequence for a given state, it does not matter which value of consequence this has, because pairs of co-monotonic actions have the same risk profile, so changing the values of equal payoffs cannot change the risk profile of an action without changing the risk profile for the other action in a similar way. The situation of the Allais paradox, therefore, does not violate the co-monotonic independence axiom, because pairs of actions are not co-monotonic: in X, we have the following preference order for states $E_1 = E_2 = E_3$, and in Y the order is $E_2 > E_1 > E_3$.

The idea of using a nonlinear function for weighting probabilities was explored in depth in behavioral studies, and many studies have worked on the idea of estimating

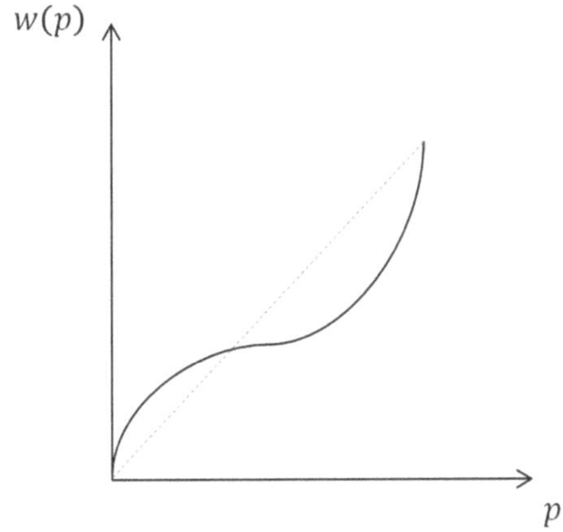

Fig. 5 Common finding for probability weighting function shape

the weighting probability function in an empirical way. A common finding is that most individuals behave according to an inverted S-curve, i.e., the function is concave for very low values of cumulative probabilities and convex for medium-high values of cumulative probabilities. Figure 5 shows this pattern. This pattern indicates that most DMs are risk prone for large gains with very low probabilities.

The approach for RDU described above is the most general form of explaining individuals' behavior that cannot be explained by Expected Utility Theory. Within this line of research, advanced studies began to emerge and new theories were developed in order to refine this model to make it even more compatible with human behavior, such as Prospect Theory (PT), which is described in the next sub-section.

4.2.2 Prospect Theory

In their work, Kahneman and Tversky (1979) describe behavioral experiments that show several decision-making situations in which most people systematically violate the axioms of Expected Utility Theory. They also mention the Allais Paradox as the main motivation for the development of Prospect Theory (PT) and identify and describe the following effects and situations that happen when rational individuals make decisions:

- Certainty effect: people tend to give more importance to consequences that are certain, rather than to merely probable consequences. The situation set out in the Allais Paradox (see the introduction to 4.2 above) is a clear example of the certainty effect: when choosing between X and Y, most people prefer X ($1 M for sure) over Y (risk to gain $5 M with 10% probability), even although Y has a higher expected monetary value than X.
- Reflection effect: Reflecting values of consequences around 0 reverts the preference order, i.e., the order of preference for negative payoffs (losses) is the opposite of the order of preference for positive payoffs (gains). For instance,

Kahneman and Tversky (1979) show the results of an experiment in which between prospects X = (4000, 0.8; 0, 0.2) and Y = (3000, 1), most people prefer Y over X; and between prospects X′ = (−4000, 0.8; 0, 0.2) and Y′ = (−3000, 1), most people prefer X′ over Y′. The authors also mention that, according to this preference pattern, most people are risk averse when dealing with gains, and risk prone when dealing with losses.

- Isolation effect: In order to simplify decision-making situations, people usually isolate components that alternatives have in common, and focus on what differentiates them. This may lead to inconsistences of preferences in choice problems. For instance, let us consider the following situations 1 and 2 (Figs. 6 and 7) (Kahneman and Tversky 1979).

In both situations above, the circles represent probability nodes, while the squares represent choice nodes. Behavioral experiments show that, in situation 1, most people prefer Y over X, and in situation 2, most people prefer X′ over Y′. However, what happens is that X = X′ and Y=Y′, because X leads to a payoff of \$4000 with probability 0.8*x*0.25 = 0.2, similar to X′; and Y leads to a payoff of \$3000 with probability 0.25*x*1 = 0.25, similar to Y′. This preference reversal happens because of the way in which the situations are represented: in situation 1, people isolate the initial part of the lottery, which may lead to a gain of nothing with 0.75 probability or

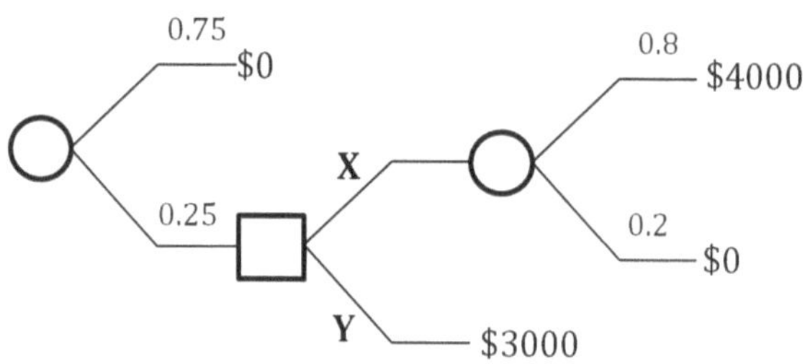

Fig. 6 Situation 1

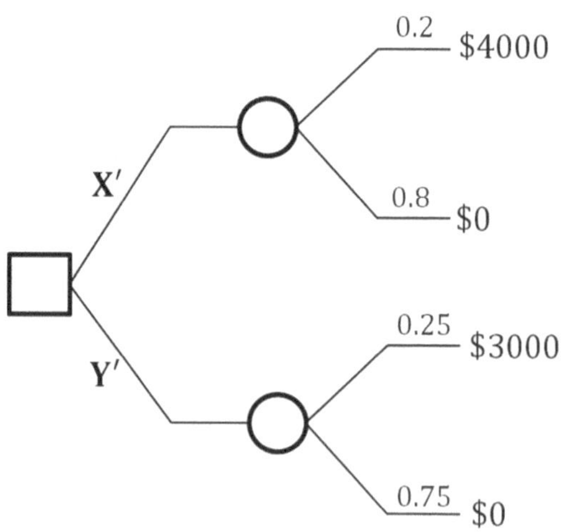

Fig. 7 Situation 2

making a choice with 0.25 probability. This behavioral pattern of preference reversal that occurs due to the dependence between events is particularly important because it shows a violation of one of the basic assumptions of Expected Utility Theory: the one that says that choices between prospects depend only on the probabilities of the final states.

Prospect Theory was developed as an extension of the RDU model in order to try to better represent these preference patterns. Prospect Theory brings a new component that is not present in RDU model: dependence on a reference point, i.e., it allows there to be different probability weighting functions for gains and for losses. As previously shown with the reflection effect, most people have risk averse behavior for gains and risk prone behavior for losses, which justifies the use of two different probability weighting functions.

The value of zero (no gain, no losses) is considered here as the reference point. Positive values are therefore considered as gains, and negative values are considered losses. Two probability weighting functions are defined: $w^{+}(p)$ for gains, and $w^{-}(p)$ for losses. The values of consequences $(x_1, x_2, \ldots, x_n)$ should be ranked relatively to the reference point (0), which is called *complete signal ranking*:

$$x_1 \geq \cdots \geq x_k \geq 0 \geq x_{k+1} \cdots \geq x_n \tag{13}$$

The values of consequences greater than the reference point are considered gains, while the values of consequences lower than the reference point are considered losses. Similarly to the rank-dependent utility model, the value of a prospect X is given by:

$$PT(X) = \sum_{i=1}^{n} \pi_i u\left(\mathrm{x}_i\right) \tag{14}$$

The Prospect Theory model differs from the RDU model in the way in which the decision weights π_i are obtained. Decision weights for gains are calculated based on $w^{+}(p)$, while decision weights for losses are calculated based on $w^{-}(p)$, as follows:

- Gains $(i \leq k) : \pi_i = w^{+}(p_1 + \ldots + p_i) - w^{+}(p_1 + \ldots + p_{i-1.})$
- Losses $(i \geq k) : \pi_i = w^{-}(p_i + \ldots + p_n) - w^{-}(p_{i+1} + \ldots + p_n)$

Therefore, Prospect Theory model can accommodate decision situations in which gains and losses are involved, throughout two weighting probability functions ($w^{+}(p)$ and $w^{-}(p)$), while in the RDU model a single probability weighting function is considered, without difference for gains and losses. When the decision situation involves only gains (without consequences that lead to losses), however, Prospect Theory and Rank-Dependent Utility both lead to the same evaluation for prospects, and this is the reason why the PT model is considered a relative generalization of the RDU model.

References

Ahn BS, Park KS (2008) Comparing methods for multiattribute decision making with ordinal weights. Comput Oper Res 35(5):1660–1670

Ahn BS, Park KS, Han CH, Kim JK (2000) Multi-attribute decision aid under incomplete information and hierarchical structure. Eur J Oper Res 125(2):431–439

Allais M (1953) Le comportement de l'homme rationnel devant le risque: critique des postulats et axiomes de l'école américaine. J Econ Soc, Econometrica, pp 503–546

Allmendinger R, Ehrgott M, Gandibleux X, Geiger MJ, Klamroth K, Luque M (2017) Navigation in multi-objective optimization methods. J Multi-Criteria Decis Anal 24(1–2):57–70

Athanassopoulos AD, Podinovski VV (1997) Dominance and potential optimality in multiple criteria decision analysis with imprecise information. J Oper Res Soc 48(2):142–150

Bana e Costa CA, Vansnick JC (1994) MACBETH—an interactive path towards the construction of cardinal value functions. Int Trans Oper Res 1(4):489–500

Belton V, Stewart T (2002) Multiple criteria decision analysis: an integrated approach. Springer Science & Business Media, Boston, MA

Berger JO (1985) Statistical decision theory and Bayesian analysis. Springer Science & Business Media, New York

Brans JP, Vincke P (1985) A preference ranking organization method: the Promethee method for multiple criteria decision making. Manag Sci 31:647–656

Coello CC, Lamont GB, Van Veldhuizen DA (2007) Evolutionary algorithms for solving multi-objective problems. Springer Science & Business Media, Boston, MA

Danielson M, Ekenberg L (2017) A robustness study of state-of-the-art surrogate weights for MCDM. Group Decis Negot 26(4):677–691

Danielson M, Ekenberg L, Larsson A, Riabacke M (2014) Weighting under ambiguous preferences and imprecise differences in a cardinal rank ordering process. Int J Comput Intell Syst 7(sup1):105–112

de Almeida AT (2013) Additive-veto models for choice and ranking multicriteria decision problems. Asia-Pacific J Oper Res 30(6):1–20

De Almeida AT, Cavalcante CAV, Alencar MH, Ferreira RJP, De Almeida-Filho AT, Garcez TV (2015) Multicriteria and multi-objective models for risk, reliability and maintenance decision analysis, International series in operations research & management science, vol 231. Springer, New York

De Almeida AT, Almeida JA, Costa APCS, Almeida-Filho AT (2016) A new method for elicitation of criteria weights in additive models: flexible and interactive tradeoff. Eur J Oper Res 250(1):179–191

Deb K, Pratap A, Agarwal S, Meyarivan T (2002) A fast and elitist multi-objective genetic algorithm: NSGA-II. IEEE Trans Evol Comput 6:182–197

Dias LC, Clímaco JN (2000) Additive aggregation with variable interdependent parameters: the VIP analysis software. J Oper Res Soc 51(9):1070–1082

Edwards W, Barron FH (1994) SMARTS and SMARTER: improved simple methods for multi-attribute utility measurement. Organ Behav Hum Decis Process 60(3):306–325

Edwards W, Miles RF Jr, Von Winterfeldt D (2007) Advances in decision analysis: from foundations to applications. Cambridge University Press, Cambridge/New York

Ehrgott M (2006) Multicriteria optimization. Springer Science & Business Media, Berlin

Ehrgott M, Gandibleux X (2002) Multiobjective combinatorial optimization. In: Multiple-criteria optimization: state of the art annotated bibliographic surveys. Springer, Boston, MA

Figueira J, Greco S, Ehrgott M (eds) (2005) Multiple criteria decision analysis: state of the art surveys. Springer, Boston/Dordrecht/London

Fonseca CM, Fleming PJ (1993) Genetic algorithms for multi-objective optimization: formulation, discussion and generalization. In: Forrest S (ed) Proceedings of the fifth international conference on genetic algorithms. University of Illinois at Urbana-Champaign, Morgan Kaufmann Publishers, San Mateo, CA

Frej EA, de Almeida AT, Costa APCS (2019) Using data visualization for ranking alternatives with partial information and interactive tradeoff elicitation. Oper Res:1–23
Jacquet-Lagréze E, Siskos J (1982) Assessing a set of additive utility functions for multicriteria decision making, the UTA method. Eur J Oper Res 10(2):151–164
Jiménez A, Mateos A, Sabio P (2013) Dominance intensity measure within fuzzy weight oriented MAUT: an application. Omega 41(2):397–405
Kahneman D, Tversky A (1979) Prospect theory: an analysis of decision under risk. Econometrica 47(2):363–391
Karasakal E, Köksalan M (2009) Generating a representative subset of the nondominated frontier in multiple criteria decision making. Oper Res 57(1):187–199
Keeney RL, Raiffa H (1976) Decision analysis with multiple conflicting objectives. Wiley, New York
Kim SH, Ahn BS (1999) Interactive group decision making procedure under incomplete information. Eur J Oper Res 116(3):498–507
Kim SH, Han CH (2000) Establishing dominance between alternatives with incomplete information in a hierarchically structured attribute tree. Eur J Oper Res 122(1):79–90
Kirkwood CW, Sarin RK (1985) Ranking with partial information: a method and an application. Oper Res 33(1):38–48
Korhonen P (2005) Interactive methods. In: Multiple criteria decision analysis: state of the art surveys. Springer, New York, pp 641–661
Korhonen P (2009) Multiple objective programming support multiple objective programming support. In: Floudas CA, Pardalos PM (eds) Encyclopedia of optimization. Springer, New York, pp 2503–2511
Korhonen P, Wallenius J (2010) Interactive multiple objective programming methods. In: Zopounidis C, Pardalos PM (eds) Handbook of multicriteria analysis, vol 9. Springer, Berlin Heidelberg, pp 263–286
Malakooti B (2000) Ranking and screening multiple criteria alternatives with partial information and use of ordinal and cardinal strength of preferences. IEEE Trans Syst Man Cybern Part A Syst Hum 30(3):355–368
Mármol AM, Puerto J, Fernández FR (2002) Sequential incorporation of imprecise information in multiple criteria decision processes. Eur J Oper Res 137(1):123–133
Mateos A, Jiménez-Martín A, Aguayo EA, Sabio P (2014) Dominance intensity measuring methods in MCDM with ordinal relations regarding weights. Knowl-Based Syst 70:26–32
Miettinen K (1999) Nonlinear multi-objective optimization. Springer, New York
Montiel LV, Bickel JE (2014) A generalized sampling approach for multilinear utility functions given partial preference information. Decis Anal 11(3):147–170
Mustajóki J, Hämäläinen RP, Salo A (2005) Decision support by interval SMART/SWING - incorporating imprecision in the SMART and SWING methods. Decis Sci 36(2):317–339
Pardalos PM, Siskos Y, Zopounidis C (eds) (2013) Advances in multicriteria analysis, vol 5). Springer, Boston, MA
Park KS (2004) Mathematical programming models for characterizing dominance and potential optimality when multicriteria alternative values and weights are simultaneously incomplete. IEEE Trans Syst Man Cybern Part A Syst Hum 34(5):601–614
Park KS, Kim SH (1997) Tools for interactive multi-attribute decision-making with incompletely identified information. Eur J Oper Res 98(1):111–123
Park KS, Kim SH, Yoon WC (1997) Establishing strict dominance between alternatives with special type of incomplete information. Eur J Oper Res 96(2):398–406
Pedrycz W, Ekel P, Parreiras R (2011) Fuzzy multicriteria decision-making: models, methods, and applications. Wiley, Chichester
Polmerol J-C, Barba-Romero S (2000) Multicriterion decision in management: principles and practice, vol 25. Springer, Boston, MA
Punkka A, Salo A (2013) Preference programming with incomplete ordinal information. Eur J Oper Res 231(1):141–150

Raiffa H (1968) Decision analysis: introductory lectures on choices under uncertainty. Addison-Wesley, London
Roy B (1996) Multicriteria methodology for decision aiding. Springer, New York
Saaty TL (1980) The analytic hierarchy process: planning; priority setting; resource allocation. McGraw-Hill International Book Company, New York
Salo AA, Hämäläinen RP (1992) Preference assessment by imprecise ratio statements. Oper Res 40(6):1053–1061
Salo AA, Hämäläinen RP (1995) Preference programming through approximate ratio comparisons. Eur J Oper Res 82(3):458–475
Salo AA, Hämäläinen RP (2001) Preference ratios in multi-attribute evaluation (PRIME)-elicitation and decision procedures under incomplete information. IEEE Trans Syst Man Cybern Part A Syst Hum 31(6):533–545
Salo A, Punkka A (2005) Rank inclusion in criteria hierarchies. Eur J Oper Res 163(2):338–356
Sarabando P, Dias LC (2010) Simple procedures of choice in multicriteria problems without precise information about the alternatives' values. Comput Oper Res 37(12):2239–2247
Savage LJ (1954) The foundations of statistics. Dover Press, New York
Slowinski R, Greco S, Matarazzo B (2012) Rough set and rule-based multicriteria decision aiding. Pesq Oper 32:213–269
Srinivas N, Deb K (1994) Multi-objective optimization using nondominated sorting in genetic algorithms. Evol Comput 2:221–248
Steuer RE (1986) Multiple criteria optimization: theory, computation, and application. Wiley, New York
Stillwell WG, Seaver DA, Edwards W (1981) A comparison of weight approximation techniques in multi-attribute utility decision making. Organ Behav Hum Perform 28(1):62–77
Vincke P (1992) Multicriteria decision-aid. Wiley, New York
Von Neumann J, Morgenstern O (1944) Theory of games and economic behavior. Princeton University Press, Princeton
Von Winterfeldt, D.; Edwards, W. Decision analysis and behavioral research, 1986.
Wakker PP (2010) Prospect theory: for risk and ambiguity. Cambridge University Press, Cambridge
Weber M (1987) Decision making with incomplete information. Eur J Oper Res 28(1):44–57
Weber M, Borcherding K (1993) Behavioral influences on weight judgments in multiattribute decision making. Eur J Oper Res 67(1):1–12
Zitzler E, Thiele L (1999) Multiobjective evolutionary algorithms: a comparative case study and the strength Pareto approach. IEEE Trans Evol Comput 3(4):257–271

Comparing Cardinal and Ordinal Ranking in MCDM Methods

Mats Danielson and Love Ekenberg

1 Introduction

One of the problems with the additive model, and other MCDA models, is that numerically precise information is seldom available, and most decision-makers experience difficulties entering realistic information when analysing decision problems. For instance, Barron and Barrett (1996a) argue that the elicitation of exact weights demands an exactness which does not exist. There are other problems, such as that ratio weight procedures are difficult to accurately employ due to response errors (Jia et al. 1998).

The utilization of ordinal or imprecise importance information to determine criteria weights is a way of handling this problem, and some authors have suggested surrogate weights, as representative numbers are assumed to represent the most likely interpretation of the preferences expressed by a decision-maker or a group of decision-makers. One such type is derived from ordinal importance information (Barron and Barrett 1996a, b; Katsikopoulos and Fasolo 2006), where decision-makers supply ordinal information on importance and the information is subsequently converted into surrogate weights corresponding to the extracted ordinal information. Often used methods are rank sum weights (RS), rank reciprocal weights (RR) (Stillwell et al. 1981) and centroid weights (ROC) (Barron 1992). For instance, Barron and Barrett (1996a) introduced a process utilizing systematic simulations to validate the selection of criteria weights when generating surrogate weights as well as true reference weights. The authors also investigated how well

M. Danielson · L. Ekenberg (✉)
Department of Computer and Systems Sciences, Stockholm University, Stockholm, Sweden

International Institute for Applied Systems Analysis, IIASA, Laxenburg, Austria
e-mail: mats.danielson@su.sel; ekenberg@iiasa.ac.at

A. T. de Almeida et al. (eds.), *Multicriteria and Optimization Models for Risk, Reliability, and Maintenance Decision Analysis*, International Series in Operations Research & Management Science 321, https://doi.org/10.1007/978-3-030-89647-8_2

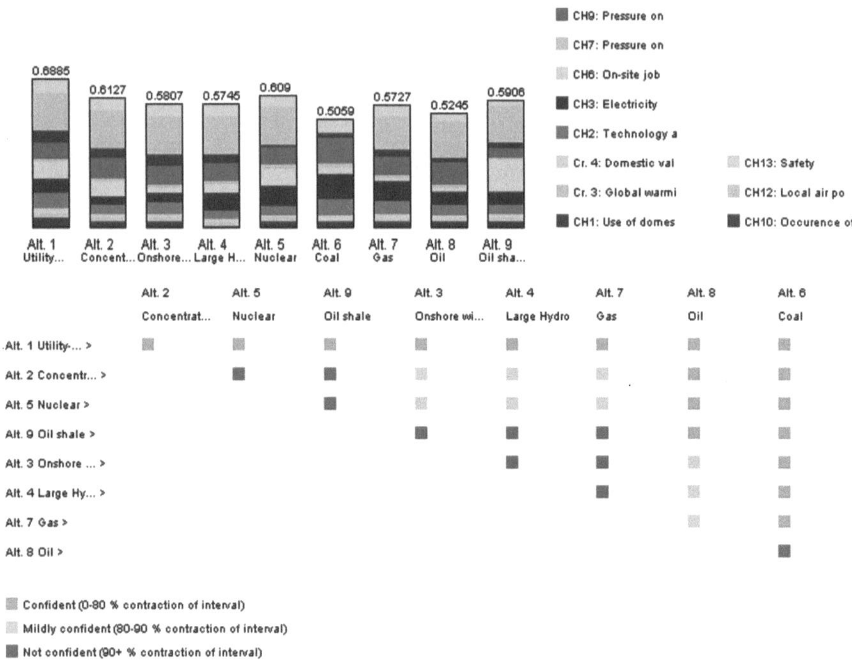

Fig. 1 The decision tool DecideIT

the result of using surrogate numbers matches the result of using the true numbers. This is heavily dependent on the distribution used to generate the weight vectors, however.

We earlier investigated various aspects of ordinal and cardinal weights in a couple of articles and compared state-of-the-art weight methods, both ordinal (ranking only) (Danielson and Ekenberg 2014; Danielson et al. 2014) and cardinal (with the possibilities to express strength) (Danielson and Ekenberg 2014, 2016a, b), in order to devise methods requiring as little cognitive load as possible. We also used these together with ranked values (utilities) and suggested a multi-stakeholder decision method that has been applied in, for example, the method of Chap. 3. This method fulfils several desired robustness properties and is comparatively stable under reasonable assumptions. Figure 1 shows the general multi-criteria multi-stakeholder tool DecideIT, combined with the CAR method from Danielson and Ekenberg (2016c) The values (utilities) of each alternative under each criterion is assessed by experts in the respective fields. Each stakeholder group will then order the criteria in terms of importance using cardinal ranking possibilities. The final evaluation then sees a combination of the stakeholders' assessments, either directly or via a set of stakeholder weights for the purpose of sensitivity analyses. The underlying method in all stages is the CAR method. An observation has been the inability of the stakeholder groups to express cardinality (similar to white cards in the Simos method), which spurred the investigation into even simpler means of expressing importance for stakeholders.

In this article, we take steps towards decreasing the cognitive load even further while still maintaining decision power and measure the effects of decreasing the load in different ways. Not least in multi-stakeholder decision situations, where different groups express different assessments (rank criteria and/or alternatives) differently, there is a need to express these assessments with simple yet powerful expressions. In that sense, this is a follow-up study to Danielson and Ekenberg (2016c) and other results.

2 Surrogate Weight Methods

The crucial issue in all these methods is how to assign surrogate weights while losing as little information as possible. Providing ordinal rankings of criteria seems to avoid some of the difficulties associated with the elicitation of exact numbers. It places fewer demands on decision-makers and is thus, in a sense, effort-saving. Furthermore, there are techniques for handling ordinal rankings with various degrees of success. A limitation of this is naturally that decision-makers usually have more knowledge of the decision situation than a pure criteria ordering, often in the sense that they have an idea regarding strengths within the importance relation information. In such cases, the surrogate weights may be an unnecessarily weak representation. Thus, we also investigate whether the methods can be extended to accommodate information regarding relational strengths while still preserving the property of being less demanding and hence more practically useful than other types of methods.

One well-known class of method is the SMART family. These were quite early suggested as methods for weight assessment from criteria rankings. The basic idea is quite simple. The criteria are ranked and then 10 points are assigned to the weight of the least important criterion (w_N). Then, the remaining weights (w_{N-1} through w_1) are given points according to the decision-maker's preferences. The overall value $E(a_j)$ of alternative a_j is then a weighted average of the values v_{ij} associated with a_j (Eq. 1):

$$E\left(a_j\right) = \frac{\sum_{i=1}^{N} w_i v_{ij}}{\sum_{i=1}^{N} w_i} \quad (1)$$

The most utilized processes for converting ordinal input to cardinal use automated procedures and yield exact numeric weights. For instance, Edwards and Barron (1994) proposed the SMARTER method for eliciting ordinal information on importance before converting it to numbers, thus relaxing information input demands on the decision-maker. An initial analysis is carried out where the weights are ordered, such as $w_1 > w_2 > ... > w_N$, and subsequently transformed to numerical weights using ROC weights. SMARTER then continues in the same manner as the ordinary SMART method.

Probably the most well-known ratio scoring method is the analytic hierarchy process (AHP), where a set of alternatives is evaluated under a criteria tree by pairwise comparisons. The process requires the same pairwise comparisons regardless of scale type. For each criterion, the decision-maker should first find the ordering of the alternatives from the best to the worst. Next, he or she should find the strength of the ordering by considering pairwise ratios (pairwise relations) between the alternatives using the integers 1, 3, 5, 7 and 9 to express their relative strengths, indicating that one alternative is equally good to another (strength = 1) or three, five, seven or nine times as good. It is also allowed to use the even integers, 2, 4, 6 and 8, as intermediate values, but using only odd integers is more common.

There are, however, some severe shortcomings of these methods, and we have in a series of articles suggested a set of alternatives. A promising candidate is the cardinal ranking (CAR) method: we have shown that it is more robust and efficient than the ones from the SMART family, AHP and many others (Danielson and Ekenberg 2014).

2.1 *Strength of Weights*

In order to make an ordinal ranking of N criteria into a stronger ranking, we add information about how much more or less important the criteria are compared to each other. Following Danielson and Ekenberg (2016c), we use the following notation for the strength of the rankings between criteria, and a suggestion for an intuitive verbal interpretation of these[1]:

$>_0$ Equally important
$>_1$ Slightly more important
$>_2$ More important (clearly more important)
$>_3$ Much more important

While being more cognitively demanding than ordinal weights, these are still less demanding than, for example, AHP weight ratios or point scores. In an analogous manner, as for ordinal rankings, decision-makers' statements can be converted into weights.

[1] Of course, this is not intended to be totally normative. Any interpretation is possible and can be formally handled in the same way.

2.2 *Preference Strength Methods*

Analogous to the ordinal weight functions above, counterparts using the concept of preference strength can be derived straightforwardly.

1. Assign an ordinal number to each importance scale position, starting with the most important position as number 1.
2. Let the total number of importance scale positions be Q. Each criterion i has the position $p(i) \in \{1,\ldots,Q\}$ on this importance scale, such that for every two adjacent criteria c_i and c_{i+1}, whenever $c_i >_{s_i} c_{i+1}$, $s_i = | p(i+1) - p(i) |$. The position $p(i)$ then denotes the importance as stated by the decision-maker. Thus, Q is equal to $\Sigma s_i + 1$, where $i = 1,\ldots,N - 1$ for N criteria.

The cardinal counterparts to the ordinal ranking methods above can then be found by using the results from Danielson and Ekenberg (2016b), where the ordinal SR weights are given by Eq. 2:

$$w_i^{\mathrm{SR}} = \frac{{}^{1}\!/_{i} + \frac{N+1-i}{N}}{\sum_{j=1}^{N} w_j^{\mathrm{SR}}} \tag{2}$$

and using steps 1–3 above, the corresponding preference strength SR weights (CSR, Eq. 3) are obtained as

$$w_i^{\mathrm{CSR}} = \frac{{}^{1}\!/_{p(i)} + \frac{Q+1-p(i)}{Q}}{\sum_{j=1}^{N} \left({}^{1}\!/_{p(j)} + \frac{Q+1-p(j)}{Q}\right)} \tag{3}$$

Using the idea of importance steps, ordinal weight methods are easily generalized to their respective counterparts. In the same manner, values (or utilities) can be assessed either ordinally (ranking only) or cardinally (additionally expressing strength).

In Danielson and Ekenberg (2016c), we combined cardinal weights with cardinal values in the CAR method and assessed the method by both simulations and a large number of real-life decision cases. The CAR method was found to outperform SMART and AHP in terms of performance and ease of use (cognitive load), but some users still wanted a method with even less cognitive load, so we tried to satisfy this while still preserving reasonable requirements of correctness.

The CAR method follows the three-step procedure presented below. Firstly, the values of the alternatives under each criterion are elicited in a way similar to the weights described above:

1A. For each criterion in turn, rank the alternatives from the worst to the best outcome.

1B. Enter the strength of the ordering. The strength indicates how strong the separation is between two ordered alternatives. Similar to weights, the strength is expressed in the notation with '$>_i$' symbols.

Secondly, the weights are elicited with a swing-like procedure in accordance with the discussion above.

2A. For each criterion in turn, rank the importance of the criteria from the least to the most important.

2B. Enter the strength of the ordering. The strength indicates how strong the separation is between two ordered criteria. The strength is expressed in the notation with '$>_i$' symbols.

Thirdly, the usual weighted overall value is calculated by multiplying the centroid of the weight simplex with the centroid of the alternative value simplex.

The same description can be used to introduce the three candidate methods, called C+O, O+C, and O+O depending on whether a cardinal or ordinal procedure is used for the representation of weights and values respectively. In the original CAR method, all the steps 1A, 1B, 2A, 2B and 3 were performed in that order. The steps in the three candidate methods that we suggest are performed as follows: In O+C, step 1B is omitted, resulting in the sequence 1A, 2A, 2B and 3 in order. In C+O, step 2B is omitted instead, resulting in the sequence 1A, 1B, 2A and 3 in order. Finally, in O+O, both steps 1B and 2B are omitted, resulting in the sequence 1A, 2A and 3 in order.

We will compare these CAR derivatives in the next section in search of a method with less cognitive load but still performing better than SMART and AHP. This is, to our knowledge, the first time ordinal and cardinal ranking methods (and combinations thereof) have been compared systematically in this way.

3 Assessment of Models for Weights

We will utilize similar techniques to those in the simulation studies described above to determine the adequacy of the methods suggested above. The assumption is that all elicitation is made relative to a weight distribution held by the decision-maker. The basic idea is that decision-makers' mindset should be reflected by the random generator for generating test vectors, but all such machinery is then dependent on the underlying distribution of the random generator, which must be considered.

In the area of MCDM, a decision-maker can generally express preferences based on scoring points, as in point allocation (PA) or direct rating (DR) methods. In PA, the decision-maker is given a point sum (e.g. 100) to distribute among the criteria. In PA, this normalization implies N–1 degrees of freedom (DoF) for N criteria. DR puts no such limit on the number of points to be allocated and the decision-maker allocates as many points as desired to each criterion. Thereafter, the points are normalized, implying N degrees of freedom for N criteria.

In the simulations below, it is important to realize which background model we are using. When following an N–1 DoF model, a vector is generated in which the components total 100%. This simulation is based on a homogenous N-variate Dirichlet distribution generator. On the other hand, following an N DoF model, a vector is generated without an initial joint restriction, only keeping components within [0%, 100%], yielding a process with N degrees of freedom. Subsequently, they are normalized so that their sum is 100%.

We will call the N–1 DoF model type of generator an N–1 generator and the N DoF model type an N generator. Depending on the simulation model used (and, consequently, the background assumption of how decision-makers assess weights), the results become different. In reality, we cannot know whether a specific decision-maker (or even decision-makers in general) adhere more to N–1 or N DoF representations of their knowledge. Both as individuals and as a group, they might use either or be anywhere in between. A reasonably robust MCDM method must therefore perform well at both endpoints of the representation spectrum and anything in between. Thus, the evaluation of MCDM methods in this paper will use a combination of both types of generator to find the most efficient and robust method.

3.1 Comparing Six MCDA Methods

We will compare the three methods, SMART, AHP and CAR, which were compared in Danielson and Ekenberg (2016c), together with three new candidate methods (O+C, C+O, and O+O)—i.e. the scaled-down versions of CAR, as described above—to look for methods that are cognitively less demanding while still yielding powerful results. Remember that the CAR method consists of cardinally ranking weights and cardinally ranking values.[2] The three candidate methods are composed as follows: O+C and O+O use ordinal SR weights while C+O and CAR use cardinal CSR weights. Furthermore, CAR and O+C use cardinally ranked values while C+O and O+O use a pure ordinal ranking of the values (such as $v_{12} > v_{14} > v_{11} > \ldots$).

3.2 Measurements

The simulations were carried out with a varying number of criteria and alternatives. There were four criteria numbers $N = \{3, 6, 9, 12\}$ and four alternatives numbers $M = \{3, 6, 9, 12\}$ in the simulation study, creating a total of 16 simulation scenarios.

[2] In the terminology of this paper, this could have been called C + C, but we retain the name by which it is more widely known.

Table 1 The winner frequency for the methods using an *N*-1 generator

N-1 DoF	SMART	AHP	CAR	O+C	C+O	O+O
3 | 3	87.9	85.2	92.4	88.5	85.6	82.7
3 | 12	71.1	82.7	85.7	77.9	74.9	70.6
6 | 6	81.2	80.2	88.2	82.5	75.9	73.3
6 | 12	73.8	81.3	85.2	78.6	73.7	70.2
9 | 9	78.4	80.9	84.7	81.0	73.3	71.4
12 | 3	86.0	77.0	88.8	88.1	82.0	81.5
12 | 12	77.3	81.5	81.9	80.1	71.8	71.0

In this and the following tables, the leftmost column contains the notation *N* | *M*, denoting a decision situation having *N* criteria and *M* alternatives

Each scenario was run ten times, each time with 10,000 trials, a total of 1,600,000 decision situations thus being generated. An *N*-variate joint Dirichlet distribution was employed to generate the random weight vectors for the *N*–1 DoF simulations as well as a standard normalized random weight generator (see Danielson and Ekenberg 2016c for details). Unscaled value vectors were generated uniformly since no significant differences were observed with other value distributions. The value vectors were then used for multiplying with the obtained weights in order to form weighted values to be compared.

The results of the simulations are shown in Table 1, which shows a subset of the results with a selection of pairs (*N*,*M*). The measure of success is the hit ratio—i.e. the number of times the highest evaluated alternative using a particular method coincides with the true highest alternative. The table thus shows the hit frequency for the three MCDA methods SMART,[3] AHP[4] and CAR, together with the three candidates.

It is clear from Table 1 that among the established methods, CAR outperforms the other methods. While CAR averages 87%, the other two well-known methods perform at around 80–81%. For example, in Table 1, CAR displays better overall ranking compared to the other methods. We know from Danielson and Ekenberg (2016c) that the other two well-known methods fare about equally, SMART being somewhat stronger when fewer alternatives are involved and AHP being somewhat stronger when more alternatives are involved. This is not surprising, since a very large amount of information is requested for AHP's pairwise comparisons when the number of criteria and alternatives increases. The gap up to CAR for both of the other methods is substantial considering the relatively high hit rate level that the methods operate at. In Table 1, using an *N*–1 generator, this can be seen where the candidate methods fare both better and worse than the established ones.

[3] SMART is represented by the improved SMARTER version by Edwards and Barron (1994).

[4] AHP weights were derived by forming quotients w_i/w_j and rounding to the nearest odd integer. Also allowing even integers in between yielded no significantly better results.

Table 2 The winner frequency for the methods using an N generator

N DoF	SMART	AHP	CAR	O+C	C+O	O+O
3 | 3	87.6	82.7	91.4	88.3	83.0	80.9
3 | 12	73.2	82.3	85.8	80.9	74.3	72.0
6 | 6	81.6	79.0	87.8	84.6	75.5	73.8
6 | 12	75.2	80.7	85.9	81.2	73.7	71.4
9 | 9	79.4	79.4	85.5	82.1	73.7	72.3
12 | 3	85.3	75.6	89.7	88.6	82.2	81.6
12 | 12	77.9	80.4	83.4	81.0	72.4	71.6

Table 3 The winner frequency for the methods using a combined generator

Combined	SMART	AHP	CAR	O+C	C+O	O+O
3 | 3	87.8	84.0	91.9	88.4	84.3	81.8
3 | 12	72.2	82.5	85.8	79.4	74.6	71.3
6 | 6	81.4	79.6	88.0	83.6	75.7	73.6
6 | 12	74.5	81.0	85.6	79.9	73.7	70.8
9 | 9	78.9	80.2	85.1	81.6	73.5	71.9
12 | 3	85.7	76.3	89.3	88.4	82.1	81.6
12 | 12	77.6	81.0	82.7	80.6	72.1	71.3

Table 4 Mean overall measurements

Total	SMART	AHP	CAR	O+C	C+O	O+O
Mean	79.7	80.6	86.9	83.1	76.6	74.6
Rank	4	3	1	2	5	6

The frequencies change somewhat in Table 2 since we employ a model with N degrees of freedom instead. Still, methods with some cardinality perform better than the pure ordinal one.

In Table 3, the N and N–1 DoF models are combined with an equal emphasis on both. The established methods yield results as expected. An interesting pattern emerges between the candidate methods. The method O+C that kept cardinality in the values perform rather well, while C+O and O+O, that only used ordinal values, perform worse than SMART and AHP regardless of using cardinal weights or not.

Table 4 shows the average of the respective columns of Table 3.

It is important that an MCDM method not only has good precision: it also needs to be robust in the sense that it performs well regardless of whether the decision-maker uses a cognitive model where the representation has N or N–1 DoF, or any combination thereof. Table 5 shows the spread in results between the N and N–1 DoF simulations, while Table 6 shows the standard deviation of these differences. These tables show that all methods are reasonably robust, the mixed cardinal/ordinal ones being a little bit less so.

We consider precision and robustness to be of equal importance to a good method. The final score for the MCDM methods—both the established ones and the candidates—are therefore computed as Final score = Mean result – Spread, thus

Table 5 Spread over different DoF

Spread	SMART	AHP	CAR	O+C	C+O	O+O
3 | 3	0.3	2.5	1.0	0.2	2.6	1.8
3 | 12	2.1	0.4	0.1	3.0	0.6	1.4
6 | 6	0.4	1.2	0.4	2.1	0.4	0.5
6 | 12	1.4	0.6	0.7	2.6	0.0	1.2
9 | 9	1.0	1.5	0.8	1.1	0.4	0.9
12 | 3	0.7	1.4	0.9	0.5	0.2	0.1
12 | 12	0.6	1.1	1.5	0.9	0.6	0.6

Table 6 Standard deviation of spread

Spread	SMART	AHP	CAR	O+C	C+O	O+O
St. Dev.	0.6	0.7	0.4	1.1	0.9	0.6
Rank	3	4	1	6	5	2

Table 7 Final scores

Final scores	SMART	AHP	CAR	O+C	C+O	O+O
Total	79.1	80.0	86.4	82.0	75.7	74.0
Rank	4	3	1	2	5	6

taking both precision and robustness into account.[5] Table 7 shows the final scores of the comparisons.

Since the CAR method performed the best in both precision and robustness, it heads the final score table, as expected. The interesting observations are made among the other candidates. One of the candidates, O + C, performs better than all other methods except CAR. It puts a considerably less demanding cognitive load on the decision-maker by requiring only cardinality in the values, not preference weights, making it a very attractive alternative even to the original (fully cardinal) CAR. Its efficiency is due to the performance of the ordinal SR weights (Eq. 3) originally developed in Danielson and Ekenberg (2014).

4 Concluding Remarks

Elicitation methods available today in MCDM are often too cognitively demanding for normal real-life decision-makers and there is a clear need for weighting methods that do not require formal decision analysis knowledge. We have investigated several methods, including state-of-the-art approaches for asserting surrogate weights with the possibility of supplying information regarding preference strength and of assigning values to consequences by ranking them. It is known from Danielson and Ekenberg (2014) that the CAR method outperforms both SMART and AHP, but all these are still considered to be difficult for some decision-makers. In the search

[5] The final score is, of course, not a percentage in the sense of Table 4, but rather a score of suitability taking both performance and robustness into account.

for a method with even less cognitive demand than CAR, three candidates were put forward. One stood the test, performing better than the benchmark methods SMART and AHP. This new method is similar to CAR in assessing values, but uses only ordinal ranking of the preference weights. The choice between cardinal and ordinal ranking in the weights has an impact on efficiency, but much less than the values. The focus on cardinally ranking weights is misplaced: it should be on values instead. This, of course, has implications for other cardinal ranking methods, such as Macbeth.

The candidate method that kept cardinality in the values performed well, while those that only used ordinal values performed worse than SMART and AHP regardless of using cardinal weights or not. This can intuitively be explained by the much greater freedom present in assigning values compared to weights, the latter being restricted by the normalization constraint and required to fall on a hyperplane with a dimension one less than the number of criteria. However, intuition alone is not a good guide to designing MCDA methods, which is why we undertook the work presented in this paper. In summary, keeping cardinal ranking in the values is a very important property of an MCDM method since using only ordinal ranking in the values instead yields methods inferior even to SMART and AHP. The choice between cardinal and ordinal ranking in the weights has an impact on efficiency and robustness, but much less than the values.

Acknowledgements This research was supported by the EU-project Co-Inform (Co-Creating Misinformation-Resilient Societies H2020-SC6-CO-CREATION-2017) and strategic grants from the Swedish government within ICT – The Next Generation.

References

Barron FH (1992) Selecting a best multi-attribute alternative with partial information about attribute weights. Acta Psych 80(1–3):91–103

Barron F, Barrett B (1996a) Decision quality using ranked attribute weights. Manag Sci 42(11):1515–1523

Barron F, Barrett B (1996b) The efficacy of SMARTER: simple multi-attribute rating technique extended to ranking. Acta Psych 93(1–3):23–36

Danielson M, Ekenberg L (2014) Rank ordering methods for multi-criteria decisions. Proceedings of the 14th Group Decision and Negotiation (GDN). Springer, Cham

Danielson M, Ekenberg L (2016a) A robustness study of state-of-the-art surrogate weights for MCDM. Group Decis Negot 7. https://doi.org/10.1007/s10726-016-9494-6

Danielson M, Ekenberg L (2016b) The CAR method for using preference strength in multi-criteria decision making. Group Decis Negot 25(4):775–797. https://doi.org/10.1007/s10726-015-9460-8

Danielson M, Ekenberg L (2016c) Trade-offs for ordinal ranking methods in multi-criteria decisions. Proceedings of GDN. Springer, Cham

Danielson M, Ekenberg L, He Y (2014) Augmenting ordinal methods of attribute weight approximation. Decis Anal 11(1):21–26

Edwards W, Barron F (1994) SMARTS and SMARTER: improved simple methods for multi-attribute utility measurement. Organ Behav Hum Decis Process 60:306–325

Jia J, Fischer GW, Dyer J (1998) Attribute weighting methods and decision quality in the presence of response error: a simulation study. J Behav Decis Mak 11(2):85–105

Katsikopoulos K, Fasolo B (2006) New tools for decision analysis. IEEE Trans Syst Man Cybern Syst Hum 36(5):960–967

Stillwell W, Seaver D, Edwards W (1981) A comparison of weight approximation techniques in multi-attribute utility decision making. Organ Behav Hum Perform 28(1):62–77

Evaluating Multi-criteria Decisions Under Conditions of Strong Uncertainty

Mats Danielson, Love Ekenberg, and Aron Larsson

1 Introduction

A variety of approaches have been suggested over the years for the evaluation of decision problems. An important category is multi-attribute utility theory (MAUT), where there are several extensively used implementations such as SMART, EXPERT CHOICE and CAR, and various varieties thereof (Danielson and Ekenberg 2016a, 2016b). In general, albeit far from always, these assume that the decision-maker can provide numerically precise decision information; in many cases, this is considered to be unrealistic in real-life decision-making, and is the reason that different interval approaches have been suggested to extend the various decision models for both multi-criteria and risk-based decision-making, such as the PRIME tool, handling multiple criteria while supporting interval-valued ratio estimates for value differences. Another approach is the preference programming method, which is an interval extension of the classical analytical hierarchy process (AHP) method (Salo and Hämäläinen, 2001) and is related to the RICH method. There are also other approaches, such as ARIADNE (Sage and White, 1984). There is a further multitude of fuzzy measurement variants of MAUT techniques. A main issue with the above approaches is that they provide very little assistance when the

M. Danielson · L. Ekenberg (✉)
Department of Computer and Systems Sciences, Stockholm University, Stockholm, Sweden

International Institute for Applied Systems Analysis, IIASA, Laxenburg, Austria
e-mail: mats.danielson@su.se; ekenberg@iiasa.ac.at

A. Larsson
Department of Computer and Systems Sciences, Stockholm University, Stockholm, Sweden

Department of Information Systems and Technology, Mid Sweden University, Sundsvall, Sweden
e-mail: aron@dsv.su.se

A. T. de Almeida et al. (eds.), *Multicriteria and Optimization Models for Risk, Reliability, and Maintenance Decision Analysis*, International Series in Operations Research & Management Science 321, https://doi.org/10.1007/978-3-030-89647-8_3

results overlap, as is usually the case in real-life decision problems. This issue will be addressed in the paper.

In the research community, there have been many suggestions as to how to handle the very strong requirements for decision-makers to provide precise information. Some main categories of approaches to remedy the precision problem are based on capacities, sets of probability measures, upper and lower probabilities, interval probabilities (and sometimes utilities), evidence and possibility theories, as well as fuzzy measures (see, for example, Dubois, 2010; Rohmer and Baudrit, 2010; Shapiro and Koissi, 2015; Dutta, 2018). The latter category seems to be used only to a limited extent in real-life decision analyses since it usually requires a significant mathematical background on the part of the decision-maker. Another reason is that the computational complexity can be problematic if the fuzzy aggregation mechanisms are not significantly simplified. This is further discussed in, for example, Danielson (2004) and Danielson and Ekenberg (2007).

In this article, we therefore suggest a method and software for integrated multi-attribute evaluation under risk, subject to incomplete or imperfect information. The software originates from our earlier work on evaluating decision situations using imprecise utilities, probabilities and weights, as well as qualitative estimates between these components derived from convex sets of weight, utility and probability measures. To avoid some aggregation problems when handling set membership functions and similar, we introduce higher-order distributions for better discrimination between the possible outcomes. For the decision structure, we use the common tree formalism but refrain from using precise numbers. To alleviate the problem of overlapping results, we suggest a new evaluation method based on a resulting belief mass over the output intervals, but without trying to introduce further complicating aspects into the decision situation.

In the next section, we briefly describe how risk and multi-criteria trees can be co-modelled trees in an integrated framework. Thereafter, we provide the conceptual model for our method and explain both the input data format and the evaluations, and how it relates to the modelling of beliefs. We finish with a real-life example.

2 Probabilistic Approaches

Probabilistic decision situations are often represented by a decision tree such as in Fig. 1.

Such a tree consists of a root node, also called a decision node, and a set of probability nodes, representing uncertainty and consequence nodes for the final outcomes. In general, the probability nodes are assigned unique probability distributions representing the uncertainties in the decision situation. When an alternative A_i is chosen, there is a probability p_{ij} that an event will occur that leads either to a subsequent event or a consequence. The consequences are assigned values v_{ijk}. The maximization of the expected value is often used as an evaluation rule. The expected value of alternative A_i in Fig. 1 is:

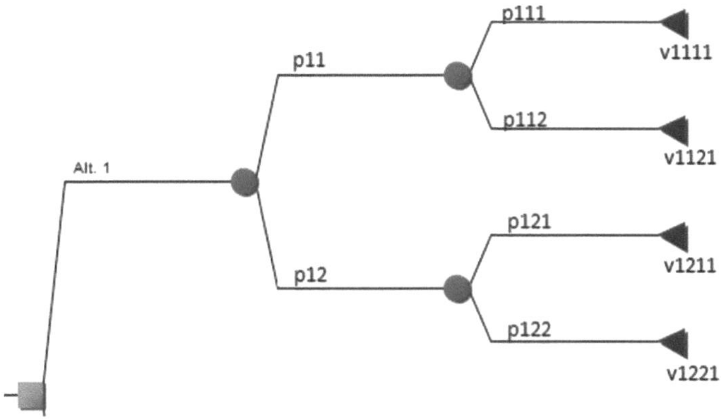

Fig. 1 A partial tree representation of the events for one alternative (Alt.1) in a decision under risk. *[The three red dots are binary events]*

$$E\left(A_i\right)=\sum_{j=1}^{2} p_{ij} \sum_{k=1}^{2} p_{ijk} v_{ijk}.$$

This is straightforwardly generalized to decision trees of arbitrary depth.

3 Multi-criteria Decision Trees

Multi-criteria decisions in the MAUT category are characterized by there being several criteria, often on different levels in a hierarchy, as in Fig. 2, where the alternatives are valued and the decision-maker assigns values to the alternatives relative to a value scale.

Normalized weights are assigned to each sub-branch in the tree and the alternatives are valued under the respective sub-criteria. A maximization of the weighted value is often used for the evaluations. In Fig. 2, the value of alternative A_i under criterion jk is v_{ijk}, while the weight of criterion jk is w_{jk}. Thereafter, the total value of alternative A_i can be calculated using

$$E\left(A_i\right)=\sum_{j=1}^{2} w_j \sum_{k=1}^{2} w_{jk} v_{ijk}.$$

The alternative with the maximum expected value is then the preferred choice.

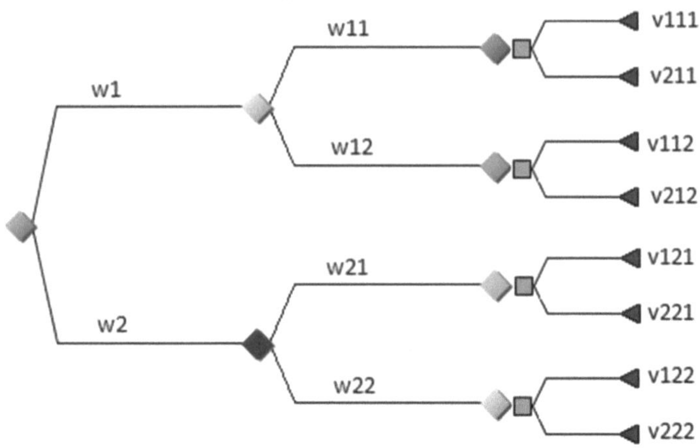

Fig. 2 A multi-criteria decision tree

4 Probabilistic Multi-criteria Hierarchies

Combining these formalisms is straightforward by calculating the value of the alternatives as expected values derived from decision trees—i.e. the valuation of the consequences can be included in the overall multi-criteria tree evaluation. Figure 3 demonstrates how this is done, where the alternatives' values under the weight w_{11} are derived from the entire underlying probabilistic decision tree.

The expected value of the tree is then calculated by

$$E\left(A_i\right)=\sum_{j=1}^{2}\left(w_j\cdot\sum_{k=1}^{2}\left(w_{jk}\cdot\sum_{m=1}^{2}\left(p_{im}\cdot\sum_{n=1}^{2}p_{imn}v_{imn}\right)\right)\right)$$

or, more generally, by

$$E\left(A_i\right)=\sum_{i_1=1}^{n_{i_0}}w_{ii_1}\sum_{i_2=1}^{n_{i_1}}w_{ii_1i_2}\cdots\sum_{i_{m-1}=1}^{n_{i_{m-2}}}p_{ii_1i_2\cdots i_{m-2}i_{m-1}}\sum_{i_m=1}^{n_{i_{m-1}}}p_{ii_1i_2\cdots i_{m-2}i_{m-1}i_m}v_{ii_1i_2\cdots i_{m-2}i_{m-1}i_m},$$

where p denotes a probability, w denotes a weight and v denotes a value.

We will formalize this in the next sections and explain how imprecision can be modelled in a combined structure.

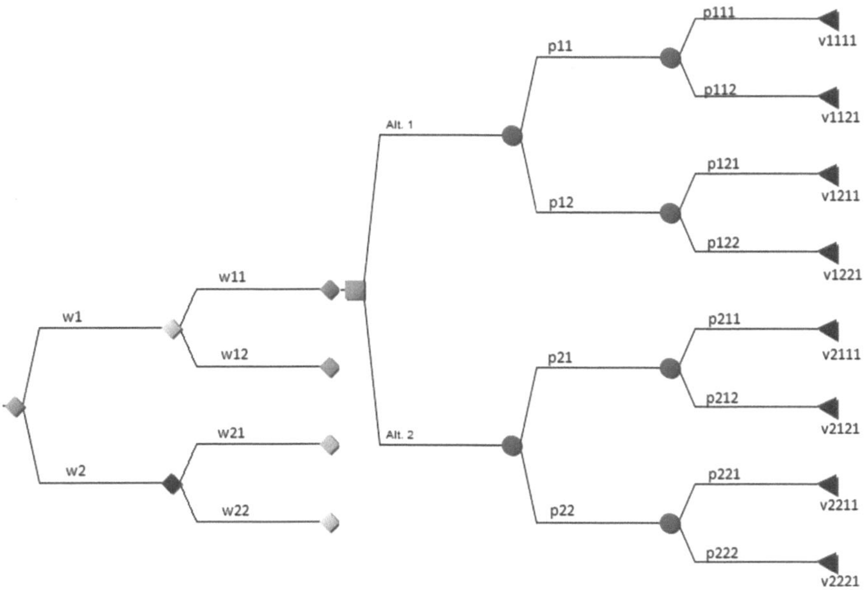

Fig. 3 Combined multi-criteria and probabilistic representations

5 Strong Uncertainty

In the type of multi-criteria decision problems we consider, we hold that strong uncertainty exists if the decision is also made under risk, with uncertain consequences for at least one criterion, in combination with imprecise or incomplete information with respect to probabilities, weights and consequences or alternative values. Decision evaluation under strong uncertainty and computational means for evaluating these models should both be capable of embracing the uncertainty in the evaluation rules and methods and provide evaluation results reflecting the effects of uncertainty for the subsequent discrimination between alternatives.

We will call our representation of a combined decision problem a multi-frame. Such a frame collects all information necessary for the model in one structure. One part of this is the concept of a graph.

Definition A graph is a structure ⟨V,E⟩ where V is a set of nodes and E is a set of node pairs. A tree is a connected graph without cycles. A rooted tree is a tree with a dedicated node as a root. The root is at level 0. The adjacent nodes, except for the nodes at level i-1, to a node at level i is at level $i + 1$. A node at level i is a leaf if it has no adjacent nodes at level $i + 1$. A node at level $i + 1$ that is adjacent to a node at level i is a child of the latter. A (sub-)tree is symmetrical if all nodes at level i have the same number of adjacent nodes at level $i + 1$. The depth of the tree is max (n | there exists a node at level n).

Definition A criteria-consequence tree T = ⟨C∪A∪N∪{r},E⟩ is a tree where

r is the root,
A is the set of nodes at level 1,
C is the set of leaves, and.
N is the set of intermediary nodes in the tree except those in A.

In a multi-frame, represented as a multi-tree, user statements can either be range constraints or comparative statements (see below); they are translated into inequalities and collected together in a value constraint set. For probability and weight statements, the same is done into a node constraint set. We denote the values of the consequences c_i and c_j by v_i and v_j respectively. Value statements are relations between value variables, and they are translated into systems of inequalities in a *value constraint set*. Probability statements are in the same manner collected in a *node constraint set*. A constraint set is said to be consistent if it can be assigned at least one real number to each variable so that all inequalities are simultaneously satisfied. Consequently, we get potential sets of functions with an infinite number of instantiations.

Definition Given a criteria-consequence tree T, let N be a constraint set in the variables $\{n_{\ldots i \ldots j \ldots}\}$. Substitute the intermediary node labels $x_{\ldots i \ldots j \ldots}$ with $n_{\ldots i \ldots j \ldots}$. N is a node constraint set for T if, for all sets $\{n_{\ldots i1}, \ldots, n_{\ldots im}\}$ of all sub-nodes of nodes $n_{\ldots i}$ that are not leaves, the statements $n_{\ldots ij} \in [0,1]$ and $\sum_j n_{\ldots ij} = 1, j \in [1, \ldots, m]$ are in N.

A probability node constraint set relative to a criteria-consequence tree then characterizes a set of discrete probability distributions. Weight and value constraint sets are analogously defined. Weight and probability node constraint sets also contain the usual normalization constraints ($\sum_j x_{ij} = 1$) requiring the probabilities and weights to total one.

Definition A multi-frame is a structure ⟨T,N⟩, where T is a criteria-consequence tree and N is a set of all constraint sets relative to T.

The probability, value and weight constraint sets thus consist of linear inequalities. A minimal requirement is that it is consistent—i.e. there must exist some vector of variable assignments that simultaneously satisfies each inequality in the system.

Definition Given a consistent constraint set X in the variables $\{x_i\}$, ${}^{X}\max(x_i) =_{\text{def}} \sup(a \mid \{x_i > a\} \cup X$ is consistent. Similarly, ${}^{X}\min(x_i) =_{\text{def}} \inf(a \mid \{x_i < a\} \cup X$ is consistent. Furthermore, given a function f, ${}^{X}\text{argmax}(f(x))$ is a solution vector that is a solution to ${}^{X}\max(f(x))$, and ${}^{X}\text{argmin}(f(x))$ is a solution vector that is a solution to ${}^{X}\min(f(x))$.

The set of orthogonal projections of the solution set is the orthogonal hull, consisting of all consistent variable assignments for each variable in a constraint set.

Definition Given a consistent constraint set X in $\{x_i\}_{i \in [1, \ldots, n]}$, the set of pairs $\langle {}^{X}\min(x_i), {}^{X}\max(x_i) \rangle$ is the orthogonal hull of the set.

The orthogonal hull is the upper and lower probabilities (weights, values) if X consists of probabilities (weights, values). The hull intervals are calculated by first finding a consistent point. Thereafter, the minimum and maximum of each variable are found by solving linear programming problems. Because of convexity, the intervals between the extremal points are feasible—i.e. the entire orthogonal hull has been determined.

6 Beliefs in Intervals

We will now extend the representation to obtain a more granulated representation of a decision problem. Often when we specify an interval, we probably do not believe in all values in the intervals equally: we may, for example, believe less in the values closer to the borders of the intervals. Additional values are nevertheless added to cover everything that we perceive as possible in uncertain situations. These additions give rise to belief distributions indicating the different strengths with which we believe in the different values. Distributions over classes of weight, probability and value measures have been developed into various models, such as second-order probability theory.

In the extended model, we introduce a focal point to each of the intervals used as parameters for belief distributions for probabilities, values and criteria weights. We can then operate on these distributions using additive and multiplicative combination rules for random variables. The detailed theory of belief distributions in this sense is described in Ekenberg and Thorbiörnson (2001), Danielson et al. (2007, 2014) and Sundgren et al. (2009).

To make the method more concrete, we introduce the unit cube as all tuples $(x_1, \ldots, x_n)$ in $[0,1]^n$. A second-order distribution over a unit cube B is a positive distribution F defined on B such that

$$\int_B F(x)\, dV_B(x) = 1,$$

where V_B is the n-dimensional Lebesgue measure on B.

We will use second-order joint probability distributions as measures of beliefs. Different distributions are utilized for weights, probabilities and values because of the normalization constraints for probabilities and weights. Natural candidates are then the Dirichlet distribution for weights and probabilities and two- or three-point distributions for values. In brief, the Dirichlet distribution is a parameterized family of continuous multivariate probability distributions. It has a probability density function given by a function of those parameters, such that $\alpha_1,\ldots,\alpha_k > 0$ depends on a beta function and the product of the parameters x_i.

More precisely, the probability density function of the Dirichlet distribution is

$$f_{dir}(p, \alpha) = \frac{\Gamma\left(\sum_{i=1}^{k} \alpha_i\right)}{\prod_{i=1}^{k} \Gamma(\alpha_i)} {p_1}^{\alpha_i - 1} {p_2}^{\alpha_2 - 1} \ldots {p_k}^{\alpha_k - 1}$$

on a set $\{p = (p_1, \ldots p_k) \mid p_1, \ldots, p_k \geq 0, \Sigma p_i = 1\}$ where $(\alpha_1, \ldots, \alpha_k)$ is a parameter vector in which each $\alpha_i > 0$ and $\Gamma(\alpha_i)$ is the Gamma function.[1]

The Dirichlet distribution is a multivariate generalization of the beta distribution and the marginal distributions of Dirichlet are thus beta distributions. The beta distribution is a family of continuous probability distributions defined on [0, 1] and parameterized by two parameters, α and β, defining the shape of the distribution.

If the distribution is uniform, the resulting marginal distribution (over an orthogonal axis) is a polynomial of degree $n - 2$, where n is the dimension of a cube B. Let all $\alpha_i = 1$, then the Dirichlet distribution is uniform with the marginal distribution[2]

$$f(x_i) = \int_{B_i^-} dV_{B_i^-}(x) = (n-1)(1-x_i)^{n-2}$$

However, we need a bounded Dirichlet distribution operating on a user-specified $[a_i, b_i]$ range instead of the general interval [0,1]. Bounded beta distributions are then derived—the so-called four-parameter beta distributions, also defined only on the user-specified range. We then define a probability or weight belief distribution as a three-point bounded Dirichlet distribution $f_3(a_i, c_i, b_i)$ where c_i is the most likely probability or weight and a_i and b_i are the boundaries of the belief with $a_i < c_i < b_i$ (Kotz and van Dorp, 2004).

For values, the generalization to a trapezoid from a triangle is analogous. We will utilize either a two-point distribution (uniform, trapezoidal) or a three-point distribution (triangular). When there is large uncertainty regarding the underlying belief distribution in values and we have no reason to make any more specific assumptions, a two-point distribution modelling upper and lower bounds (the uniform or trapezoid distributions) is preferred. On the other hand, when the modal outcome can be estimated, the beliefs are more congenially represented by three-point distributions. Because triangular distributions are less centre-weighted than other three-point distributions, the risk of underestimation is less, which is why there are no particular reasons to use any other distribution for real-life decision purposes.

[1] The details of this are provided in any standard textbook in Bayesian statistics, such as Kendall and Stuart (1969).

[2] For a more elaborated treatment of these properties and the suitability for representing second-order properties, see, for example, Ekenberg and Thorbiörnson (2001) and Ekenberg et al. (2005).

7 Evaluation Steps

We will use a generalization of the ordinary expected value for the evaluation—i.e. the resulting distribution over the generalized expected utility is

$$E\left(A_i\right)=\sum_{i_1=1}^{n_{i_0}} w_{ii_1}\sum_{i_2=1}^{n_{i_1}} w_{ii_1i_2}\cdots\sum_{i_{m-1}=1}^{n_{i_{m-2}}} p_{ii_1\,i_2\cdots i_{m-2}\,i_{m-1}}$$

$$\sum_{i_m=1}^{n_{i_{m-1}}} p_{ii_1i_2\cdots i_{m-2}i_{m-1}i_m} v_{ii_1i_2\cdots i_{m-2}i_{m-1}i_m1},$$

given the distributions over random variables p and v. There are only two operations of relevance here, multiplication and addition.

Let G be a distribution over the two cubes A and B. Assume that G has a positive support on the feasible distributions at level i in a general decision tree, as well as on the feasible probability distributions of the children of a node x_{ij} and assume that $f(x)$ and $g(y)$ are the marginal distributions of G(z) on A and B, respectively. Then the cumulative multiplied distribution of the two belief distributions is $H(z) = \iint_{\Gamma_x} f(x)g(y)dxdy = \int_0^1\int_0^{z/x} f(x)g(y)dxdy = \int_z^1 f(x)G(z/x)\,dx$ where G is a primitive function to g, $\Gamma_z = \{(x,y) \mid x\cdot y \le z\}$, and $0 \le z \le 1$.

Let $h(z)$ be the corresponding density function. Then

$$\mathrm{h}(z)=\frac{d}{dz}\int_z^1 f(x)G\left(z/x\right)dx=\int_z^1 f\frac{f(x)g\left(z/x\right)}{x}dx.$$

The addition of the products is the standard convolution of two densities restricted to the cubes. The distribution h on a sum $z = x + y$ associated with the belief distributions $f(x)$ and $g(y)$ is therefore given by

$$\mathrm{h}(z)=\frac{d}{dz}\int_0^z f(x)g\left(z-x\right)dx.$$

Then we can obtain the combined distribution over the generalized expected utility.

As in most of risk and decision theory, we assume that a large number of events will occur and a large number of decisions will be made. In business administration, this is called the principle of going concern. In such an operating environment, the expected value becomes a reasonable decision rule and, at the same time, the belief distributions over the expected values tend to normal distributions or similar. But the resulting distributions will be normal only when the original distributions are symmetrical, which of course is not usually the case for beta and triangular distributions. The result then will instead be skew-normal. Thus, we use a truncated skew-normal distribution, generalizing the normal distribution by allowing for non-zero skewness and truncated tails. We can then conveniently represent truncated (skew-)normal distributions as probability distributions of (skew-)normally distributed random variables that are bounded. Assume that a distribution X has a normal distribution within the interval (a, b). Then X, $a < X < b$, has a truncated normal distribution and its probability density function is given by a four-parameter expression that tends to normality as the intervals are widened (see, for instance, Loeve, 1977).

8 Real-Life Decision Example

In the following, we will illustrate the approach with an example derived from a real-life decision problem. The example is modelled and evaluated using the DecideIT[3] tool version 3.0, which, among other features, implements the above approach to handling strong uncertainty. Consider a pulp mill company that wishes to evaluate whether to rebuild or possibly exchange its recovery boiler.[4] The decision problem is viewed as two sequential decisions. The first decision is to what extent the boiler will be enhanced, and three different alternatives are considered: (1) do nothing; (2) rebuild boiler in order to secure deliveries; and (3) replace existing recovery boiler.

The second decision concerns what to do with the power turbine exploiting the pressure from the boiler in order to produce electricity (since a new boiler can allow for more powerful turbines). Furthermore, the existing turbine would need to be revised within one year were it not replaced. For this sub-decision, four alternatives were evaluated: (1) revise and use the existing turbine; (2) replace with a smaller 70 kg/s turbine; (3) replace with a bigger 80 kg/s turbine; and (4) replace with a 100 kg/s turbine (which is only feasible with a new boiler).

The alternatives are evaluated based on the following set of criteria:

Cr. 1. Discounted cash flow with weight variable w_1. Assessed on a monetary scale.
Cr. 2. Initial cash drain with weight variable w_2. Assessed on a value scale $[-10, 0]$.
Cr. 3. Internal environment with weight variable w_3. Assessed using comparisons.
Cr. 4. External environment with weight variable w_4. Assessed using comparisons.
Cr. 5. Delivery dependability with weight variable w_5. Assessed using comparisons.
Cr. 6. Room for a production increase with weight variable w_6. Assessed on a value scale [0, 10].

[3] DecideIT is supplied by www.preference.nu

[4] The boiler is the part of a pulp mill where chemicals left from the cooking of wood are recovered and reformed. This generates heat, which is used both in the process and to produce electricity.

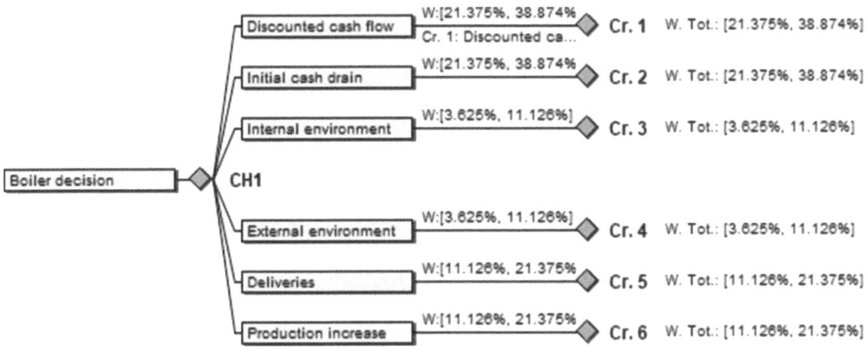

Fig. 4 The criteria tree in DecideIT

The criteria weights were provided as comparisons:

$$\begin{aligned} w_1 - w_2 &= 0 \\ w_2 - w_5 &> 0 \\ w_5 - w_6 &= 0 \\ w_6 - w_4 &> 0 \\ w_4 - w_5 &= 0 \end{aligned}$$

This constraint set essentially says that Cr. 1 and Cr. 2 are most and equally important, followed by Cr. 5 and Cr. 6, which are of equal importance and, in turn, more important than Cr. 3 and Cr. 4, also being of equal importance. The resulting orthogonal weight hull for each criterion is shown within brackets in Fig. 4. Cr. 1 is connected to a decision tree shown in Fig. 5 according to the approach in Larsson et al. (2005).

Cr. 1 was assessed through discounted cash flow analysis (EBITA), using a risk-free discount rate with a ten-year time frame, providing a net present value for each consequence node C1 to C17 in Fig. 5, where interval values for each consequence within brackets are shown in kSEK. The cash flows were based upon unit margins of paper production and power production, together with annual estimated production. Since the estimates were uncertain, interval statements were used. This way of modelling risk in discounted cash flow analysis can be labelled risk-adjusted net present value since the risk is modelled by means of probabilities for different consequences, each associated with a net present value, as opposed to incorporating risk in the discount rate (see Aven, 2011).

For the first boiler alternative, keeping the boiler, there was an initial sub-decision regarding a choice between a new 70 kg/s turbine, a new 80 kg/s turbine, or keeping the existing turbine but with more frequent revisions. The chance nodes in the tree reflect whether or not the old turbine will break down during its final year of operation. The probability of the existing turbine breaking down while awaiting a new turbine was assessed to lie within the interval [2%, 10%]. For the second

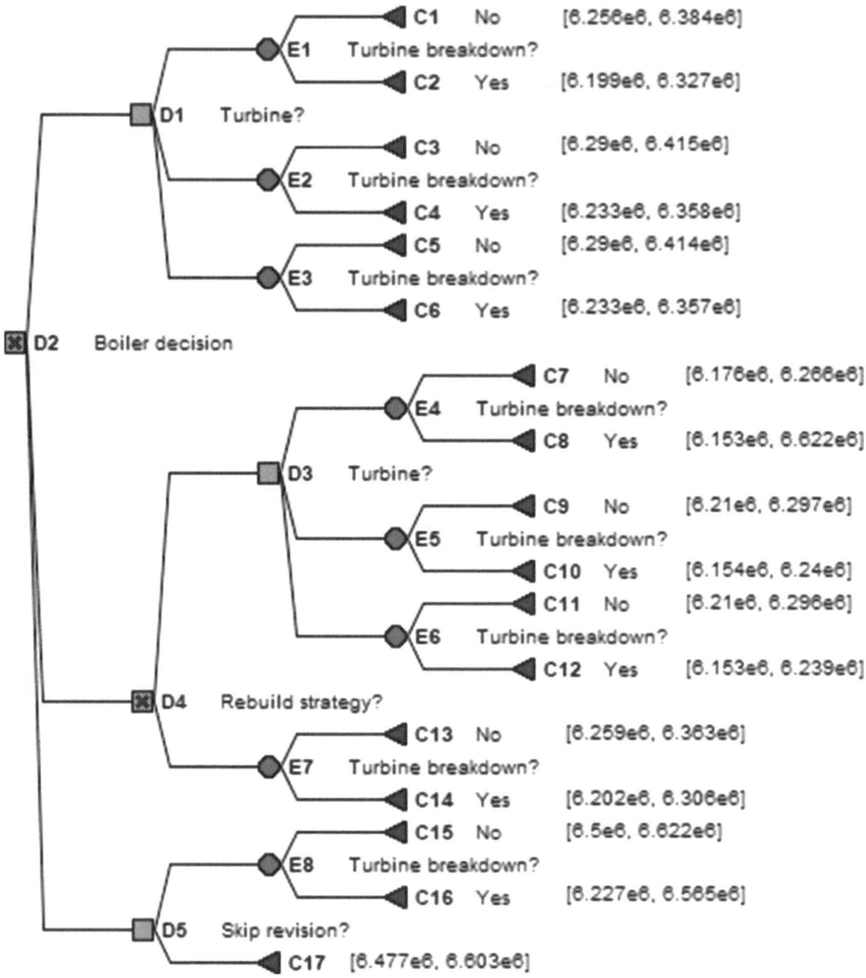

Fig. 5 Decision tree for the discounted cash flow criterion

alternative, the action of rebuilding the boiler can either be done to secure the deliveries only or to additionally enable increased power production by utilizing a more powerful turbine.

For the third alternative, acquiring a new recovery boiler together with a new 100 kg/s turbine, the existing turbine needed to be in use for two years instead of only one year due to increased planning and installation time. This resulted in the breakdown probability of the old turbine being estimated to be higher compared to the other two boiler alternatives, at [10%, 20%], which is the probability for consequence C16 in Fig. 5. The discounted cash flow analysis strongly supports the alternative of enabling increased power production if rebuilding the boiler (Table 1).

Table 1 Alternative values or rankings per criterion. Interval values within brackets

	A_1: Do nothing	A_2: Rebuild boiler	A_3: New boiler
Discounted cash flow[a]	$[6.28\times10^6, 6.41\times10^6]$	$[6.25\times10^6, 6.36\times10^6]$	$[6.48\times10^6, 6.60\times10^6]$
Initial cash drain	0	$[-6, -5]$	-10
Internal environment	$A_1 > A_2$ and $A_1 > A_3$		
External environment	$A_3 > A_2$ and $A_2 > A_1$		
Deliveries	$A_3 > A_2$ and $A_2 > A_1$		
Production increase	0	[5, 7]	10

[a] *Expected value interval*

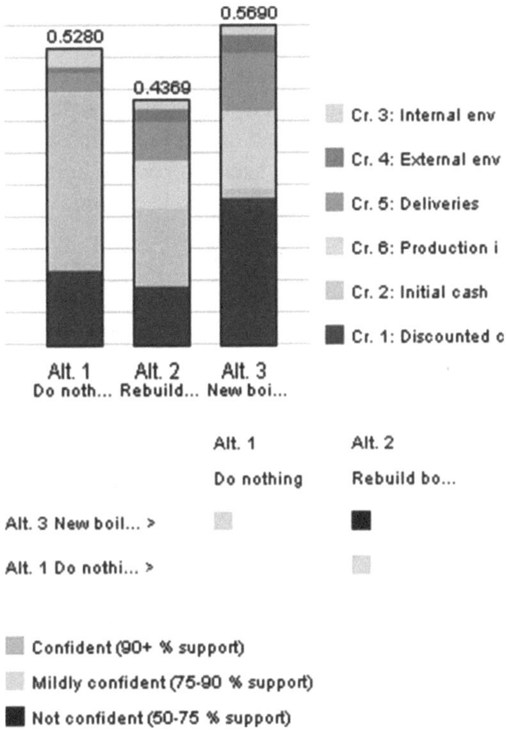

Fig. 6 Main decision evaluation result

For the above multi-criteria decision problem modelled in DecideIT, a main decision evaluation window is shown in Fig. 6, consisting of bar charts of stacked centroid part-worth values for the criteria for each alternative. The part-worth value φ_{il} for alternative A_i under criterion l is simply given by $\varphi_{il} = {}^c w_l \cdot {}^c v_{il}$, where ${}^c w_l$ and ${}^c v_{il}$ are the centroid weights for criterion l and the centroid alternative value for alternative A_i under criterion l. The height of each bar is then the sum $\varphi_{i1} + \varphi_{i2} + \ldots + \varphi_{in}$, where n is the number of direct sub-criteria.

In addition, the results of an embedded a priori sensitivity analysis are presented in the main evaluation window in a table of pairwise comparisons between all three alternatives, done according to the approach above, enabling investigation of the belief support for the given ranking of Alt. 3 being the most preferred alternative, followed by Alt. 1. In this way, the evaluation windows provide an informative decision evaluation in the presence of strong uncertainty. The main outcome is the resulting belief distribution of the combined input belief distributions over the expression E(Alt. 3) – E(Alt. 1) and the support where this expression is positive is 89%. It would thus be unreasonable to select Alt. 3 over Alt. 1.

9 Concluding Remarks

In classic decision theory, a decision-maker is expected to assign precise numerical values to the different decision components such as weights, probabilities and values. However, in real-life problems, this requirement is too strong in many situations and some kind of representation and evaluation mechanism is important. Many candidates have been suggested, such as sets of probability measures, upper and lower probabilities, as well as interval weights, probabilities and utilities enabling a more realistic representation of the input sentences. In these contexts, higher-order analyses can add information, enabling further discrimination between alternatives. Decision trees can still be utilized to represent the decision structure, where the various estimates can be done by intervals and qualitative assessments. However, much is accomplished by enhancing this with an evaluation method based on a belief mass interpretation of the various data. We have discussed here how multi-criteria and probabilistic trees can be viewed in an integrated framework and the effects of employing second-order information in decision trees. We have also demonstrated an implementation of the theory on a real-life decision problem and how the multiplicative and additive effects strongly influence the resulting distribution over the expected values. The result is a method that can offer considerably more discriminative power when selecting alternative options.

Acknowledgements This research was supported by the EU-project Co-Inform (Co-Creating Misinformation-Resilient Societies H2020-SC6-CO-CREATION-2017) and partly funded by strategic grants from the Swedish government within ICT – The Next Generation. A much shorter earlier version of this paper appeared at the conference ISIPTA'19 in Gent, Belgium.

References

Aven T (2011) Foundations of risk analysis, 2nd edn. Wiley

Danielson M (2004) Handling imperfect user statements in real-life decision analysis. Int J Inf Technol Decis Mak 3(3):513–534

Danielson M, Ekenberg L (2007) Computing upper and lower bounds in interval decision trees. Eur J Oper Res 181:808–816

Danielson M, Ekenberg L (2016a) A robustness study of state-of-the-art surrogate weights for MCDM. Group Decis Negot 7. https://doi.org/10.1007/s10726-016-9494-6

Danielson M, Ekenberg L (2016b) The CAR method for using preference strength in multi-criteria decision making. Group Decis Negot 25(4):775–797. https://doi.org/10.1007/s10726-015-9460-8
Danielson M, Ekenberg L, Larsson A (2007) Belief distribution in decision trees. Int J Approx Reason 46(2):387–407
Danielson M, Ekenberg L, Larsson A, Sundgren D (2014) Second-order risk constraints in decision analysis. Axioms 3:31–45
Dubois D (2010) Representation, propagation, and decision issues in risk analysis under incomplete probabilistic information. Risk Anal 30(3):361–368
Dutta P (2018) Human health risk assessment under uncertain environment and its SWOT analysis. Open Public Health J 11:72–92
Ekenberg L, Thorbiörnson J (2001) Second-order decision analysis. Int J Uncertainty Fuzziness Knowledge-Based Syst 9(1):13–38
Ekenberg L, Thorbiörnson J, Baidya T (2005) Value differences using second-order distributions. Int J Approx Reason 38(1):81–97
Kendall MG, Stuart A (1969) The advanced theory of statistics. Volume 1: distribution theory, 3rd edn. Griffin
Kotz S, van Dorp JR (2004) Beyond beta: other continuous families of distributions with bounded support and applications. World Scientific Press, Singapore
Larsson A, Johansson J, Ekenberg L, Danielson M (2005) Decision analysis with multiple objectives in a framework for evaluating imprecision. Int J Uncertainty Fuzziness Knowledge-Based Syst 13(5):495–509
Loeve M (1977) Probability theory. Graduate texts in mathematics. Volume 45, 4th edn. Springer
Rohmer J, Baudrit C (2010) The use of the possibility theory to investigate the epistemic uncertainties within scenario-based earthquake risk assessments. Nat Hazard, Springer 56(3):613–632
Sage AP, White CC (1984) ARIADNE: a knowledge-based interactive system for planning and decision support. IEEE Trans Syst Man Cybern 14(1):35–47
Salo AA, Hämäläinen RP (2001) Preference ratios in multi-attribute evaluation (PRIME) – elicitation and decision procedures under incomplete information. IEEE Trans Syst Man Cybern 31(6):533–545
Shapiro AF, Koissi MC (2015) Risk assessment applications of fuzzy logic, Casualty Actuarial Society, Canadian Institute of Actuaries, Society of Actuaries
Sundgren D, Danielson M, Ekenberg L (2009) Warp effects on calculating interval probabilities. Int J Approx Reason 50(9):1360–1368

A Framework for Building Multicriteria Decision Models with Regard to Reliability, Risk, and Maintenance

Adiel Teixeira de Almeida and Lucia Reis Peixoto Roselli

1 Introduction

Many multicriteria decision-making/aiding (MCDM/A) problems are to be found in organizations. The solutions found for these problems can produce different impacts on the organization's strategies.

Therefore, in order to support DM's evaluation of MCDM/A problems, MCDM/A building models were developed with a view to their being guides that offers solutions.

The main focus of this chapter is to present and discuss some issues that are raised by using MCDM/A building models, including some that deal with some problems in the RRM (risk, reliability, and maintenance) context.

2 Building Multicriteria Decision Models

An MCDM/A building model offers a formal and simplified representation of an MCDM/A problem. It consists of structured steps to represent the problem in line with the DM's preferences during the decision-making process. According to Box and Draper (1987) all models are wrong since they are simplifications of the "real world", but some models are useful as they make it possible to describe, study, and analyze problem situations. The key is to evaluate how wrong a model can be, i.e., to identify the point after which it is no longer useful.

A. T. de Almeida (✉) · L. R. P. Roselli
CDSID – Center for Decision Systems and Information Development, Universidade Federal de Pernambuco, Recife, Brazil
e-mail: almeida@cdsid.org.br; lrpr@cdsid.org.br

A. T. de Almeida et al. (eds.), *Multicriteria and Optimization Models for Risk, Reliability, and Maintenance Decision Analysis*, International Series in Operations Research & Management Science 321, https://doi.org/10.1007/978-3-030-89647-8_4

One of the first MCDM/A building models was developed by Simon in 1960. This model proposed three steps to solve problems. The first one was the intelligence step, which is related to identifying future conflict situations in an organization. The second was the design step, which was about constructing the model by formalizing important aspects presented in the problem. Finally, the last step was the choice step, which sought to indicate the solution to the problem.

Moreover, two further steps can be integrated into this model, namely the review step, which is used to review the definitions made in the previous step and the implementation of the solution step (Polmerol and Barba-Romero 2000).

Currently, there are many MCDM/A building models in the literature, such as Roy (1996), Polmerol and Barba-Romero (2000), Belton and Stewart (2002), and de Almeida et al. (2015).

In Belton and Stewart (2002) the building model developed had five steps. The first was about identifying the problem, which is equivalent to the intelligence step in the Simon model. The second and third steps dealt with structuring the problem and constructing the decision model; these steps are equivalent to the design step in the Simon model. The fourth and five steps made a recommendation and implemented it.

In de Almeida et al. (2015), their decision model had twelve steps, which are aggregated into three major phases. The initial phase is the preliminary phase, during which problems are structured. The next phase is preference modeling, which is about choosing an adequate MCDM/A method that will be used to solve the problem. Lastly, the finalization phase is used for review and to implement the solution.

Based on these models, it can be seen that they present structured steps to formally represent the problem based on a DM's preferences expressed during the process. According to Guitouni and Martel (1998), no building model will ever be perfected to characterize all decision-making problems. Thus, for each problem a decision model should be constructed to consider the DM's preferences.

It is while the model is being built that the MCDM/A method that would be the most appropriate for solving the MCDM/A problem is indicated. Therefore, these methods deal with real problems, which formalize the problem by following some well-structured steps with a view to producing a solution that can be applied to solve this problem.

In this context, according to Keisler and Noonan (2012), problems are present in the "real world" and are transferred to the "model world". In the "model world", these problems are structured, processed, and the solution found. Then, this solution is returned to the "real world" to be implemented.

Moreover, the models constructed are particular for each specific MCDM/A problem. In other words, for each preference expressed by the DM in the steps, the model is shaped for the specific problem. As illustrated in Fig. 1, at the beginning of the process, there are many possible models, but during the steps for selecting a model, assumptions are made, sets of approaches are selected and simplifications are introduced, resulting in some models being eliminated.

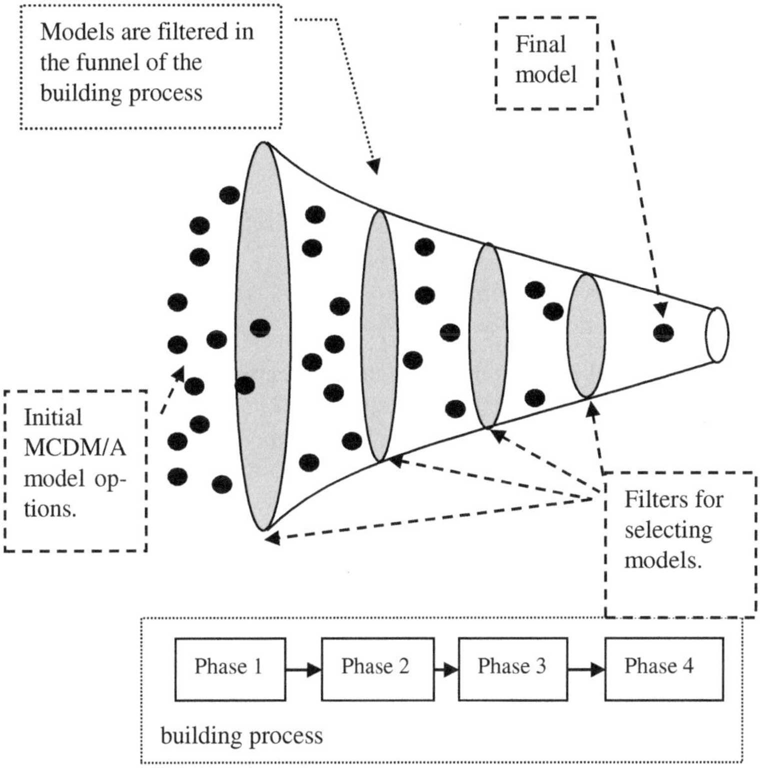

Fig. 1 Selecting the model in a funnel of the building process. (Adapted from de Almeida et al. 2015)

To conclude this topic, to support the building model process, problem structuring methods (PSM) can be used (Rosenhead and Mingers 2004; Eden 1988; Eden and Ackermann 2004; Ackermann and Eden 2001; Franco et al. 2004). According to Eden (1988) problem structuring seeks to build a formal representation for the problem, and this includes identifying objective and subjective factors of the decision-making process.

Among PSM methods, the value focus thinking (VFT) approach (Keeney 1992) aims to investigate the DM's values in order to guide the decision process. In this approach, DMs need to address two issues, namely, deciding what he/she wants for the decision situation, i.e., what his/her objectives for the problem are, and evaluating how he/she will achieve these objectives, which are represented by the alternatives that may be the solutions for the problem.

Thus, based on the answer to these two questions, this approach presents a structured way of thinking about the decision-making process and the DM's subjective judgments.

3 A Framework for Building Multicriteria Models in RRM

In this section, the building model presented in de Almeida et al. (2015) is discussed in order to highlight important concepts of the MCDM/A approach and is improved on by including steps presented in RRM decision situations. This model has four phases, namely:

- Phase 1 – Preliminary phase
- Phase 2 – Probabilistic Modeling phase
- Phase 3 – Preference Modeling and choice of MCDM/A method phase
- Phase 4 – Finalization phase

The first phase, called the preliminary phase, integrates four steps of this building model, and seeks to define the problem situation. Thus, the steps which comprise this phase are characterized to present the basic elements of an MCDM/A problem, such as: problem objectives, attributes associated with each objective, and the alternatives.

The second phase, called probabilistic modeling, consists of three steps that are used to define important elements present in probabilistic problems. This phase was included in this adapted building model, based on that of de Almeida et al. (2015), in order to provide a structured process to evaluate the RRM problem.

The third phase also has three steps. This phase is responsible for modeling the DM's preferences with regard to the elements presented in the previous steps and is an important phase in the decision model. At the end of this phase, the building model has been defined, as illustrated in Fig. 1. Moreover, it is at the end of this phase that model indicates the appropriate MCDM/A method that should be applied to find the solution to the problem.

The fourth phase, called the finalization phase, has four steps and is responsible for presenting a recommendation for the MCDM/A problem. In this phase, the MCDM/A method will identify to produce a recommendation for the problem. Thus, this recommendation will be tested, reviewed, and implemented for the problem situation. The framework for the model set out in this chapter and based on de Almeida et al. (2015) is illustrated in Fig. 2.

Compared to Simon's model, this model does not have the intelligence phase but its steps are broadly equivalent to the phases of Simon's model as follows: steps 1 to 10 to the design phase, 11 to the choice phase, while 12 and 13 are equivalent to the review step that was added to Simon's model, and, similarly, 14 is equivalent to the implementation step that was added to Simon's model.

However, note that a review step is not performed only in steps 12 and 13. It is present in the whole model, as a procedure that prompts successive refinements (Ackoff and Sasinieni 1968). These refinements permit returning to previous steps to review the preferences expressed and definitions made. Moreover, because it is possible to make refinements, some steps can be evaluated in a simplified way, and then later reviewed after more information becomes available from the successive steps. These refinements are identified by the dashed arrows between each of the steps.

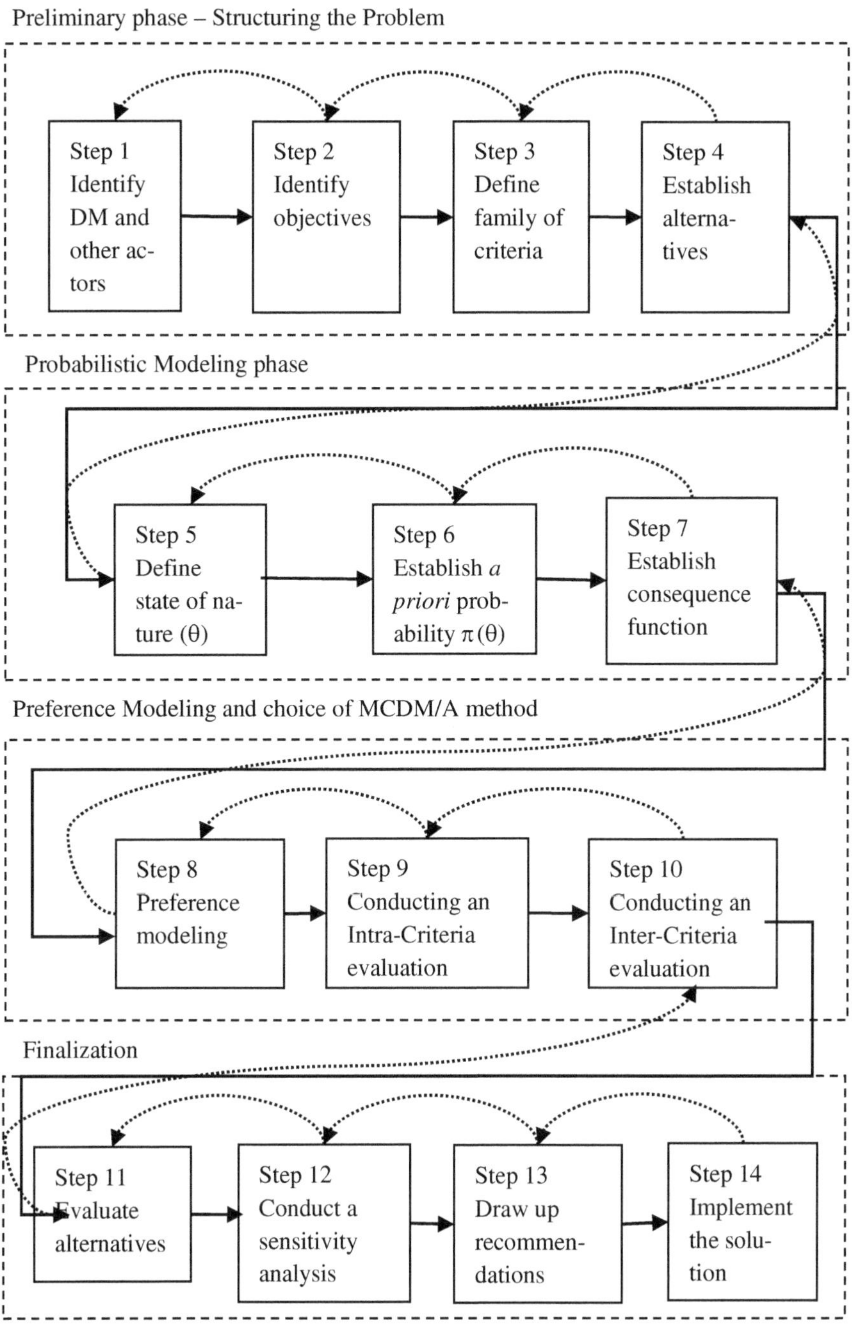

Fig. 2 Framework for building an MCDM/A model. (Adapted from de Almeida et al. 2015)

3.1 Step 1 – Identify DM and Other Actors in the Decision-Making Process

In MCDM/A problems, the main figure is the decision maker (DM). The DM is the person who is responsible for the decision. The whole building model is based on the preferences that the DM expresses for the problem situation.

Decision problems may involve only one DM, when this is an individual decision, or more than one DM, when a group decision needs to be taken. The focus of this model is on individual decision, but for group decision adaptations to the model can be made.

In addition to the DM, other actors may be present in the decision scenario. Therefore, it is important to identify these actors and their role in the decision-making process. The other actors are: the analyst, the client, one or more experts, and the stakeholders.

The analyst has knowledge about the MCDM/A approach; his/her role is to provide methodological support to the DM throughout the decision-making process. The analyst must interact with the DM during all the steps that are followed to find the adequate MCDM/A method, based on the DM's preferences.

The client can be considered a close advisor to the DM, who may deputize temporarily for the DM when the DM is absent. The client does not express his/her preferences, but only communicates the DM's preferences to the analyst.

The expert has factual information about the behavior of some variables which are not under the control of the DM. He/she should not declare his/her preferences, but only give factual information to help the DM acquire a fuller understanding of the problem scenario.

Stakeholders represent a group of people who may be affected by the decision; they do not participate in the decision-making process but can influence DM's preferences by reinforcing the importance of certain themes to them.

3.2 Step 2 – Identify Objectives

In MCDM/A problems, multiple objectives are present and the DM wishes to meet the whole set of objectives. Thus, the second step in the framework is to identify the objectives of the problem.

During this step, the DM must identify which objectives are the bases of interest for his decision. The reason why the problem must be solved is so that these objectives can be met. Based on these objectives, the DM will express his/her preferences.

Since the identification of objectives impacts all the future steps, this step can be considered the most important in the framework. If the definition of the objectives is incomplete or vague, potential problems will arise in future stages of the model.

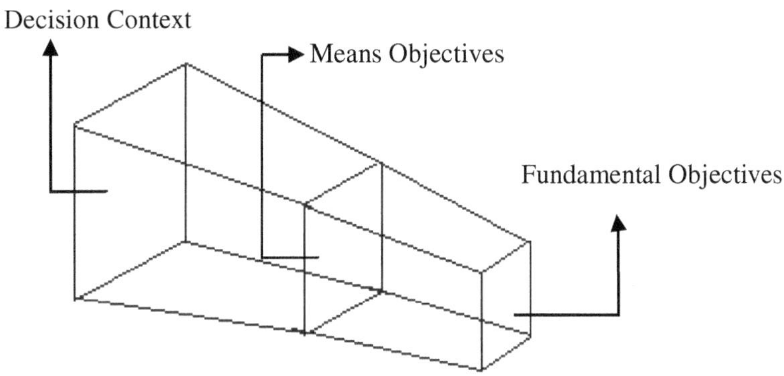

Fig. 3 Hierarchical structure of objectives. (Adapted from Keeney 1992)

The value focus thinking (VFT) approach, developed by Keeney (1992), can be used to support this step because it presents several relevant comments for the process of correct assessment of objectives.

According to the VFT, the process of finding the right objectives is not an easy task and some gimmicks can be used to assist this process such as using wish lists. This theory also classifies objectives into two categories: fundamental objectives and means objectives. Fundamental objectives are those that underlie the problem: being able to identify and achieve them, representing the reason for solving the problem. Means objectives are those that lead to fundamental objectives.

Besides the conceptual separation of such objectives, their hierarchical structure can be developed which facilitates understanding their relationship to each other, as illustrated in Fig. 3. Therefore, defining the set of objectives is a relevant step which is important as it provides a complete understanding of the problem situation. This usefully supports the subsequent steps, namely identifying the criteria and the alternatives.

3.3 Step 3 – Define Family of Criteria

For each objective identified, some criteria must be established to represent it. A criterion can be considered as a function that measures the level of achievement that some alternative obtains in the objective. According to Keeney (1992), criteria are characterized by the degree to which their related objective is successfully met.

Attributes are characterized as the lowest level to which a fundamental objective can be broken down and seek to measure the performance level of a given objective for a given situation (Keeney 1992).

At this stage of the model, criteria should be established in a non-redundant, exhaustive, and coherent form for all objectives (Roy 1996). Also, criteria must have three properties: they must be measurable, operational, and understandable.

Their meaning is understood to be as follows: measurable means that criteria have to represent the objectives in detail; operational means that criteria should provide a common basis for value judgment; and as to understandable, it is assumed that criteria cannot be ambiguous when evaluating the alternatives (Keeney 1992).

As for identifying objectives, if criteria are in disagreement with these properties and definitions, future problems will arise in the subsequent steps of the model, which may lead to the use of an inconsistent MCDM/A method, and consequently, an unrepresentative solution may be indicated.

In addition, according to Keeney (1992), three types of attributes can be observed: natural attributes, constructed attributes, and proxy attributes. This classification depends on the values that will be presented within each criterion.

Natural attributes have the same interpretation for all DMs and they are clearly defined independently of the decision context. Examples include: price, distance, and duration. Constructed attributes are used when it is not possible to use natural ones. However, they are only suitable for the context of a specific decision. An example is when a subjective assessment needs to be used and a scale can be constructed to represent the alternative assessments in the criterion. Finally, proxy attributes are used in the latter case as an indirect measurement associated with the objective.

Moreover, the criteria can be deterministic or probabilistic. In problems in which information about consequences is known to be certain, i.e., the evaluation of each alternative in the specific criterion is represented by a constant level of performance; this criterion can be characterized as a deterministic criterion.

On the other hand, in a problem where information about consequences is probabilistic, the evaluation for each alternative in a specific criterion is based on information that might use a probability density function (PDF). For these problems, the probabilistic modeling phase has to be considered.

RRM decision problems require a probabilistic modeling phase, although in some cases, simplifications can be made in order to represent probabilistic consequences as deterministic indices, which in general can be some statistic of the PDF (e.g., means, percentiles, etc.).

3.4 Step 4 –Establish Alternatives

To establish problem alternatives, the first evaluation that must be made is about identifying the characteristics of the alternatives that will be used in the problem. To identify these characteristics, three questions need to be answered:

- Is the set of alternatives discrete or continuous?
- Are the alternatives stable or can they change throughout the process (Vincke 1992)?
- Can the problem be solved by choosing one alternative as a solution, and excluding the rest of them or by combining alternatives?

After defining these characteristics, the problematic adopted in the problem has to be defined. This concerns how the DM intends to evaluate the set of alternatives. Some types of problematic are:

- Choice Problematic: this is used when the DM desires to reduce the initial set of alternatives to a smaller subset.
- Ranking problematic: this is used when the DM desires to rank the alternatives from best to worst.
- Sorting Problematic: this is used when DM desires to classify alternatives into previously defined categories.
- Description Problematic: this is used when DM desires to describe alternatives.
- Portfolio Problematic: this problematic finds a combination of a subset of alternatives that maximizes the objectives and is limited by constraints.

Finally, alternatives for the problem can be generated, alternatives already presented in the environment can be used or new alternatives can be created. The VFT methodology emphasizes that the DM must create alternatives, and not only accept those that already exist and that are available to him/her when the problem occurs.

For each criterion an alternative is given an outcome (or consequence), which will be evaluated in the MCDM/A approach. The consequences can be deterministic or probabilistic. Deterministic consequences are those for which an exact value can be defined as the evaluation of the alternative in the criterion. Probabilistic consequences are used when problems are in an uncertain scenario. In this case, the evaluation of an alternative in a specific criterion is based on a probability distribution which represents this criterion.

Therefore, when this step is concluded a consequence matrix can be obtained. The consequence matrix for decision problem presents the evaluation of each alternative in each criterion.

3.5 Step 5 – Define State of Nature

This step will deal with problems in which some variables (state of nature) are not under the control of the DM, and thus cause random changes in the consequences matrix. The State of Nature is a typical ingredient in the traditional Decision Theory approach (Raiffa 1968; Berger 1985; Edwards et al. 2007; Goodwin and Wright 2004).

In these cases, the presence of the experts is very important since they give factual information about such variables to the DM. For example, in problems where the failure mode has to be evaluated, an expert's knowledge about the situation can be useful to support the DM in obtaining the evaluation of each alternative in the specific criterion.

Some precautionary measures should be taken in this step. For example, for the state of nature, the analyst has to consider a probabilistic modeling of such information. Also, experts have to supply only factual information about these variables, since it is inappropriate to include preference information from experts in the decision model.

3.6 Step 6 – Establish a Priori Probability

For these problems, *a priori* information about the state of nature (θ) is characterized as an important element which should be defined in order to construct the model. This quantification can be provided by using probability distributions of θ, π(θ), called *a priori* probability distributions (Berger 1985).

Therefore, as stated in the previous phase, an expert's knowledge about the problem scenario can be used to quantify the *a priori* probability distribution π(θ). Some procedures to develop the elicitation of expert's prior knowledge are set out in the literature.

Keeney and von Winterfeldt (1991) proposed the following steps to elicit *a priori* probabilities:

- Identify and select the problem
- Identify and select experts
- Discuss and refine the problem
- Train experts to provide the elicitation, evaluating the reason to perform the elicitation
- Conduct the elicitation process
- Analyze the results
- Solve disagreements
- Document the results

One of the elicitation procedures is the equiprobable intervals method (Raiffa 1968). This method is based on developing equal intervals of probability based on estimating the most likely value of the state of nature (θ) given some probabilities. This method follows some steps:

- Define the range of the minimum and the maximum values of the state of nature based on the value of an event that is unlikely to occur, with a probability of 0.001, and an event that is likely to occur, with a probability of 0.999.
- Development of equal intervals of probability in order to define other values of state of nature, the third value defined is the intermediate value with a probability of 0.5.
- Repeat the step again dividing the intervals into equal parts and estimating values of the state of nature. This will give the values of state of nature, a probability of 0.25 and 0.75.
- After some points have been defined, a consistency test should be performed with the expert to confirm if the values estimated are consistent.

- Finally, having defined the points, a statistical analysis can be performed in order to discover the probability distribution which best fits the points.

Therefore, these problems can be presented in a risk scenario. In these cases, it is appropriate to conduct the probabilistic modeling phase in order to formalize the problem and to support the DM's understanding of the problem. If no probabilities are obtained, then an uncertain scenario is considered.

In general, for problems presented in a risk scenario, Bayesian Decision Theory (Berger 1985) is used to support the decision process. On the other hand, for problems in an uncertain scenario, it is recommended such procedures as MaxMin or MinMax be used (Raiffa 1968; Berger 1985).

3.7 Step 7 – Establish Consequence Function

As is well known, the expected utility function $[E_\theta\ u(a)]$ of an alternative *a* is given by Eq. 1 as follows:

$$E(a) = \int \pi(\theta)\, u(\theta, a)\, d\theta \tag{1}$$

where:

$u(\theta,a)$ is the utility of alternative *a* when the state of nature is θ.

Then, one can obtain the utility $u(a)$ using the *a priori* probability $\pi(\theta)$.

The utility $u(\theta,a)$ is obtained by Eq. 2

$$u(\theta, a) = \int P(x|\theta, a)\, u(x) dx \tag{2}$$

where:

$P(x|\theta,a)$ is the consequence function.

$u(x)$ is the utility function of x which is obtained by preference modeling as dealt with in steps 8, 9, and 10.

The focus of this step is the consequence function, which associates the consequence to the state of nature and the chosen alternative. It is the probability $P(x|\theta,a)$ of obtaining *x* given θ and the alternative *a* (Berger 1985).

In general, $P(x|\theta,a)$ is obtained based on statistical data analysis or assumptions with regard to its behavior, as illustrated in de Almeida and Souza (2001), in a problem of service supply selection for maintenance, in which the consequence is the time to repair and θ is μ, the parameter of $f(x)$ which is assumed to be an exponential probability function. This is the probabilistic model, which is the usual case in RRM.

3.8 Step 8 – Preference Modeling

Preference modeling is the first step for the third phase of the decision model. This phase presents higher interaction between the analyst and the DM, where the flexibility presented in the model allows not only reviews of the previous steps, but these three steps to be integrated.

This phase plays an important role in building the model, and has to be developed with care, because at the end of it the MCDM/A method is defined that will be used to solve the problem. Modeling preferences with the DM is one of the main steps within the decision-making process.

Based on this step the DM's preference structure will be characterized. A preference relationship system or preference structure is represented by a collection of preference relations applied to the set of alternatives, which is constructed based on exhaustive and not exclusive comparisons.

Thus, some of the main preference structures of a DM are: Structure (P, I); Structure (P, Q, I); Structure (P, Q, I, R). Thus, based on these structures, the preference relations are:

- Indifference (I): for DM there are clear reasons for declaring equivalence between two alternatives.
- Strict Preference (P): for DM there are clear reasons to justify that one alternative is preferable to another.
- Weak preference (Q): for DM there is no clear reason for declaring either indifference or strict preference. Therefore, the DM's preference lays between P and I relations.
- Incomparability (R): for DM there are no reasons to justify any of the other three relationships. Incomparability is useful when DM is unable or unwilling to establish comparisons between two alternatives.

Thus, in this step, it is necessary to evaluate which Preference Structure best represents the DM's preferences for the problem. For example, Structure (P, I) should be used when the DM can define relations for each comparison of consequences. Thus, for this structure, the property of Ordenability, which is related to the possibility of providing comparisons for each pair, is the first that will be tested. Therefore, based on the agreement of this property, the transitivity property should be tested, where if x, y, and z are consequences and x P y and y P z, consequently x P z.

On the other hand, the structure (P, I, Q, R) allows DM to have doubts about the comparisons between the alternatives, and therefore the DM may remain undecided between two relations, such as Q, or may not be willing to express his/her preferences over some pair, such as R.

Moreover, in this step, one more important consideration that must be taken into account concerns the rationality considered by the DM in the problem, which can be: compensatory or non-compensatory. The terms compensatory and non-compensatory are associated with studies by Fishburn (1976).

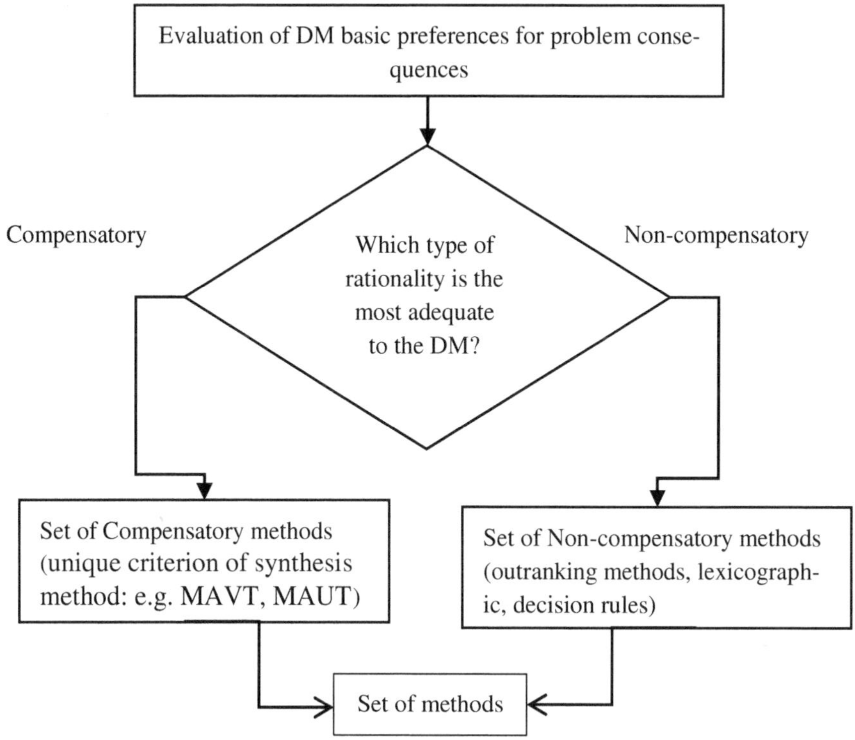

Fig. 4 Evaluation of compensatory and non-compensatory rationality. (Adapted from de Almeida et al. 2015)

Compensatory rationality exists when a worse performance of an alternative in the *criterion i* can be compensated by a higher performance of the same alternative in the *criterion j*. For this rationality, the trade-offs between the consequences are performed.

Non-compensatory rationality is the opposite of the previous one, when compensations between performances are not relevant for the DM. In this case, the difference in performance between two consequences is not relevant for the DM. The information that is relevant to him/her is which alternative wins over the criterion, even if the difference between them is very small.

Depending on the structure defined and the rationality that the DM presents, at the end of this step, a set of coherent MCDM/A methods is pre-selected, according to Fig. 4.

Figure 4 presents a flowchart to illustrate this step. From this figure, it can be seen that the estimation about compensatory or non-compensatory rationality is very important since this is used to define what family of MCDM/A methods is indicated and therefore will be pre-selected.

According to de Almeida et al. (2015), MCDM/A methods are characterized as a methodological formulation or a theory, which has an axiomatic structure. These methods are generic and can be applied in different problem situations in order to help find a solution.

Regarding compensatory rationality, a unique criterion of synthesis method (Roy 1996) is recommended to be applied, where the most usual is the additive aggregation based on the MAVT (Multi-Attribute Value Theory) or MAUT (Multi-Attribute Utility Theory) (Keeney and Raiffa 1976). The additive aggregation combined the criteria and generates a global value for each alternative. For non-compensatory rationality, it is recommended that outranking methods be used (Roy 1996). These methods make pairwise comparisons between the alternatives, as commented in the first chapter of this book.

However, the careful definition of the rationality, based on the DM's preferences for the problem situation, is not even considered. In inappropriate cases, a familiar MCDM/A model is selected for use before all the DM's preferences have been evaluated, i.e., at the beginning of the building model.

According to Wallenius (1975), in general, DMs do not feel comfortable about using decision models which they consider are difficult. In the same context, Bouyssou et al. (2006) commented that heuristics can be suggested to facilitate solving the problem. Therefore, in these cases, the analyst should be alert and ensure that the method used to characterize the DM's preferences was appropriate, and therefore presents a recommendation which bring benefits to the decision situation.

3.9 Step 9 – Conducting an Intra-Criterion Evaluation

Intra-criterion evaluation is the evaluation of each alternative in each criterion, assigning a marginal utility function. Within the intra-criterion evaluation, an important concept is the scale and the scale transformation. For utility function an interval scale is considered, in which the utility zero is assigned to the worst consequence.

Sometimes the marginal utility function may be constructed over a consequence expressed as a verbal scale. A widely used quantitative verbal scale is the Likert scale (1932).

Depending on the pre-selected family of MCDM/A methods, the form of evaluating the intra-criterion will be developed in different ways.

As to compensatory rationality, where the methods of unique criterion of synthesis are adequate, the evaluation of each alternative in each criterion is represented by a value function for deterministic consequences or a utility function for probabilistic consequences. The value or utility functions can be linear or non-linear.

To construct the value function, few procedures are presented in Belton and Stewart (2002), it being simpler to model the problem in this case. On the other hand, in order to elicit the utility functions, the DMs behavior regarding risk has to be investigated. When the DM is considered risk averse or risk prone, the utility function is non-linear. For the DM who is neutral to risk, the utility function is linear.

Regarding non-compensatory rationality, where it is appropriate to use outranking methods, this step is conducted in another way. If the preferences for consequences which were expressed for each criterion are ordered, there is no need to conduct further evaluation in this step, but, the threshold estimation is characterized as being part of the intra-criteria evaluation. On the other hand, when probabilities are assigned to consequences, then a utility function might be applied, incorporating the DM's attitude to risk. This would make necessary an integration between marginal utility function with outranking methods, as already done (de Almeida, 2005; de Almeida, 2007; Brito et al., 2010).

3.10 Step 10 – Conducting an Inter-Criteria Evaluation

The last step of this phase is the inter-criteria evaluation. Inter-criterion information allows the quantitative criteria to be combined in an aggregation process. This step involves elicitation procedures to obtain the criteria weights (de Almeida et al. 2015)

Different mechanisms of aggregations are presented in the literature; the mechanism selected depends on the MCDM/A method that will be used.

As to methods of unique criterion of synthesis, scale constants (k_j) are used to aggregate the criteria. Scale constants do not represent how important the criteria are to DMs and cannot be directly determined. They represent the ratio between criteria, considering the set of consequences present in each one of them. The main differences between the MCDM/A methods presented in this classification are in the elicitation procedure applied to obtain the scale constants.

An example of an elicitation procedure, for deterministic consequences, is the tradeoff procedure (Keeney and Raiffa 1976), which presents a robust axiomatic structure which seeks indifference points to formulate (n-1) equalities, where n is the number of criteria in the problems. These equalities are used to find the exact values of scaling constants. The FITradeoff method (de Almeida et al., 2016) uses the same robust axiomatic structure with some advantages, needing only partial information from the DM, as mentioned in the first chapter of this book.

For probabilistic consequences, MAUT (Multi-Attribute Utility Theory) or prospect theory could be applied, as explained in the first chapter of this book.

As for outranking methods, the weights are defined directly by the DM. They represent the level of importance that each criterion in the problem has for the DM. The weights are normalized so that they sum to one.

At the end of this step, the decision model has been built and an appropriate MCDM/A method is indicated to solve the problem. In other words, the end of this step represents the end of the funnel, illustrated in Fig. 1. The next phase deals with applying the method procedure, testing the robustness of the solution, reviewing the decision-making process and implementing the recommendation.

3.11 Step 11 – Evaluate Alternatives to Find a Solution

This step is the first step of the finalization phase. In it, the algorithm of the MCDM/A method selected is processed and presents the solution to the problem. The MCDM/A method selected is not personalized for the problem, since it is generic and can be applied to many different situations.

On the other hand, the decision model built, which is implemented in order to indicate the adequate MCDM/A method, is personalized for each problem, and constructed based on the DM's preferences which were expressed in the previous step.

3.12 Step 12 – Conduct a Sensitivity Analysis

Sensitivity analysis is a relevant step which aims to test the robustness of the decision model. Thus, after the sensitivity analysis, the recommendation found in the last step will be confirmed or reevaluations will be indicated for the building model.

The sensitivity analysis is characterized as being used to change problem inputs in order to analyze how these changes impact the recommendation made for solving the problem. In other words, this step verifies if the recommendation found in step 11 is sensitive to variations in the data of the problem, such as the consequence matrix and the criteria weights.

Sensitivity analysis can be conducted in two ways: individually, by changing one parameter at a time, and simultaneously by changing several parameters of the decision model.

With regard to the former, changes to the values of the scale constants (or weights) or of some consequences can be made. An example of variation can be generated by applying a percentage change of 10% to the nominal value, thereby generating values that are higher or lower than the original ones.

In MCDM/A problems, the values of consequences can be generated by considering some approximations since it is quite difficult to have access to all the data accurately. Therefore, it should be interesting to modify the values of consequences in order to test the robustness of the model since this model presents approximations. Many modifications can be performed to test the robustness of the building model.

For a complete evaluation, several changes must be done simultaneously. The Monte Carlo simulation is an approach used for simultaneous sensitivity analysis. In this approach, a random variation of data is applied to test the decision model. Thus, the solution found to each problem created is compared to the initial recommendation found in step eleven.

To test the robustness of the model, the frequency of changes in the initial recommendation is calculated after conducting a large number of simulations.

Moreover, to complement this analysis, statistical hypothesis tests can be applied in order to evaluate the significance of these changes (Daher and de Almeida 2012).

Therefore, this step is important because based on its results, it is confirmed if the model built can be used formally to represent the MCDM/A problem and to find the representative solution for it, or if the model has to be reevaluated, and thus to return to some previous step in order to review the preferences expressed. It is worth mentioning that the approximations provided in some steps of the decision model can be reevaluated based on the impact that they can cause for the recommendation found. This places the responsibility on the DM for determining whether to keep these approximations or to revise them in earlier steps of the model.

3.13 Step 13 – Draw Up Recommendation

In this step the recommendation found in step 11 and tested in step 12 is presented to the DM, especially with regard to it degree of accuracy investigated in the last step. If the recommendation is favorable for the DM, the implementation of this recommendation can be made, i.e., the solution can be applied in the real problem situation.

If the recommendation and its analysis of robustness are not favorable for the DM, the decision model must be reviewed in order to identify steps where the DM's preferences were not coherent or have changed during the process, and to identify steps where approximations made have impacted the recommendation found. As already stated, there is no right model being possible to DM review the previous assumptions made.

3.14 Step 14 – Implement the Solution

Finally, after the solution is found and accepted by the DM, it must be implemented. Brunsson (2007) presented important matters related to the implementation process, and emphasized that the implementation step depends on the decision situation and the decision model built.

As a result of the magnitude of the decision problem, the implementation process can be a complex process, and take more time to do than does the process for building the decision model. In this case, changes can occur in the problem scenario thereby modifying consequences and producing new solutions for the problems. In this case, should be interesting for DM to review the decision model build in order to update the problem elements and preferences expressed.

4 Conclusions

This chapter presents a framework for building decision models in the RRM context. This framework presented well-structured steps to support the DM in the evaluation of MCDM/A problems.

The framework developed was adapted from de Almeida et al. (2015) and had three phases. The preliminary phase aims to present important elements of the problem. The preference modeling phase deals with modeling the DM's preferences regarding the elements defined, and the finalization phase is when the recommendation found for the problem is identified and tested.

In this framework, an additional phase was included in order to improve the earlier framework. This new phase was the probabilistic modeling phase which has important features to support the DM when he/she is dealing with probabilistic problems.

Therefore, the framework developed in this chapter can be used to formalize MCDM/A problems in order to present the adequate recommendation. It is important to highlight that building models are always wrong since they are a simplification of problem reality, but some of them are necessary to represent the problem elements and support the DM to solve them following a rational process (Box and Draper 1987).

References

Ackermann F, Eden C (2001) SODA – Journey making and mapping in practice. In: Rosenhead J, Mingers J (eds) Rational analysis in a problematic world revisited, 2nd edn. Wiley, pp 43–61

Ackoff RL, Sasinieni MW (1968) Fundamentals of operations research. Wiley, New York, p 455

Belton V, Stewart TJ (2002) Multiple criteria decision analysis: an integrated approach. Springer, Boston, MA

Berger JO (1985) Statistical decision theory and Bayesian analysis. Springer, New York

Bouyssou D, Marchant T, Pirlot M, Tsoukis A, Vincke P (2006) Evaluation and decision models with multiple criteria: stepping stones for the analyst. Springer, New York

Box G, Draper N (1987) Empirical model-building and response surfaces. Wiley, New York

Brito AJ, de Almeida AT, Mota CMM (2010) A multicriteria model for risk sorting of natural gas pipelines based on ELECTRE TRI integrating Utility Theory. Eur J Oper Res 200:812–821

Brunsson N (2007) The consequences of decision-making. Oxford University Press, New York

Daher S, de Almeida A (2012) The use of ranking veto concept to mitigate the compensatory effects of additive aggregation in group decisions on a water utility automation investment. Group Decis Negot 21(2):185–204

de Almeida AT (2005) Multicriteria modeling of repair contract based on utility and ELECTRE I method dependability and service quality criteria. Ann Oper Res 138:113–116

de Almeida AT (2007) Multicriteria decision model for outsourcing contracts selection based on utility function and ELECTRE method. Comput Oper Res 34:3569–3574

de Almeida AT, Souza FMC (2001) Gestão da Manutenção: na Direção da Competitividade (Maintenance Management: toward Competitiveness). Editora Universitária da UFPE, Recife

de Almeida AT, Ferreira RJP, Cavalcante CAV (2015) A review of multicriteria and multi-objective models in maintenance and reliability problems. IMA J Manag Math (forthcoming)

de Almeida AT, de Almeida J, Costa APCS, De Almeida-Filho AT (2016) A new method for elicitation of criteria weights in additive models: flexible and interactive tradeoff. Eur J Oper Res 250:179–191
Eden C (1988) Cognitive mapping. Eur J Oper Res 36(1):1–13
Eden C, Ackermann F (2004) SODA. The principles. In: Rosenhead J, Mingers J (eds) Rational analysis for a problematic world revisited, 2nd edn. Wiley, Chichester
Edwards W, Miles RF Jr, Von Winterfeldt D (2007) Advances in decision analysis: from foundations to applications. Cambridge University Press, Cambridge
Fishburn PC (1976) Noncompensatory preferences. Synthese 33:393–403
Franco LA, Cushman M, Rosenhead J (2004) Project review and learning in the construction industry: embedding a problem structuring method within a partnership context. Eur J Oper Res 152(3):586–601
Goodwin P, Wright G (2004) Decision analysis for management judgment. Wiley, London
Guitouni A, Martel JM (1998) Tentative guidelines to help choosing an appropriate MCDA method. Eur J Oper Res 109(2):501–521
Keeney RL (1992) Value-Focused Thinking: a path to creative decision-making. Harvard University Press, London
Keeney RL, Raiffa H (1976) Decisions with multiple objectives: preferences and value trade-offs, Wiley series in probability and mathematical statistics. Wiley, New York
Keeney RL, von Winterfeldt D (1991) Elicitation probabilities from experts in complex technical problems. IEEE Trans Eng Manag 38:191–201
Keisler JM, Noonan PS (2012) Communicating analytic results: a tutorial for decision consultants. Decis Anal 9:274–292
Likert R (1932) A technique for the measurement of attitudes. Arch Psychol 140:1–55
Polmerol J-C, Barba-Romero S (2000) Multicriterion decision in management: principles and practice. Kluwer, Boston
Raiffa H (1968) Decision analysis: introductory lectures on choices under uncertainty. Addison-Wesley, London
Rosenhead J, Mingers J (eds) (2004) Rational analysis for a problematic world revisited, 2nd edn. Wiley, Chichester
Roy B (1996) Multicriteria methodology for decision aiding. Springer, New York
Simon HA (1960) The new science of management decision. Harper & Row Publishers, Inc, New York
Vincke P (1992) Multicriteria decision-aid. Wiley, New York
Wallenius J (1975) Comparative evaluation of some interactive approaches to multicriterion optimization. Manag Sci 21(12):1387–1396

Part II
MCDM/A Models for Risk Decision Analysis

A Participatory MCDA Approach to Energy Transition Policy Formation

Mats Danielson, Love Ekenberg, Nadejda Komendantova, Ahmed Al-Salaymeh, and Leena Marashdeh

1 Energy Transition

Projections show that energy demand in Jordan will increase during the next decades, largely due to population growth, migration dynamics in the region, an increase in the quality of life, and the increasing electricity needs for the desalination of water and cooling of buildings due to climate change, both requiring large amounts of energy. For example, energy demand forecasts for Jordan show an annual increase of 5% of Jordan's primary energy demand and 6% of Jordan's electricity demand annually by the year 2020 (Komendantova et al. 2017). Due to the lack of energy resources, the question of how to cover energy demand is a constant challenge in Jordan. The country is heavily dependent on imports of energy, largely from fossil fuels. It therefore also suffers from fluctuation in energy prices, which increases the Jordanian national debt and affects its national economy. According to the Ministry of Energy and Mineral Resources (MEMR), Jordan imports over 95% of its energy needs. This situation will become even more acute if the annual primary energy demand growth of 7% is considered.

M. Danielson · L. Ekenberg (✉)
Department of Computer and Systems Sciences, Stockholm University, Stockholm, Sweden

International Institute for Applied Systems Analysis, IIASA, Laxenburg, Austria
e-mail: mats.danielson@su.se; ekenberg@iiasa.ac.at

N. Komendantova
International Institute for Applied Systems Analysis, IIASA, Laxenburg, Austria
e-mail: komendan@iiasa.ac.at

A. Al-Salaymeh · L. Marashdeh
Mechanical Engineering Department, Faculty of Engineering and Technology, University of Jordan, Amman, Jordan
e-mail: salaymeh@ju.edu.jo

A. T. de Almeida et al. (eds.), *Multicriteria and Optimization Models for Risk, Reliability, and Maintenance Decision Analysis*, International Series in Operations Research & Management Science 321, https://doi.org/10.1007/978-3-030-89647-8_5

One of the crucial developments in the Jordanian energy sector in 2017 was the completion of the Aqaba terminal. The goal of this project was to secure the supply of crude oil and oil products to Jordan. The terminal has storage capacities for crude oil, oil products and liquefied petroleum gas. The Logistic Company for Jordan's Oil Facilities was established in the year 2016 as the operator and manager of this project. The costs of crude oil and oil products imports reached JOD 1333 million in 2016. In general, the year 2016 witnessed a decrease of around 21% in the consumption of oil products due to a decrease in the demand for oil products used in electricity generation and large imported quantities of natural gas. The oil shale sector also experienced significant development in that year. Jordanian decision-makers consider this energy source to be strategically important, considering the fact that Jordan has the fourth largest oil shale reserve in the world, exceeding 70 billion tons. In 2017, the Jordanian government signed several memoranda of understanding and gave concessions for local and international companies to invest in the area of oil shale, including in-situ retorting and direct burning to generate electricity. In 2017, the natural gas sector also experienced some development. The National Petroleum Company signed a production-sharing agreement with the IPG Company to develop the Risha field. Two liquefied natural gas (LNG) agreements were also signed between NEPCO and Shell International Company to expand the use of natural gas in power plants and industries. In 2017, the green corridor project saw ongoing grid expansion and reinforcement plans; this will continue, as NEPCO's plan is that it will contribute to the upgrading of the national grid capacity to assimilate 1200 MW of renewable energy projects in the southern area of Jordan. It is expected that the project will be completed by the end of 2018.

The deployment of new technologies, along with higher use of existing technologies, which are needed to cover energy demand and to diversify energy supply, will lead to an energy transition in Jordan and a transformation of the Jordanian energy system. Energy transition in Jordan will be and already is a complex process, which has political, social, economic and technical dimensions. Therefore, a holistic, inclusive and comprehensive governance approach to energy transition is essential. The process of substituting one energy source with another, and one technology with another, can result in significant socio-technical changes which might lead to many frictions and conflicts. This process will lead not only to technological change but also to a socio-technological transition process, which will be combined with shifts in generation and distribution technologies, business models, governance structures, consumption patterns, values and worldviews. For a sustainable implementation of this process, new forms of governance are needed.

Various incentives paved the way for technology transfer in the MENA region. However, they failed due to a variety of factors, amongst them social and public acceptance (Komendantova et al. 2017). Other reasons included the governance of this transition, which was based on a top-down framework of national renewable energy master plans elaborated by the MENA governments. The realization of these plans was far behind the settled targets, mainly because energy transition roadmaps underestimated the intricacy of managing transformative change towards sustainable energy systems (Brand 2015). There are several examples and good practices from Europe, such as the *Energiewende* (energy transition) in Germany or energy transition through climate and energy models in Austria. However, plans for energy transition in the MENA region should consider completely different energy market structures, stakeholder networks and societal aspirations towards energy, climate and environmental policies in the region (FES 2015). Therefore, careful consideration of stakeholders' views, concerns and conflicting priorities is required when considering a sustainable energy transition and transformation of the energy system, as well as for compromise-oriented energy governance solutions.

The transformation of energy systems often faces risks and boundaries regarding the implementation of climate change mitigation policies, which are connected with decision-making processes (Patt 2015). These boundaries include not only technological and economic factors, but also human factors, including conflicting views of the risks and benefits of different technologies, as well as social and public acceptance, and willingness to use technology and to pay for it (Komendantova et al. 2018). Today, public interest in energy infrastructure is different from what it was half a century ago when the existing infrastructure was built. The existing energy infrastructure was perceived as a driver for socio-economic development. Nowadays, people want to participate in the decision-making process on technologies that affect their communities. Participation in decision-making processes is often perceived as a democratic principle of the inclusiveness of people (Beierle and Cayford 2002). The lack of opportunity to exercise this right leads to protests, delays in the implementation of projects, and even the cancellation of projects because of public protests or actions of stakeholders who were not included in the decision-making process (Kunreuther et al. 1994).

International legislation also lays down the right to participate. The Aarhus Convention requires the involvement of stakeholders in decision-making processes on infrastructure projects and the provision of clear and transparent information about how to get involved. However, often there are limits to participation; the fact that energy transition is a topic heavily dominated by technological and economic content hinders effective public participation (Devine-Wright 2012). Different views on participatory governance exist. Some argue that complex decision-making processes on critical infrastructures, such as energy, should be left in the hands of experts and scientists. Public participation is reserved as a method for evaluating this decision-making process and its outcomes (Rowe and Frewer 2000). Others argue that participation is very beneficial because it brings additional knowledge of stakeholders at the national level (Hänlein 2015), such as the knowledge of local areas, which might be limited (Jasanoff 1998. There is also evidence that integrating the views of all stakeholders—and not only those of specialized experts—can enhance the legitimacy of decision-making processes and build trust (Renn 2008). Evidence from energy generation and transmission projects in Europe shows that decision-making processes along the so-called 'decide-announce-defend' (DAD) model, where the decision is taken by the national government, aided by experts and then implemented through a top-down approach, is no longer feasible (Wolsink 2000; Komendantova and Battaglini 2016). The DAD model often leads to conflicting opinions, as well as protests which delay implementation and may even lead to the cancellation of projects (Wolfsink 2012). Discussions on the framework of the so-called NIMBY (not-in-my-backyard) concept often end up simply identifying factors of acceptance, which is a more passive attitude towards a top-down decision-making process where a person cannot change anything. Nowadays, many scientists argue that NIMBY is a misleading concept to understand local objections and concerns. One flaw of the concept is that it does not involve local knowledge to improve the results of decision-making processes (Batel and Devine-Wright 2015). There is also the need to understand how engagement and participation can go beyond a discussion of a project's details and shape a discussion about centralized and decentralized energy transition, as this is a complex topic where human factors play a significant role. Understanding is needed about how participatory governance works in different countries and how centralization or decentralization of decision-making shapes the process of stakeholders' involvement in the discussion of energy transition issues (Komendantova et al. 2015).

Even though a significant part of existing literature on participatory governance research is focused on Europe, there is also evidence of the advantages of a participatory approach for other countries. For instance, Xavier et al. (2017) studied the implications of human factors for the transformation of the energy sector in South Africa. Having analysed several infrastructure projects, the authors express the need to incorporate public participation within the project cycle and to institutionalize it as a part of the whole decision-making process. They also find that existing conflicts in stakeholders' views and opinions can be mitigated through engagement and different methods of multi-criteria discussion.

Yazdanpanah et al. (2015) looked at human factors of energy transition in Iran that influence the willingness to use renewable energy sources. Applying the theory of planned behaviour, the authors identify the main factors as moral norms, attitudes and perceived behavioural control, which is also connected with the possibility to influence decision-making processes.

Thus, it is necessary to develop compromise solutions to mitigate the risk that differences in views about electricity generation technologies needed for energy transition will turn into conflicting opinions. These human factors include perceptions of different risks connected with technological deployment, and views about the benefits and impacts generated by different technologies. To accomplish this, we have developed a multi-stakeholder multi-criteria approach to assess the relevance of Jordan's electricity generation technologies against a set of criteria under uncertainty which we will present in the following. The next section describes the criteria used and the stakeholder groups. Section 3 describes the stakeholder workshop set-up and the resulting criteria rankings, as well as some methodological considerations. Section 4 demonstrates the decision methodology and Sect. 5 provides the results from the different workshops. Finally, Sect. 6 concludes the chapter.

2 Criteria and Stakeholders

Each Jordanian technology which is considered in the national energy planning was evaluated against a set of criteria. Altogether, there were eleven criteria, including 20 indicators, nine of which were quantitative and eleven qualitative. Data for quantitative indicators were collected from national and international statistical databases, reports and projects. Data for qualitative indicators were collected from surveys of stakeholders in Jordan.

The criteria were divided into two sets:

- **Contribution to national energy policy targets**, such as to secure reliable and affordable power supply. This included such criteria as decreasing dependence on foreign resources, climate change mitigation, domestic industry development, technology and knowledge transfer, as well as affordable electricity system costs.
- **Sensitivity to local conditions and impacts on local communities**. This included aspects of land and water resources, on-site job creation, air pollution and health, hazardous waste and safety issues.

The project team selected eleven out of initially 32 relevant criteria. These were then discussed during the stakeholder workshops to see whether the stakeholders agreed with the criteria definitions, whether the criteria were relevant for the case countries and whether stakeholders would recommend any further criteria. The stakeholders' reactions confirmed the robustness of the selected criteria and their definitions, which were also communicated to stakeholders during the workshops.

- ***Use of domestic energy sources.*** The dependence on foreign energy imports can be decreased by tapping into domestic resources that are either available today or could be exploited in the mid to long term. Two indicators are relevant here: a) the current domestic potential of each technology's energy carrier to decrease energy import dependence; and b) the future domestic potential of each technology's energy carrier to decrease energy import dependence by 2040/50.
- ***Global warming potential.*** The technology should contribute to the mitigation of climate change. This criterion is based on the indicator total lifecycle GHG emissions (CO_2-eq) per generated kWh.
- ***Domestic value chain***. The technology should have a high potential to use components and services provided by domestic industries throughout the entire value chain. This criterion is based on the indicator existing potential for the integration of domestic industries to manufacture a significant share of components and provide essential services during the manufacturing, construction and installation (MCI) and operation and maintenance (OM) phases of the technology.
- ***Technology and knowledge transfer***. Based on existing policies, the technology should have a high potential to benefit from technology and knowledge transfer to stimulate future domestic value added in electricity generation. This criterion is based on the indicators a) effectiveness of educational policies to foster skill development and R&D; and b) effectiveness of industrial policies to enhance

industry linkages between domestic and foreign firms geared towards horizontal technology transfer.

- ***Electricity system cost***. The electricity system cost of the technology should be as low as possible so as not to constitute a burden for Jordan's overall budget. This criterion is based on the indicators a) electricity-generation cost measured as levelized cost of electricity (LCOE) in €/MWh; and b) estimated additional integration cost at increasing penetration levels based on uncertainty/variability and distance/location.
- ***On-site job creation***. The technology should have a high potential to create direct on-site jobs over the entire lifetime of the power plant. This criterion is based on the indicators a) MCI: the average amount of labour in FTE person-years per MW; and b) OM: the average amount of labour in FTE permanent jobs per MW.
- ***Pressure on local land resources***. The technology should cause minimal additional pressure on valuable land resources regarding the amount and value of required land to avoid the deprivation of any locally relevant livelihood resources. This criterion is based on the indicators a) land requirement: the area of land directly required by the technology at the site of its deployment in ha/MW; and b) land value: the importance of the land surrounding typical project sites for providing livelihood resources and services to adjacent communities.
- ***Pressure on local water security***. The technology's water consumption should be appropriate to the local water risk context and cause minimal pressure on local water security. This criterion is based on the indicators a) average operational water consumption of each technology measured in l/MWh; and b) average water risk at typical project sites of each technology based on the Water Risk Index (https://www.wri.org/).
- ***Occurrence and manageability of non-emission hazardous waste***. The disposal of non-emission hazardous waste produced during the operation of the technology and the risk stemming from national waste management capabilities should be low to minimize adverse consequences on human health and the environment. This criterion is based on the indicators a) disposal of non-emission hazardous waste; and b) potential national capabilities to manage the disposal of the respective types of non-emission hazardous waste.
- ***Local air pollution and health***. The amount of air pollutants (NO_x, SO_2 and PM) emitted by the technology should be low to minimize pressure on local air quality and health risks for people in adjacent communities. This criterion is based on the indicators a) air pollutants (SO_2, NO_x, and $PM_{2.5}$) emitted by O&M activities of power plants in kt/MWh; and b) premature deaths by $PM_{2.5}$/MWh of electricity produced.
- ***Safety***. Severe accidents from the construction, operation and maintenance of electricity-generating technologies, as well as during the transport and storage of resources and equipment, should be minimized to reduce accidents resulting in fatalities within and outside power plants. This criterion is based on the indicators a) historical immediate fatalities from severe accidents during transport and storage of resources and equipment, and operation and maintenance activities of power plants, per unit of electricity (MWh) produced (hereafter referred

to as 'normalized fatalities'); and b) potential of regulatory and operational emergency preparedness, and response capabilities of the private and public sector to mitigate and manage the risk of catastrophic accidents with maximum and severe consequences during the construction and operation phase of each technology (hereafter referred to as 'normalized fatalities').

Six groups of different stakeholders were involved in the MENA-SELECT workshops. These groups represent the most relevant stakeholders for energy policy in Jordan, which are: policymakers, finance and industry, academia, young leaders, national and local NGOs, as well as civil society and local communities. These groups include the following stakeholders who participated in different events of stakeholders' dialogue organized in the framework of the MENA-SELECT project, such as workshops and surveys.

2.1 *Policymakers*

This group represents decision-makers in the Jordanian government and representatives of relevant organizations, who are responsible for developing and implementing energy policy in Jordan. The participants were from the Ministry of Energy and Mineral Resources, the Ministry of Water and Irrigation, the Amman Chamber of Industry, the Ministry of Public Works, the National Electric Power Company and the Jordan Press Foundation/Business section.

- The **Ministry of Energy and Mineral Resources** (MEMR) is the overarching legislative authority on energy-related issues in Jordan and, as such, lays down the goals and political framework conditions for the development of the energy market.
- The **Ministry of Water and Irrigation** (MWI) is responsible for the implementation of the Energy Efficiency and Renewable Energy Policy for the Jordanian water sector by rehabilitating different systems, installing new systems, and renewable energy projects that include different programmes such as solar energy systems for administrative buildings in the water sector, the utilization of hydropower potential to power the water sector, the utilization of biofuel potential in wastewater facilities, and large-scale renewable energy-based power generation for the water sector on available lands (MWI 2015).
- The **National Electric Power Company** (NEPCO) is responsible for the construction, planning, development, operation, maintenance and management of the control systems, the electric transmission and interconnection networks, as well as for management of the processes of purchasing, transmitting, control and selling the electric power in Jordan and to neighbouring countries. It also conducts the planning studies in this regard. The company provides services, consultancy and studies related to electric power to various parties inside and outside Jordan.

- The **Amman Chamber of Industry** (ACI) is a non-profit organization which represents the industrial sector in Jordan. The ACI forms and develops a framework to crystallize the industrial point of view of its members in relation to economic issues in general and industrial issues in particular. To this effect, the chamber cooperates with the ministries and relevant government economic planning, especially with regard to industry, in coordination with the Jordan Chamber of Industry. Within the framework of ACI's strategy and plans, it aims to promote the use of renewable energy and reduce energy costs for the factories.
- The **Ministry of Public Works and Housing** (MPWH) aims to provide new government buildings that are environmentally friendly and energy-saving.
- The **Jordan Press Foundation** is the owner of the Al-Rai newspaper and is a shareholding company responsible for media coverage.

2.2 Finance and Industry

The participants of this group represented energy and environment companies, engineering companies, banks and factories, represented by the following companies: GREENVIRO for renewable energy, control and communication, Al-Masar Engineering Company, Control and Communications Company (CCC), Arab Bank, Greenplans Environmental Consultations, Petra Elevators Company and Qatrana Cement Company.

- The **banks** in Jordan are involved in energy projects through signed agreements with the Jordan Renewable Energy and Energy Efficiency Fund (JREEEF) to finance renewable energy projects.
- Jordan's **industrial sector** is composed mainly of the mining and quarrying and manufacturing subsectors. Large-scale industries operate primarily in the field of phosphate and potash mining, the industrial production of cement, fertilizers and refined petroleum. The industrial sector's energy consumption represented about 16% of the total energy consumed in Jordan in 2016 (MEMR 2016).
- **Al-Masar Engineering** is a company specializing in the design and implementation of solar energy systems to generate electricity, store system electricity in batteries or connect them to the network. Al-Masar Engineering is one of the first companies to use energy-saving heating pumps to heat water for home use.
- **GREENVIRO** is a Jordan-based company that was established in 2013. The company offers customized energy consumption consultations, energy-saving system designs, energy-saving products, and the design and installation of solar energy systems.
- The **Control and Communications Company** (CCC) was established in 1990. It provides markets with industrial control systems, file tracking, access control, time attendance, fire alarms and other systems that could be used in the field of renewable energy systems.

- **Greenplans Environmental Consultations Ltd. Co.** is an engineering consulting firm specializing in environmental engineering and consulting services that cover the areas of water, environment, waste and energy (renewable energies and energy efficiency) related to engineering and environmental projects, industries, facilities and development zones.
- **Petra Elevators** is a company that provides and designs an innovative range of elevators, lifts and other technical devices for smooth riding comfort, preciseness and reliable speed control.
- The **Qatrana Cement Company** was established in 2007 with a total investment of 500 million US dollars. The Qatrana cement plant is located 80 km south of Amman. A 30-megawatt power plant, which runs on coal, will be constructed to supply energy to the Al Manaseer cement factory in Qatraneh. It is planned that the project will be operational by 2025.

2.3 Academia

This group represents researchers and academics in the field of energy. The participants were faculty members and researchers from the University of Jordan, the King Abdullah II Design and Development Bureau (KADDB), Al-Zaytoona University, the Applied Science University, and the German Jordanian University. For example, the KADDB is an independent government entity within the Jordanian Armed Forces (JAF) aiming at becoming a global defence and security research and development hub in the region. The Bureau's scope of work includes defence design and development, testing and evaluation, technology incubation in the Kingdom and defence technology training.

- The **University of Jordan** is a public university located in Amman. It is Jordan's largest and leading institution of higher education.
- The **Al-Zaytoona University** is a private university and includes six faculties, encompassing 19 undergraduate specializations and one graduate programme.
- The **Applied Science University** is a private university located in Amman, Jordan. It was established in 1991 as the largest private university in Jordan in terms of campus area and the number of student enrolments.
- The **German Jordanian University** is a public university in Madaba, Jordan. It offers more than 20 programmes to about 5000 students, primarily from Jordan. The university was modelled on the German applied science model, characterized by a focus on putting knowledge into practice and promoting knowledge transfer. It aims to play a significant role in promoting links between Jordan and Europe, particularly Germany. By taking advantage of the best educational practices in both Jordan and Germany, the university has positioned itself as a leader in its field.

2.4 *Young Leaders*

This group represents the graduate students in the field of energy, as well as young employees at energy and engineering companies such as Green Essence, KEPCO KPS IPP3 power plant and GREENVIRO.

- Green Essence specializes in renewable energy systems and is an authorized dealer for Suntech, the global leading PV panel manufacturer ranked as the largest manufacturer in the world. It can be compared to the leading German inverter manufacturers, such as SMA.
- KEPCO KPS IPP3 power plant is located on a greenfield site at Al Manakher, 30 km from the Jordanian capital Amman. It is the world's biggest tri-fuel power plant with an installed capacity of 573 MW. The plant is designed to use natural gas and heavy fuel oil (HFO) as its main fuels and light fuel oil (LFO) as the backup fuel.

2.5 *Civil Society and NGOs*

This group of stakeholders represents national non-governmental organizations in the field of energy, environment and engineering. The participants were from the Centre for Energy Services, the Renewable Energy Establishments Society, the Jordan Engineers Association (JEA), the Jordan Environment Society (JES), the Jordan Energy Chapter and the Sanibel Society for the Environment.

- The **Centre for Energy Services** is an integrated centre for energy, renewable energy and energy efficiency. It provides a range of training and advisory services through a qualified team to build capacity in this sector.
- The **Jordan Engineers Association** is a trade union of engineers in Jordan, and it is the largest trade union in the country.
- The **Jordan Environment Society** was established in 1988 as a non-profit non-governmental organization. It is the largest NGO in Jordan in its field. The objectives of the JES include, but are not limited to, protecting the environment and its basic elements such as water, air, soil and wildlife.
- The **Jordan Energy Chapter** is partnered with the American Energy Engineers Association, which is a non-profit professional society with over 18,000 members in more than 100 countries. Its mission is "to promote the scientific and educational interests of those engaged in the energy industry and to foster action for sustainable development".
- The **Sanibel Society for the Environment** is a non-profit organization concerned with environmental protection.

2.6 *Local Communities*

This group represents the local community from different cities in North and South Jordan. The participants were employees from the Ministry of Municipal Affairs in Alsalt city, Madaba city and Zarqa city, and interested citizens from Amman and Madaba.

The large-scale power-generating projects such as the Hussein Thermal Power Station and the first nuclear power plant are under development and construction in Zarqa city. The cities of Amman, Alsalt and Madaba host a number of small- and large-scale renewable energy projects.

The following workshops with different groups of stakeholders took place:

- Civil society and NGOs, 7 November 2016.
- Finance and investment, 9 November 2016.
- Academia, 10 November 2016.
- Future decision-makers, 12 November 2016.
- Local communities, 13 November 2016.
- Political decision-makers, 15 November 2016.
- Final workshop with mixed groups of stakeholders, 28 February 2017.

The following organizations participated in the workshops:

- *Academia*: Al Balqa Applied University, Mutah University, University of Jordan, Applied Science University, German Jordanian University, American University.
- *Local communities*: Greater Amman Municipality, Salt community, Ministry of Municipal Affairs, Municipality of Al Zarqa, Municipality of Madaba.
- *Civil society and NGOs*: Energy Services Centre, Renewable Energy Establishments Society, Jordan Engineer Association, Jordan Environment Society, EDAMA, Sanibel Society for Environment, Jordan Press Foundation.
- *Private sector*: Arab bank, GREENVIRO for renewable energy, Control and Communication Company, Al-Masar Engineering Company, Greenplans, Qatrana Cement, NEPCO.
- *Government*: Ministry of Public Works, Ministry of Water and Irrigation, Ministry of Municipal Affairs, Amman Chamber of Industry, Ministry of Energy and Mineral Resources, Parliament.

3 Criteria Ranking

One of the problems with most models for criteria ranking is that numerically precise information is seldom available, and most decision-makers experience difficulties entering realistic information when they analyse the challenges of decision-making. Several attempts have been made to resolve this issue. Methods allowing for less demanding ways of ordering the criteria, such as ordinal rankings or interval approaches for determining criteria weights and values of alternatives, have been suggested, but the evaluation of these models is sometimes quite complicated and difficult for decision-makers to accept.

The problem is eliciting stakeholder information. Different elicitation formalisms have been proposed by which a decision-maker can express preferences. Such formalisms are sometimes based on scoring points, as in point allocation (PA) or direct rating (DR) methods. See chapter "Comparing Cardinal and Ordinal Ranking in MCDM Methods" for a general discussion of such methods. Simos proposed a simple procedure using a set of cards, trying to indirectly determine numerical values for criteria weights (Simos 1990a, b). The Simos method is, however, a bit different from the methods discussed above. It is a relatively simple method to express criteria hierarchies easily while introducing some cardinality if needed. It has been widely applied and has been well-received by real decision-makers. When this method is used, a group of decision-makers is provided with a set of coloured cards with the criteria written on them. They are also given a set of blank cards. Then they are asked to rank the coloured cards from the least important to the most important, where criteria of equal importance are grouped together. Furthermore, the decision-makers are asked to place the blank cards between the coloured cards to express preference strengths. Then the surrogate numbers can be computed. A constant value difference, u, between two consecutive cards is assumed here. A blank card between two consecutive coloured cards signifies a difference of $2 \times u$, two white cards represent a difference of $3 \times u$, etc.

However, one problem with the Simos method is that it is not robust when the preferences are changed (Scharlig 1996) and it has some other contra-intuitive features, such as that it only picks one of the weight vectors satisfying the model, while there can, of course, be an infinite number of them. Furthermore, because the weights are determined differently depending on the number of cards in the subsets of equally ranked cards, the differences between the weights also change in an uncontrolled way when the cards are reordered. This is why Figueira and Roy (2002) suggested a revised version, where there is a more robust proportionality when these blank cards are used. This is accomplished by asking the decision-makers to state how many times more important the most important criterion or criteria group is compared to the least important. This addition seemingly solves some problems but introduces the complication that the decision-maker has to reliably and correctly estimate a proportional factor, z, between the largest and the smallest criteria weights.

We therefore used a variant of the Simos method for elicitation purposes and kept the card ranking part while changing the evaluation significantly compared to the Simos method and its revisions. At that point, the participants already knew the criteria well from the previous sections of the workshops. The key challenge in our workshops was to elicit a collective ranking. Most methods for ranking and weighting deal with individuals; we had to do it as a group effort. This was the main reason to opt for card-ranking through a silent negotiation, not the calculation behind it.

Each criterion was written on a coloured card and arranged horizontally on a table. Each of the participants then successively ranked the cards from the least

Table 1 Card semantics

Equal level of cards	Equally important
No blank card	Slightly more important
One blank card	More important (clearly more important)
Two blank cards	Much more important
Three blank cards	Very much more important

important to the most important by moving the cards to a vertical arrangement, where the highest-ranked criterion was put on top and so forth. If two criteria were considered to be of equal importance, they were put on the same level. This process went on for four rounds, where the number of moves for each round was 8, 5, 3 and 2. Furthermore, the first and third rounds were concluded by an open discussion before the following round. The ranking procedure lasted 120 min or until a final ranking was achieved that the participants found acceptable.

It is true that the decreasing number of moves can be disputed and is a weak point of the method since it induces/forces the participants to act strategically in relation to the information gained during the process. When this method is used, therefore, potential conflicts must come into the open and be dealt with. In some cases, by working with a set of final rankings in the evaluations, it shows whether the differences are of importance or not. After the first ordinal ranking was finalized, the participants were asked to introduce preference strengths in the ranking by introducing the blank cards during three additional rounds (with three, two and one move). The number of white cards (i.e. the strength of the rankings between criteria) was also interpreted verbally (Table 1).

The final rankings of the six workshops were handed to the representatives of each stakeholder group during the final workshop after 2 months, where the exercise was also repeated with this group. They could there present each ranking and its rationale to the other participants during an introductory presentation round.

3.1 Ranking of Different Criteria

The ranking of different criteria during the six workshops with homogeneous groups of stakeholders showed that electricity system costs are perceived as an important criterion by all groups of stakeholders. Safety and global warming potential are also perceived as important criteria. Safety has the highest importance for decision-makers and is also important for local communities, future decision-makers, and finance and investment. Global warming potential is important for local communities and for finance and investment. Global warming potential was a contested criterion, being perceived as the least important by academia. At the same time, domestic value chain integration was perceived as the least important criterion by almost all stakeholder groups, excluding academia and decision-makers. Non-emission hazardous waste was the least important criteria for civil society, academia, future decision-makers and current decision-makers. Pressure on local

Table 2 Ranking of criteria by different stakeholder groups

Stakeholders	Use of domestic energy sources	Global warming potential	Domestic value chain integration	Technology and knowledge transfer	Electricity system costs	On-site job creation	Pressure on land resources	Pressure on local water security	Non-emission hazardous waste	Local air pollution and health	Safety
Young leaders	Moderate-low importance	Moderate-low importance	Least importance	Moderate importance	High importance	Moderate importance	Least importance	Moderate importance	Least importance	Moderate-low importance	High importance
National NGOs	Moderate-low importance	Moderate-low importance	Least importance	Moderate-low importance	High importance	Moderate-low importance	Least importance	Moderate-low importance	Least importance	Least importance	Moderate-low importance
Local communities	Least importance	High importance	Least importance	Least importance	High importance	Least importance	Least importance	Moderate-low importance	Least importance	Moderate-low importance	High importance
Academia	Moderate importance	Least importance	Moderate-low importance	Moderate importance	High importance	Moderate importance	Least importance	Moderate importance	Least importance	Moderate importance	Moderate-low importance
Finance/Industry	Least importance	High importance	Least importance	Least importance	High importance	Least importance	Least importance	Moderate-low importance	Least importance	Moderate-low importance	High importance
Policy-makers	Moderate importance	Least importance	Moderate-low importance	Least importance	Moderate importance	Least importance	Moderate-low importance	Least importance	Least importance	Least importance	High importance
Compromise	Moderate-low importance	Least importance	Least importance	Moderate-high importance	Moderate-high importance	High importance	Least importance	Moderate-high importance	Moderate-low importance	High importance	Moderate-low importance

land resources was ranked as the least important criterion for academia, and pressure on local water resources was ranked as the least important criterion for decision-makers. Table 2 shows different criteria and their importance for the six stakeholder groups.

- The ranking of the criteria by **civil society and NGOs** showed that the electricity system cost is by far the most important criterion and weighs more than one third in the decision-making process, the ten remaining criteria together making less than 70%. During the round of open discussion after the ranking exercise, the following arguments were discussed and the following criteria were debated as the most important: electricity system costs, socio-economic impacts and safety. Safety was discussed in light of further efforts needed to develop safety regulations for existing and emerging technologies. The question was also how to include safety regulations in the national legislative framework.
- For **finance and investment** stakeholders, the main focus of discussion was on the safety of electricity generation. Safety seems to be one of the most important criteria; however, the implementation of safety measures will invariably lead to higher energy costs. A further concern of participants was that safety-monitoring authorities in Jordan lack the power to reinforce safety regulations. The responsibility level of stakeholders should be increased in order to guarantee the safety of power plant operations. Participants believed that technologies such as nuclear power technologies will be transferred from more experienced countries; therefore the know-how and guarantees for safety will be also transferred.
- During the open discussion among **academia**, a serious debate took place between participants who thought that electricity system cost was the most important criterion and participants who had a common understanding that safety is the most important criterion. In their opinion, safety and security were vital.
- Among **future decision-makers,** electricity system costs were considered the most important criterion, especially in conditions of limited budget and budget deficit in Jordan. However, there was no consensus on this criterion among participants. Other participants strongly objected that values are more important than costs. Safety and transfer of knowledge were considered crucial for the implementation of safety regulations. The risk of climate change impacts was also closely connected with safety issues.
- Representatives of **local communities** intensively debated what is more important, safety or impacts on human health and on locally available resources such as water and land. Also, electricity costs play a significant role for local communities. There was no common opinion on these criteria among different communities, and the participants did not come to a compromise solution on which criterion was the most important.
- The discussion after ranking among **decision-makers** was short: participants agreed on the ranking and on the importance of the safety criterion. The aspect of safety was picked up again during the ranking of the procedural criteria, where it was agreed that safety should remain a top priority.

3.2 Procedural and Output Justice

- Representatives of **civil society and NGOs** argued that every infrastructure project should be combined with the implementation of participation procedures, namely in the framework of environmental impact assessment. These procedures should guarantee that the local community benefits from the project. Further sensibilization of the population is needed to guarantee sufficient levels of awareness and knowledge about the project and about possibilities to participate. Two factors were intensively debated: access to information, and whether or not it could be considered as participation, and compensation, namely who should be responsible, how it should be organized and who should be compensated.
- The main focus of discussion among **finance and investment** was about the access to information and meaningful participation. While some participants argued that access to information should be a prerequisite for meaningful participation, others argued that participation produces access to information. However, a consensus was reached that benefit sharing should come after the access to information and meaningful participation, and that compensation should be the least important criterion. This was mainly because of the participants' perception that compensation is only due after a disaster has occurred.
- **Academia** intensively debated whether or not providing information is part of stakeholders' involvement and inhabitants' engagement. It was agreed that conditions for engagement should be provided during all phases of decision-making processes rather than compensating for adverse impacts of a non-inclusive and non-transparent decision-making process.
- Some participants among **future decision-makers** argued that participation in decision-making processes should be prioritized. Participants initially had different ideas on what access to information meant. It was argued that access to information reduces fear and enables participation; therefore, access to information should be the most important criterion. It was agreed that access to the information criterion should precede the participation criterion in the decision-making process. Furthermore, it was decided that benefits should be ranked third, especially if the technology creates benefits for the entire society, the state and the nationwide economy.
- The aspects of involvement and participation were intensively discussed by **local communities**. It was agreed that community involvement in decision-making processes is the most important criterion, and should go much beyond simply informing or providing information, though the availability of clear and transparent information is a necessary requirement. It was also agreed that compensation is the last criterion and that projects should generally be deployed to leave communities as better places afterwards rather than simply compensating them for impacts from the projects.
- **Decision-makers** agreed that awareness-raising measures are a first step. Therefore, clear and transparent information should be available to stakeholders and inhabitants to guarantee public and social support for infrastructure deployment.

4 Decision Evaluation

A common approach to solve decision problems with multiple criteria is to specify a set of criteria that represent the relevant aspects of a problem and then define a weight function over the criteria set. Value functions are then defined for the alternatives to each attribute. It is common here to use a weight function over the attribute set using fixed numbers on a normalized scale. The criteria weights thus describe each criterion's significance in the specific decision context. Value functions of the alternatives are defined in a similar way. Thereafter, the overall score of each alternative is calculated by aggregating the various components.

We have followed this general approach in the evaluation process. The performance of the different electricity generation technologies was estimated from a larger expert survey. Together with the surrogate weights, the experts therefore provided the decision base for the multi-criteria analysis. Using a weighted aggregation principle, we combined the multiple criteria and stakeholder preferences with the valuation of the different technology options under the criteria surrogate weights. The results of the process were (i) a detailed analysis of each technology's performance compared with the other technologies and (ii) a sensitivity analysis to test the robustness of the result. The resulting multi-criteria decision trees look like Fig. 1.

During the process, we considered the entire range of values as the alternatives presented across all criteria, as well as how plausible it was that an alternative outranked the remaining ones, thus providing a robustness measure. Because of the complexity of these calculations, we used the state-of-the-art MCDA software DecideIT for the analysis, which allows for imprecision of the kinds that exist in this analysis (see, for example, Danielson and Ekenberg 2007).

4.1 Encoding of Criteria Weights

One of the central issues in such situations is how to assign weights while avoiding too much information loss and preserving correctness in the weight assessments. Using criteria ordinal rankings usually avoids some of the elicitation difficulties that appear when limited to precise numbers only. Techniques for ordinal rankings are, however, quite different regarding their accuracy, and decision-makers usually also have usable knowledge of decision situations expressed in criteria orderings (see, for example, Danielson and Ekenberg 2016a, b; Danielson et al. 2014), information that should also be used. The so-called surrogate weights based only on ordering may thus be too weak a representation. In the analyses, we have therefore included information regarding relational strengths. The analytical part of the project evaluation consists of translating the rankings to surrogate weights, evaluating them by applying the cardinal ranking (CAR) method, and then using

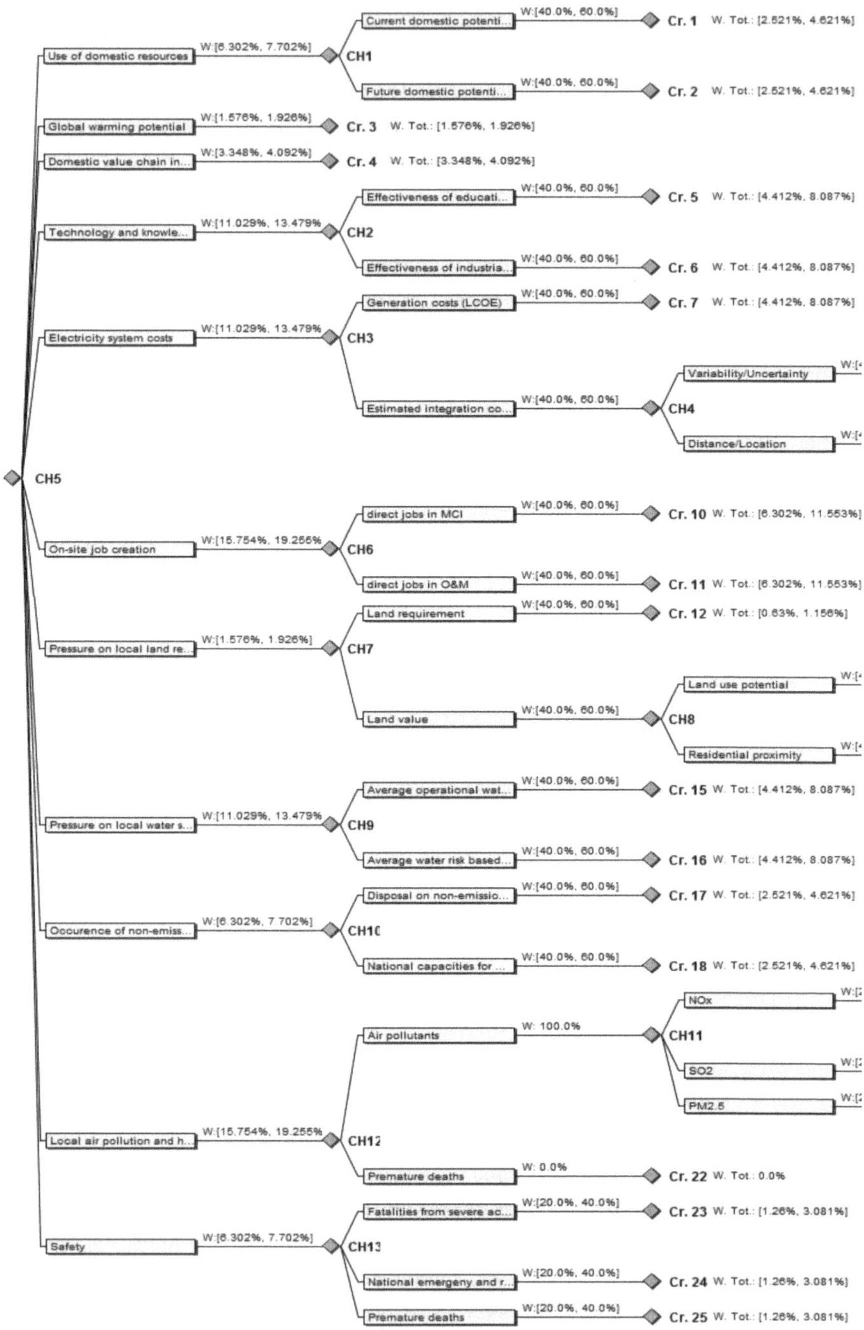

Fig. 1 A multi-criteria tree for the final workshop evaluation

these values in the DecideIT software which is designed to solve this type of problem under uncertainty. Information loss is thereby limited.

The criteria weight generation is further described in chapter "Comparing Cardinal and Ordinal Ranking in MCDM Methods", but the general idea is the following:

- Assign an ordinal number to each importance scale position, starting with the most important position as number 1.
- Let the total number of importance scale positions be Q. Each criterion i has the position $p(i) \in \{1,\dots,Q\}$ on this importance scale, such that for every two adjacent criteria c_i and c_{i+1}, whenever $c_i >_{s_i} c_{i+1}$, $s_i = | p(i+1) - p(i) |$. The position $p(i)$ then denotes the importance as stated by the decision-maker. Thus, Q is equal to $\Sigma s_i + 1$, where $i = 1,\dots,N-1$ for N criteria.

In chapter "Comparing Cardinal and Ordinal Ranking in MCDM Methods", we argue that the best method for cardinal ranking with properties similar to Simos cards is to use what is called CSR weights, expressed as:

$$w_i^{\mathrm{CSR}} = \frac{1/_{p(i)} + \frac{Q+1-p(i)}{Q}}{\sum_{j=1}^{N}\left(1/_{p(j)} + \frac{Q+1-p(j)}{Q}\right)} \tag{1}$$

which were consequently employed in this study. Based on the weightings of each stakeholder group, expressed as CSR weights, and observations made during the workshops, the analysis of potential conflict lines and commonalities between the different stakeholder preferences was facilitated through negotiation.

4.2 *Aggregating the Components*

One of the problems with most models for criteria ranking is that numerically precise information is seldom available. We have solved this in part by introducing surrogate weights as described above. This is only a part of the solution, however, since the elicitation can still be uncertain and the surrogate weights might not be a fully adequate representation of the preferences involved, which is, of course, a risk with all kinds of aggregation. To allow for analysis of how robust the problem is to changes in the input data, we also introduced intervals around the surrogate weights and around the values of the technology options. Thus, in this elicitation problem, the possibly incomplete information is handled by allowing the use of intervals, where ranges of possible values are represented by intervals in combination with pure orderings without the use of surrogate weights at all if the latter turn out to be inadequate.

There are thus several approaches to elicitation in MCDM problems, and one way of partitioning the methods into categories is in terms of how they handle imprecision in weights and values, such as fixed numbers, comparative statements, representing orderings or intervals.

Computationally, methods using fixed numbers are very easy to solve, while systems of relational or interval constraints normally require more elaborate optimization techniques. On the other hand, if the model only accepts fixed numbers, we impose constraints that might severely affect the decision quality. If we allow for imprecision in terms of intervals and relations, we usually get a more realistic representation of the problem. These can be represented by interval statements, for instance, such as $w_i \in [y_i - a_i, y_i + b_i]$, where $0 < a_i \leq 1$ and $0 < b_i, \leq 1$, or comparative statements, such as $w_i \geq w_j$. Systems of such equations can be solved, and aggregations of decision components in these formats can be optimized, by using the methods from Danielson and Ekenberg (1998, 2007). The disadvantage here is that many decision-makers perceive these methods as difficult to understand and accept because of complex computations and loss of user transparency.[1]

The decision information can be considered as constraints in the multi-dimensional solution space formed by all decision variables, which are collected as linear constraints to the solution sets of the spaces spanned by the weight and value variables, respectively. To further aid in the modelling of the problem, the orthogonal hull concept is introduced, indicating which parts of the statements are consistent with the information given so far. This then becomes the projection of the constrained spaces onto each variable axis, and can thus be seen as the meaningful interval boundaries for the decision situation. The same type of input is used for the components involved—i.e. alternative values v and weights w_j, although the normalization constraints $\boldsymbol{\Sigma}\ w_j = 1$ must not be violated in the weight case.

All input into the model was subject to consistency checks performed by the DecideIT tool. The calculations are based on the weighted sum of the alternative values under the criteria and sub-criteria aggregated for the entire decision problem. For instance, in a three-level tree such as the current one, this becomes

$$V(A_s) = \sum w_i \sum w_{ij} \sum w_{ijk} v_{ijk}(A_s),$$

where $v_{ijk}(A_s)$ is the value of alternative A_s under sub-criteria ijk. Given this, we then calculate the strength of alternatives as a mean for further discriminating the alternatives. The strength simply denotes the difference in weighted value—i.e. the expression $V(A_i) - V(A_j)$ for the difference between alternatives A_i and A_j. In this way, we can readily calculate the maximum and minimum difference between the alternatives.

An important feature of this process is the sensitivity analysis. This analysis attempts to highlight what information is the most critical for the obtained results and must therefore be subject to careful additional consideration. It also highlights which of the assessments are too imprecise to be of any assistance in the discrim-

[1] This should be kept in mind here, as always when working with aggregation methods of whatever kind, and this should affect how the elicitation mechanisms and software tools are used.

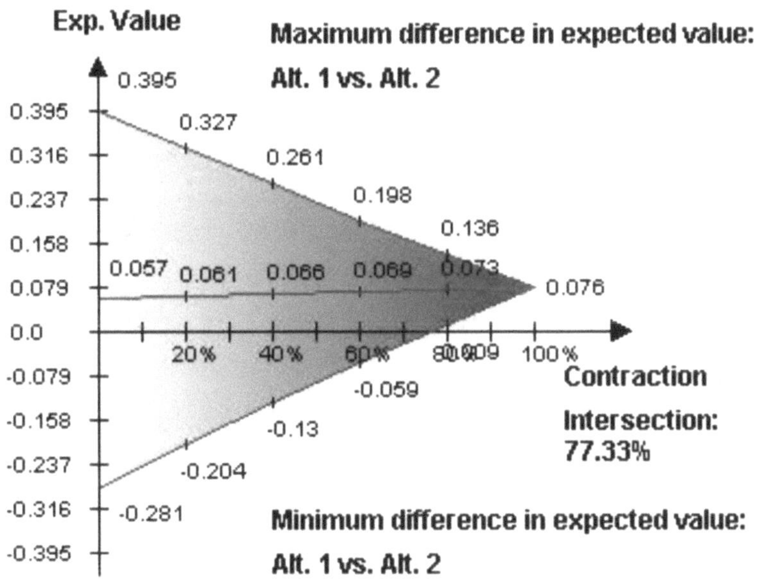

Fig. 2 Alternative comparison between total expected values of alternatives utility PV and CSP respectively

ination of alternatives and thus should be made more accurate, thereby triggering and facilitating iteration in the process. The embedded sensitivity analysis, called the concept of contraction, is performed by reducing the widths of the intervals (contraction) for the values and weights in the analysis model of the decision problem. The concept's idea is to shrink the orthogonal hull while studying the stability of the maximum strength at different contraction levels. The level of contraction is indicated as a percentage so that, for a 100% level of contraction, all orthogonal hull intervals have been reduced to their respective focal points. The contraction can be seen as cutting the hull from the extreme points (having lower reliability or a lower degree of belief towards the focal point), increasing the lowest permitted degree of belief. When dealing with interval statements only this is quite simple; it is more complicated when comparative constraints are involved.

Thus, the programme calculates the result of assigning all possible values, given the estimated interval and relations. We can then see all the possible weighted values given the background information. Figure 2 shows how two of the strategies, utility PV and CSP, relate to each other given the information provided during the final workshop.

Another good tool for studying how sensitive the result is for error estimates is to use sensitivity analyses. These are used to investigate how stable the choice of a strategy is when the input data change. Here we primarily investigate the limits within which the probabilities and values must stay for the decision not to change.

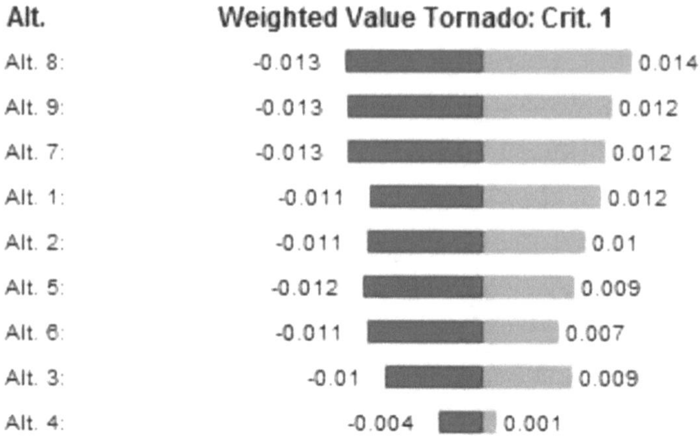

Fig. 3 A tornado diagram showing how much changes to the criterion 'current domestic potential of the energy carrier' affects the respective alternatives

This is done by letting the input values vary between possible and realistic values and investigating how these fluctuations affect the outcome. Thus, the values are systematically varied up and down. See Fig. 3.

5 Trade-Offs Between Technologies

Analyses and the application of DecideIT software allowed identifying what the preferences of stakeholders in terms of criteria mean for the most preferable technology. In the figures, all of the pairwise comparisons are shown simultaneously. A red square means that the difference between the alternatives is not significant when contracting the information bases, while a yellow or green square indicates that the difference is either significant or highly significant. Furthermore, the figures show how large an impact the respective criteria have on the values of the alternatives. For instance, in Fig. 4, the criterion electricity costs has a significant impact on most of the alternative technologies, while the criterion pressure on local resources has a lower impact.

The results for the **civil society and NGOs** group show that alternative 1 (utility-scale photovoltaic) is the preferred alternative, meaning that solar radiation converted into electricity by the photovoltaic effect is the most favourable technology, slightly better than coal and nuclear, followed by gas, large-scale hydro, oil shale, concentrated solar power (CSP, which concentrates solar radiation onto a receiver and then converts it into thermal energy), onshore wind and oil. There is strong confidence that oil is considered much worse than most of the technologies and that coal, nuclear and gas are considered better than onshore wind, CSP, oil shale and large-scale hydro. A significant role in these results is played by electricity

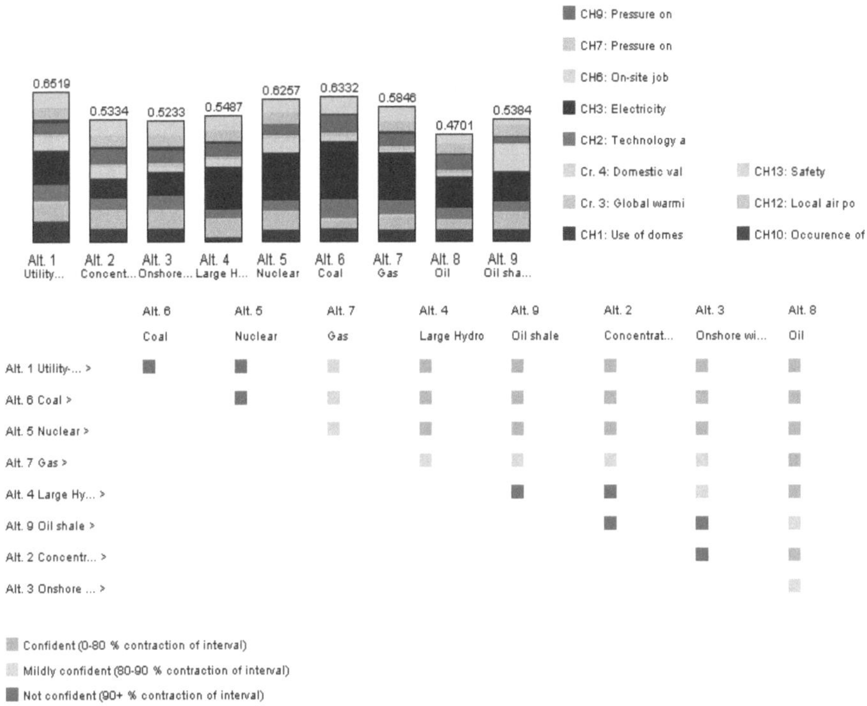

Fig. 4 Preferences of civil society and NGOs stakeholder group

systems costs criterion, which was considered one of the most important criteria and pushed coal, gas, nuclear and PV to the top of the ranking. See Fig. 4.

Utility PV was considered the most favourable technology by the **finance and investment** stakeholder group. Utility PV was slightly better than nuclear and large-scale hydro, followed by onshore wind, CSP, gas, coal, oil shale and oil. There is strong confidence that oil is worse than all technologies except oil shale and that utility PV, nuclear and large-scale hydro are better than all other technologies. Global warming potential is improving the positions of most of the technologies except coal, oil and gas. Electricity systems costs are pushing up coal, nuclear, gas and large-scale hydro. See Fig. 5.

For **academia,** the utility PV was definitely considered the most favourable technology, followed by nuclear, oil shale, coal, gas, CSP, onshore wind, large-scale hydro and oil. There is strong confidence that oil is worse than all technologies except large-scale hydro, onshore wind and CSP. Utility PV was considered a much better technology than all technologies except oil shale. Nuclear is much better than onshore wind, large-scale hydro and oil. Local air pollution plays a role in these results and pulls back coal. Electricity systems costs are pushing up nuclear, coal and gas. On-site job creation is considered important for oil shale. At the same time, pressure on water resources pulls this technology down. See Fig. 6.

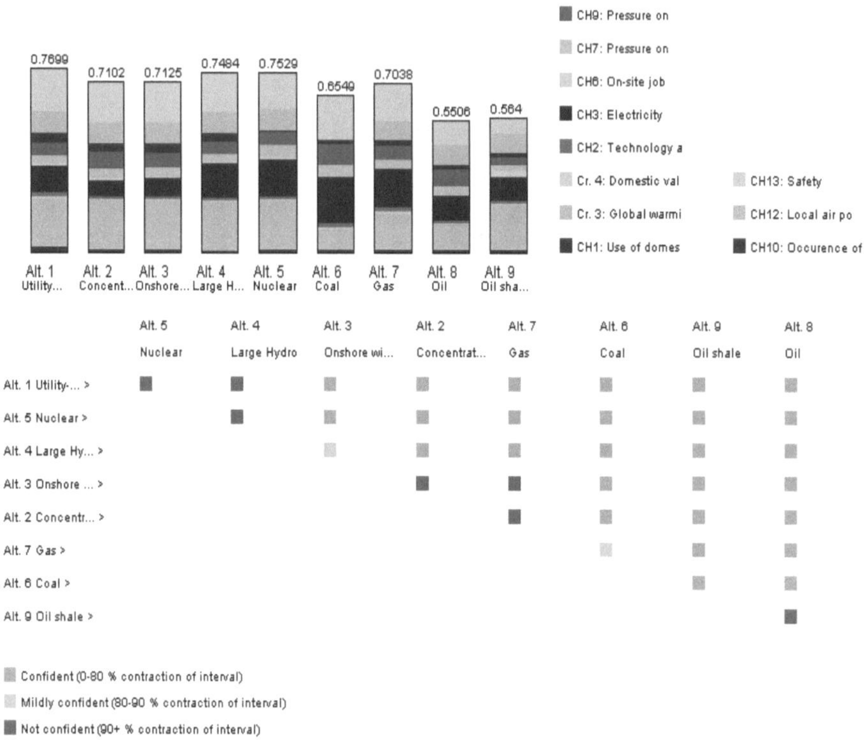

Fig. 5 Preferences of finance and investment stakeholder group

The results in the group of **future decision-makers** show that utility PV is considered the most favourable technology, slightly better than coal and nuclear, followed by gas, large-scale hydro, CSP, onshore wind, oil shale and oil. There is strong confidence that oil is worse than almost all technologies except oil shale, and that utility PV, coal and nuclear are better than all other technologies. Safety is an important criterion for stakeholders in this group. Also, electricity systems costs are pushing up nuclear, coal and gas, as well as PV. See Fig. 7.

In the group of **local community** representatives, utility PV is considered the most favourable technology, slightly better than coal and gas, followed by nuclear, onshore wind, large-scale hydro, CSP, oil, and oil shale. There is a strong confidence that oil shale is worse than all technologies and that utility PV and coal are better than onshore wind, large-scale hydro, CSP, oil, and oil shale. The safety criterion plays a significant role in all technologies and is pulling down oil and oil shale. Electricity systems costs are pushing up nuclear, coal and gas. Pressure on local water resources is reducing the positions of oil shale and nuclear. The availability of domestic resources is reducing the position of large-scale hydro. See Fig. 8.

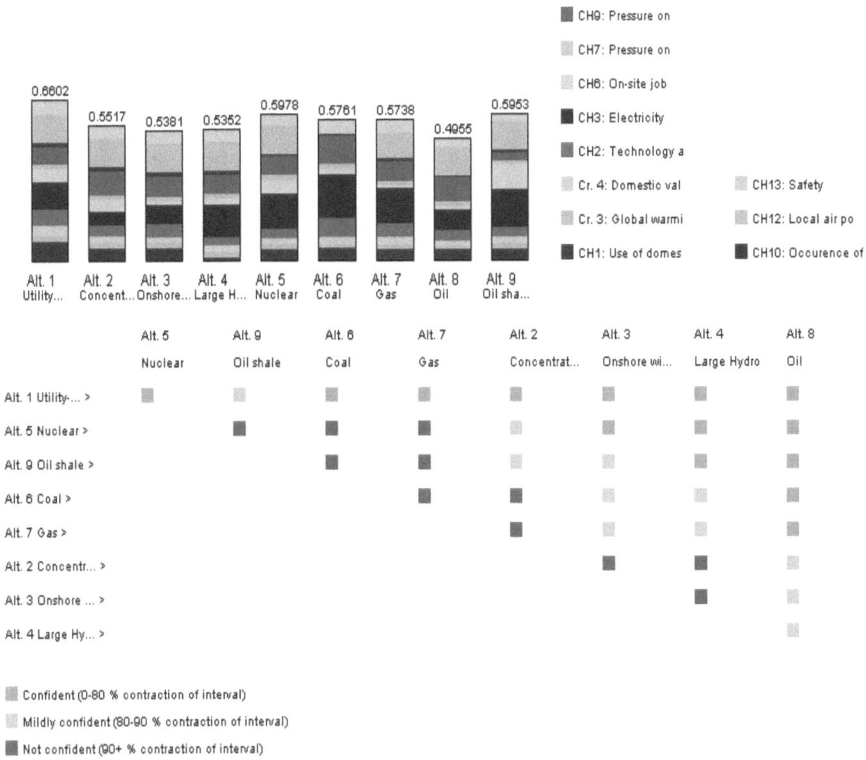

Fig. 6 Preferences of the academia stakeholder group

Utility PV is considered as the most favourable technology by **decision-makers**, followed by oil shale, nuclear, coal, gas, large-scale hydro, onshore wind, CSP and oil. See Fig. 9.

There is strong confidence that oil is worse than most of the technologies, except CSP, onshore wind and large-scale hydro, and that utility PV is better than gas, large-scale hydro, onshore wind, CSP and oil. Electricity systems costs are considered an important criterion which pushes up nuclear, gas and coal. Local air pollution reduces the positions of gas, and the availability of domestic resources criterion reduces the position of large-scale hydro. Pressure on water resources is reducing the positions of oil shale and nuclear.

During the first round of the final workshop, to which representatives from different stakeholder groups were invited, utility PV was considered the most favourable option, followed by nuclear, gas and coal, as well as CSP, large-scale hydro, onshore wind, oil shale and oil. There is strong confidence that oil is worse than utility PV, nuclear, gas and coal, and that nuclear is better than CSP, large-scale hydro, onshore wind, oil shale and oil. Electricity systems costs play an important role and are pushing up nuclear, coal and gas. Local air pollution reduces the

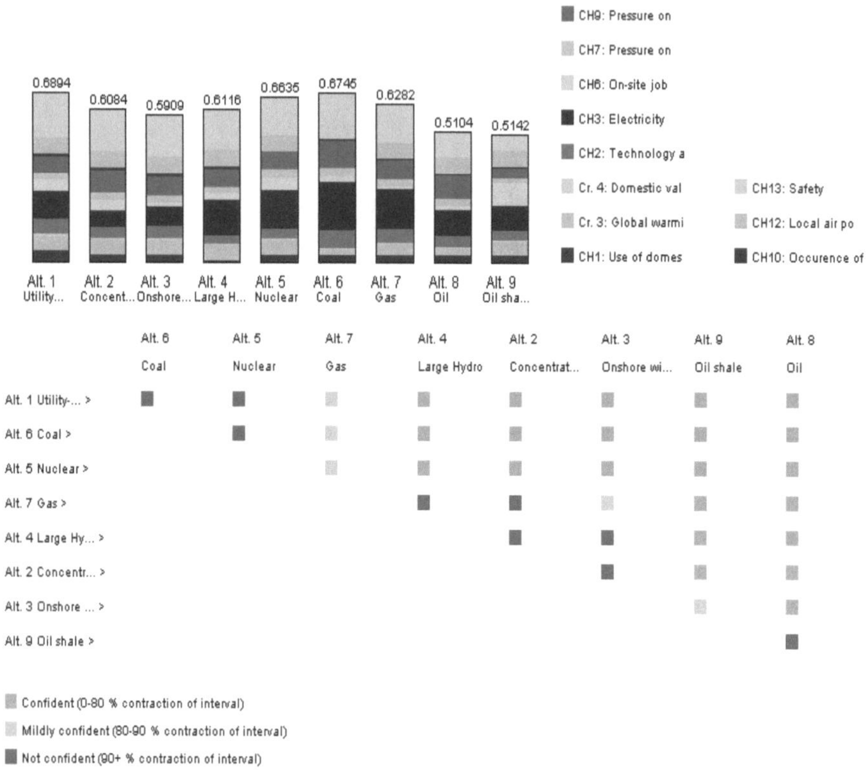

Fig. 7 Preferences of young leaders/future decision-makers

position of coal. On-site job creation is important for oil shale. Pressure on water resources reduces the positions of oil shale and nuclear. See Fig. 10.

The final ranking during this workshop showed that utility PV is definitely the most preferable option, followed by CSP, nuclear, oil shale, onshore wind, large-scale hydro, gas, oil and coal. There is strong evidence that coal is the least preferable option except for oil, that utility PV is better than all other options, and that CSP, nuclear and oil shale are better than oil and coal. Local air pollution plays an important role for all technologies and is pushing down coal. On-site job creation is important for oil shale. Pressure on water resources is reducing the positions of oil shale and nuclear. Electricity systems costs are less important than in the previous round but still play a role, together with local air pollution, on-site job creation and pressure on water resources. See Fig. 11.

Global warming potential was another criterion with high polarization of opinions, local community representatives ranking the criterion high, and academia and policymakers ranking it low. Pressure on local water security and non-emission hazardous waste were two criteria which received a low ranking but where the positions of stakeholders were homogeneous. National NGOs and academia ranked

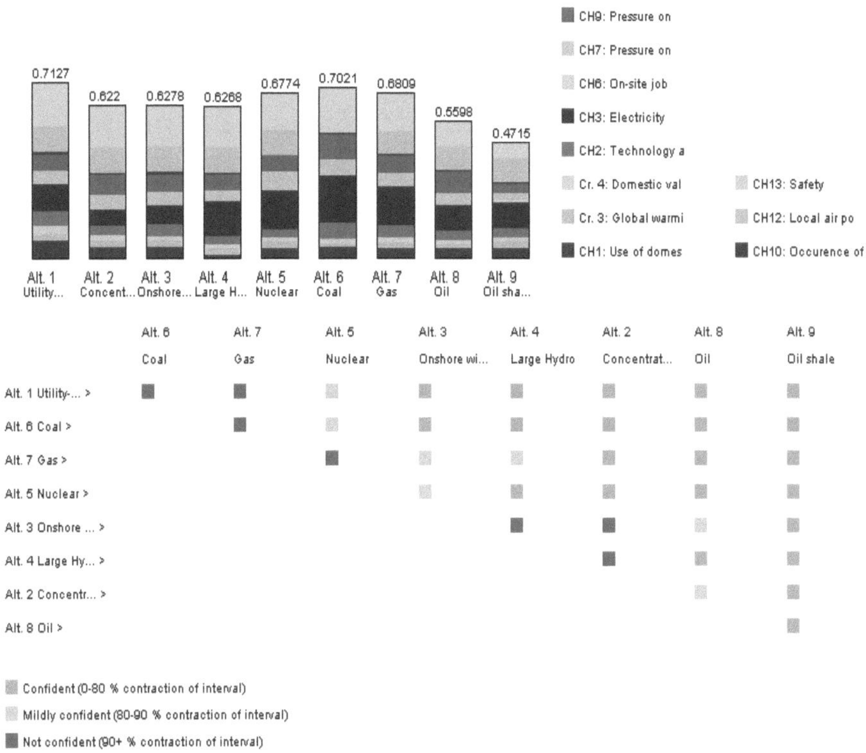

Fig. 8 Preferences of the local community stakeholder group

technology and knowledge transfer and on-site job creation significantly higher than local community and policymakers. The use of domestic energy sources was also a criterion with high polarization of opinions: it was ranked high by decision-makers, national NGOs and academia, and received a low ranking from finance and industry, and from local communities.

6 Conclusions

We have used a new multi-stakeholder multi-criteria approach to assess the relevance of Jordan's electricity generation technologies against a set of criteria under uncertainty, reflecting environmental, social and economic components of sustainable development. The performance of different electricity generation technologies was estimated based on a large expert survey. Together with the surrogate weights, they provided the decision base for the multi-criteria analysis. The multiple criteria and stakeholder preferences were combined with the valuation of the different technology options under the criteria surrogate weights. The results of the evalua-

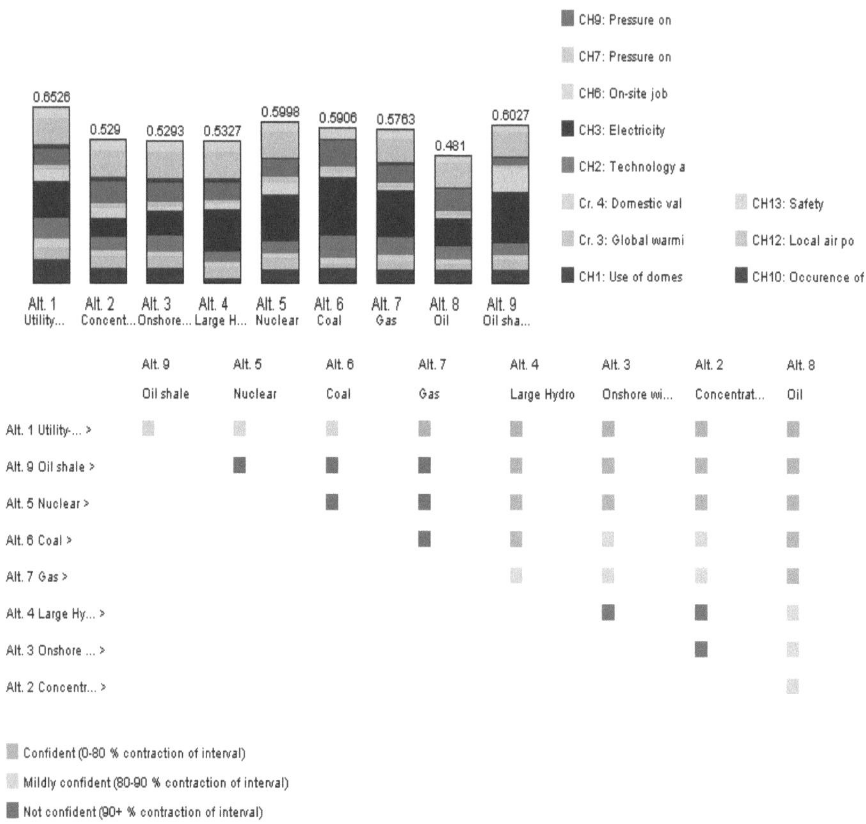

Fig. 9 Preferences of the political decision-makers stakeholder group

tions are: (i) a detailed analysis of the performance of each technology compared with the other technologies; and (ii) a sensitivity analysis to test the robustness of the result. The overall results show that currently the discourse in Jordanian society is dominated by economic rationality, such as electricity costs, supported by concerns about safety during operation and maintenance of electricity generation power plants. The results also show a strong desire of all stakeholder groups to have an opportunity for engagement in decision-making processes on energy transition, rather than purely compensating local communities for the installation of electricity generation and transmission technologies. The discourse about energy transition in Jordan is strongly dominated by energy security concerns. In almost all group rankings, the safety of energy generation and the affordability of electricity prices were ranked as top priorities. The criteria, which are relevant for the social and environmental impacts of technologies, were moved by participants from the middle or the bottom of the ranking. It seems that concerns about climate change mitigation do not belong to the dominant discourse, as the criterion on climate change mitigation was frequently ranked at the bottom of the list. One stakeholder

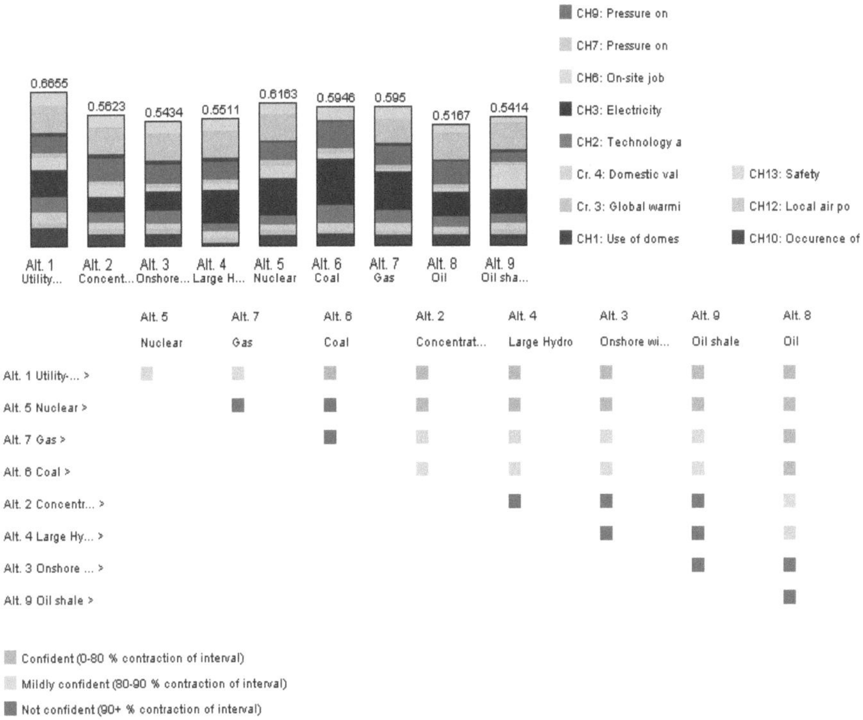

Fig. 10 Results of the first round of the final workshop

group, the local communities, ranked global warming potential at the top of the ranking, probably because people on the ground are feeling the direct impacts of climate change. However, while evaluating renewable energy technologies, the most frequent positive characteristic was ‘clean’ and ‘with little impact on the environment’. It seems that there is a certain level of awareness about environmental protection issues; however, the level of awareness about climate change risks and the need for climate change mitigation is lower.

Comparison of visions of the environmental, social and economic future of Jordan showed that the young people have the most optimistic approach. For instance, they did not identify any negative tendencies. Among economic factors, the positive expectations connected with investment in new technologies and reduction of dependency on energy imports were mentioned most frequently. The positive expectations about social development are connected with the creation of employment opportunities and the generation of further knowledge. In general, there was a perception that the environmental future of Jordan is positive. Among negative tendencies, the possible increase in electricity costs was named most frequently. In the social area, this is the destruction of traditional values and of traditional family structure. In environmental areas, the most frequent concerns were about water scarcity.

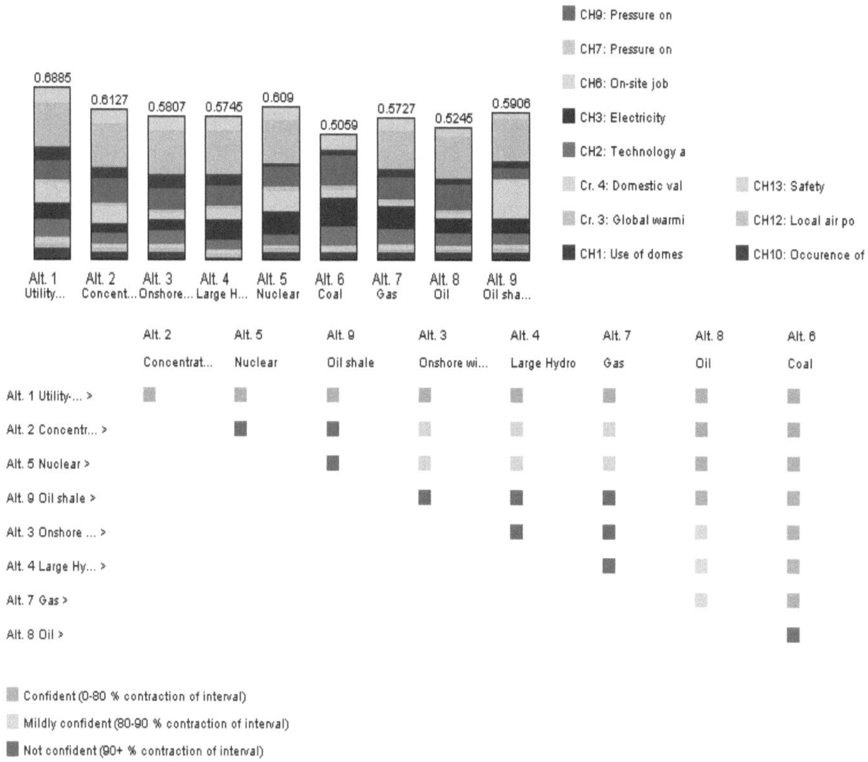

Fig. 11 Results of the second (last) round of the final workshop

Discussing procedural and output justice, the majority of stakeholders had the opinion that compensation for the deployment of infrastructure was the least favourable criterion and that further efforts are necessary to facilitate the engagement of stakeholders and laypeople in decision-making processes on energy transition. Providing possibilities for participation in decision-making processes was considered the most important criterion among the four criteria of procedural and output justice.

Solar, nuclear and oil are the three energy generation technologies most discussed in the Jordanian media. However, attitudes to these technologies are quite different. Solar is perceived mostly positively, with PV being a top priority technology. At the same time, CSP does not enjoy the same high level of support as PV. Nuclear was often considered as the second or third most favourable technology. However, opinions here are strongly polarized and several stakeholders are strongly opposed to nuclear. Even though oil is discussed frequently in media, it is considered the least favourable technology by all stakeholder groups. Shale oil is considered much more positively, mainly due to the Jordanian resources, aspirations for technology transfer and impulses for socio-economic development which are connected with the deployment of this technology. The strong recommendation from stakeholders during almost all workshops was to add oil shale technology as one of the most discussed in Jordan. In some stakeholder groups, such as academia or local communities, there was also a recommendation to add waste to energy technology, with major arguments focusing on its positive features such as the possibilities of reducing the costs of waste disposal, clean technology and the potential to create green jobs.

Overall, solar radiation converted into electricity by the photovoltaic effect (utility PV) remains the most favourable technology. It was ranked as the top priority in the frames of all stakeholder groups; furthermore, during the final ranking with the mixed group of stakeholders, utility PV was ranked at the top of the list. During the individual ranking, stakeholders ranked PV as the most favourable technology. Other solar technology, such as CSP, is ranked significantly lower. The main reason is the high investment costs of this technology.

Acknowledgements The work in this chapter was supported by the German Federal Ministry for Economic Cooperation and Development (BMZ) in funding the Middle East North Africa Sustainable Electricity Trajectories (MENA-SELECT) project as well as by the EU-project Co-Inform (Co-Creating Misinformation-Resilient Societies H2020-SC6-CO-CREATION-2017) and strategic grants from the Swedish government within ICT – The Next Generation.

References

Batel S, Devine-Wright P (2015) A critical and empirical analysis of the national-local 'gap' in public responses to large-scale energy infrastructures. J Environ Plan Manag 58(6):1076–1095

Beierle T, Cayford J (2002) Democracy in practice: public participation in environmental decisions. RFF Press, Washington, DC

Brand B (2015) The integration of renewable energies into the electricity systems of North Africa. Schriftenreihe technische Forschungsergebnisse. Volume 20. Hamburg, Verlag Dr. Kovac

Danielson M, Ekenberg L (1998) A framework for analysing decisions under risk. Eur J Oper Res 104(3):474–484

Danielson M, Ekenberg L (2007) Computing upper and lower bounds in interval decision trees. Eur J Oper Res 181(2):808–816

Danielson M, Ekenberg L (2016a) A robustness study of state-of-the-art surrogate weights for MCDM. Group Decis Negot 7. https://doi.org/10.1007/s10726-016-9494-6

Danielson M, Ekenberg L (2016b) The CAR method for using preference strength in multi-criteria decision making. Group Decis Negot 25(4):775–797. https://doi.org/10.1007/s10726-015-9460-8

Danielson M, Ekenberg L, He Y (2014) Augmenting ordinal methods of attribute weight approximation. Decis Anal 11(1):21–26

Devine-Wright P (2012) Explaining 'NIMBY' objections to a power line: the role of personal, place attachment and project-related factors. Environ Behav 45:761–781

FES (2015) Renewable energy transitions in Jordan and the MENA region. Amman: Friedrich Ebert Stiftung. ISBN: 978-9957-484-58-3

Figueira J, Roy B (2002) Determining the weights of criteria in the ELECTRE type methods with a revised Simos' procedure. Eur J Oper Res 139:317–326

Hänlein R (2015) Public participation and transparency in power grid planning. Recommendations from the BESTGRID Project. Handbook – Part 1. Germanwatch, Bonn

Jasanoff S (1998) The political science of risk perception. Reliab Eng Syst Saf 59(1):91–99
Komendantova N, Battaglini A (2016) Beyond decide-announce-defend (DAD) and not-in-my-backyard (NIMBY) models? Addressing the social and public acceptance of electric transmission lines in Germany. Energy Res Soc Sci 22:224–231
Komendantova N, Vocciante M, Battaglini A (2015) Can the BestGrid process improve stakeholder involvement in electricity transmission projects? Energies 8:9407–9433. https://doi.org/10.3390/en8099407
Komendantova N, Irshaid J, Marashdeh L, Al-Salaymeh A, Ekenberg L, Linnerooth-Bayer J (2017) Country fact sheet Jordan: energy and development at a glance. 2017: Background paper. Middle East North Africa Sustainable Electricity Trajectories (MENA-SELECT) project funded by the Federal Ministry for Economic Cooperation and Development (BMZ)
Komendantova N, Riegler M, Neumueller S (2018) Of transitions and models: community engagement, democracy, and empowerment in the Austrian energy transition. Energy Rev Soc Sci 39:141–151
Kunreuther H, Linnerooth-Bayer J, Fitzgerald K (1994) Siting hazardous facilities: lessons from Europe and America. Wharton Risk Manage Decision Process Center
MEMR (2016) Annual report 2016. Amman, Ministry of Energy and Mineral Resources
MWI (2015) Energy efficiency and renewable energy policy for the Jordanian water sector. Ministry of Water and Irrigation, Amman
Patt A (2015) Transforming energy: solving climate change with technology policy. Cambridge University Press
Renn O (2008) Risk governance. Coping with uncertainty in a complex world. Earthscan, London
Rowe G, Frewer L (2000) Public participation methods: a framework for evaluation. Sci Technol Human Values 25:13–29
Scharlig A (1996) Pratiquer electre et PROMETHEE un complement à decider sur plusieurs critères. Collection Diriger L'Entreprise. Presses Polytechniques et Universitaires Romandes, Lausanne
Simos J (1990a) Evaluer l'impact sur l'environnement: une approche originale par l'analyse multicriteere et la negociation. Presses Polytechniques et Universitaires Romandes, Lausanne
Simos J (1990b) L'evaluation environnementale: un processus cognitif neegociee. Doctoral thesis, DGF-EPFL, Lausanne
Wolfsink M (2012) The research agenda on social acceptance of distributed generation in smart grids: renewable as common pool resources. Renew Sustain Energy Rev 16(1):822–835
Wolsink M (2000) Wind power and the NIMBY-myth: institutional capacity and the limited significance of public support. Renew Energy 21(1):49–64
Xavier R, Komendantova N, Jarbandhan V, Nell D (2017) Participatory governance in the transformation of the South African energy sector: critical success factors for environmental leadership. J Clean Prod 154:621–632
Yazdanpanah M, Komendantova N, Ardestani RS (2015) Governance of energy transition in Iran: investigating public acceptance and willingness to use renewable energy sources through socio-psychological model. Renew Sust Energ Rev 45:565–573

A Proposition of a Multidimensional HAZOP Analysis (MHAZOP) to Support a Decision-Making Process

Thalles V. Garcez and Marcelo Hazin Alencar

1 Introduction

Over the centuries, the idea of risk control has been discussed by different thinkers of different historical Ages, and has, for example, been associated with divine control, or even with random occurrences (Bernstein 1998). This understanding has evolved over the centuries to the present day. Today, according to Aven (2018), the term risk is regarded as a fundamental issue in the search for viable decisions when discussing solutions to real-world problems such as those related to technology, health, safety, or climate change. de Almeida et al. (2016a) point out that in the last decades there has been an increase in the number of scientific research studies developed in different contexts regarding risk management. This transformation is due to many factors such as contractual requirements, norms and regulations, competition between organizations, and society having access to information. As a consequence, the development of models applied to risk management has taken place with a view to providing more robust results for those involved in the process. This arises from identifying, eliminating, or mitigating the risk of a single undesirable event or a set of such events. These transformations seek to incorporate a broader and different view as to the use of techniques, while always aiming to detect potential risks in order to reduce or eliminate accidents, occupational diseases, and impacts on the environment, and thereby seeks to improve the quality of life of people at work and in society.

According to Dunjó et al. (2010), the techniques used for Process Hazard Analysis (PHA) are conducted so as to identify hazards in new or existing processes.

T. V. Garcez (✉) · M. H. Alencar
Universidade Federal de Pernambuco, REASON - Research Group on Risk Assessment and Modelling in Environment, Assets, Safety, Operations and Nature, Recife, PE, Brazil
e-mail: thalles.garcez@ufpe.br; marceloalencar@insid.org.br

A. T. de Almeida et al. (eds.), *Multicriteria and Optimization Models for Risk, Reliability, and Maintenance Decision Analysis*, International Series in Operations Research & Management Science 321, https://doi.org/10.1007/978-3-030-89647-8_6

PHAs are designed not only to ensure safe projects and system operations, but also to complement risk assessments and to make safety devices fit for purpose. Therefore, this type of analysis can be considered as the basis for the safety of processes and for risk management programs. Among techniques of PHA analysis, the HAZOP (Hazard and Operability) study stands out worldwide. Its aim is to study the hazards associated with operating and maintenance problems, while taking the opinions of experts and deviations arising from project conditions into account (Baybutt 2015a, 2015b; de Almeida et al. 2015). Basically, HAZOP identifies how a process can deviate from the specifications of a project. This is identified by means of structured and sequential procedures, which include combining predetermined keywords and parameters that describe the processes, activities, or operations that a system or set of equipment performs. Moreover, HAZOP seeks to identify the ways in which these potential hazards are controlled or how any consequences arising from hazards are mitigated.

Although HAZOP is an efficient approach to hazard assessment, it has some limitations, such as: it is a qualitative approach, and thus does not provide a quantitative analysis of the results; it is a heuristic approach in which the focus is mainly on the brainstorming developed by the team responsible for the study, and therefore, there is no-one who has a primary role in the decision-making process, i.e., no one person is designated as the decision-maker (DM); when deviations are generated, the inductive/deductive starting point is counter-intuitive, and thus does not consider, for example, compound deviations (Baybutt 2015b; Guo and Kang 2015).

Given the "limitations" of the classic HAZOP approach, a structured framework, called Multicriteria HAZOP (MHAZOP), is proposed in order to contribute to the risk management process. It takes hazard scenarios, multiple consequence dimensions, and the DM's preferences into account.

One of the main results of HAZOP analysis is that it identifies many hazards. Therefore, given the limited resources available, it becomes a major challenge for managers to take action to address all risks (Othman et al. 2016). Furthermore, the DM's preferences will be incorporated into the MHAZOP decision model in order to support the choice of the alternative, and by doing so, the multiple criteria will be analyzed simultaneously (de Almeida et al. 2015). Therefore, MHAZOP will establish a ranking of potential hazards, thus enabling response actions related to the risk management process to be implemented in a structured way so as to eliminate or mitigate existing risks.

2 HAZOP Study

HAZOP is a structured and systematic analysis of a planned or existing product, process, procedure, or system, the main objective of which is to identify risks to people, equipment, the environment, and/or organizational objectives. It is a qualitative technique that is based on the use of guide-words, and is carried out by a multidisciplinary team that should always seek to provide a solution for dealing with risk (IEC 31010 2009).

HAZOP can be divided into four main stages (IEC 61882 2001): definition, preparation, examination, and documentation. Its development is carried out by using questions based on using guide-words in the project/process being analyzed. These guide-words drive the study group's reasoning in order to identify the main deviations from the intent of the project/process. Risk events are assumed to be caused by deviations from the design or operational failures. This search process should prompt group ideas and discussions so as to maximize the chances of a more thorough analysis of the system. Subsequently, the causes and the consequences of these possible undesirable events are identified, and, finally, mitigating actions are suggested that aim to eliminate or minimize risk.

To facilitate the HAZOP process, the system under analysis is divided into nodes, which may have different "sizes" depending on the complexity of the system and the severity of the hazards identified. Therefore, it is emphasized that this selection of the size of the nodes is a subjective decision which depends on the objectives that the DM has to reach, i.e., what applying HAZOP is expected to achieve.

The final results of HAZOP analysis identify a large number of hazards, so they need to be prioritized. According to Othman et al. (2016), in practice this prioritization is based on DMs' experience of making evaluations based on their deductive judgment, which often only takes into account aspects related to safety and costs. For this problem, Othman et al. (2016) presented a structured methodology to incorporate prioritizing hazards into HAZOP analysis using an analytic hierarchy process (AHP), called HAZOP-AHP. Besides prioritizing the hazards identified from the HAZOP assessment, they also provide a means for the assessors to quantitatively analyze the hazards with the appropriate countermeasures to be taken.

Ramzan et al. (2009) present a systematic procedure based on an Extended HAZOP methodology and the MCDA PROMETHEE technique for the case study of a distillation column. The decision analysis included generating alternatives for safe design; analysis of alternatives based on the extended HAZOP methodology to identify hazards and generate alternatives; and, the economic module for estimating both fixed and operating cost and calculating extended costs (risk cost).

Unlike previous HAZOP proposals, MHAZOP analyzes the concept of risk from the point of view of Decision Theory, in which the multidimensionality of the consequences is dealt with. This considers the possible damage in multiple dimensions and incorporates the DM's preference structure, thus enabling the DM, based on his/her perceptions and behavior, to obtain differentiated information so that risks in productive systems can be better managed.

Lastly, the existence of means of detecting and dealing with prioritized deviations must be analyzed and so too must means of preventive and mitigating safeguards be investigated in order to make recommendations and suggest improvement actions for each deviation. Finally, a person responsible for correcting the process is appointed and the effectiveness of the implementation is monitored.

3 MCDM Approach

Multicriteria Decision Making (MCDM) is a set of methods and techniques developed to support organizations and individuals to solve decision problems. MCDM considers the DM's preference structure and involves value judgment. The DM's preferences will be incorporated into the decision model in order to support choosing the alternative, and while doing so, multiple criteria will be analyzed simultaneously (de Almeida et al. 2015).

According to Munda (2008), from the operational point of view, the greatest strength of multicriteria decision-making methods is their ability to analyze issues that are characterized as conflicting, from different points of view, thereby allowing the evaluation of the problem to be analyzed in an integrated way.

One important aspect of the problem that MDCM considers is the DM's rationality. This can be compensatory or non-compensatory. Compensatory rationality allows the overall performance of an alternative to be obtained by trading off values. Single-criterion MCDM methods reflect compensatory rationality.

Multi-Attribute Utility Theory (MAUT) incorporates the issue of tackling problems which have multiple objectives (Keeney and Raiffa 1976), by aggregating utility functions, and does so by considering the DM's preference structure. The multi-attribute utility function is estimated in accordance with the domain of consequences, for which a structured protocol is used that is based on an axiomatic structure.

According to Keeney and Raiffa (1976), the evaluation process of a multi-attribute utility function has five steps: introduction of the terminology and idea, identification of the relevant hypotheses of independence, evaluation of the conditional utility function, evaluation of the scale constants, and the consistency check.

Choosing the MAUT method is justified because it has a well-structured protocol, which is supported by a very solid and consistent axiomatic structure for decisions which involve several criteria. In addition, according to Brito and de Almeida (2009), it is in the probabilistic modeling stage that the uncertainties are inserted within the axiomatic structure, which allows a more consistent approach to be taken with regard to applying MAUT in multicriteria decision problems under situations of uncertainty. This step of probabilistic modeling complements the modeling of the DM's preference structure.

In addition, MAUT takes the DM's behavior in relation to risk into account. Thus, the DM may be prone to risk or risk neutral for a given situation in each of the dimensions of consequence analyzed (Keeney and Raiffa 1976).

4 Multidimensional HAZOP Analysis

Multicriteria decision-making approaches can be incorporated into the risk management process in order to provide a more structured decision-making process in the context under analysis. Different multicriteria methods have been applied in recent years, thus providing a multidimensional analysis of risk. In this section, a multicriteria decision model called Multidimensional HAZOP Analysis (MHAZOP) is proposed. This model integrates Multi-Attribute Utility Theory (MAUT) with HAZOP. The model is used to conduct a multidimensional analysis of risks, based on the steps set out in the framework which is shown in Fig. 1. The framework was constructed by combining the steps of the classic HAZOP (Macdonald 2004) with the approach of the multidimensional risk analysis, thereby seeking a vision of systemic risk management.

4.1 *Identifying the Decision Maker (DM)*

During the first step of the framework, who the DM will be is defined. The DM plays the central role in the decision process. He/she has the authority to take the decision and is accountable for it and therefore his/her preference structure is adopted. Therefore, it is important to emphasize that the DM's preferences should reflect the organization's interests and objectives, and also those of the senior managers who will be held responsible for any and all consequences of this decision.

The DM may be influenced by other actors, such as analysts, clients, experts, and stakeholders. The analyst gives methodological support to the DM by structuring and building the decision model. Stakeholders try to influence the DM's behavior so as to obtain a satisfactory result for themselves or those whom they represent. In general, these stakeholders are affected by the DM's decisions. Experts are actors who have specialized knowledge of some part of the system, who provide realistic information that is incorporated into the decision model. According to de Almeida et al. (2015), experts may be relevant for decision problems in the context of Risk, Reliability, and Maintenance (RRM), since this requires many probabilistic issues to be modeled, which experts have experience of.

What should also be emphasized is that in several situations it will be necessary to include the preferences of several DMs, i.e., there will be a need for more than one DM. In other words, a group decision will be made (Kilgour et al. 2010). However, this framework places the emphasis on decision problems for which there is only one DM.

As the proposed model seeks to be a tool that aids the management of risks, the DM chosen can be appointed from among managers who have key functions within the company. Among such functions are those that serve risk management by supplying the main information that is used to help a DM carry out his/her functions. These managers may well include the maintenance manager, the manager responsible for health and safety at work and in the environment, or the operations and production manager.

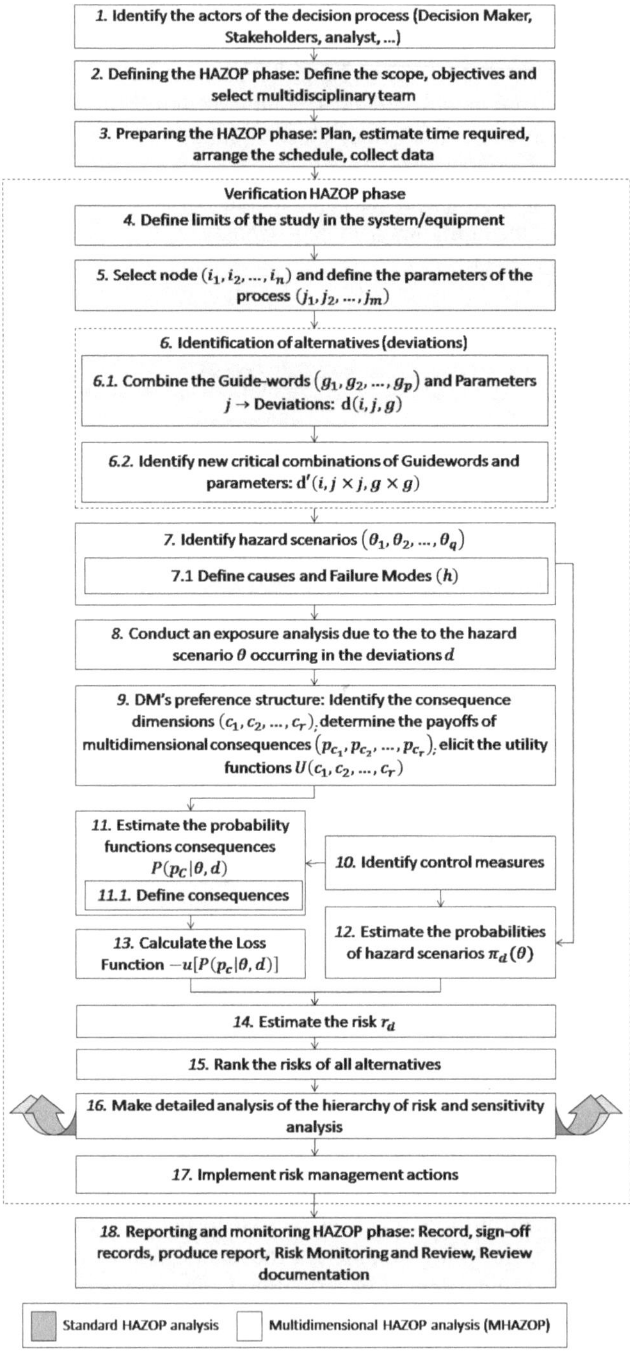

Fig. 1 Multidimensional HAZOP framework

4.2 Definition HAZOP Phase: Define the Scope, Objectives and Select Multidisciplinary Team

The first step of HAZOP is the definition stage. This sets the scope of the main objectives previously defined by the DM and the multidisciplinary team. During this stage, a qualitative analysis of the multidimensional risks involved in the processes of a given company is conducted with a view to proposing measures that prevent and/or mitigate risks. According to Baybutt (2015a), drawing up a statement of the purpose, scope, and objectives of these measures is essential to ensure that HAZOP studies are focused and complete.

Why a HAZOP study is conducted? Generally, this is for a combination of reasons which include meeting the regulatory requirements; the requirements of the industry in general and of the company in particular; and to reduce legal liabilities, as part of a post-incident investigation; the insurance company's requirements, etc. HAZOP also helps to guarantee that the result is consistent with what the study set out to do.

The scope of HAZOP defines what should be included in the study, as well as what should not be included. Items that can be considered include: boundaries (limits) of the process; equipment, procedures, control systems, etc.; support systems; modes of operation (states of the process during its life cycle such as startup, normal operation, and shutdown); external events (natural events such as lightning strikes, events induced by humans, cascade events and failures in utilities and support systems, etc.) (Baybutt 2015a).

The objectives define what is to be considered, specifically, the types of hazards and consequences. In some situations, there may be objectives which overlap in the definition of their scope for different views of practitioners.

Hence, identifying the purpose, scope, and objectives is an important step of the decision process because these have an impact on all the steps in this process. Furthermore, they may influence even the process of establishing the set of alternatives. The approaches of Problem Structuring Methods (PSM) are very useful for conducting this step (Keeney 1992; Rosenhead and Mingers 2001).

HAZOP is a technique performed by teamwork, in which it is preferable for people to have different functions and multidisciplinary knowledge, such as managers and operators of production, maintenance, design, procurement, software, etc. Therefore, having multidisciplinary teamwork allows the maximum use of the experiences and different competences of those belonging to the group and thus for them to achieve an understanding of the problems of different areas and interfaces of the system under analysis.

4.3 HAZOP Preparation Phase: Plan, Estimate Time Required, Arrange the Schedule, and Collect Data

The HAZOP preparation phase includes activities such as establishing how to plan the study to be conducted, collecting data, choosing the method of registration, estimating the time required to undertake the study, and defining the schedule of activities to be carried out. It is worth pointing out that data collection can be supported by other surveys of risk analysis in ventures similar to the one analyzed, of technical norms and of regulations on the subject.

4.4 Define Limits of the Study in the System/Equipment/Process

This stage defines what systems/equipment and production processes will enter the HAZOP study. In order to do this, the level of analysis, which will establish the depth of the analysis in the system (system, subsystems, components, processes, etc.) should be defined as should the criteria which will be used to select the systems/processes, i.e., the priority systems/processes. This should be conducted in accordance with the set of objectives that the DM has defined, which may involve objectives related to the impact on safety, environment, operation, cost, etc. Therefore, since multiple objectives will be considered, MCDM models are appropriate for tackling this decision problem (de Almeida et al. 2015).

After setting priorities, these systems/processes should be identified and described, and their limits should be defined. This step is essential to avoid redundant analysis of the same system/process in different stages of the HAZOP study, or even to avoid creating gaps in HAZOP because evaluating some part of the system/process was overlooked.

4.5 Selecting Nodes ($i_1, i_2, \ldots, i_n$) and Identifying the Parameters of the Process ($j_1, j_2, \ldots, j_m$)

To start the HAZOP study, the lines (nodes or circuits) ($i_1, i_2, \ldots, i_n$) in the flowchart of the process under analysis should be defined. Each node corresponds to a subsystem/piece of equipment in which the activities and tasks developed in that area will be detailed. This procedure can avoid excluding a risk that should be analyzed.

The parameters of the process ($j_1, j_2, \ldots, j_m$) refer to the process variables that are being evaluated, which will then be evaluated in relation to the intention of the original design conditions, for example, temperature, pressure, flow, density, etc.

Table 1 Examples of basic guide-words

Guide-word	Meaning
No or none	Negation, absence
More	Quantitative increase
Less	Quantitative decrease
Also	Qualitative modification/addition
Part of	Qualitative modification/subtraction
Reverse	Opposite direction
Other than	Complete substitution

4.6 *Identifying Alternatives (Deviations)*

4.6.1 Combining the Guidewords ($g_1, g_2, \ldots, g_p$) and Parameters j → Deviations: d(i, j, g)

Basically, the HAZOP analysis investigates how an *i* node of a plant, sector, or piece of equipment can deviate d from the intention of the design. For this purpose, guide-words *g* (Table 1) are used to evaluate the project variables *j*. Therefore, HAZOP thoroughly investigates each succeeding step of a process in order to discover all possible deviations from normal operating conditions, thus serving as a reminder that operability is as important as identifying hazards. Subsequently, HAZOP investigates the causes and consequences of this deviation from design, and offers suggestions so that such deviations do not occur.

The guide-words are compared with the parameters of the process (temperature, pressure, level, etc.), thereby generating the possible problems to be studied, as shown in Table 2.

4.6.2 Identify New Critical Combinations of Guide-Words and Parameters: d(i, j × j, g × g)

One of the criticisms of traditional HAZOP is that it does not consider the possible occurrence of compound deviations. Therefore, this stage of the proposed model seeks to identify these possible compound deviations that can produce critical combinations for the system under analysis. In other words, in addition to the simple combination of the guide-words and the process parameters for node (i, j, g) in order to determine each deviation $d(i, j \times j, g \times g)$, it is interesting to evaluate the combination of the occurrence $(g_1, g_2, g_3 \ldots, g_1 \times g_2, g_1 \times g_3, \ldots, g_2 \times g_3, \ldots, g_1 \times g_2 \times g_3, \ldots)$ with multiple parameters of the process $(j_1, j_2, j_3, \ldots, j_1 \times j_2, j_1 \times j_3, \ldots, j_2 \times j_3, \ldots, j_1 \times j_2 \times j_3, \ldots)$, as exemplified in Table 3.

This combination is justified because of some systems in which the occurrence of a simple deviation d(i, j, g) may not have serious consequences, and therefore they may be minimized in the risk management process. This occurs mainly in

Table 2 Deviation formed by combining parameters and a guide-word

Parameter (variables)	Guide-word	Deviations
Flow	No or none	No flow
	Less	Less flow
	More	More flow
	Reverse	Reverse flow
	Also	Contamination
Pressure	Less	Low pressure
	More	High pressure
Temperature	Less	Low temperature
	More	High temperature
Level	Less	Low level
	More	High level
Viscosity	Less	Low viscosity
	More	High viscosity
Reaction	No or none	No reaction
	Less	Incomplete reaction
	More	Uncontrolled reaction
	Reverse	Reverse reaction
	Also	Secondary reaction

situations in which deviations can generate possible hidden failures, or are of little consequence, but the occurrence of multiple failures has failures with catastrophic consequences. For example, the deviation "No flow $d(G_1 \times P_1)$" may not be critical for a given process, i.e., it may not generate great consequences, should the deviation happen. However, the occurrence of the deviation formed by "No flow & High Temperature $d(G_1 \times P_1; G_3 \times P_2)$" can generate catastrophic consequences. Therefore, it is essential to evaluate the multiple combinations of the guide-words and multiple parameters of the process.

4.7 Identify Hazard Scenarios ($\theta_1, \theta_2, \ldots, \theta_q$)

The identification of the hazard scenarios is related to states of nature (θ), defined in Decision Theory. The state of nature corresponds to one of the basic ingredients of Decision Theory and it consists of factors in the system that are not under the DM's control and may change randomly, thus influencing the outcomes of the decision process (Raiffa 1968; Berger 1985; Edwards et al. 2007; de Almeida et al. 2015).

This step consists of defining all possible hazard scenarios $\Theta = \{\theta_1, \ldots, \theta_q\}$ resulting from operational failures defined previously by the deviations $d(i,j \times j, g \times g)$. The resulting hazard scenarios do not define the mode of failure or accidental causes, but the phenomena or accidents associated with the deviation, which are influenced by the failure mode and factors adjacent to the occurrence of the deviation, such as, for example, there being immediate ignition or delated

Table 3 New critical combinations of guide-words and parameters: $d(i, j \times j, g \times g)$: word-guides (no or none; less; more) and process parameters (flow; temperature)

Parameters (variable)	**Guide-word**	Flow (P1): No or none (G1)	Flow (P1): Less (G2)	Flow (P1): More (G3)	Temperature (P2): No or none (G1)	Temperature (P2): Less (G2)	Temperature (P2): More (G3)
Flow (P1)	No or none (G1)	No flow $d(G_1 \times P_1)$	-	-	$d(G_1 \times P_1)$	No flow & Low Temperature $d(G_1 \times P_1; G_2 \times P_2)$	No flow & High Temperature $d(G_1 \times P_1; G_3 \times P_2)$
	Less (G2)	-	Less flow $d(G_2 \times P_1)$	-	$d(G_2 \times P_1)$	Less flow & Low Temperature $d(G_2 \times P_1; G_2 \times P_2)$	Less flow & High Temperature $d(G_2 \times P_1; G_3 \times P_2)$
	More (G3)	-	-	More flow $d(G_3 \times P_1)$	$d(G_3 \times P_1)$	More flow & Low Temperature $d(G_3 \times P_1; G_2 \times P_2)$	More flow & High Temperature $d(G_3 \times P_1; G_3 \times P_2)$
Temperature (P2)	No or none (G1)	$d(G_1 \times P_1)$	$d(G_2 \times P_1)$	$d(G_3 \times P_1)$	-	-	-
	Less (G2)	$d(G_1 \times P_1; G_2 \times P_2)$	$d(G_2 \times P_1; G_2 \times P_2)$	$d(G_3 \times P_1; G_2 \times P_2)$	-	Low Temperature $d(G_2 \times P_2)$	-
	More (G3)	$d(G_1 \times P_1; G_3 \times P_2)$	$d(G_2 \times P_1; G_3 \times P_2)$	$d(G_3 \times P_1; G_3 \times P_2)$	-	-	High Temperature $d(G_3 \times P_2)$

ignition, whether or not there is confined space, half-life of toxicity of a certain product, etc. In addition to these hazard scenarios, one should consider the normal operating scenario θ_N, in which all operations are perfectly normal, and no type of hazard occurs. Thus, in this case, it is considered that no hazard scenarios occur.

There are several techniques that can help this step, such as Preliminary Risk Analysis (PRA), Failure Modes and Effects Analysis (FMEA), and Event Tree Analysis (ETA).

FMEA is a bottom-up approach that is used to analyze all potential failure modes of a system component and is widely used in industry as a means of identifying, ranking, and mitigating the failure modes of components (Whiteley et al. 2016; Du et al. 2016; Akbarzade Khorshidi et al. 2016; Selim et al. 2016). The criticality determined by FMEA is traditionally established by calculating the risk priority number (RPN) that is obtained by the product of its severity, occurrence, and detectability (Lolli et al. 2015; Zhou and Thai 2016).

Ruijters and Stoelinga (2015) point out that HAZOP and FMEA are similar tools in the sense that they both list the possible causes of a failure. The major difference between these tools is that FMEA considers failure modes of the components of a system, whereas HAZOP considers abnormalities of a process.

Event Tree Analysis represents systems by means of diagrams where these systems include the combination of equipment and actions required to obtain data for the purpose of the study. Event trees are constructed horizontally, beginning with an initial event that describes a scenario or situation where the system is required. Heravi and Charkhakan (2015) emphasize that ETA is a technique that is commonly applied to identify the consequences that may arise when these potentially dangerous initial events occur, and thereby provides a way to describe a sequence of probabilistic events along with their probabilities and impacts.

4.8 Conduct an Exposure Analysis of Objects Due to the Hazard Scenario θ Occurring in the Deviations d(i,j × j, g × g)

In this step the objects that are exposed to the impacts due to the occurrence of a certain hazard scenario θ will be analyzed. For each combination of hazard scenario θ and deviation $d(i,j \times j, g \times g)$, mathematical models and numerical applications should be applied in the various surrounding characteristics of the objects exposed to the source of the hazard. Therefore, it is desired, by using a quantitative approach, to estimate the possible impacts on the various dimensions of consequences such as losses to structures and properties, the environment, and the health and safety of people, for which there may be, depending on the mathematical complexity, probabilistic modeling that incorporates the dynamic view of the system, or a simplistic approach that incorporates the deterministic view of modeling consequences.

The modeling of the exposure analysis will estimate the hazard zone, which can be determined by the area or diameter or distance from the source of the hazard (Garcez and de Almeida 2014a, b). In the context of natural gas pipelines, for example, a hazard zone is a region in which the physical effects of a hazard exceed critical thresholds, thus inducing negative impacts on people, property, and the environment (Dziubiński et al. 2006; Brito and de Almeida 2009; Brito et al. 2010). Jo and Ahn (2002) use the thermal radiation intensity of $15kW/m^2$ as the critical threshold.

4.9 DM's Preference Structure

According to Cox LA (2012), the application of utility functions rather than simple risk formulae – consisting of terms such as exposure, probability, and consequence – allows a DM's risk attitudes to be taken into account, thereby improving the effectiveness of the decision-making process at reducing risks. Cox Jr (2009) discusses many issues related to the decision process in the risk context, including the limitations of risk assessment using risk matrices and a normative decision framework.

An important issue to be evaluated when modeling the DM's preferences is the assessment of rationality regarding compensation among criteria. Firstly, it is necessary to evaluate with the DM her/his basic preference properties and preference system. Secondly, the type of rationality that is the most adequate to the DM is established, i.e., a compensatory or a non-compensatory approach. This answer guides the choice of the MCDM/A method (de Almeida et al. 2015).

For each objective previously established by the DM, a criterion or attribute has to be proposed. In the RRM context, the criteria will represent different dimensions of consequences $(c_1, c_2, \ldots, c_r)$ in the assessment risk model. To Roy (1996), the criteria cannot have redundancy; they must be exhaustive, since all objectives have to be present and represented by the criteria; and they have to be consistent, in the sense that the DM's preferences over the criteria have to be coherent with the global evaluation of consequences. A structured view for building criteria is shown by Keeney (1992).

After identifying the consequence dimensions $(c_1, c_2, \ldots, c_r)$,the pay-offs of multidimensional consequences $\left(p_{c_1}, p_{c_2}, \ldots, p_{c_r}\right)$ should be determined, estimating the possible dimensions impacts, resulting from hazard scenarios (θ) and the occurrence of deviations $d(i,j \times j, g \times g)$, in a hazard zone already estimated previously.

According to Brito and de Almeida (2009), the traditional representation of risk with probabilities or product of probabilities by consequence values does not reflect the aversion of people in relation to harmful events with low probability and great catastrophic consequences. This requires an approach that considers the DM's preferences. Thus, utility functions can be used on the consequences to incorporate the DM's preference to the risk of losses due to accidents (Garcez and de Almeida 2014a).

Therefore, the DM's utility functions need to be elicited. This step consists of eliciting the utility function related to the utility of different performance of outcomes in the criterion $(c_1, c_2, \ldots, c_r)$. Keeney and Raiffa (1976) described a five-step multi-attribute utility elicitation procedure: Introduction to terminologies and ideas; Identification of independence assumptions; Evaluation of the conditional utility functions; Evaluation of scale constants; and Verification of the consistency and reiteration. This procedure identifies the DM's behavior regarding risk, which may be classified into: neutral, averse, or prone to risk. For a neutral risk behavior, the $u(c_r)$ is a linear function, and the others are non-linear functions. In the elicitation procedure, a utility function is obtained on a scale of 0 to 1. It should be observed that the utility function is given in an interval scale (de Almeida et al. 2015).

The elicitation of the utility function occurs over a closed range of consequences, where the maximum value is limited to a null consequence (i.e., there is no impact) and the minimum value is the greatest of the estimated consequences for each $d(i, j \times j, g \times g)$. It is worth mentioning that although discrete and enumerable values (number of victims, for example) can be considered, the consequence sets can be considered continuous for the purposes of estimating the utility function.

Another step of the elicitation procedure of the multi-attribute utility function is that of verifying the independence and additive independence between the attributes, i.e., to verify if there is preferential independence between the sets of consequences in the dimensions analyzed $(c_1, c_2, \ldots, c_r)$. Given that the DM's additive independence is verified, the additive utility function can be described by:

$$U(c_1, \ldots, c_r) = k_{c_1} U(c_1) + \cdots + k_{c_r} U(c_r) \tag{1}$$

where $U(c_1) \ldots U(c_r)$ are the one-dimensional utilities for the different risk dimensions, and the constants $k_{c_1} \ldots k_{c_r}$ are constants of scales estimated by elicitation processes based on comparing lotteries of payoffs, and $\sum_1^r k_{c_i} = 1$, as described by Keeney and Raiffa (1976).

Regarding the compensatory approach, the meaning of the "weights", normally called scale constants k_c, does not involve only the importance of the criteria. Their elicitation is related to the scales of the value function in each criterion (Vincke 1992; Belton and Stewart 2002). For probabilistic consequences, using MAUT, there are very well-structured elicitation procedures for obtaining the scale constants used to aggregate the utility functions of the criteria (Keeney and Raiffa 1976).

There are many elicitation procedures in the literature for eliciting scale constants (Weber and Borcherding 1993). Among these are the tradeoff and the swing procedures (Keeney and Raiffa 1976; Edwards and Barron 1994). There is also a flexible elicitation procedure for additive model scale constants (FITradeoff) that is proposed by de Almeida et al. (2016b), who use partial information to perform dominance tests based on a linear programming problem.

4.10 Identify Control Measures

This step of the model seeks to identify control measures that can be implemented to eliminate/mitigate existing risks. These measures may be associated, for example, to the safety engineering of the project, management control processes, provision of warning labels or maintenance activities to prevent failures occurring, as well as skills-giving/training actions. Examples are shutdown systems, relief/exhaust systems, fire protection systems, smoke detectors, and evacuation procedures.

4.11 Estimate the Consequences of Probability Functions $P(p_C|\theta, d(i,j \times j, g \times g))$

The criterion or attribute may be considered in two ways, regarding its variability and uncertainty: it may be deterministic or probabilistic. A deterministic criterion is assumed to have a constant level of performance or fixed outcome. A probabilistic criterion has a consequence x, which is a random variable and is specified in terms of its probability density function (pdf): $f(x)$. If a criterion is a random variable, with a not relevant variability it may be assumed to be deterministic. In this case, it is assumed that the standard deviation is so small, that the mean of the variable may represent the consequence x (de Almeida et al. 2015).

Regarding uncertainty, a DM may regard a criterion or an attribute as being ambiguous in the representation of its value function and therefore fuzzy numbers (Pedrycz et al. 2010) could be used to represent them. In this case, a fuzzy approach may be considered for the decision model, which may influence the choice of the MCDM/A method.

Several uncertainties are present in the modeling of the consequence function. Such uncertainties are desirable because it becomes unlikely that all the multidimensional consequences that can occur due to a deviation d occurring can be defined in a deterministic way. Therefore, in this step what is desired is to estimate a probability distribution on the possible values of the payoffs $P(p_C|\theta, d(i,j \times j, g \times g))$, given that a hazard scenario and the deviation may have occurred.

4.12 Estimate the Probabilities of Hazard Scenarios $\pi_d(\theta)$

In this step, an estimate is made of the probabilities $\pi_{d(i,j \times j, g \times g)}(\theta_{1, \ldots, q})$ of the hazard scenarios θ stipulated in the previous step and associated with each deviation $d(i,j \times j, g \times g)$. Risk analysis allows the failures in the system to be anticipated, thereby helping to identify potential causes and possible consequences. Such anticipation can be achieved by analyzing accidents that have previously occurred in facilities, both in the company's own facilities and in similar companies;

records in the specialized literature or held in international databases; laboratory tests; and physical/chemical resistance tests, for the purposes of simulation. This type of analysis allows a statistical (frequentist) evaluation of the causes and local conditions that favor the occurrence of deviations.

Another alternative for obtaining these probabilities is by taking advantage of an expert's a priori knowledge. The analyst can apply an elicitation procedure to obtain the $\pi(\theta)$. This procedure is usually applied to an expert on the behavior of θ (de Almeida et al. 2015). Due to the problem of archives generally containing limited or incomplete data, these two ways for estimating probabilities can be combined in an attempt to take advantages of each vision, to do which complementary forms, the frequentist approach (historical), and the Bayesian approach are used.

4.13 Calculate the Loss Function $-u[P(p(c)|\,\theta, d())]$

According to Decision analysis, risk is considered to be the expected value of the loss, defined as:

$$L(\theta, \mathrm{d}) = -u\left[P\left(p\left(c_1, \ldots, c_r\right)|\theta, d\left(i, j \times j, g \times g\right)\right)\right] \tag{2}$$

Know, as:

$$u\left(P\left(p\left(c_1, \ldots, c_r\right)|\theta, d\left(i, j \times j, g \times g\right)\right)\right) = E_p\left[p\left(c_1, \ldots, c_r\right)\right]$$
$$= \int\limits_{p \in P} P\left(p\left(c_1, \ldots, c_r\right)|\theta, d\left(i, j \times j, g \times g\right)\right) u\left(p\left(c_1, \ldots, c_r\right)\right) dp\left(c_1, \ldots, c_r\right), \tag{3}$$

then,

$$L\left(\theta, d\left(i, j \times j, g \times g\right)\right) = -u\left(P\left(p\left(c_1, \ldots, c_r\right)|\theta, d\left(i, j \times j, g \times g\right)\right)\right)$$
$$= -\int\limits_{p \in P} P\left(p\left(c_1, \ldots, c_r\right)|\theta, d\left(i, j \times j, g \times g\right)\right) u\left(\mathrm{p}\left(c_1, \ldots, c_r\right)\right) d\mathrm{p}\left(c_1, \ldots, c_r\right) \tag{4}$$

4.14 Estimate Risk $r_{d(i,j \times j, g \times g)}$

In this step, an estimate is made of the risks associated with a given deviation $d(i,j \times j, g \times g)$ evaluated in the study. Calculating the risk is based on the expected value of the loss. Therefore, the losses associated with each hazard scenario θ and deviations $d(i,j \times j, g \times g)$ are added, in the different dimensions of the consequences tackled $(c_1, c_2, \ldots, c_r)$, multiplied by the probabilities of the hazard scenarios $\pi_{d(i,j \times j, g \times g)}(\theta_{1, \ldots, q})$. The loss associated with a θ_N scenario of normality is also added to the risk calculation:

$$r_{d(i,j\times j,g\times g)} = E_\theta \left[L\left(\theta, d\left(i,\ j \times j, g \times g\right)\right)\right]$$

$$= \left(\sum_\theta \sum_c \pi\left(\theta\right) L\left(\theta, d\left(i, \mathrm{j} \times j, g \times g\right)\right)\right) + \pi\left(\theta_N\right)(-1) \tag{5}$$

Considering that the risk values of each dimension correspond to the negative of the utility of the distribution of consequences, and since utilities are on a scale of 0 to 1, and losses on a scale of -1 to 0, then the risks will be on a scale from -1 to 0. Thus, the closer the risk value is to -1, the safer is the deviation $d(i,j \times j, g \times g)$, and consequently the closer to the value 0, the greater the risk of the deviation.

4.15 *Rank the Risk from All Alternatives r_d*

Having computed the risk values for each deviation $d(i,j \times j, g \times g)$, the alternatives analyzed can be placed in decreasing order, thus forming a ranking of decreasing risks (where the most critical risk is the one placed first in the ranking and the risk placed last in the ranking is the least critical one of those considered in the study). A differential of this approach when compared to other traditional approaches used for risk management is that a more structured analysis is used which, besides aggregating the multiple criteria, takes into account the DM's preferences in relation to the risk. Thus, a more robust result is obtained than, for example, would be the case by using the RPN (Risk Priority Number) in FMEA (Failure Mode and Effect Analysis) or even subjective nominal scales. These are sometimes inserted into the HAZOP worksheet to assess the level of occurrence and impact consequences of the deviations under analysis. The ranking obtained by using the multidimensional analysis proposed in this section can help DMs to allocate resources, since these are usually limited and scarce. Therefore, it would be better to allocate resources so as to prioritize deviations $d(i,j \times j, g \times g)$, by adopting preventive and mitigating measures, when managing maintenance risks.

4.16 *Conduct a Detailed Analysis of the Hierarchy of Risk and Sensitivity Analysis*

Sensitivity analysis is a tool of the utmost importance for evaluating uncertainties when conducting probabilistic risk analysis. A more detailed analysis of the final hierarchy obtained from the model proposed in this section can be seen after conducting an overall sensitivity analysis of the results. This level of detail of the study with respect to the risk hierarchy arises from the robustness analysis of the process, based on varying the parameters of the model and on the input data. According to de Almeida et al. (2015), this analysis can indicate whether the aspects of the study model and input data are considered robust or sensitive in order to verify if any steps should be reevaluated due to some hypotheses or input data. Moreover, in the case of the ordering problem, what is evaluated is whether some alternative(s) of the ranking established have undergone a variation in its/their position in the ranking. If this is observed, an analysis can be made of the frequency at which this occurs, in which alternatives this is observed, besides the relevance of the change(s) in positions. According to Medeiros et al. (2015, 2016, 2017), sensitivity analysis provides the DM with a more assertive recommendation based on the analysis of the variations that were observed throughout the simulation, and also with information about uncertainties of different groups of parameters of the model.

4.17 Implement Risk Management Actions

Finally, after the DM has received the recommendation and accepted the proposed solution, then, a start can be made on implementing it. This may be either: simple and immediate or complex and time consuming. Implementation of the risk management actions may be as complex as conducting the process that led to decision-making and may take much more time to accomplish than the decision process itself.

The effectiveness of the risk elimination/mitigation actions should be re-assessed by recalculating the risk values. Before doing so, there is a need for a period of time deemed necessary to elapse only after which the return on implementing the risk management the actions can be judged satisfactory. From that point on, the values associated with the risks analyzed should be recalculated. If the value associated with a given risk is reduced, it is understood that the actions implemented were effective. Otherwise, other alternatives for action to combat risk must be sought in order to obtain a lower risk value than the existing one, based on the calculations of the model.

4.18 Reporting and Monitoring HAZOP Phase: Record, Sign-Off Records, Produce Report, Monitor and Review Risk and Review Documentation

After the conclusion of the last step, if no return to revise previous steps is necessary, then, finalization is tackled in this step by analyzing the final results and producing the report for the DM, with the final recommendations.

A good report indicates to the DM the extent to which the solution can be trusted. The DM should be advised on the nature of the models. The DM should understand that there is no right model and the usefulness of the model is the main issue to be evaluated (de Almeida et al. 2015).

5 Final Remarks

The use of multicriteria modeling to quantitatively assess the consequences of the undesired events considered in a given multidimensional risk analysis is proposed in this paper with a view to guaranteeing an improvement in the process by generating results that can assure managers prioritize actions adopted to preventing and mitigating the risks. Therefore, what is sought is to support managerial actions in the context of uncertainty by using MAUT as a decision support method which is implemented in a way that is associated with HAZOP as shown in the proposed model. Incorporating the DM's preferences is highlighted as a differential of this approach. This is done in order to incorporate subjective aspects based on the DM's perceptions and behavior in the modeling.

Acknowledgments This paper is part of a research study funded by the Fundação de Amparo à Ciência e Tecnologia do Estado de Pernambuco (FACEPE) and the National Council for Scientific and Technological Development – CNPq.

References

Akbarzade Khorshidi H, Gunawan I, Ibrahim MY (2016) Applying UGF concept to enhance the assessment capability of FMEA. Qual Reliab Eng Int 32:1085–1093. https://doi.org/10.1002/qre.1817

Aven T (2018) An emerging new risk analysis science: Foundations and implications. Risk Anal 38:876–888. https://doi.org/10.1111/risa.12899

Baybutt P (2015a) The importance of defining the purpose, scope, and objectives for process hazard analysis studies. Process Saf Prog 34:84–88. https://doi.org/10.1002/prs.11682

Baybutt P (2015b) A critique of the Hazard and operability (HAZOP) study. J Loss Prev Process Ind 33:52–58. https://doi.org/10.1016/j.jlp.2014.11.010

Belton V, Stewart VBTJ (2002) Multiple criteria decision analysis: An integrated approach. Springer

Berger J (1985) Statistical decision theory and Bayesian analysis (Springer Series in Statistics), 2nd edn. Springer

Bernstein PL (1998) Against the gods: The remarkable story of risk. Wiley

Brito AJ, de Almeida AT (2009) Multi-attribute risk assessment for risk ranking of natural gas pipelines. Reliab Eng Syst Saf 94:187–198. https://doi.org/10.1016/j.ress.2008.02.014

Brito AJ, de Almeida AT, Mota CMM (2010) A multicriteria model for risk sorting of natural gas pipelines based on ELECTRE TRI integrating utility theory. Eur J Oper Res 200:812–821. https://doi.org/10.1016/j.ejor.2009.01.016

Cox LA Jr (2009) Risk analysis of complex and uncertain systems. Springer Science & Business Media

Cox Louis Anthony (Tony) J (2012) Evaluating and improving risk formulas for allocating limited budgets to expensive risk-reduction opportunities. Risk Anal 32:1244–1252. https://doi.org/10.1111/j.1539-6924.2011.01735.x

de Almeida AT, Cavalcante CAV, Alencar MH et al (2015) Multicriteria and multiobjective models for risk, reliability and maintenance decision analysis. Springer, Cham

de Almeida AT, Alencar MH, Garcez TV, Ferreira RJP (2016a) A systematic literature review of multicriteria and multi-objective models applied in risk management. IMA Journal of Management Mathematics dpw021. https://doi.org/10.1093/imaman/dpw021

de Almeida AT, de Almeida JA, Costa APCS, de Almeida-Filho AT (2016b) A new method for elicitation of criteria weights in additive models: flexible and interactive tradeoff. Eur J Oper Res 250:179–191. https://doi.org/10.1016/j.ejor.2015.08.058

Du Y, Lu X, Su X et al (2016) New failure mode and effects analysis: an evidential downscaling method. Qual Reliab Eng Int 32:737–746. https://doi.org/10.1002/qre.1753

Dunjó J, Fthenakis V, Vílchez JA, Arnaldos J (2010) Hazard and operability (HAZOP) analysis. A literature review. Journal of Hazard Mater 173:19–32. https://doi.org/10.1016/j.jhazmat.2009.08.076

Dziubiński M, Frątczak M, Markowski AS (2006) Aspects of risk analysis associated with major failures of fuel pipelines. J Loss Prev Process Ind 19:399–408. https://doi.org/10.1016/j.jlp.2005.10.007

Edwards W, Barron FH (1994) SMARTS and SMARTER: Improved simple methods for multi-attribute utility measurement. Organ Behav Hum Decis Process 60:306–325. https://doi.org/10.1006/obhd.1994.1087

Edwards W, Miles RF Jr, Von Winterfeldt D (2007) Advances in decision analysis: from foundations to applications. Cambridge University Press

Garcez TV, de Almeida AT (2014a) A risk measurement tool for an underground electricity distribution system considering the consequences and uncertainties of manhole events. Reliab Eng Syst Saf 124:68–80. https://doi.org/10.1016/j.ress.2013.11.007

Garcez TV, de Almeida AT (2014b) Multidimensional risk assessment of manhole events as a decision tool for ranking the vaults of an underground electricity distribution system. IEEE Transactions on Power Delivery 29:624–632. https://doi.org/10.1109/TPWRD.2013.2273083

Guo L, Kang J (2015) An extended HAZOP analysis approach with dynamic fault tree. J Loss Prev Process Ind 38:224–232. https://doi.org/10.1016/j.jlp.2015.10.003

Heravi G, Charkhakan MH (2015) Predicting change by evaluating the change implementation process in construction projects using event tree analysis. J Manag Eng 31:04014081. https://doi.org/10.1061/(ASCE)ME.1943-5479.0000325

IEC 31010 (2009) ISO.IEC 31010:2009 – Risk management – Risk assessment techniques.

IEC 61882 (2001) IEC 61882: Hazard and operability studies (HAZOP studies) – Application guide.

Jo Y-D, Ahn BJ (2002) Analysis of hazard areas associated with high-pressure natural-gas pipelines. J Loss Prev Process Ind 15:179–188. https://doi.org/10.1016/S0950-4230(02)00007-4

Keeney RL (1992) Value-focused thinking: A path to creative decisionmaking. Harvard University Press

Keeney RL, Raiffa H (1976) Decision with multiples objectives preferences and value trade-offs. Wiley

Kilgour DM, Eden C, Rennecker JA et al (2010) Handbook of group decision and negotiation. Springer, Dordrecht

Lolli F, Ishizaka A, Gamberini R et al (2015) FlowSort-GDSS – A novel group multi-criteria decision support system for sorting problems with application to FMEA. Expert Syst Appl 42:6342–6349. https://doi.org/10.1016/j.eswa.2015.04.028

Macdonald D (2004) Practical – Hazops, trips and alarms. Elsevier

Medeiros C, Alencar MH, Garcez TV, de Almeida AT (2015) Global sensitivity analysis based on a Monte Carlo simulation: a study based on a natural gas system. In: Safety and reliability of complex engineered systems. CRC Press, pp 2571–2576

Medeiros CP, Alencar MH, de Almeida AT (2016) Hydrogen pipelines: enhancing information visualization and statistical tests for global sensitivity analysis when evaluating multidimensional risks to support decision-making. Int J Hydrogen Energy 41:22192–22205. https://doi.org/10.1016/j.ijhydene.2016.09.113

Medeiros CP, Alencar MH, de Almeida AT (2017) Multidimensional risk evaluation of natural gas pipelines based on a multicriteria decision model using visualization tools and statistical tests for global sensitivity analysis. Reliab Eng Syst Saf 165:268–276. https://doi.org/10.1016/j.ress.2017.04.002

Munda G (2008) Social multi-criteria evaluation for a sustainable economy. Springer, Berlin, p 210

Othman MR, Idris R, Hassim MH, Ibrahim WHW (2016) Prioritizing HAZOP analysis using analytic hierarchy process (AHP). Clean Technol Environ Policy. https://doi.org/10.1007/s10098-016-1104-4

Pedrycz W, Ekel P, Parreiras R (2010) Fuzzy multicriteria decision-making. Wiley, Ltd, Chichester

Raiffa H (1968) Decision analysis: Introductory lectures on choices under uncertainty. Addison-Wesley

Ramzan N, Naveed S, Feroze N, Witt W (2009) Multicriteria decision analysis for safety and economic achievement using PROMETHEE: a case study. Process Saf Prog 28:68–83. https://doi.org/10.1002/prs.10263

Rosenhead J, Mingers J (2001) Rational analysis for a problematic world revisited: Problem structuring methods for complexity, uncertainty and conflict, Second Edition. Wiley

Roy B (1996) Multicriteria methodology for decision aiding. Springer US, Boston, MA

Ruijters E, Stoelinga M (2015) Fault tree analysis: a survey of the state-of-the-art in modeling, analysis and tools. Computer Science Review 15–16:29–62. https://doi.org/10.1016/j.cosrev.2015.03.001

Selim H, Yunusoglu MG, Yılmaz Balaman Ş (2016) A dynamic maintenance planning framework based on fuzzy TOPSIS and FMEA: application in an international food company. Qual Reliab Eng Int 32:795–804. https://doi.org/10.1002/qre.1791

Vincke P (1992) Multicriteria decision-aid. Wiley, Bruxelles

Weber M, Borcherding K (1993) Behavioral influences on weight judgments in multiattribute decision making. Eur J Oper Res 67:1–12. https://doi.org/10.1016/0377-2217(93)90318-H

Whiteley M, Dunnett S, Jackson L (2016) Failure mode and effect analysis, and fault tree analysis of polymer electrolyte membrane fuel cells. Int J Hydrogen Energy 41:1187–1202. https://doi.org/10.1016/j.ijhydene.2015.11.007

Zhou Q, Thai VV (2016) Fuzzy and grey theories in failure mode and effect analysis for tanker equipment failure prediction. Saf Sci 83:74–79. https://doi.org/10.1016/j.ssci.2015.11.013

Multidimensional Risk Evaluation in Natural Gas Pipelines: Contributions from Sensitivity Analysis and Risk Visualization to Improving the Management of Risk

Francisco Filipe Cunha Lima Viana, Marcelo Hazin Alencar, Rodrigo José Pires Ferreira, and Adiel Teixeira de Almeida

1 Introduction

As concern about how to manage risks has grown, so too has the use of structured procedures to identify, eliminate, or mitigate risks in situations of uncertainty. This is because pressure from society and market competition has made operations under risk even more difficult, and therefore, more studies have been published in areas such as impacts on the environment, finance, occupational health, accidents caused by Nature, terrorism, supply chains, medicine, and industrial production (de Almeida et al. 2015).

Besides, the concepts underpinning what risk is have been frequently debated over the years, a common notion is based on means of probabilities and expected values (Aven and Renn 2009). Generally, risk is regarded to arise when an event of a forecast value is subject to a probability behavior in which the consequence is uncertain (Kaplan and Garrick 1981; Aven 2011; Goerlandt and Kujala 2014).

The intention of estimating risk is to establish at what point it becomes unacceptably high so that mitigation actions can be designed and implemented in order to inhibit the severity of future impacts. From this perspective, to draw up mitigating actions requires proposing, evaluating, and selecting measures to alleviate risks (Meyer et al. 2009).

In some systems, the impacts on the population, the environment, and organizations are extremely negative and take years to be re-established. The challenge is thus to gauge the capacity that the system has to recover from the losses, i.e., its resilience to the risk.

F. F. C. L. Viana · M. H. Alencar (✉) · R. J. P. Ferreira · A. T. de Almeida
CDSID – Center for Decision Systems and Information Development, Universidade Federal de Pernambuco, Recife, PE, Brazil
e-mail: marceloalencar@cdsid.org.br; rodrigo@cdsid.org.br; almeida@cdsid.org.br

A. T. de Almeida et al. (eds.), *Multicriteria and Optimization Models for Risk, Reliability, and Maintenance Decision Analysis*, International Series in Operations Research & Management Science 321, https://doi.org/10.1007/978-3-030-89647-8_7

For the context of losses, the impacts caused by an accidental scenario occurring adversely affect different perspectives (dimensions), thus causing the performance of an organization's objectives to vary (de Almeida et al. 2017). Therefore, the literature also contains studies on how best to evaluate multidimensional risk and some of these will be discussed in this chapter, taking into account the context of transporting natural gas along pipelines.

2 Evaluating Multidimensional Risk

Multicriteria approaches play an important role for risk management since the concepts associated with the methods are able to deal with the uncertainties as well as to understand the conflicts between the objectives (Medeiros et al. 2017). Therefore, there is a need to consider whether more than one objective should be included in the approach taken to a risk management problem.

De Almeida et al. (2015) list some of the loss dimensions considered for the context of risks:

- Human losses: this involves damage to life, to people physically affected, caused by an accidental scenario. An estimate can be made of the likely number of fatalities or of injuries to people in a given accidental scenario.
- Environmental losses: this consider the impacts on the natural environment and biodiversity of the areas affected.
- Financial loss: this is associated with losses in revenues, and other financial losses such as reimbursing customers for services not provided, damages to property, fines, and indemnities.
- Operational losses: this refers to damage to facilities, everyday materials, consumables, and everything related to the production system studied.

Establishing tradeoffs between conflicting objectives is complex for decision-making. For example, how can different preferences of different decision-makers for bearing losses be reconciled when such losses include the impact on financial objectives, the impact on human lives or on the environment. MCDM/A methods include some which offer an appropriate procedure that makes tradeoffs possible (Keeney and Raiffa 1976).

The studies developed by Brito and de Almeida (2009) and Alencar and de Almeida (2010) prove that it is possible to estimate risks using multiple dimensions by adopting a well-structured approach to decision-making. The authors incorporate the subjective judgments of the DM into the model and do not only address data.

For this situation, it should be noted that taking note of the consequence with respect to data and information of the system (facts) makes the concept of risks based on evidence and facts, important for specialists. This knowledge, according to Aven (2016b) must be free of non-epistemic values in the first instance. However, the DM must nevertheless make value judgments at a later stage of implementing in the model. In the second stage, the DM needs to look beyond the facts, and include value-based considerations from other sources of information. Thus, an activity is considered safe when judgments based on facts and on values are obtained. In the light of practice, de Almeida et al. (2015) present the steps to run a multicriteria model for application to a context where there is risk and for analyzing the nuances involved in decision-making under uncertainty.

Additionally, de Almeida et al. (2017) explain that the consensus of running an industrial activity is directly dependent on society's approval since society is subject to possible consequences of that activity. Thus, the risk management process should not be conducted in isolation, but rather should take into account the varying different judgments of different DMs. By combining these judgments, strategies can be formulated and selected that estimate the consequences at the start of the evaluation process and set priorities on how resources should be allocated with a view to avoiding or mitigating adverse impacts that may arise in accidental scenarios (WMO 2006; Aye et al. 2016).

Aggregating judgments may imply different points of view that will conflict with each other. A multidimensional approach has been considered an efficient approach to aggregating points of views for decision purposes (Medeiros et al. 2016). Therefore, modeling risks requires the different contexts of a problem to be assessed, i.e., to take into account that a problem must consider different dimensions such as the human, social, financial, and technical ones (Garcez and De Almeida 2014). Some studies have endeavored to use multidimensional perspectives (Brito and de Almeida 2009; Brito et al. 2010; Alencar and de Almeida 2010; Lins and de Almeida 2012; Garcez and de Almeida 2014) so as to estimate risk instead of using just a single dimension.

The importance of having a multiple vision of objectives is aligned with real world risk issues, since objectives set for daily practices never lead to isolated conclusions. On this, Shields et al. (2015) argue that at some stage of assessing risk, the isolated analysis of different criteria always leads to the same decision. Thus, a multicriteria approach incorporates multidimensionality to represent that there are several objectives in a decision-making process.

3 The Importance of Sensitivity Analysis (SA) for Evaluating Risk

One of the problems commonly found in the context of risk assessment concerns how to obtain the data for analysis, or even that data are lacking. This is because information about risk is not very easy to get or, most of the time, has not been considered worth recording. For example, repeated incidents in daily operations may be an input to more realistic models but may not be computed in daily processes. Medeiros et al. (2017) corroborate this view and therefore recommend that risk should be evaluated in more holistic contexts, which relate risks to more systematic perspectives. In short, the evaluation should contain more than simply specifying what operational contexts are, but rather should indicate how other aspects of the system are related to the operational level. This means, first, to find processes or organizations that affect the conditions of risk and then to explore the interactions between them. To illustrate this, Daher et al. (2015) point out that in order to establish strategies that reduce risk it is important to gather information about imminent threats, the vulnerability of critical assets, and the likelihood of risks and consequences.

In a decision-making context, sensitivity analysis (SA) is used to focus on the importance of evaluating the uncertainties of a parameter (or a set of such parameters) that affect the result of a numerical model (Saltelli et al. 2008; Pianosi and Wagener 2015). This analysis plays a significant role in understanding how the final recommendation to the DM is impacted by such variations (Alzbutas et al. 2014; Medeiros et al. 2016). Generally, SA can be classified by varying the parameters: locally, where a chosen parameter is varied; or in a global way, observing how the varying several sources of uncertainties modifies the results (Markert et al. 2014; Silva Monte et al. 2015; Borgonovo et al. 2016, 2018b; Borgonovo 2017).

One of the widely used methods for SA is the Monte Carlo Simulation (MCS), in which a parameter is assumed to follow a probability distribution (Shields et al. 2015) to generate several replications of a numerical model. The model is calculated by varying a specific input from the initial input case to the shifting input case studied (Plischke et al. 2013; Borgonovo et al. 2016). Then, at each replication (in a one-at-a-time method), a sample of the parameter is simulated, in some cases using some replication strategies (Pasman and Rogers 2012) and thus initial data to run the model are obtained.

In the context of risks, some studies have shown why it is important to explore the sensitivity stage of risk assessment models (Saha et al. 2016; Yeo et al. 2016; Yu et al. 2018). Such an analysis helps managerial staff understand the impacts of different scenarios at the operational status level. Therefore, the information given by SA increases awareness about dominant influences on risk. In other words, this shows how managers could better control each of the variables in order to reduce risk (Ahmadi et al. 2015).

For the MCDM/A context, SA raises a concern about the final recommendation to the DM. In this view, it is important to understand how varying the parameters of the problem, and the DM's preferences expressed by the parameters of the MCDM/A model, directly affect the decision. Some studies in the literature develop SA by applying methodologies such as Monte Carlo Simulation (Gómez-Delgado and Tarantola 2006; Medeiros et al. 2016, 2017; Yu et al. 2018).

When DMs are confident that the risk assessment model produces consistent results, they can set and monitor strategies on a sounder basis and plan actions that focus more accurately on tackling negative impacts that may arise from accidental scenarios. As noted by Medeiros et al. (2017), planning these actions includes optimizing the allocation of resources such as money, time, personnel, technology, and safety equipment. In general, resource allocation is of great relevance for the processes of assessing, monitoring, and controlling risk (Daher et al., 2015).

Before the results from risk models are applied to guide a real world policy-making, an effective sensitivity analysis should be developed so that the assessment is designed to cover all possible scenarios regarding the management of risk (Borgonovo et al. 2016). As an example, Marangoni et al. (2017) introduce a study that seeks a fuller understanding of future scenarios that may arise because of climate change due to CO_2 emissions. They regard doing so as key to designing hedging strategies when policies are being drawn up. They model uncertainty with respect to several CO_2 emission scenarios so as to assess future emissions from energy combustion. Thereby, the study reveals key factors for such changes which were identified because of interaction across parameters, which is a determinant to develop adequate policies.

For comparison, interactions between parameters are also observed by López-Benito and Bolado-Lavín (2017) in their study of a natural gas pipeline. A combination of different parameters is checked in order to identify the most relevant input that affects drops in pressure and temperature when natural gas is transported. It turns out that dependence among parameters arises in two situations: when physical conditions are impossible operationally, and when some parameters indicate the technological and economic infeasibility of a proposed action.

On a conceptual discussion, Goerlandt et al. (2017) emphasize the importance of delineating the scope of methodological approaches in SA and pay great attention to the analysis of validity and validation of quantitative risk in the literature. Limitations in this area are explicit and there is a scarcity of studies that cover the need to conduct studies on refining risk measures. Borgonovo (2017) states that numerical models that deal with uncertainty in parameters should be tested so as to reflect combined implications. Some reasons to justify the need to use SA in risk analysis models are given below:

(a) Variation in assumptions and input quantities. Risk managers denote assumptions due to cautionary thinking, i.e., sometimes estimates are made that are higher than the right estimate. In practice, this point indicates that the parameters of the models are changed frequently (Aven 2016a).
(b) Probabilistic behavior. The natural uncertainties of some parameters may affect the results of the analysis, thus stressing the need to delineate the observation of probabilistic risk in greater detail (Borgonovo 2017).
(c) Risk evidence reasoning. Depending on the type of concept used to assess risk, different types of presumption evidence emerge to characterize it. On the one hand, parameters may be observable, based on objective facts (Medeiros et al. 2017). On the other hand, the idea of risk is understood as a way to formalize judgments, being not observable or based on subjective viewpoints, that is, a subjectivity view of probability (Aven 2011; Goerlandt et al. 2017; Medeiros et al. 2017).

Zio (2018) emphasizes the importance of SA in risk assessment given that a variety of combinatorial sets of events, scenarios, and conditions needs to be observed because some lead to critical and unsafe conditions.

3.1 Sensitivity Analysis and Evaluating Multidimensional Risk

In the context of multidimensional risks, Medeiros et al. (2017) investigate the sensitivity of the parameters of the risk assessment model for natural gas pipelines, based on MAUT (Multi-attribute Utility Theory). Medeiros et al. (2016) developed an SA study that supported assessing multidimensional risk in the context of transporting hydrogen transport as proposed by Alencar and de Almeida (2010). In these studies, the authors discuss how the parameters of the model alter the results of ranking the risk zones of the pipeline. An MCS structured process (Fig. 1) is conducted in order to obtain several results from the multicriteria model applied so that further analysis about the parameters can be made.

Initially, the DM conducts an in-depth analysis of the initial ordering of the risks of the sections, r_0. According to Medeiros et al. (2017), this analysis aims to associate the DM's thinking with regard to possible changes in the risk assessment variables. The intention is to analyze the quality of information from the original ranking on how to prioritize the sections in terms of improvements, maintenance, inspections, and other benefits. Thus, to better guide this stage some questions can be raised, such as:

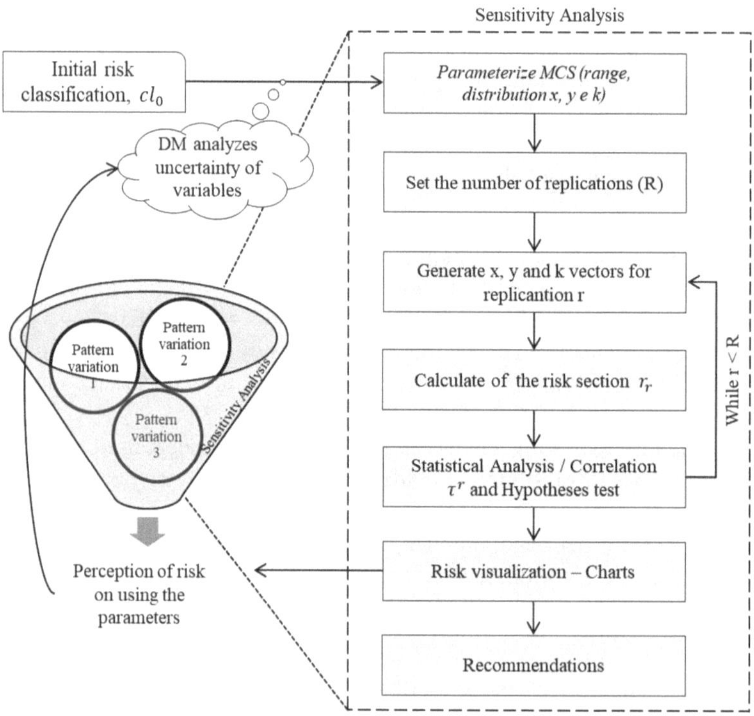

Fig. 1 Framework of the MCS for sensitivity analysis. (Adapted from Medeiros et al. 2017)

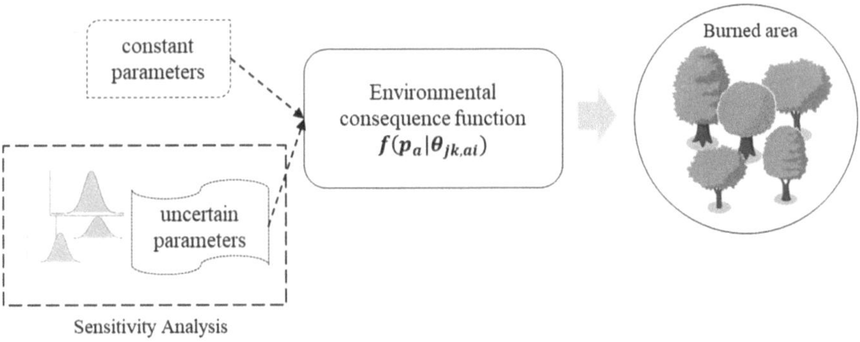

Fig. 2 Sensitivity analysis for the environmental consequence function

- If the percentage probability of a given scenario occurring is increased, will the recommendations on how best to prioritize sections remain the same?
- If the number of people injured were larger, what changes in the ordering could occur?
- What is the confidence level of a recommendation?

This step is investigative and may depend on operational technical factors, as it is subject to observing the characteristics of the section in the pipeline. The DM will evaluate to what extent the parameters of the risk assessment model lie in the uncertainty domain. Therefore, the intention is to define all the parameters that will affect the consequence function. As an example, Fig. 2 depicts how the environmental consequence function is obtained by examining the influence of the two types of parameters: the constant parameters, which are those that present determined values; and the uncertain parameters, those that change over time, which also are the input for the SA.

Then, MCS must be parametrized by setting the parameters that will vary in the simulation process over a given probability. Medeiros et al. (2017) follow a sequence of steps that guide the MCS:

Step 1: Calculate the risk, based on the original values of the parameters. Set the original ranking, r_0.
Step 2: Specify the parameters (or a group of them – patterns), the ranges, and the probability density function (PDF) for each parameter.

As to the parameters of the risk assessment model, it is known that N is the number of sections of the pipeline, and S is the number of accident scenarios of the model. Let X be the input vector such that $X = [X_i, X_{i+1} \ldots X_N]$ represents all the characteristics of each area a_i of the pipeline, and let Y be the input vector that defines the characteristics of the scenario such that $Y_{jm} = [Y_m, Y_{m+1} \ldots Y_M]$, i being the index corresponding to each section, m the scenario, and j the failure module (rupture or hole). For each of the parameters, a PDF, $f_x(x)$ and $f_y(y)$, is assigned in order to simulate the vector of random numbers.

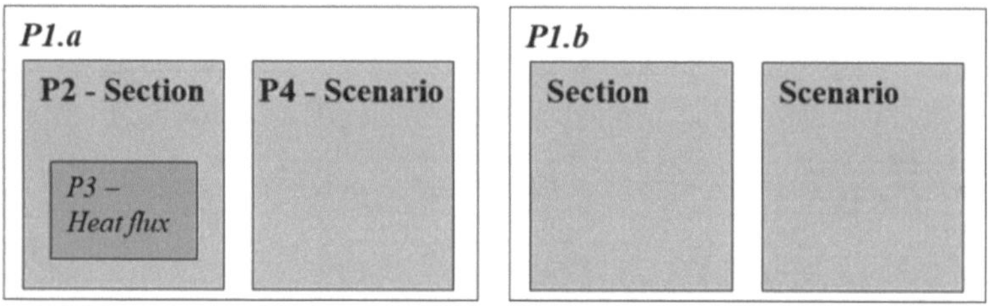

Fig. 3 Patterns simulated. (Adapted from Medeiros et al. 2017)

As the distribution of each parameter is unknown, the triangular and uniform distributions are used in order to generate the data. Fishman (2013) considers that these distributions as efficient for the context of generating data.

SA can be used differently depending on the approach used. When a significant amount of data needs to be dealt with, identifying patterns, correlations, and outliers plays an important role in the analysis (Medeiros et al. 2016). Thus, uncertainty inherent in the problem analyzed is assessed thoroughly.

From a different perspective, Medeiros et al. (2017) structured a simulation process in view of different patterns of variation, which consisted of generating comparisons between the variation of groups of some specific parameters of the model and their original values.

These patterns are defined based on similar characteristics so as to find the set of sensitive parameters and understand their behavior, as shown in Fig. 3. For each pattern, random numbers are generated regarding the PDF assigned.

Step 3: Set the number of replications (R). This should be as large as necessary to achieve the expected change.

Step 4: Obtain the sample for the random variable X and/or with respect to its marginal PDF function. For each replication denote these samples as $[X_i, X_{i+1} \ldots X_N]^r$ and $[Y_m, Y_{m+1} \ldots Y_M]^r$, which r is the index of the replication.

Based on the distribution that the DM has chosen, the random number generator creates the vector X and Y at each replication. For uniform distribution to be generated, two parameters are required (a, b), which represent the maximum and minimum values at which a parameter varies. In the case of the triangular distribution, the DM estimates the values of three parameters, namely two extremes and the most probable value, the latter being the original value of the study. For both uniform and triangular distribution, the interpretation of the parameters takes into account the inherent uncertainty, i.e., the minimum and maximum values that the parameter can reach.

Step 5: Calculate the risk associated with each section. Obtain the ranking of the sections in relation to the total risk of replication r.

Step 6: Determine the Kendall τ correlation coefficient for each r.

The correlation measure, τ, is used to indicate the robustness of the original ranking, represented by r_0, relative to the simulated one, r_r, in order to give the DM the best recommendation. In each replication, a comparison is made between the natural ranking (ascending) and the ranking generated in the replication for the n individuals. If the ranking does not change, the natural order is preserved, then the individual is valued with $+1$, otherwise, -1. The value of τ is the ratio of the sum of all current and maximum possible total individuals – for further details see Siegal (1956). Then, by comparing the various values of τ obtained and the level of significance inherent in the data generated, inferences are made about the robustness of the risk assessment model.

Medeiros et al. (2017) indicate that the dispersion of τ can be analyzed by its median, maximum, and minimum values, mode and standard deviation calculated from the replications. In addition, the coefficient value can infer the traceability of the variations of the simulated parameters at different values of significance. Thus, by reversing the process, information can be generated about the accuracy of the simulated vector of the parameters.

As for this latter approach of τ, the coefficient calculated is used to verify the null hypothesis (H0) that there is no correlation between the original and simulated rankings. If H0 is rejected, then the disorder is caused by a range of variation that causes the original ranking to be similar. Subsequently, the significance test is validated by the value of z and its critical value. Thus, the analysis indicates the level of confidence of the results determined from each replication, thereby generating more information for the DM.

3.2 Visualization of Risk for Sensitivity Analysis

In view of the variety of information presented to obtain an ordering recommendation, the DM needs to analyze different levels of uncertainty of the parameters, and how the variation in such parameters modifies the results significantly. To be more specific, MCS covers an amount of data regarding the simulation of inputs and the outputs of the model.

Payne (1976) points out that increasing the amount of information indicates the variability of responses and decreases the quality of choices as well as making the DM feels more confident. Medeiros et al. (2016) note that information must be summarized so that irrelevant or redundant information is not generated which would confuse the DM and affect the final decision. Therefore, a risk system that involves a great deal of information should ensure the rapid and clear perception of risk in a way that supports mitigating losses, as illustrated in Fig. 3 which shows that using parameters to perceive risks adds to understanding outputs from the model.

In addition to the amount of information, the underlying complexity of some operations leads to the need to deal with risk information as efficiently as possible. To illustrate this, Mittal et al. (2017) enhance the understanding of risk involved in

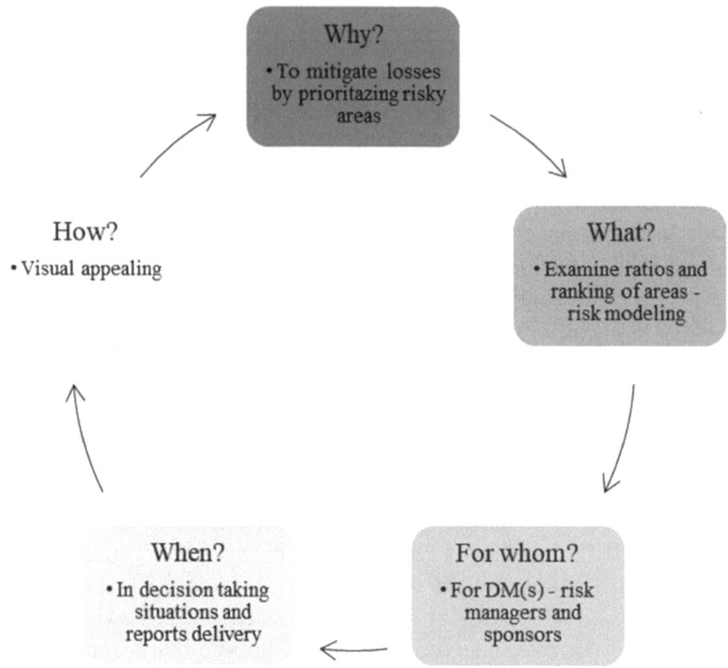

Fig. 4 Risk visualization process for a hydrogen pipeline (Adapted from Medeiros et al. 2016)

an oil and gas operation by using graphs, reports, and 3D visuals of smoke, fire, and explosions.

De Almeida et al. (2015) mention that being able to visualize risk provides support to process information in the stages of risk management that deal with matters such as identifying, analyzing, assessing, communicating, and reducing risks.

Based on the framework addressed by Eppler and Aeschimann (2009), Medeiros et al. (2016) assess risks in the context of hydrogen pipelines to exemplify how risk visualization is best used. The scope of the framework seeks to show why, what, for whom, when, and how risk-related information needs to be displayed – see Fig. 4.

For the sensitivity analysis presented, some visual approaches can be taken that use a back-to-front approach to highlight the risk associated with the uncertainties. In other words, the output is produced graphically in such a way that a detailed analysis of the ranking of sections should be prioritized.

Medeiros et al. (2016) depict risk information using a variety of visual elements as a means to create knowledge of risk. That is, the information on the quantitative output generated is displayed in charts and tables as a means to synthetize the simulation of the parameter, thereby framing the sensitivity of the ranking. The following items prompt discussion of such approaches:

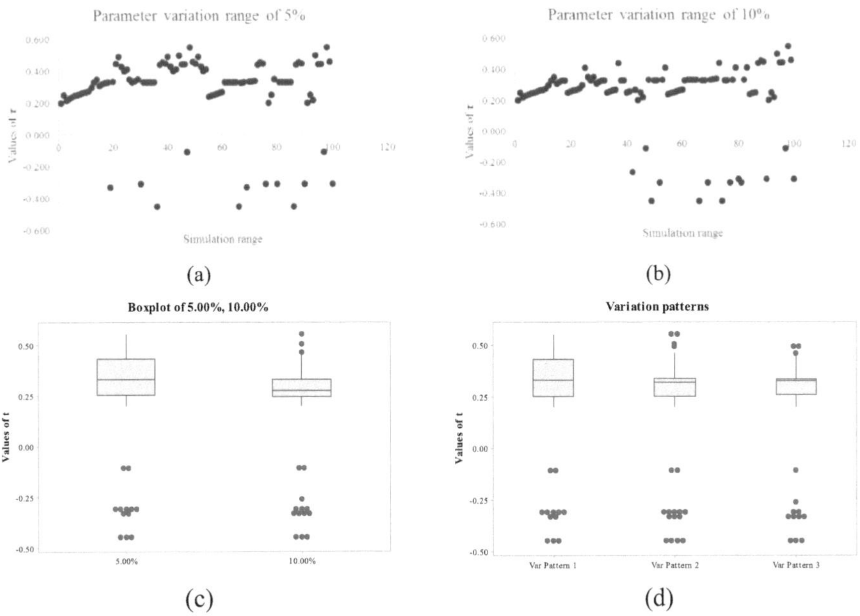

Fig. 5 Dispersion of τ τ values for ranges and patterns

- The dispersion of the correlated index τ adds information about the values of the variations and patterns obtained in the Monte Carlo simulation.

Data concentrated greater than 0 shows a positive correlation with the original ranking in Fig. 5a, despite there being notable negative values in many of the total number of simulations. Figure 5c and d consolidates the aspects of a broad dispersion for different ranges of variation. Thus, the perspective given by both charts is complementary. While the first makes a wild extension of the values simulated, the second sets out a summarized view of the dispersion given by different variations of range in the scatter plot. Medeiros et al. (2017) take the same approach of making pattern-structuring variation instead of ranges. In both applications, the DM is prompted to think about how uncertainties in the parameters (or a group of them) produce the original ranking, which Borgonovo et al. (2018a) regard as a relevant distinction between decision and value sensitivity;

- Sections that are not prioritized in the original ranking but eventually because of the uncertainty behavior come to change its position are also crucial to DM perception. Such observation is taken to evaluate each individual section making a comparison to its original ranking and the percentage change over others positioning ranking.

The information of each individual section variation is displayed in Fig. 6a. For example, section a_4 is initially ranked in the first position. Additionally, its variation

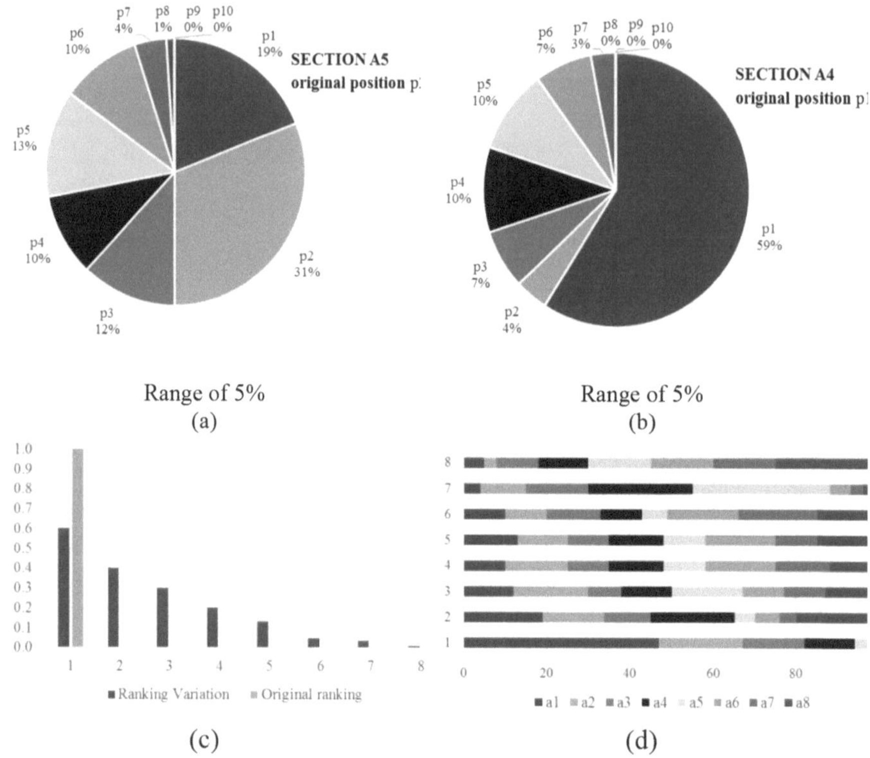

Fig. 6 Variation of sections and positions. (Adapted from Medeiros et al. 2016, 2017)

regarding other positions is not quite sufficiently preferable. In contrast, section a_5, shown in Fig. 6b, varies in the second position 31% of the time in the range of 5%. A similar evaluation of Fig. 6c is possible. The DM verifies the behavior of a given section ranked in a position, represented by the light gray bar, and its variation along the positions shown in the darker gray bars. A broader perspective is given in Fig. 6d, in which the variation of each section is depicted throughout the ranking. The results refer to those simulated data that showed a correlation with the confidence level.

Combined analysis can also be performed as means of comparing the percentage change and position variation over a variety of intervals of uncertainties – see Medeiros et al. (2016) and Walls et al. (2016) for further details.

Based on the analysis of the studies discussed in this chapter, it is observed that they provide additional information to a decision-making process and thereby contribute to managing risk in gas pipeline systems better. Additionally, the DM can judge the alternatives considered in the decision problem more clearly and this leads to better decision-making.

References

Ahmadi A, Moridi A, Han D (2015) Uncertainty assessment in environmental risk through Bayesian networks. J Environ Info 25:46–59. https://doi.org/10.3808/jei.201500294
Alencar MH, de Almeida AT (2010) Assigning priorities to actions in a pipeline transporting hydrogen based on a multicriteria decision model. Int J Hydrogen Energy 35:3610–3619. https://doi.org/10.1016/j.ijhydene.2010.01.122
Alzbutas R, Iešmantas T, Povilaitis M, Vitkutė J (2014) Risk and uncertainty analysis of gas pipeline failure and gas combustion consequence. Stoch Environ Res Risk Assess 28:1431–1446. https://doi.org/10.1007/s00477-013-0845-4
Aven T (2011) A risk concept applicable for both probabilistic and non-probabilistic perspectives. Saf Sci 49:1080–1086. https://doi.org/10.1016/j.ssci.2011.04.017
Aven T (2016a) On the use of conservatism in risk assessments. Reliab Eng Syst Saf 146:33–38. https://doi.org/10.1016/j.ress.2015.10.011
Aven T (2016b) Risk assessment and risk management: review of recent advances on their foundation. Eur J Oper Res 253:1–13. https://doi.org/10.1016/j.ejor.2015.12.023
Aven T, Renn O (2009) On risk defined as an event where the outcome is uncertain. J Risk Res 12:1–11. https://doi.org/10.1080/13669870802488883
Aye ZC, Jaboyedoff M, Derron MH et al (2016) An interactive web-GIS tool for risk analysis: A case study in the Fella River basin, Italy. Nat Hazards Earth Syst Sci 16:85–101. https://doi.org/10.5194/nhess-16-85-2016
Borgonovo E (2017) Sensitivity analysis: an introduction for the management scientist.
Borgonovo E, Hazen GB, Plischke E (2016) A common rationale for global sensitivity measures and their estimation. Risk Anal 36:1871–1895. https://doi.org/10.1111/risa.12555
Borgonovo E, Cillo A, Smith CL (2018a) On the relationship between safety and decision significance. Risk Anal 38:1541–1558. https://doi.org/10.1111/risa.12970
Borgonovo E, Morris MD, Plischke E (2018b) Functional ANOVA with multiple distributions: implications for the sensitivity analysis of computer experiments. SIAM-ASA J Uncertainty Quantificat 6:397–427
Brito AJ, de Almeida AT (2009) Multi-attribute risk assessment for risk ranking of natural gas pipelines. Reliab Eng Syst Saf 94:187–198. https://doi.org/10.1016/j.ress.2008.02.014
Brito AJ, de Almeida AT, Mota CMM (2010) A multicriteria model for risk sorting of natural gas pipelines based on ELECTRE TRI integrating Utility Theory. Eur J Oper Res 200:812–821. https://doi.org/10.1016/j.ejor.2009.01.016
Daher SFD, Alencar MH, de Almeida AT (2015) Recent patents on industrial risk management. Recent Pat Comput Sci 8(2):144–151
de Almeida AT, Cavalcante CAV, Alencar MH, et al (2015) Multicriteria and multiobjective models for risk, Reliability and Maintenance Decision Analysis
de Almeida AT, Alencar MH, Garcez TV, Ferreira RJP (2017) A systematic literature review of multicriteria and multi-objective models applied in risk management. IMA J Manag Math 28:153–184. https://doi.org/10.1093/imaman/dpw021
Eppler MJ, Aeschimann M (2009) A systematic framework for risk visualization in risk management and communication. Risk Manag 11:67–89. https://doi.org/10.1057/rm.2009.4
Fishman G (2013) Monte Carlo: concepts, algorithms, and applications. Springer Science & Business Media
Garcez TV, de Almeida AT (2014) Multidimensional risk assessment of manhole events as a decision tool for ranking the vaults of an underground electricity distribution system. IEEE Trans Power Deliv 29:624–632. https://doi.org/10.1109/TPWRD.2013.2273083
Goerlandt F, Kujala P (2014) On the reliability and validity of ship-ship collision risk analysis in light of different perspectives on risk. Saf Sci 62:348–365. https://doi.org/10.1016/j.ssci.2013.09.010
Goerlandt F, Khakzad N, Reniers G (2017) Validity and validation of safety-related quantitative risk analysis: a review. Saf. Sci. 99:127–139

Gómez-Delgado M, Tarantola S (2006) GLOBAL sensitivity analysis, GIS and multi-criteria evaluation for a sustainable planning of a hazardous waste disposal site in Spain. Int J Geogr Inf Sci 20:449–466. https://doi.org/10.1080/13658810600607709

Kaplan S, Garrick BJ (1981) On the quantitative definition of risk. Risk Analysis 1:11–27. https://doi.org/10.1111/j.1539-6924.1981.tb01350.x

Keeney RL, Raiffa H (1976) Decision with multiple objectives: preferences and value trade-offs, Wiley Seri. Wiley, New York

Lins PHC, De Almeida AT (2012) Multidimensional risk analysis of hydrogen pipelines. Int J Hydrogen Energy 37:13545–13554. https://doi.org/10.1016/j.ijhydene.2012.06.078

López-Benito A, Bolado-Lavín R (2017) A case study on global sensitivity analysis with dependent inputs: the natural gas transmission model. Reliab Eng Syst Saf 165:11–21. https://doi.org/10.1016/j.ress.2017.03.019

Marangoni G, Havlik P, Keppo I et al (2017) Sensitivity of projected long-term CO2 emissions across the shared socioeconomic pathways. Nat Clim Chang 7:113–117. https://doi.org/10.1038/nclimate3199

Markert F, Melideo D, Baraldi D (2014) Numerical analysis of accidental hydrogen releases from high pressure storage at low temperatures. Int J Hydrogen Energy 39:7356–7364. https://doi.org/10.1016/j.ijhydene.2014.02.166

Medeiros CP, Alencar MH, de Almeida AT (2016) Hydrogen pipelines: enhancing information visualization and statistical tests for global sensitivity analysis when evaluating multidimensional risks to support decision-making. Int J Hydrogen Energy 41:22192–22205. https://doi.org/10.1016/j.ijhydene.2016.09.113

Medeiros CP, Alencar MH, de Almeida AT (2017) Multidimensional risk evaluation of natural gas pipelines based on a multicriteria decision model using visualization tools and statistical tests for global sensitivity analysis. Reliab Eng Syst Saf 165:268–276. https://doi.org/10.1016/j.ress.2017.04.002

Meyer V, Scheuer S, Haase D (2009) A multicriteria approach for flood risk mapping exemplified at the Mulde river, Germany. Nat Hazards 48:17–39. https://doi.org/10.1007/s11069-008- 9244-4

Mittal V, Borges V, Shaba K (2017) Advanced risk visualization for the offshore industry. In: Walls L, Revle M, Bedford T (eds) European safety and reliability conference, Glasgow, Scotland, 2016. Risk, reliability and safety: innovating theory and practice. Taylor and Francis, London, p 2983

Pasman HJ, Rogers WJ (2012) Risk assessment by means of Bayesian networks: a comparative study of compressed and liquefied H2 transportation and tank station risks. Int J Hydrogen Energy 37:17415–17425. https://doi.org/10.1016/j.ijhydene.2012.04.051

Payne JW (1976) Task complexity and contingent processing in decision making: An information search and protocol analysis. Organ Behav Hum Perform 16:366–387. https://doi.org/10.1016/0030-5073(76)90022-2

Pianosi F, Wagener T (2015) A simple and efficient method for global sensitivity analysis based on cumulative distribution functions. Environ Model Softw 67:1–11. https://doi.org/10.1016/j.envsoft.2015.01.004

Plischke E, Borgonovo E, Smith CL (2013) Global sensitivity measures from given data. Eur J Oper Res 226:536–550. https://doi.org/10.1016/j.ejor.2012.11.047

Saha N, Ahmed MB, Ngo HH et al (2016) Industrial metal pollution in water and probabilistic assessment of human health risk. J Environ Manage 185:70–78. https://doi.org/10.1016/j.jenvman.2016.10.023

Saltelli A, Ratto M, Andres T et al (2008) Global sensitivity analysis: The primer. Wiley

Shields MD, Teferra K, Hapij A, Daddazio RP (2015) Refined stratified sampling for efficient Monte Carlo based uncertainty quantification. Reliab Eng Syst Saf 142:310–325. https://doi.org/10.1016/j.ress.2015.05.023

Siegal S (1956) Nonparametric statistics for the behavioral sciences. McGraw-Hill

Silva Monte MB, de Almeida T, Filho A (2015) A reliability-based approach to maximize availability in a water supply system. IEEE Lat Am Trans 13:3807–3812. https://doi.org/10.1109/TLA.2015.7404912

Walls L, Revie M, Bedford T (eds) (2016) Risk, reliability and safety: Innovating theory and practice. CRC Press, Taylor & Francis Group, 6000 Broken Sound Parkway NW, Suite 300, Boca Raton. 33487-2742

WMO (2006) Social aspects and stakeholder involvement in integrated flood management: Associated Programme on Flood Management

Yeo C, Bhandari J, Abbassi R et al (2016) Dynamic risk analysis of offloading process in floating liquefied natural gas (FLNG) platform using Bayesian network. J Loss Prev Process Ind 41:259–269. https://doi.org/10.1016/j.jlp.2016.04.002

Yu X, Liang W, Zhang L et al (2018) Risk assessment of the maintenance process for onshore oil and gas transmission pipelines under uncertainty. Reliab Eng Syst Saf 177:50–67. https://doi.org/10.1016/j.ress.2018.05.001

Zio E (2018) The future of risk assessment. Reliab Eng Syst Saf 177:176–190. https://doi.org/10.1016/j.ress.2018.04.020

Multidimensional Decision-Making Process for Managing Flood Risks in Postmodern Cities: Challenges, Trends, and Sharing Insights to Construct Models That Deal with Climate Changes

Lucas Borges Leal da Silva, Marcelo Hazin Alencar, and Adiel Teixeira de Almeida

1 Introduction: Characterizing Flooding in the Urban Context

Modern societies undergo dynamic interactions between people, Nature, and its resources. Although these interactions are essential for promoting social and economic development, strategic policies that seek to improve the quality of life in urban areas can cause an imbalance between these actors and can have serious adverse impacts on everyday life.

In fact, nowadays, public administrations face new challenges in order to adapt human life to alarming trends (and also to their consequences) such as: climate changes (IPCC 2018), an increase in the extent and frequency of natural hazards (Neumayer et al. 2014), threats to the supply of food and water (Gharehgozli et al. 2017; Srinivasan et al. 2012), inadequate distribution of energy (Richard and Eugene 2014), and migratory crises (Trost et al. 2018).

As to the occurrence of extreme events in urban areas, Mitchell (1993) pointed out more than 25 years ago how important it is to improve response techniques to natural disasters that had been occurring with ever-greater frequency. In particular, he highlights flooding events since they occur in many different regions all over the world. Therefore, decision-makers (DMs) seek to improve how the risks from floods are managed, while recognizing that this activity involves different factors. However, just how complex can it be to model this decision problem?

L. B. L. da Silva · M. H. Alencar (✉) · A. T. de Almeida
CDSID – Center for Decision Systems and Information Development, Universidade Federal de Pernambuco, Recife, PE, Brazil
e-mail: marceloalencar@cdsid.org.br; almeida@cdsid.org.br

A. T. de Almeida et al. (eds.), *Multicriteria and Optimization Models for Risk, Reliability, and Maintenance Decision Analysis*, International Series in Operations Research & Management Science 321, https://doi.org/10.1007/978-3-030-89647-8_8

1.1 Justifying Flood Risk Management (FRM) Practices and Research

The effects of climate change on cities have had implications for urban governments. Over many years, their planning for local development has included the need to draw up, implement, and amend procedures for managing risk that effectively addresses dangers posed by imminent climate change. Recent studies alert to the need to develop tools that assess the vulnerability of urban spaces and that classify risks in order to combat extreme events, floods being the most recurrent of these (CRED-UNISDR 2015).

Bearing these matters in mind, urban floods cannot be managed individually so responses to the impact of potential flooding are complex since they are interlinked with political, socio-economic and environmental issues. Thus, to understand some features of urban flood management, an integrated methodology should be developed (and constantly improved) in which spatial-temporal relations are not only defined but it is also made clear that these should be regularly monitored and re-examined.

This should provide clarity regarding the concepts of vulnerability and resilience, the main point being to incorporate in risk modeling the way the interactions among differing spatial scales occur (Lei et al. 2014). Some insights and methods from the literature may well provide the basis for drawing up an integrated methodology in order to improve routine decision-making processes in the context of public administration.

Therefore, this chapter discusses what cities worldwide can do to respond to climate change and therefore seeks to reframe their role. Naturally, this highlights the vulnerability of a city's critical infrastructure and the need for strategic planning. This shows that public managers now face a decision problem with multiple objectives that may conflict with each other. As to mitigating the impact of floods, what technical-scientific institutions need to do is to search for innovative solutions that minimize these impacts. This involves examining social, economic, and environmental questions.

Therefore, it is necessary to understand how the growth in urbanization has reconfigured postmodern societies and has thus added to the complexity of flood management.

1.2 Urbanization Processes: A New Challenge for Postmodern Societies

That urban areas have expanded so much is an unavoidable phenomenon that has arisen as a result of major changes in economic activity, and therefore in what have become complex inter-relationships between the industrial, commercial, and service sectors, and between these sectors and the markets they serve. They have had to adapt their activities to the new demands from the market and society. This includes that they must bear in mind that urbanization to a large extent is about seeing to it that the spatial layout is dynamic (Chen and Frauenfeld 2016). This need for dynamism has led the world's most powerful cities to enhance their critical infrastructures, their emergency plans, their support services, and their strategic planning. These are being designed and, when necessary, adapted to reduce the impact of major flooding and this is making these cities safer.

However, it is worth endeavoring to understand how the postmodern configuration of societies contributes to increasing the vulnerability in urban areas to natural disasters such as floods and landslides. In this context, Da Silva et al. (2018) pointed out that society's dependence on exploiting natural resources has led to this having a huge adverse impact on the environment and has thus contributed to causing climate change.

Moreover, the United Nations reports that nearly 80% of the world population of 11 billion people will live in urban conurbations by 2050 (UN 2019). This puts a spotlight on hazardous events that may occur due to the inter-play relationship between urbanization, climate change, and natural catastrophes.

Hodgkins et al. (2019) analyzed historical trends in the annual peak flows in the United States by basin type, namely those that have been minimally altered, those that are regulated, and those that have been urbanized. For urbanized basins, their statistical analysis concluded that the magnitude of this trend was significantly correlated with the extent to which the basin had been urbanized. They claimed that their analysis showed that increases in the volume and frequency of flooding undoubtedly reveal the adverse influence of humans.

Saadi et al. (2018) analyzed trends in demand for urban land and the implication that this leads to greater flooding, for which they used a Land-Use Model System. The results from this were based on three urban expansion scenarios for 2030: business-as-usual; limited expansion; and extreme expansion. They established three classes of city, according to their degree of urban density. Their results revealed increases in flood damage in a range at least from 15% to 30% (business-as-usual scenario). Furthermore, the Land-Use Model System showed that the more that land-use had expanded, the higher the levels of flooding that had occurred (see Fig. 1).

Since floods are closely related to urbanization, Bae and Chang (2019) demonstrated the historical influence of land use so as to show that factors regarding damage from floods have changed over time. They reinforced that assertion by concluding that population density was the common factor that best explained flood damage in different degrees with respect to the advance in urbanization, no matter how rapidly this occurred. Therefore, they suggested the need for different flood management strategies, as urbanization increases, in order to minimize flood damage.

As a result of the increases in urbanization, it is even more commonplace for public managers to have to engage on complex decision-making, including on assessing what the potential damage from flooding might be. Thus, they seek to structure decision-making processes jointly and collaboratively within their city

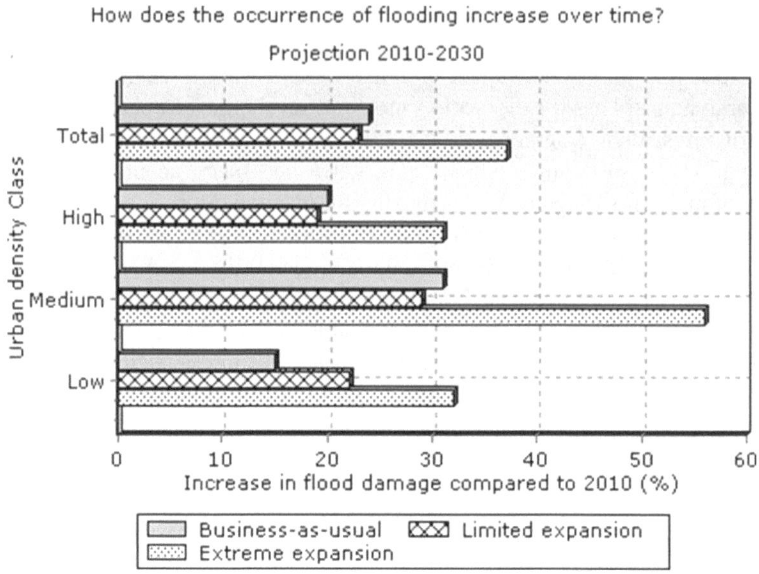

Fig. 1 Predicted increase in flood damage due to future urbanization for the horizon 2030 compared to the baseline situation. (Adapted from Saadi et al. 2018)

government in order to implement efficient actions that seek to mitigate floods and to have controls that seek to predict and minimize flooding.

When this is done, then all concerns about floods that are likely to be raised are considered by managers during decision-making processes. Thus, all public managers have to plan, finance, implement, and engage on actions that meet the needs of urban spaces and to manage these activities efficiently. This results in improving the overall quality of life of urban dwellers, especially in the context of natural hazards.

However, there are multiple impacts that floods can cause that need to be addressed during this process, which requires social, human, economic, institutional, and environmental questions to be integrated (Gigović et al. 2017; Meyer et al. 2009a; Ashley et al. 2005). That is why some researchers, on seeking approaches that guide DMs' decision-making, have applied multicriteria methods to model decision problems from a multidimensional point of view.

This chapter gives an overview of multicriteria modeling and shares some insights into how best to manage flood risks in urban areas that take multiple (and possibly conflicting) objectives into account. Moreover, such modeling seeks to contribute toward aiding public policy to prioritize preventive actions to combat potential disasters. It does so by prioritizing risks, selecting projects for portfolios and suggesting how to improve communications to and between citizens at risk from flooding.

From this perspective the chapter focuses on a multidimensional approach to assessing risks from flooding in order to aid decision modeling in the urban context. All sections summarize the potential benefits of these modeling proposals for managing, controlling, and mitigating the consequences of flooding and for taking emergency actions to combat flooding. Furthermore, multicriteria studies are presented and insights for modeling are shared and briefly discussed throughout the chapter. In the next section, this chapter will focus on clarifying how climate changes led to extreme events that impact our lives in urban communities.

2 Impact of Flooding on Everyday Life and Climate Change Effects: A Starting Point for Constructing Decision Problems

The issue of climate change has gained international projection recently with regard to forecasting the climate based on either changes in the frequency of short-term extreme weather events or in their intensity. Thus, hot flashes, heavy rainfall, floods, droughts, and other natural disasters have been of great interest to researchers because of their huge impact, not only on the environment, but also on people. Disasters result in high monetary costs and, often, in the loss of human lives.

A special report from the Intergovernmental Panel on Climate Change (Hoegh-Guldberg et al. 2018) reinforces this concern due to climate change intensifying the behavior of natural disasters. According to IPCC, global warming of nearly 0.5% which was caused by human activity was detected by analyzing trends in the intensity and frequency of some climate and weather extremes over time spans.

Instead of keeping to the previous goal of a maximum average rise in the global temperature of 2 °C by 2100 (IPCC 2012), the new report proposed limiting the increase in global warming to 1.5 °C. This new goal implies 420 million fewer people being frequently exposed to extreme heatwaves, and about 65 million fewer people being exposed to exceptional heatwaves, assuming constant vulnerability.

In this context, public managers find it helpful to incorporate climate variables into flood management decision problems, so that their contribution to the analysis of risks from floods includes measurements that suggest alternatives in order to anticipate problems and impacts that may arise in cities in the future.

In fact, an integrated policy between public and private technical-scientific institutions considering climate change effects is essential in order to foster innovative solutions. This policy should include (in the agenda for cities) important topics such as land use, consumerism, urban violence, etc. Most ordinary people and most organizations consider that environmental problems should be addressed and reiterate the priority of linking the three spheres of sustainability: human development, economic growth, and conservation of the environment (Ramaswami et al. 2016) in order to prevent and/or mitigate the regularly occurring impacts of flooding.

The impact of global change and the consequent changes in climate on water, food, forests, lakes, and other features of the environment have been intensively discussed in the literature (Şen 2018). Therefore, risk management encompasses procedures for drawing up both quantitative and/or qualitative estimates in order that results include not only the probability of occurrence of dangerous events but also their consequences.

With particular regard to the public administration context, Lööf and Nabavi (2013) comment that it is the responsibility of governments to establish, implement, and manage measures that can mitigate the direct and indirect risks of climate change, thereby making the urban environment more resilient and less vulnerable. These concepts are closely related to the relationship between the high adaptability of cities and governments being able to draw on valuable decision-making expertise (Hoegh-Guldberg et al. 2018).

However, before listing the most common decision problems which DMs face in this area, we must understand not only how the interaction between climate change and the occurrence of floods impacts and changes the urban dynamics but also the interaction between people in cities. Fig. 2 shows a summary of the principal aspects of the diverse forms of harm that potentially can be caused by flooding.

First of all, adverse impacts on humans have been widely studied over the years. For example, Jonkman et al. (2009) focused their efforts on using preliminary data to analyze where fatalities increase. They also produced a qualitative description of the majority of victims of the catastrophic flooding caused by Hurricane Katrina in 2005 in New Orleans (U.S.A.): nearly 60% of fatalities were over 65 years old. The authors compare this event with historical flood events and affirm that there is an empirical relationship between mortality and the characteristics of such floods and that the evidence shows that the overall mortality rate for cities that have suffered such events is around 1%.

Moreover, Alderman, Turner, and Tong (2012) pointed out that the relationship between flood events and human health is deeper than people usually think. They conducted a systematic literature review and noted assessed recent epidemiological evidence on the impacts of floods on human health. They found there was an increased risk of outbreaks of diseases such as leptospirosis, hepatitis E, gastrointestinal illnesses, and, particularly, in areas where hygiene is poor and people live in makeshift homes. Moreover, they also list other health problems that stem from floods such as epidemics/pandemics, post-traumatic (mental and physical) illnesses, and the fact that floods have a negative impact on the preservation of culture and on interpersonal relationships.

On the other hand, Priori, Alencar, and De Almeida et al. (2017) analyzed how climate changes and intense floods may undermine the major critical infrastructure of urban centers, such as the energy supply, logistics, transportation, communications, drainage, etc. They focused on an adaptation system using Value-Focused Thinking (VFT), a method for structuring problems.

Moreover, financial systems and the general economy of societies, whether global or local, can be severely affected by extreme events (Surminski and Eldridge 2017; McGrath et al. 2019; Neumayer et al. 2014; Heo and Heo 2019). Therefore,

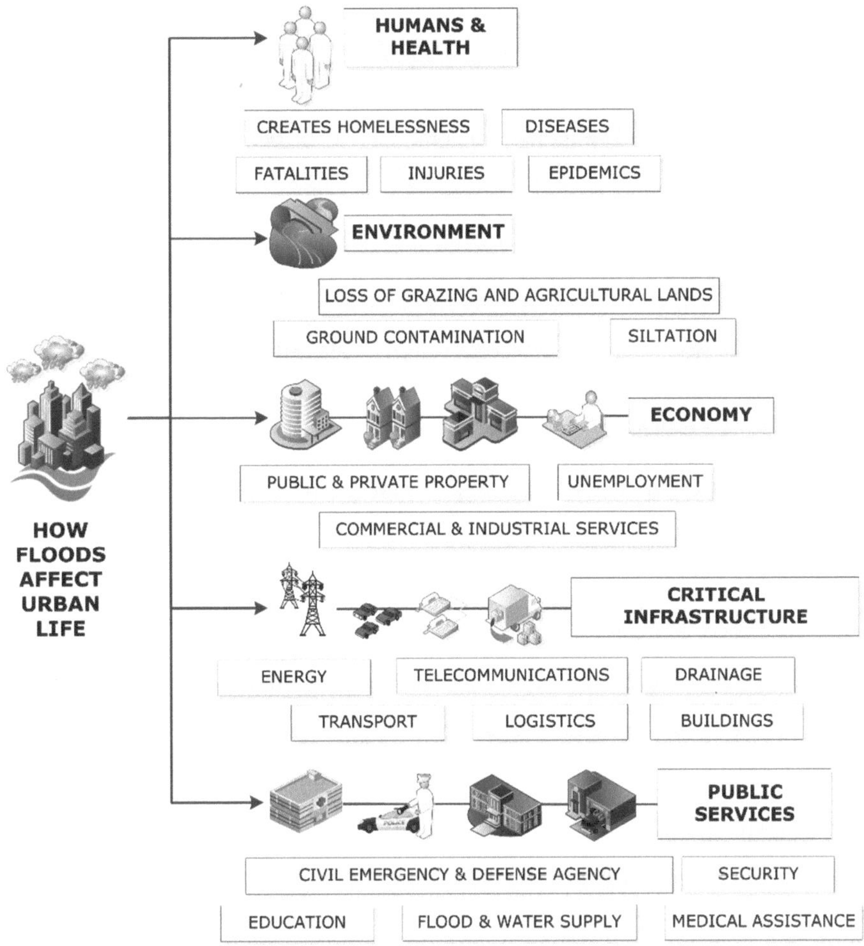

Fig. 2 Scheme of potential forms of damage in urban areas caused by floods that result from climate change

a great deal of research and discussion can be found in the literature that measures economic and financial damage to public and private assets, the impact of this on the provision of public services (education, security, emergency defense, and civil agency, health), and future impacts of extreme events on local trade and employment as a whole.

The keywords shown in Fig. 2 exemplify a variety of potential forms of flood damage in urban area due to climate change. Therefore, tools need to be developed that aid assessing and, subsequently, categorizing and assessing levels of vulnerability and risk due to climate change. These can then be used to adapt all societies to change.

In this context, a multicriteria approach can aid the decision process. It is essential that the scientific foundations that underpin periodic actions are very robust and up-to-date. This applies not only to addressing the speed and extent of climate change, but also to assessing its impacts and risks and the actions needed to mitigate these. In this context, it is important to point out that the behavior of precipitation is neither standardized nor can it be controlled by human action.

3 MCDA/M Approach to FRM in Urban Areas

The results of this analysis for decision-making are used in Flood Risk Management while the literature shows that there are different approaches to helping public and private entities estimate risks in a reasonable way.

Cost-benefit analysis is a traditional economic technique that holistically assesses whether the expected benefits from implementing a risk-reducing action outweigh its costs. It is the most common approach to FRM. However, Samuels and Gouldby (2009) criticize in their report the quantification in monetary terms and aggregation into a single value, as this makes it impossible to associate factors of different natures that are also affected by risk. Elements of Decision Theory are also used for risk assessment (Cuellar and McKinney 2017), as are the methodologies of hydro-meteorological analysis which aim at continuous improvement (Patra et al. 2016).

However, the policies used to mitigate floods often take multiple strategic objectives into account that frequently conflict with each other and seek to integrate especially the social, economic, and environmental dimensions. This is why some researchers, when seeking approaches to support DMs' decision-making, have applied multicriteria methods.

Multicriteria methodology has been used in several risk management contexts, especially for Natural Hazards. A systematic review of the literature by de Almeida et al. (2017) identified research trends in dealing with multidimensional models, which take a DM's preference structures into account.

3.1 Multicriteria Aid for Risk Decision Problems: A Brief Description of the State-of-Art

Due to the exposure and vulnerability of urban spaces in contemporary society and the imperfect condition of the critical infrastructure of most cities, flooding affects the environment adversely and has the potential to cause heavy economic and social losses, including the possibility of multiple deaths (Yamashita et al. 2015). Therefore, choosing a multicriteria method is often appropriate in order to deal with such particularities.

Since the decision-making process is a key success factor of any organization, MCDM/A may be used. It is an approach based on a DM's preference structure and involves value judgments of multiple objectives which often conflict with each other, yet need to be dealt with simultaneously.

Despite the diversity of current methods, three basic characteristics identify a multicriteria problem: a discrete set of alternatives with at least two criteria and there being a DM (De Almeida et al. 2015).

From this point of view, decision-making processes accept the subjectivity that is involved by establishing preference relations between all sets of alternatives.

Several papers led to great advances in modeling problems with multicriteria aggregation methods, as these models were shown to be applicable to situations that reflected real life more and more. For instance, Koksalan, Wallenius, and Zionts (2011) and Miles Jr. (2007) analyzed the evolution, history, and perspectives of these methods.

It is worth noting that there are several studies in the literature to manage risks from natural hazards and these include a range of methodologies for constructing how best to make assessments. Some applications which focused on risks in the environment are also found (Brito et al. 2010; Alencar and De Almeida 2010; Meyer et al. 2009b; Cuellar and McKinney 2017).

This chapter, however, seeks to focus on the multidimensional approach as a risk management tool. This allows different situations to be combined in the same assessment, which consider – partially or wholly – their relative importance and mode. This approach is therefore recommended for flood risk analysis and will be explained in detail below.

To do so, risk management, risk assessment, and risk analysis take into account issues such as mathematical procedures, computational tools, and a variety of approaches and decision support models (De Almeida et al. 2015) in which the context of decision-making can be related to:

Quantitative risk analysis for prioritizing/classifying vulnerable areas: here, the main objective is to define what locations managers have to consider in order to minimize the impacts of flooding (Fadlalla et al. 2015; Xiao et al. 2018; Gigović et al. 2017); Allocating resources for strategic actions to prevent flood damages and also with regard to the limitations of human, financial and technical resources (Karamouz et al. 2018); and Strategic planning of urban policies to mitigate flooding in critical infrastructures due to climate changes (Huong and Pathirana 2013; Priori et al. 2017).

With this in mind, several studies in the literature seek to improve quantitative and qualitative risk estimation procedures that characterize the hydrological behavior of an extreme event.

3.2 The Role of Uncertainty in Evaluating Flood Risks

These advances in modeling enabled researchers to become involved in different areas of knowledge such as water resources (Godskesen et al. 2017), research and development (R&D) projects (Karasakal and Aker 2017), the electricity sector (Cucchiella et al. 2017), construction (Miniotaite 2017), the financial sector (Ferreira et al. 2018), risk management (Medeiros et al. 2017), and the development of new methods (De Almeida et al. 2016).

Especially in risk management problems, de Brito and Evers (2015) noted that the greatest interest is in applying MCDM/A applications to flood risk management. Their study is a systematic review of more than a thousand papers that apply MCDM/A to flood-related problems, in order to provide an overall picture of what has motivated researchers from 37 different countries over the past two decades.

Future changes in the climate worldwide are widely discussed in the literature and Di Baldassarre et al. (2016) state such changes are another source of uncertainty. Furthermore, they pointed out that it can be a challenge to calibrate flood risk models because sometimes DMs do not have all rainfall data or have lost data so these circumstances are an additional source of uncertainty.

In the context of MCDA/M, Keeney and Raiffa (1976) studied the mathematical association between uncertainty and utility by using a consistent formulation of the parameters needed to model the problem. Characteristics concerning multidimensional methods for assessing uncertainty are set out in Sect. 5.

Given that this is a particular decision problem to do with managing flood risks, Sects. 4 and 5 will briefly present some approaches developed by the authors, as a result of applying multicriteria models in real practice.

4 A Multimethodology Framework for Multicriteria Assessment of Flood Risk

Da Silva et al. (2018) developed a multimethodology framework to support decision-making to understand the risks from floods and to integrate the dimensions of risk, but also, whenever a DM faces uncertainty, so as to use this data to prioritize and also manage strategic portfolios of projects to face this problem.

The framework proposed by the authors is based on a multidimensional evaluation for risk categorization, which has already been developed, to implement actions to avoid risk. This way, when this risk can be mitigated, the framework then leads to a way to a multicriteria model that will be used to manage projects that combat risk. These actions are held in a portfolio and include strategic alternatives that perform effectively in order to minimize the impacts of flooding.

The framework presented in Fig. 3 is divided into 5 steps, and for each one of them this approach defines a list of activities as well as clarifies the roles of decision-making actors. This aims to maximize the benefits of this approach.

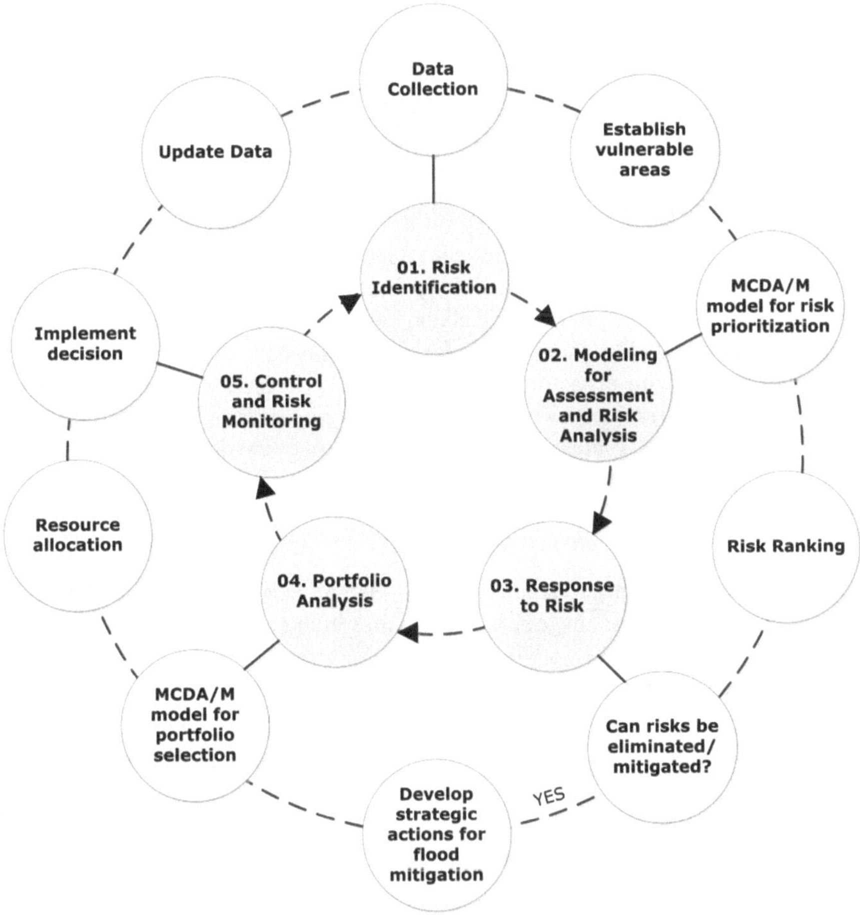

Fig. 3 Framework proposal conducts the multidimensional flood risk analysis. (Adapted from Da Silva et al. 2018)

The initial step, Risk Identification, comprises a stage of understanding the problem. Here, DM is supported by analysts to characterize flooding problem, collecting important parameters which will be used as input for the risk modeling (given its inherent probabilistic character). This stage models the problem using the data collected and it quantifies the flood risk in the area of study by considering the relation among natural, social, and economic aspects of floods.

Thus, in the next step, Modelling for assessment and risk analysis and Classification, a MCDA/M model is used to analyze and determine this measure of risk, using methods that are suitable to the problem of prioritizing risk in an urban area. There are several procedures that have been presented in the literature to support DMs when they are constructing a model, such as the de Almeida et al. (2015) suggestion that divides the procedure into 3 phases – in successive refinements:

1. A preliminary phase, in which the actors must characterize the problem by explicating the DMs, the objectives, the criteria used to model these objectives, the space of consequences and the problematic, for example;
2. The modeling of preferences phase is developed by modeling DMs' preferences and then choosing a suitable MCDM/A method; and
3. The finalization phase, regarding the evaluation of alternatives and recommendation to DM, and finally the decision implementation.

It is worth to know that the authors put a spotlight on building MCDM/A models to represent real problems, once models can be regarded as a creative process, for which DMs need to have an intellectual and cultural background that enables them to be aware of and understand the complexity of this task.

By interacting with all actors of this process, DMs increase their perceptions about objectives, criteria, space of actions. This greatly enriches the model and therefore enhances the evaluation and implementation of the decision.

In the context of MCDM/A methods (De Almeida et al. 2015), categorizing risk based on the disaster management cycle allows DMs to respond to risk in three ways in step 3, the Response to Risk:

- Accept: to implement an action, it would take an excessive amount of time to prepare a strategy to manage risk or it requires high-costs to deal with it, so it is better to accept the risk.
- Transfer the risk to a third party or parties that can manage the outcome. Broadly speaking, outsource or share risk with them.
- Eliminate/mitigate the risk: action is taken to reduce the causes of threats wherever possible.

Therefore, if the assessed risk can be minimized/mitigated, then potential flood control alternatives can be established, while taking the critical infrastructures that are affected by it into account. Step 4, Portfolio Analysis, is used to prompt discussion among public administration, companies, scientific communities, and citizens to help DMs to generate a portfolio that will be analyzed at a later stage.

As to the Portfolio Analysis step, first of all, the DM, assisted by the analyst, compiles a set of projects or programs and other activities that they consider will support effective management of the decision-making problem. The portfolio problem consists of choosing, within a set of actions, a subset that best meets an organization's objectives and does not exceed its constraints. This results in prioritizing projects and then implementing them.

Next, based on the results from the risk analysis of the delimited area, a portfolio should be constructed. This involves selecting from the projects already identified those that maximize satisfaction in the following dimensions: the social (such as determining the level of risk to which the population is potentially exposed), the economic (reducing losses), and Nature (preserving the environment).

However, it is known that these objectives often conflict with each other, which prompts the need to apply multicriteria decision support methods to this problem.

In addition, the benefits derived from all actors interacting with each other are similar to the first MCDM/A model (step 2), which contributes to better results.

Finally, Step 5, Control and Risk Monitoring, is the last step of the framework and it implement a monitoring plan to control risk parameters in the area of study. It comprises the record of the risks to be addressed, as well as the real performance of the actions selected and developed to mitigate floods. Thus, this step fosters continuous improvement and is used (Da Silva et al. 2018):

- to design and execute projects;
- to assess the risk and its implications;
- to understand the problem and to model the DM's preferences more coherently;
- to learn of opportunities and to exploit these in order to benefit affected populations.

It is worth to know that this is a learning process that can be implemented periodically in order to assess flood risk in urban areas. Thereafter, DMs assisted by other actors of decision-making process update the framework and generate knowledge for improving new input data. Thus, the framework proposed by the authors seeks to analyze carefully the risk behavior through the implementation of strategic actions, according to the simplifications, specifications, and constraints of the model.

5 Multicriteria Model for Prioritizing Flood Risks Using Decision Theory and MultiAttribute Utility Theory

In flood control and management in urban areas, Van Wesenbeeck et al. (2016) point out that flood risk adaptation planning is a global response to climate change. Therefore, engineering responses to higher levels of unwanted events require a heavy investment of financial, human, and material resources to try to ensure that cities become more resilient and less vulnerable to such disasters.

The methodology is based on how decision models are constructed, according to the classical approach of operational research (Da Silva et al. 2019). Thus, the proposed model will construct the problem in stages and will be associated with a georeferencing platform (GIS) in order to integrate current risk management practices in large urban centers with innovative ideas that are often suggested by multidimensional risk analysis.

The activities related to each phase are described below. It is important to consider that the methodology will be improved, revised, and expanded during the lifecycle of a project, the need for which will be verified during the research.

- Phase 1: Preliminary

The preliminary phase of the model seeks to identify and list the efforts of the actors involved in the process in order to conceive the multidimensional problem in a clear and grounded way, in which:

- The DM understands the main objectives of applying the model, represents the preferences of the professionals involved in flood mitigation, establishes criteria, alternatives – defined by zones or areas – among other data.
- Analysts, along with potential facilitators, contribute factual information about the problem so that they all understand the multidimensional environment and the methodology involved in the process, with a view to determining the likelihood and veracity of recommendations made by the model.
- Preferences will be integrated with data from management and mitigation processes already practiced in public administrations (such as hydrological modeling for flood forecasting, as well as having access to georeferencing platforms to characterize the study areas).

- Phase 2: Natural Disaster Characterization

Karamouz, Nazif, and Falahi (2013) state that hydrological variables and events are generally investigated by analyzing their observation records. However, many characteristics of these processes seem to vary in a way that cannot be deterministically analyzed, and for which hydrological modeling currently seeks to cover scenarios of hazard occurrence, both in the intensity and duration of flooding.

Regarding the hazardous environment, this phase seeks to:

- estimate the likelihood of hazard scenarios occurring, as this particularity has a direct impact on the step that calculates the associated risk;
- identify factors not controlled by the DM, i.e., those relevant parameters that behave like the State of Nature.

This will be of the utmost importance in order for the mathematical model of performance aggregation to approximate reality and for the later phase to be completed.

Keeney and Raiffa (1976) establish a set of procedures to determine utility functions such that they represent the DM´s behavior besides risk, and these explain his/her value judgments (or preferences) about each dimension analyzed here.

Since flooding is caused by expected or unexpected precipitation, the intensity of which the city's critical infrastructure is unable to bear, it is assumed that the decision problem needs to incorporate rainfall behavior in order to undertake further analysis of possible scenarios arising from natural disasters.

In this context, it is important to point out that the behavior of precipitation can be neither standardized nor controlled by human action and should therefore be considered as a factor that the DM cannot control regarding the scenarios arising from floods.

Thus, there are underlying probabilistic mechanisms and the best one can do is to characterize them properly so that rational inferences can be made. Several

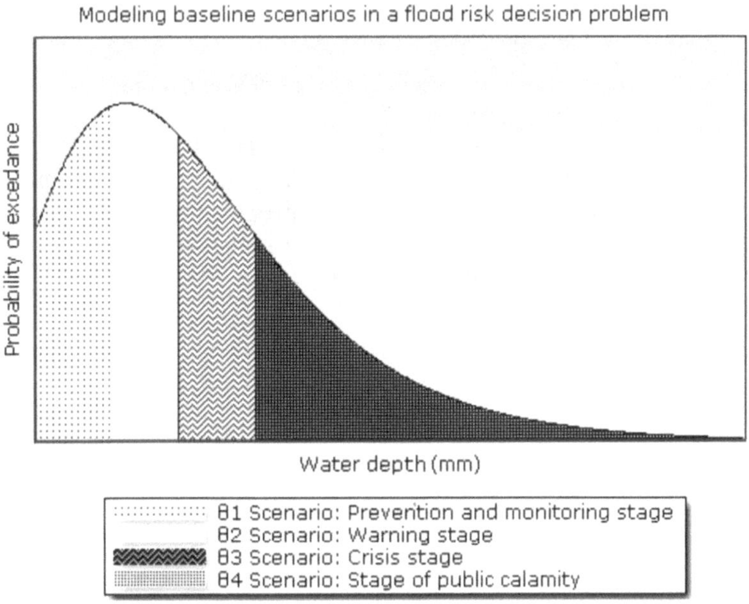

Fig. 4 Modeling scenarios on considering that the depth of flood waters is a state of nature

Table 1 Scenarios description for obtaining a priori probabilities on considering that the depth of flood waters is a state of nature

Scenario	Description	A priori probability Π_{θ_i}
θ1(h1 mm)	Prevention and monitoring stage.	$\pi_{\theta_1} = \Pi < h2$
θ2(h2 mm)	Warning stage.	$\pi_{\theta_2} = h2 < \Pi < h3$
θ3(h3 mm)	Crisis stage.	$\pi_{\theta_3} = h3 < \Pi < h4$
θ4(h4 mm)	Stage of public calamity.	$\pi_{\theta_4} = \Pi > h4$

hydrological parameters can be set in order to construct a scenario of a probabilistic nature due to objective data having been collected; therefore, each relevant scenario should be modeled as a state of nature in the decision problem. The most significant hydrological parameter is the depth of flood waters. This might be the key factor for defining four baseline scenarios in this model (see proposal in Fig. 4 and Table 1) according to a proper probability density function that represents flood frequency in the case study.

The DM establishes the parameters of division between categories of rainfall intensity, depending on the local characteristics of the behavior of rainfall and its particularities.

Assuming that all scenarios for the state of nature occur in a random and independent way, the calculation of a consequence function as a PDF adds uncertainty to the modeling. Thus, the probabilities of the consequences are such that:

$$P_{dim}\left(x|\,\theta_j, a_i\right) = f_{dim}\left(x|\,\theta_j, a_i\right), \tag{1}$$

where $P_{dim}(x|\,\theta_j, a_i)$ means the probability of occurrence in dimension dim whose alternative a_i happens for scenario θ_j.

As a consequence of Equation 1, the expected loss can be calculated by combining the probability of the consequences and the utility function (Equation 2). Here the loss is of the alternative a_i whose scenario θ_j has happened is analyzed for each dimension individually. Furthermore, Berger (1985) argues that if losses are related to risks, they must be considered as the negative of the utility function, thus justifying the negative sign of Equation (2).

$$L_{\theta_j}(a_i) = -\int_x P\left(x|\,\theta_j, a_i\right) u(x)dx, \tag{2}$$

Equation (3) estimates the unidimensional risk. It is calculated by summing all the expected losses values related to possible scenarios, as shown in Equation (3).

$$r_{dim}(a_i) = \sum_{\theta} \pi\left(\theta_j, a_i\right) L_{\theta_j}(a_i) \tag{3}$$

- Phase 3: Multidimensional Risk Determination

The mathematical formulation of the risk will then be constructed according to the chosen (or filtered) model. It should be applied to each dimension assessed in the evaluation context, as the main contribution of this phase is to obtain expected values of the occurrence of unwanted events, by aggregating dimensions and hazard scenarios Berger (1985). In addition, this formulation seeks to understand while using existing practices which parameters or indices are analyzed in order to prioritize one risk over another.

Keeney and Raiffa (1976) pointed out that the ratios between the attributes of the problem must be calculated in order to aggregate the required information into a unified result. Questions about the DM's preferences regarding lotteries with hypothetical performances are put, and a set of equations is obtained to calculate the scale constants K_{dim}.

Thus, the overall risk of an alternative a_j is given by aggregating unidimensional risks using the respective scale constants.

The global risk represents, therefore, the expected values of occurrence of the unwanted events, and adds dimensions and the danger scenarios, as mentioned by Berger (1985).

As a result, the model gives the DM a risk ranking based on Utility Theory, for the most critical areas by assigning priorities to risks.

Alencar and de Almeida (2010) presented another type of result analysis that can be conducted in this step. It is based on the interval scale of the utility functions compared with increments (ratios) of risk regarding alternatives of lesser priority in the ranking. Equation (4) shows how this parameter is calculated.

$$ratio(a_j)_{\beta i} = \frac{r_{global}(a_j)_{\beta i} - r_{global}(a_j)_{\beta i+1}}{r_{global}(a_j)_{\beta i+1} - r_{global}(a_j)_{\beta i+2}} \quad (4)$$

where $r_{global}(a_j)_{\beta i}$ means the risk from the alternative a_j whose position is β_i. So, the increment is calculated using the relative difference between overall risks, the positions of which are adjacent.

- Phase 4: Multidimensional Risk Mapping

Again with the help of the GIS (georeferencing) platform, the decision support system which is used as a tool to apply this methodology will use the results obtained from the previous step, thereby allowing a graphical visualization of the study area, with the respective alternatives evaluated (in this case, geographical areas) with their respective risk assessments (either by rating or ranking, depending on the issues that will be defined throughout the project).

This graphical visualization allows constant interaction between the outcomes to promote a learning process, the benefits of which will extend throughout the local flood management and mitigation process, as well as to create the opportunity for public managers to redefine their policy priorities. This visualization also prompts projects that benefit the most urgent demands to be prioritized efficiently.

After prior prioritizing flood risks, a monitoring plan can be drawn up to design, plan, and execute important actions (projects) that will help the DM to control impacts and to avoid the damages that may otherwise arise. This stage involves the joint participation of many specialists, researchers, and managers, and they can understand how important and necessary it is to have structural and non-structural measures to combat flooding.

6 Open Issues & Insights for Multicriteria Models in FRM context

In order to apply multidimensional models based on Sects. 4 and 5, when structuring of any kind of problem in the context of flood risk management, insights can help to achieve the goals of applying multicriteria methodologies.

It should be noted that flexibility is powerful when analyzing a model for flood mitigation and control, since it can be used in any city in the world that faces a similar flood problem.

Many factors – such as unplanned urbanization, irregular occupation of river-banks, cutting down the riparian forest, etc. – contribute to inflicting harm on the financial, environmental, and social dimensions.

Therefore, we must detect not only the social impacts related to these factors – such as mortality, disease, epidemics, and pandemics – but also economic ones –

damage to property and the infrastructure and the need for emergency plans, for example. In practice, this chapter puts a spotlight on policies for flood management that often take into account multiple strategic objectives.

As a possible source of insights into how best to build MCDA/M models, we cite a survey of relevant attributes shown in Fig. 2 to evaluate possible hierarchical relationships between them in order to structure the decision problem criteria. For this case, although the assumption of hierarchical attributes may better represent the problem for DMs, some studies point to the need to establish an adequate elicitation of the parameters that represent the weights of each attribute, since bias in the process has frequently been observed (Pöyhönen et al. 2001; Pöyhönen and Hämäläinen 1998).

As a tool for applying the modeling proposals, as seen in the previous section, a decision support system (DSS) is a powerful tool in which procedures sequenced by the model are instrumented to perform the entire decision process, from receiving and storing data, processing them and manipulating them (Pressman 2010).

We must point out that the system developed for this study considers the user's cognitive style i.e., the way in which he/she observes and analyzes the data. To this end, the DSS was conceived to meet several functional aspects, as commented on by (Pressman 2010).

This is why an important trend in developing Decision Support Systems for the flood risk environment is the insertion of GIS tools as illustrated in Fig. 5.

Early studies by Mennecke (1997) show the GIS platform integrated with other technologies and they asserted that this integration was as an essential tool for reducing or eliminating bias, for instance. Some benefits of this powerful tool are highlighted by Mennecke and West Jr. (2001) include: increases our knowledge about the resources available in a given area; it increases our knowledge about the resources available in a given area; it facilitates formulating and evaluating different alternative strategies, by answering what if questions about policies, and the analysis and distribution of resources; it reduces the time taken to prepare reports, graphs and maps, which improves the effectiveness of the geographic information used in policy analysis and when evaluating planning options; it improves future planning research by making the data already available existing guidelines and establish guidelines for collection, storage and processing of the new data to be captured; it improves order response time information that is generated by managers and planners for making the information more affordable; it produces new information due to its ability to manipulate data previously available, thanks to the data manipulation capability via computer; it facilitates model development dynamics to support planning; and it allows more appropriate use of the human resources available for collecting and analyzing data – it has already been seen that the costs of these resources are high – by eliminating redundancies and data overlaps and efforts.

Finally, some researchers have devoted themselves to incorporating prospective theory into modeling a DM's preferences (in terms of utility) (Farrow and Scott 2013). Thus, Liu, Fan, and Zhang (2014) focus on emergency response to a disaster, considering DMs' psychological behavior such as reference dependence, aversion to loss, and judgmental distortion to calculate values of potential response results

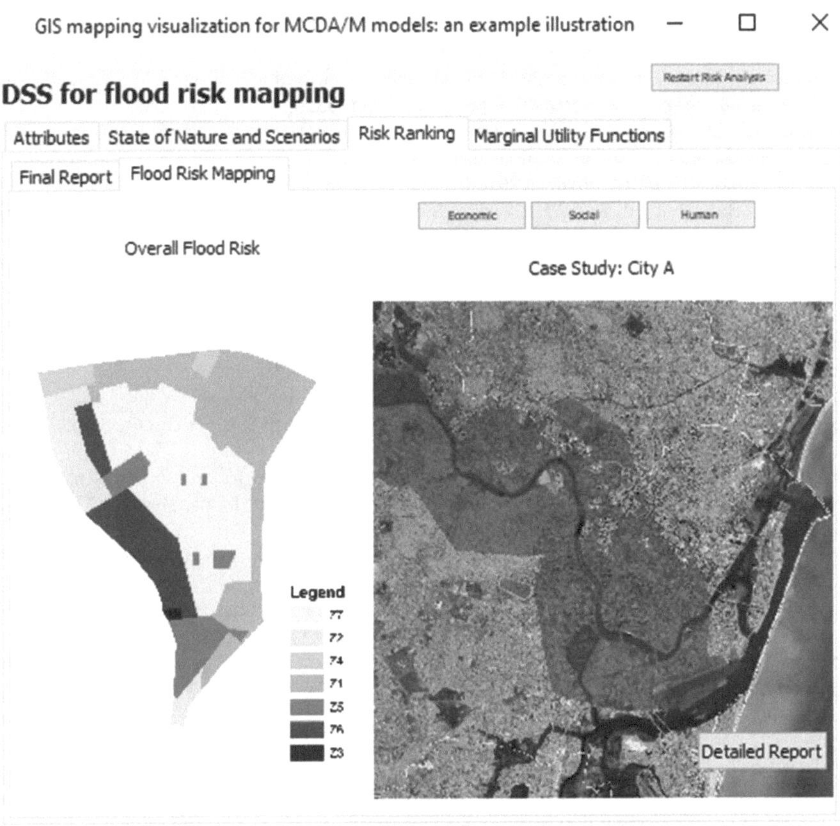

Fig. 5 Example illustration to incorporate GIS-based DSS tools into flood risk problems

concerning each criterion. They proposed a cumulative prospect theory (CPT) to solve risk decision-making problems in emergency response. An insight for multicriteria modeling can be obtained by analyzing the feasibility and validity of the method.

On the other hand, all methodologies and trends pointed out here may help the calculation of flood risk as a catastrophic event, since it seeks to analyze the multiple dimensions involved in the process. Therefore, the use of the risk ranking support model makes it simpler to search for a better compromise solution given the multidimensional nature of a natural disaster.

In addition, this approach can help to prioritize mitigation actions effectively, as it considers the relationship between forms of flood impact, not just a hydrological analysis to measure natural impacts (a practice usually adopted in urban administrations) and considering different types of problem, usually faced by managers in everyday life.

References

Alderman K, Turner LR, Tong S (2012) Floods and human health: a systematic review. Environ Int 47:37–47. https://doi.org/10.1016/j.envint.2012.06.003

Alencar MH, De Almeida AT (2010) Assigning priorities to actions in a pipeline transporting hydrogen based on a multicriteria decision model. Int J Hydrogen Energy 35(8):3610–3619. https://doi.org/10.1016/j.ijhydene.2010.01.122

Ashley RM, Balmforth DJ, Saul AJ, Blanskby JD (2005) Flooding in the future – predicting climate change, risks and responses in urban areas. Water Sci Technol 52(5):265–273

Bae S, Chang H (2019) Urbanization and floods in the Seoul metropolitan area of South Korea: what old maps tell us. Int J Disast Risk Reduc 37(May). https://doi.org/10.1016/j.ijdrr.2019.101186

Berger JO (1985) Statistical decision theory and bayesian analysis. Edited by Springer-Verlag New York: Springer-Verlag. https://doi.org/10.1007/978-1-4757-4286-2

Brito M d, Evers M (2015) Review article: multi-criteria decision making for flood risk management: a survey of the current state-of-the-art. Nat Hazards Earth Syst Sci Discuss 3:6689–6726. https://doi.org/10.5194/nhessd-3-6689-2015

Brito AJ, De Almeida AT, Mota CMM (2010) A multicriteria model for risk sorting of natural gas pipelines based on ELECTRE TRI integrating utility theory. Eu J Oper Res 200(3):812–821. https://doi.org/10.1016/j.ejor.2009.01.016

Chen L, Frauenfeld OW (2016) Impacts of urbanization on future climate in China. Climate Dynam 47(1):345–357. https://doi.org/10.1007/s00382-015-2840-6

CRED-UNISDR (2015) "The human cost of weather related disasters: 1995–2015." https://www.unisdr.org/files/46796_cop21weatherdisastersreport2015.pdf. Accessed 29 Sept 2019

Cucchiella F, Gastaldi M, Trosini M (2017) Investments and cleaner energy production: a portfolio analysis in the Italian electricity market. J Clean Prod 142:121–132. https://doi.org/10.1016/j.jclepro.2016.07.190

Cuellar AD, McKinney DC (2017) Decision-making methodology for risk management applied to Imja lake in Nepal. Water (Switzerland) 9(8):14–16. https://doi.org/10.3390/w9080591

Da Silva LBL, Palha RP, Alencar MH, De Almeida AT (2018) A multidimensional risk evaluation framework for managing floods in urban areas. In: Haugen S, Barros A, Gulijk CV, Kongsvik T, Vinnem J (eds) Safety and Reliability: Safe Societies in a Changing World. Proceedings of ESREL 2018 – European Safety and Reliability Conference, 2018, Trondheim, 1st edn. CRC Press, London, pp 2763–2770

Da Silva LBL, Humberto JS, Lima LES, Alencar MH, Almeida JA, De Almeida AT (2019) Multicriteria modelling for managing flood risks in urban areas. In: Beer M, Zio E (eds) 29th European Safety and Reliability Conference (ESREL), Hannover. Research Publishing Services, Singapore, pp 3714–3722

De Almeida AT, Cavalcante CAV, Alencar MH, Ferreira RJP, de Almeida-Filho AT, Garcez TV (2015) Multicriteria and multiobjective models for risk, reliability and maintenance decision analysis. International Series in Operations Research & Management Science, vol 1a. Vol. 231. Springer International Publishing, Cham. https://doi.org/10.1007/978-3-319-17969-8

De Almeida AT, Almeida JA, Costa APCS, de Almeida-Filho AT (2016) A new method for elicitation of criteria weights in additive models: flexible and interactive tradeoff. Eu J Oper Res 250(1):179–191. https://doi.org/10.1016/J.EJOR.2015.08.058

De Almeida AT, Alencar MH, Garcez TV, Ferreira RJP (2017) A systematic literature review of multicriteria and multi-objective models applied in risk management. IMA J Manag Math 28(October):153–184. https://doi.org/10.1093/imaman/dpw021

Di Baldassarre G, Brandimarte L, Beven K (2016) The seventh facet of uncertainty: wrong assumptions, unknowns and surprises in the dynamics of human–water systems. Hydrol Sci J 61(9):1748–1758. https://doi.org/10.1080/02626667.2015.1091460

Fadlalla R, Elsheikh A, Ouerghi S, Elhag AR (2015) Flood risk map based on GIS, and multi criteria techniques (case study Terengganu Malaysia). J Geogr Inf Syst no. August:348–357. https://doi.org/10.4236/jgis.2015.74027

Farrow S, Scott M (2013) Comparing multistate expected damages, option price and cumulative prospect measures for valuing flood protection. Water Resour Res 49(5):2638–2648. https://doi.org/10.1002/wrcr.20217

Ferreira L, Borenstein D, Righi MB, de Almeida-Filho AT (2018) A fuzzy hybrid integrated framework for portfolio optimization in private banking. Expert Syst Appl 92:350–362. https://doi.org/10.1016/j.eswa.2017.09.055

Gharehgozli A, Iakovou E, Chang Y, Swaney R (2017) Trends in global e-food supply chain and implications for transport: literature review and research directions. Res Transp Bus Manag 25:2–14. https://doi.org/10.1016/j.rtbm.2017.10.002

Gigović L, Pamučar D, Bajić Z, Drobnjak S (2017) Application of GIS-interval rough AHP methodology for flood hazard mapping in urban areas. Water (Switzerland) 9(6):1–26. https://doi.org/10.3390/w9060360

Godskesen B, Hauschild M, Albrechtsen H, Rygaard M (2017) ASTA – a method for multi-criteria evaluation of water supply technologies to assess the most sustainable alternative for Copenhagen. Sci Total Environ 618:399–408. https://doi.org/10.1016/j.scitotenv.2017.11.018

Heo BY, Heo WH (2019) Economic analysis of disaster management investment effectiveness in Korea. Sustainability (Switzerland) 11(11):1–13. https://doi.org/10.3390/su11113011

Hodgkins GA, Dudley RW, Archfield S, Renard B (2019) Effects of climate, regulation, and urbanization on historical flood trends in the United States. J Hydrol 573(August 2018):697–709. https://doi.org/10.1016/j.jhydrol.2019.03.102

Hoegh-Guldberg O, Jacob D, Taylor M, Bindi M, Brown S, Camilloni I, Diedhiou A et al. (2018) "Chapter 3: Impacts of 1.5°C Global Warming on Natural and Human Systems. In: Global Warming of 1.5 °C.", 175–311. https://www.ipcc.ch/site/assets/uploads/sites/2/2019/02/SR15_Chapter3_Low_Res.pdf. Acessed 29 Sept 2019.

Huong HTL, Pathirana A (2013) Urbanization and climate change impacts on future urban flooding in Can Tho City, Vietnam. Hydrol Earth Syst Sci 17(1):379–394. https://doi.org/10.5194/hess-17-379-2013

IPCC (2012) Managing the Risks of Extreme Events and Disasters to Advance Climate Change Adaptation. https://doi.org/10.1017/CBO9781139177245

IPCC (2018) Global Warming of 1.5 °C – SR15. https://www.ipcc.ch/sr15/ Accessed 29 Sept 2019.

Jonkman SN, Maaskant B, Boyd E, Levitan ML (2009) Loss of life caused by the flooding of New Orleans after Hurricane Katrina: analysis of the relationship between flood characteristics and mortality. Risk Anal 29(5):676–698. https://doi.org/10.1111/j.1539-6924.2008.01190.x

Karamouz M, Nazif S, Falahi M (2013) Hydrology and hydroclimatology: principles and applications. CRC Press, Boca Raton. https://doi.org/10.1201/b13771

Karamouz M, Rasoulnia E, Olyaei MA, Zahmatkesh Z (2018) Prioritizing investments in improving flood resilience and reliability of wastewater treatment infrastructure. J Infrastruct Syst 24(4):04018021. https://doi.org/10.1061/(ASCE)IS.1943-555X.0000434

Karasakal E, Aker P (2017) A Multicriteria sorting approach based on data envelopment analysis for R&D project selection problem. Omega (United Kingdom) 73:79–92. https://doi.org/10.1016/j.omega.2016.12.006

Keeney RL, Raiffa H (1976) Decisions with multiple objectives: preferences and value trade-offs. John Wiley and Sons, New York

Koksalan M, Wallenius J, Zionts S (2011) Multiple criteria decision making: from early history to the 21st century. World Scientific Books, World Scientific Publishing Co. Pte. Ltd., n 8042

Lei Y, Wang J, Yue Y, Zhou H, Yin W (2014) Rethinking the relationships of vulnerability, resilience, and adaptation from a disaster risk perspective. Nat Hazards 70(1):609–627. https://doi.org/10.1007/s11069-013-0831-7

Liu Y, Fan Z, Zhang Y (2014) Risk decision analysis in emergency response: a method based on cumulative prospect theory. Comput Oper Res 42:75–82. https://doi.org/10.1016/j.cor.2012.08.008

Lööf H, Nabavi P (2013) Increasing returns to smart cities. Reg Sci Policy Pract 5:255–262. https://doi.org/10.1111/rsp3.12008
McGrath H, Kotsollaris M, Stefanakis E, Nastev M (2019) Flood damage calculations via a RESTful API. Int J Disast Risk Reduc 35(July 2018):101071. https://doi.org/10.1016/j.ijdrr.2019.101071
Medeiros CP, Alencar MH, de Almeida AT (2017) Multidimensional risk evaluation of natural gas pipelines based on a multicriteria decision model using visualization tools and statistical tests for global sensitivity analysis. Reliab Eng & Syst Safe 165(March):268–276. https://doi.org/10.1016/j.ress.2017.04.002
Mennecke BE (1997) Understanding the role of geographic information technologies in business: applications and research directions. J Geogr Inf Decis Anal 1:44–68
Mennecke B, West L Jr (2001) Geographic information systems in developing countries: issues in data collection, implementation and management. J Glob Inf Manag 9(4):44–54. https://doi.org/10.4018/jgim.2001100103
Meyer V, Haase D, Scheuer S (2009a) A multicriteria flood risk assessment and mapping approach. Flood Risk Manag Res Pract 2016(May):1687–1694. https://doi.org/10.1201/9780203883020.ch200
Meyer V, Haase D, Scheuer S (2009b) A multicriteria flood risk assessment and mapping approach. https://doi.org/10.1201/9780203883020.ch200
Miles RF Jr (2007) The emergence of decision analysis. In: Edwards W, Miles RF Jr, von Winterfeldt D (eds) Advances in decision analysis: from foundations to applications. Cambridge University Press, Cambridge. https://doi.org/10.1017/CBO9780511611308
Miniotaite R (2017) Multicriteria analysis of assembling buildings from steel frame structures. IOP Conf Ser: Mater Sci Eng 245(2). https://doi.org/10.1088/1757-899X/245/2/022077
Mitchell JK (1993) "Natural hazard predictions and responses in very large cities." In Prediction and Perception of Natural Hazards: Proceedings Symposium, 22–26 October 1990, Perugia, Italy, edited by J Nemec, J M Nigg, and F Siccardi, 29–37. Dordrecht: Springer Netherlands. https://doi.org/10.1007/978-94-015-8190-5_4.
Neumayer E, Plümper T, Barthel F (2014) The political economy of natural disaster damage. Glob Environ Chang 24(1):8–19. https://doi.org/10.1016/j.gloenvcha.2013.03.011
Patra JP, Kumar R, Mani P (2016) Combined fluvial and pluvial flood inundation modelling for a project site. Procedia Technol 24:93–100. https://doi.org/10.1016/j.protcy.2016.05.014
Pöyhönen M, Hämäläinen RP (1998) Notes on the weighting biases in value trees. J Behav Decis Mak 11(2):139–150. https://doi.org/10.1002/(sici)1099-0771(199806)11:2<139::aid-bdm293>3.3.co;2-d
Pöyhönen M, Vrolijk H, Hämäläinen RP (2001) Behavioral and procedural consequences of structural variation in value trees. Eu J Oper Res 134(1):216–227. https://doi.org/10.1016/S0377-2217(00)00255-1
Pressman RS (2010) Software engineering. A practitioner's approach, 7th edn. McGraw-Hill, New York. https://doi.org/10.1017/CBO9781107415324.004
Priori L, Alencar MH, de Almeida AT (2017) Adaptations to possible climate change impacts: problem structuring based on VFT methodology. In: Filho WL (ed) Innovation in Climate Change Adaptation. Springer International Publishing, Cham, pp 145–157. https://doi.org/10.1007/978-3-319-25814-0_11
Ramaswami A, Russell A, Culligan PJ, Karnamadakala RS, Kumar E (2016) Meta-principles for developing smart, sustainable, and healthy cities. Science 352:940–943
Richard AS, Eugene DC (2014) Understanding the global energy crisis: West Lafayette. Purdue University Press
Saadi I, Bruwier M, Mustafa A, Peltier Y, Archambeau P, Erpicum S, Orban P, Dassargues A, Dewals B, Pirotton M, Teller J, Cools M (2018) Development trajectory of an integrated framework for the mitigation of future flood risk: results from the FloodLand project. Transp Lett 10(5):243–256. https://doi.org/10.1080/19427867.2016.1256120

Samuels P, Gouldby B (2009) "Language of risk – project definitions."http://www.floodsite.net/html/partner_area/project_docs/T32_04_01_FLOODsite_Language_of_Risk_D32_2_v5_2_P1.pdf Acessed 29 Sept 2019.
Şen Z (2018) Climate change impact on floods. In: Flood modeling, prediction and mitigation. Springer International Publishing, Cham, pp 337–379. https://doi.org/10.1007/978-3-319-52356-9_8
Srinivasan V, Lambin EF, Gorelick SM, Thompson BH, Rozelle S (2012) The nature and causes of the global water crisis: syndromes from a meta-analysis of coupled human-water studies. Water Resour Res 48(10):1–16. https://doi.org/10.1029/2011WR011087
Surminski S, Eldridge J (2017) Flood Insurance in England – an assessment of the current and newly proposed insurance scheme in the context of rising flood risk. J Flood Risk Manag 10(4):415–435. https://doi.org/10.1111/jfr3.12127
Trost M, Wanke EM, Ohlendorf D, Klingelhöfer D, Braun M, Bauer J, Groneberg DA, Quarcoo D, Brüggmann D (2018) Immigration: analysis, trends and outlook on the global research activity. J Glob Health 8(1):1–11. https://doi.org/10.7189/jogh.08.010414
United Nations – UN (2019) World population prospects 2019. https://population.un.org/wpp/ Accessed 29 Sept 2019.
Van Wesenbeeck BK, Boer W, Narayan S, Van der Star WRL, De Vries MB (2016) Coastal and riverine ecosystems as adaptive flood defenses under a changing climate. Mitig Adapt Strat Glob Chang 1–8. https://doi.org/10.1007/s11027-016-9714-z
Xiao Y, Yi S, Tang Z (2018) A spatially explicit multi-criteria analysis method on solving spatial heterogeneity problems for flood hazard assessment. Water Resour Manag 32(10):3317–3335. https://doi.org/10.1007/s11269-018-1993-6
Yamashita S, Watanabe R, Shimatani Y (2015) Smart adaptation to flooding in urban areas. Procedia Eng 118:1096–1103. https://doi.org/10.1016/j.proeng.2015.08.449

Part III
MCDM/A Models for Reliability and Maintenance Decision Analysis

Multicriteria Decision Model to Support Maintenance Planning in Sewage Systems

Alexandre Ramalho Alberti and Cristiano Alexandre Virgínio Cavalcante

1 Introduction

Protection systems, such as automotive airbags, military defence systems, isolation valves, and defibrillators, remain inactive during normal operating periods of the main production system, and their operation is required only during the occurrence of specific demands, typically emergency events. In protection systems, the transition from the operational condition to the failure state is not immediately verified, as there is no interruption of the main production process, which results in hidden failures that can only be identified through inspections or during demand events, when the system is required to work and does not fulfil its function. Unmet demands can have very negative effects, which justifies the concern with appropriate maintenance planning for such systems (Vaurio 1999; Jia and Christer 2002; Cavalcante et al. 2011).

Availability is a key indicator to evaluate the performance of a protection system, as it indicates how much the system is able to contain demand events. To guarantee a satisfactory level of availability, inspection policies have been proposed as good alternatives (Jia and Christer 2002). Inspections have the sole objective of obtaining information about the state of the system without affecting its condition, but in different situations, inspections can result in obtaining incorrect information or even

A. R. Alberti (✉)
RANDOM – Research Group on Risk and Decision Analysis in Operations and Maintenance, Universidade Federal de Pernambuco, Recife, Brazil
e-mail: a.r.alberti@random.org.br

C. A. V. Cavalcante
INSID – National Institute of Information and Decision Systems, Universidade Federal de Pernambuco, Recife, Brazil
e-mail: c.a.v.cavalcante@random.org.br

A. T. de Almeida et al. (eds.), *Multicriteria and Optimization Models for Risk, Reliability, and Maintenance Decision Analysis*, International Series in Operations Research & Management Science 321, https://doi.org/10.1007/978-3-030-89647-8_9

interfering negatively with the system condition. The intensity with which errors occur depends on the quality of the resources involved in the execution of the maintenance policy (Alberti et al. 2018). These aspects of inspection quality have been observed and incorporated into mathematical models presented in previous works.

Berrade et al. (2012, 2015) observe that in certain situations, there may be errors in the identification of the failed state of protection systems: false positives can lead to an early renewal of the system, while the occurrence of false negatives increases system vulnerability, increasing the likelihood of occurrence of unmet demands. The authors consider a hybrid inspection and preventive replacement policy, which can compensate the negative impacts of imperfect inspections on system availability.

Alberti et al. (2018), in a study of isolation valves used in water distribution networks, verify that the system's deterioration and failure process can be represented by a two-stage failure model, where a defective state (where the system is operational but exhibits deviations from its normal operating conditions) can be identified before a failure occurs. In this context, the authors verified that errors regarding the identification of the failure state are negligible, but there may be errors in the identification of the defective state (both false positives and false negatives). In an application of the model, the authors show that, even with significant probabilities of errors in the identification of the defective state, it is worth implementing a maintenance policy that determines the renewal of the system based on the identification of defects or failures, and not only based on the identification of failures.

Alberti et al. (2018) also consider that an inspection can lead to defect induction in the protection system. In an application they verified that, even if the probability of defect induction is low, the impact on system performance is quite significant, which may lead to the recommendation of an inspection-free simple age-based replacement policy. Scarf and Cavalcante (2012) present a model for critical systems (which have pronounced failures) that also considers this possibility, while presents a model that considers the possibility of failure induction.

Another important factor that has been studied is the variation in the quality of component replacement, which may be related to variation in the manufacturing quality or even the reuse of a component through its recycling and/or variation in the installation service quality. Scarf et al. (2009) model this aspect considering that the component used in the replacement comes from a heterogeneous population composed of weak items, with low reliability and susceptible to early failures, and strong items that present late failures, so that the probability distribution of the time until the defect arrival can be estimated as a mixture of the characteristic distributions of these subpopulations, considering their proportions. Berrade et al. (2012, 2015) and Alberti et al. (2018) share the same notion in models applied to protection systems.

Mathematical models are developed from certain simplifying assumptions and are not able to consider all aspects of reality, but they can be very useful tools to obtain guidelines for decision-making in maintenance.

This chapter presents a mathematical model for a hybrid inspection and preventive maintenance policy applied to a protection system whose process of deterioration and failure can be modelled considering the delay-time concept (Christer 1999). The model has similarity to the model presented in recent work of both consider the possibility of errors in the identification of defective and failed states, but different assumptions are considered in their development. The model presented here was developed in an innovative way, so it would be possible to more accurately calculate the cost rate and the rate of unmet demands resulting from the adoption of a maintenance policy. Its development was motivated by a case study on shut-off valves used in sewage collection systems.

A multicriteria model based on the multi-attribute value theory (MAVT) (Keeney and Raiffa 1976) is presented, considering two criteria: the expected cost rate and the rate of unmet demands (which can be considered as a risk measure). Depending on the scenario, the losses in non-financial dimensions (human, environmental, etc.) are proportional to the rates of unmet demands, and the multicriteria model allows the appropriate treatment of the characteristic multidimensionality of the problem in certain situations. The model can be used for the definition of a maintenance policy and even for the evaluation of investment scenarios in improving the quality of maintenance, as demonstrated in the presented application.

The remainder of this chapter is organized as follows: the next section briefly presents the context of a case study that motivated the development of this work. In Sect. 4, the mathematical model for the maintenance policy for a protection system considering the possibility of errors in inspections is developed, and in Sect. 5, a framework for the construction of the multicriteria model and obtaining results is presented. In Sect. 6, a numerical application is presented using simulated data, and finally, in the last section, conclusions are presented.

2 Motivation

The development of the model presented in this chapter was motivated by a case study on shut-off valves used in sewage collection and transportation systems. It is worth mentioning that although the model has been motivated by a specific context, it can be applied in other contexts, as long as the model is consistent with the observed conditions.

Urban sewage networks are essential public infrastructures whose performance has an impact on community health, pollution control and economic and environmental sustainability of cities, so it is important to establish proper maintenance and rehabilitation plans for these systems (Baah et al. 2015; Diogo et al. 2018).

Sewage collection and transport networks are usually formed by branched, buried and open channel (i.e. under atmospheric pressure) pipes with gravity flow, which transport the sewage from the collection points to the treatment plants. The design of the system shall ensure the slope necessary for the flow of the sewage to occur by gravity and with the speed necessary to keep the sewers clean and transport materials with the wastewater. When it is not possible to maintain the necessary slope (e.g. due to geographical limitations), lift stations (or pumping stations) can be used to pump the fluid to higher altitudes and to enable new gravity routes (Grigg 2012). Other models of sewage collection systems are possible, for example, with networks that operate under pressure, but these systems are not of interest for this work.

Shut-off valves are used to interrupt the flow of sewage, a procedure necessary for the maintenance of system components such as pipes in general, as well as pumps and other types of valves used in lift stations (Humes and Stolberg 2006). Because of their operational characteristics, shut-off valves can be characterized as protection systems, and the demands are situations that require the interruption of the flow in a specific area.

While in water distribution networks, which operate under pressure, the failure of isolation valves can be verified without errors through downstream pressure gauges (Alberti et al. 2018), in the context of sewage collection networks that operate in open channel conditions, the failure of shut-off valves may not be identified during an inspection when there is greater dependence on the inspector's perception. Moreover, as is common in mechanical equipment (Alberti et al. 2018), prior to failure, shut-off valves have an apparent defective state, which may be characterized with greater difficulty in their actuation or small leakage, and the defect is also subject to misclassification errors.

The occurrence of unmet demands in this context can have quite negative effects from an economic and environmental point of view. The maintenance of equipment in a lift station is an example of a demand event: when one or more pumps of a lifting station fail, the non-operation of the shut-off valve makes it impossible to stop the arrival of sewage in the station well, which hinders the execution of the maintenance plan and may lead to the well overflowing in the case of long delays. Well overflow can have negative environmental impacts, with the risk of contamination of sources of drinking water and inconvenience caused to nearby communities, in addition to the financial impact to the company. Figure 1 gives a good idea about how these consequences can happen. Thus, a multicriteria approach can be very useful for the evaluation of maintenance policies for shut-off valves.

3 Notation

The notation presented below is used throughout this chapter.

- Decision Variables:

  M – maximum number of inspections until the preventive replacement.
  T – time interval between two consecutive inspections.

- Notation for the development of the maintenance policy's mathematical model:

  X time until defect arrival – non-negative random variable.

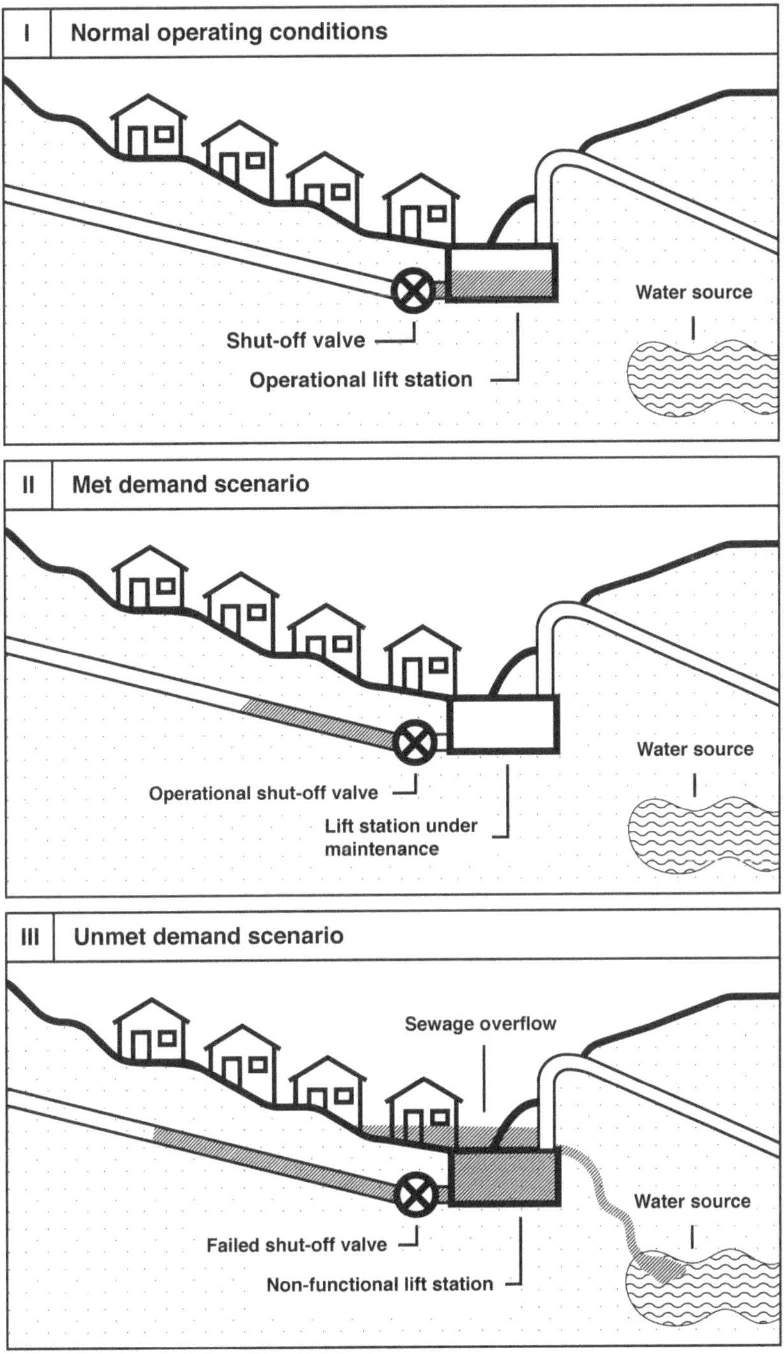

Fig. 1 Graphical representation of the consequences of failure in a shut-off valve from a sewage collection networks

H – sojourn time in the defective state (delay-time) – non-negative random variable.
Z – time between the occurrences of two consecutive demands – non-negative random variable.
f_x, F_x, R_x – for X, functions of the probability density, cumulative probability distribution and reliability, respectively.
f_h, F_h, R_h – for H, functions of the probability density, cumulative probability distribution and reliability, respectively.
f_z, F_z, R_z – for Z, functions of the probability density, cumulative probability distribution and reliability, respectively.
μ – rate of demands.
p – proportion of weak items in the component population.
w – probability of false positives during inspection.
q_1 – probability of false negatives during inspection for defective state.
q_2 – probability of false negatives during inspection for failed state.
c_v – cost of an inspection.
c_p, c_d, c_f – replacement costs when the component is good, defective and failed, respectively.
C_{UD} – cost of an unmet demand.
t_v – time required to perform an inspection.
t_p, t_d, t_f – *time* for performing the replacement when the component is good, defective and failed, respectively.
t_{ud} – time to *normalize* the system when an unmet demand occurs.
EC, EL – expected cost and expected length of a renewal cycle, respectively.
ρ – probability of a renewal cycle ending in an unmet demand.
$C\infty$ – expected cost per unit of time in the long run.
λ – rate of unmet demands.

- Notation for multicriteria model development:

A – alternative/solution – in this context, a maintenance policy, which is defined by the combination of decision variables T and M.
Aj – performance of alternative A for criterion j.
V – multi-attribute value.
Vj – unidimensional value for criterion j.
kj – scale constant for criterion j.
nc – number of criteria.
c, r – indexes for the cost (cost rate) and risk (rate of unmet demands) criteria, respectively.

4 The Proposed Mathematical Model

To support the evaluation of maintenance policies in contexts such as the one presented in this study, a delay-time model was developed for a protection system subject to demands that occur according to a homogenous Poisson process.

We consider a single component protection system composed of a component and a socket, which together provide an operational function (Ascher and Feingold 1984). The component deteriorates over time, and the operational status of the system depends on its state: if the component is in a good or defective state, the system is operational, and if the component fails, the system is unable to fulfil its function. With the component replacement, the system as a whole is renewed.

A hybrid inspection and preventive replacement policy (MT policy) is proposed, similar to the one proposed by Vaurio (1999), which guides the performance of M inspections with a time interval T between the beginning of two consecutive inspections. When an inspection indicates that the component is defective or failed, it is replaced with a new unit, and after the M-th inspection, the component is replaced regardless of its state. It is a flexible policy format, and special cases are the pure inspection policy ($M = \infty$) and the simple age-based replacement policy ($M = 1$).

It is also considered that the inspection is subject to errors in the identification of the component condition (false positives and false negatives). Because there is no difference regarding the action recommended in cases of defect or failure indication, only one type of false positive is considered: the component is in good state, but the inspection indicates the opposite, which leads to its early replacement. Regarding false negatives, two situations can be verified: the component is in the defective state and the inspection indicates that it is good (false negative type 1), or the component is in the failed state and the inspection indicates that it is still good (false negative type 2). The probabilities of false negatives type 1 and type 2 are expected to be different because the abnormal conditions of the system tend to become more evident when it is failed.

The relationships between the possible component states, the possible results of the inspection and their probabilities are presented in Table 1.

4.1 Model Assumptions

For the construction of the model the following assumptions are considered:

Table 1 Consequence matrix for the decision problem

	Component's state			
		Good	Defective	Failed
Inspection outcome	Negative	$1-w$	q_1	q_2
	Positive	w	$1-q_1$	$1-q_2$

1. The maintenance policy is rescheduled at the beginning of each renewal cycle.
2. A component's replacement is performed when an inspection indicates a defective or failed state, when an unmet demand occurs, or at the M-th inspection, whichever occurs first.
3. At $(M.T)$, an inspection is also carried out to check the component's state and the information is recorded.
4. The distributions F_x, F_h and F_z are known and statistically independent.
5. At an inspection performed when the component is in good state, there is a probability w of a false positive occurring.
6. At an inspection performed when the component is in the defective state, there is a probability q_1 of a false negative occurring.
7. At an inspection performed when the component is in the failed state, there is a probability q_2 of a false negative occurring.
8. The time and cost of maintenance actions are constant and known.
9. The time spent on maintenance actions is not configured as downtime for the protection system. Here, it is considered that during the maintenance actions, there is an interruption in the process of arrival of demands or that demands are met through alternative solutions.
10. Demands occur according to a homogeneous Poisson process with a known rate. Hence, f_z is an exponential distribution with a characteristic parameter μ.
11. The time required to perform an inspection t_v is much smaller than T ($t_v << T$).

The assumption 9 is particularly true for examples where, for the maintenance of the protection system, it is necessary to stop the operation of the main system, as is the case for shut-off valves: to maintain the valve, it is necessary to interrupt the flow through it, which means an interruption of the process of the arrival of demands. Emergency brakes are another example where this condition applies. Moreover, this assumption is a good approximation when $t_p, t_d, t_f, t_{ud} << T$.

The mathematical models for protection systems presented in the literature are based on a characterization of the system's renewal cycle according to the decision variables of the maintenance policy, with a penalty for the resulting downtime (Vaurio 1999; Cavalcante et al. 2011; Berrade et al. 2012, 2015; Alberti et al. 2018). In contrast to these models, the model presented in this chapter considers the replacement of the component in the case of unmet demands, which is a more realistic scenario. Once assumptions 9 and 10 were considered, it was possible to develop an analytical model considering this change.

4.2 Model Development

The model was developed from the enumeration of all possible renewal cycle scenarios that may occur under the presented conditions (represented graphically in Fig. 2). Figure 2 depicts an exhaustive and mutually exclusive set of scenarios

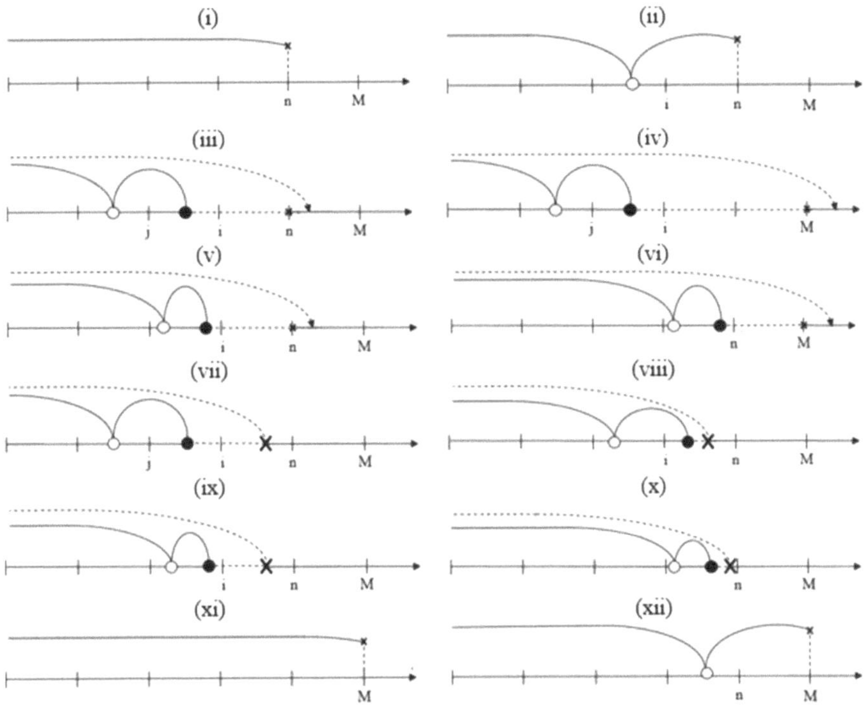

Fig. 2 Graphical representation of all the possible renewal cycle scenarios ○ defect arrival, • failure, — demand arrival process

such that the sum of the probabilities of occurrence of these scenarios must be equal to 1 regardless of the values of the decision variables. For each scenario, we present the calculation of the probability of its occurrence ($P_{scenario}$) and, from this result, the expressions to calculate the expected values for a renewal cycle of the following measures are derived: the cost ($EC_{scenario}$) and length of the renewal cycle ($EL_{scenario}$). After the scenarios are detailed, the expressions are grouped to model the performance of the maintenance policy as a function of the decision variables M and T.

4.2.1 Mathematical Development of the Scenarios

Scenario 1 ($M > 1$) – the component is replaced after a false positive at the n-th inspection ($n < M$):

$$P_1\left(T, n\right) = (1 - w)^{n-1}.w.R_x(n.T) \tag{1}$$

$$EC_1(T,M) = \sum_{n=1}^{M-1} \left[\left(n.c_v + c_p \right).P_1(T,n) \right] \tag{2}$$

$$EL_1(T,M) = \sum_{n=1}^{M-1} \left[\left(n.T + t_p \right).P_1(T,n) \right] \tag{3}$$

Scenario 2 ($M > 1$) – the defect arrives at the i-th interval between inspections and is identified at the n-th inspection, before the failure ($i \leq n < M$):

$$P_2(T,n) = \sum_{i=1}^{n} \left[(1-w)^{i-1}.{q_1}^{n-i}.(1-q_1).\int_{(i-1).T}^{i.T} f_x(x).R_h(n.T-x)\,dx \right] \tag{4}$$

$$EC_2(T,M) = \sum_{n=1}^{M-1} \left[(n.c_v + c_d).P_2(T,n) \right] \tag{5}$$

$$EL_2(T,M) = \sum_{n=1}^{M-1} \left[(n.T + t_d).P_2(T,n) \right] \tag{6}$$

Scenario 3 ($M > 2$) – the defect arrives at the j-th interval between inspections, the failure occurs at the i-th interval and is identified at the n-th inspection, before the occurrence of a demand ($j < i \leq n < M$):

$$P_3(T,n) = \sum_{i=2}^{n} \sum_{j=1}^{i-1} \left[\begin{array}{l} (1-w)^{j-1}.{q_1}^{i-j}.{q_2}^{n-i}.(1-q_2). \\ \int_{(j-1).T}^{j.T} f_x(x).\int_{(i-1).T-x}^{i.T-x} f_h(h).R_z(n.T-x-h)\,dh\,dx \end{array} \right] \tag{7}$$

$$EC_3(T,M) = \sum_{n=2}^{M-1} \left[\left(n.c_v + c_f \right).P_3(T,n) \right] \tag{8}$$

$$EL_3(T,M) = \sum_{n=2}^{M-1} \left[\left(n.T + t_f \right).P_3(T,n) \right] \tag{9}$$

Scenario 4 ($M > 1$) – the defect arrives at the j-th interval between inspections, the failure occurs at the i-th interval and the replacement occurs at the M-th inspection, before the occurrence of a demand ($j < i \leq M$):

$$P_4(T,M) = \sum_{i=2}^{M}\sum_{j=1}^{i-1}\left[\begin{array}{l}(1-w)^{j-1}.{q_1}^{i-j}.{q_2}^{M-i}.\\ \int_{(j-1).T}^{j.T} f_x(x).\int_{(i-1).T-x}^{i.T-x} f_h(h).R_z(M.T-x-h)\,dhdx\end{array}\right] \quad (10)$$

$$EC_4(T,M) = \left(M.c_v + c_f\right).P_4(T,M) \quad (11)$$

$$EL_4(T,M) = \left(M.T + t_f\right).P_4(T,M) \quad (12)$$

Scenario 5 ($M > 1$) – the defect arrives at the i-th interval between inspections, the failure occurs before the subsequent inspection and is identified at the n-th inspection, before the occurrence of a demand ($i \le n < M$):

$$P_5(T,n) = \sum_{i=1}^{n}\left[\begin{array}{l}(1-w)^{i-1}.{q_2}^{n-i}.(1-q_2).\\ \int_{(i-1).T}^{i.T} f_x(x).\int_{0}^{i.T-x} f_h(h).R_z(n.T-x-h)\,dhdx\end{array}\right] \quad (13)$$

$$EC_5(T,M) = \sum_{n=1}^{M-1}\left[\left(n.c_v + c_f\right).P_5(T,n)\right] \quad (14)$$

$$EL_5(T,M) = \sum_{n=1}^{M-1}\left[\left(n.T + t_f\right).P_5(T,n)\right] \quad (15)$$

Scenario 6 (any M) – the defect arrives at the n-th interval between inspections, the failure occurs before the subsequent inspection, and the replacement occurs at the M-th inspection, before the occurrence of a demand ($n \le M$):

$$P_6(T,M) = \sum_{n=1}^{M}\left[\begin{array}{l}(1-w)^{n-1}.{q_2}^{M-n}.\\ \int_{(n-1).T}^{n.T} f_x(x).\int_{0}^{n.T-x} f_h(h).R_z(M.T-x-h)\,dhdx\end{array}\right] \quad (16)$$

$$EC_6(T,M) = \left(M.c_v + c_f\right).P_6(T,M) \quad (17)$$

$$EL_6(T,M) = \left(M.T + t_f\right).P_6(T,M) \quad (18)$$

Scenario 7 ($M > 2$) – the defect arrives at the j-th interval between inspections, the failure occurs at the i-th interval and is not detected before a demand that occurs at the n-th interval ($j < i < n \le M$):

$$P_7(T,n) = \sum_{i=2}^{n-1}\sum_{j=1}^{i-1}\left[\begin{array}{l}(1-w)^{j-1}.{q_1}^{i-j}.{q_2}^{n-i}.\\ \int_{(j-1).T}^{j.T} f_x(x).\int_{(i-1).T-x}^{i.T-x} f_h(h).\int_{(n-1).T-(x+h)}^{n.T-(x+h)} f_z(z)dzdhdx\end{array}\right] \tag{19}$$

$$EC_7(T,M) = \sum_{n=3}^{M}\left\{\left[(n-1).c_v + c_f + C_{UD}\right].P_7(T,n)\right\} \tag{20}$$

$$\begin{array}{l}EL_{7jin}(T,j,i,n) = (1-w)^{j-1}.{q_1}^{i-j}.{q_2}^{n-i}.\\ .\int_{(j-1).T}^{j.T} f_x(x).\int_{(i-1).T-x}^{i.T-x} f_h(h).\int_{(n-1).T-(x+h)}^{n.T-(x+h)} f_z(z).\left(x+h+z+t_f+t_{UD}\right)dzdhdx\end{array} \tag{21}$$

$$EL_7(T,M) = \sum_{n=3}^{M}\sum_{i=2}^{n-1}\sum_{j=1}^{i-1} EL_{7jin}(T,j,i,n) \tag{22}$$

Scenario 8 ($M > 1$) – the defect arrives at the i-th interval between inspections, the failure occurs at the n-th interval and a demand occurs before the subsequent inspection ($i < n \leq M$):

$$P_8(T,n) = \sum_{i=1}^{n-1}\left[\begin{array}{l}(1-w)^{i-1}.{q_1}^{n-i}.\\ \int_{(i-1).T}^{i.T} f_x(x).\int_{(n-1).T-x}^{n.T-x} f_h(h).F_z(n.T-x-h)\,dhdx\end{array}\right] \tag{23}$$

$$EC_8(T,M) = \sum_{n=2}^{M}\left\{\left[(n-1).c_v + c_f + C_{UD}\right].P_8(T,n)\right\} \tag{24}$$

$$\begin{array}{l}EL_{8in}(T,i,n) = (1-w)^{i-1}.{q_1}^{n-i}.\\ \int_{(i-1).T}^{i.T} f_x(x).\int_{(n-1).T-x}^{n.T-x} f_h(h).\int_{0}^{n.T-(x+h)} f_z(z).\left(x+h+z+t_f+t_{UD}\right)dzdhdx\end{array} \tag{25}$$

$$EL_8(T,M) = \sum_{n=2}^{M}\sum_{i=1}^{n-1} EL_{8in}(T,i,n) \tag{26}$$

Scenario 9 ($M > 1$) – the defect arrives at the i-th interval between inspections, the failure occurs before the subsequent inspection and is not identified before a demand that occurs at the n-th interval ($i < n \leq M$):

$$P_9(T,n) = \sum_{i=1}^{n-1} \left[\begin{array}{l} (1-w)^{i-1}.{q_2}^{n-i}. \\ \int_{(i-1).T}^{i.T} f_x(x). \int_0^{i.T-x} f_h(h). \int_{(n-1).T-(x+h)}^{n.T-(x+h)} f_z(z)dzdhdx \end{array} \right] \tag{27}$$

$$EC_9(T,M) = \sum_{n=2}^{M} \left\{ \left[(n-1).c_v + c_f + C_{UD} \right].P_9(T,n) \right\} \tag{28}$$

$$\begin{array}{l} EL_{9\ in}(T,i,n) = (1-w)^{i-1}.{q_2}^{n-i}. \\ \int_{(i-1).T}^{i.T} f_x(x). \int_0^{i.T-x} f_h(h). \int_{(n-1).T-(x+h)}^{n.T-(x+h)} f_z(z). \left(x+h+z+t_f+t_{UD} \right) dzdhdx \end{array} \tag{29}$$

$$EL_9(T,M) = \sum_{n=2}^{M} \sum_{i=1}^{n-1} EL_{9\ in}(T,i,n) \tag{30}$$

Scenario 10 (any M) – the defect arrives at the n-th interval between inspections, the failure and later a demand occur before the subsequent inspection ($n \leq M$):

$$\begin{array}{l} P_{10}(T,n) = (1-w)^{n-1}. \\ \int_{(n-1).T}^{n.T} f_x(x). \int_0^{n.T-x} f_h(h).F_z(n.T-x-h)\,dhdx \end{array} \tag{31}$$

$$EC_{10}(T,M) = \sum_{n=1}^{M} \left\{ \left[(n-1).c_v + c_f + C_{UD} \right].P_{10}(T,n) \right\} \tag{32}$$

$$\begin{array}{l} EL_{10n}(T,n) = (1-w)^{n-1}. \\ \int_{(n-1).T}^{n.T} f_x(x). \int_0^{n.T-x} f_h(h). \int_0^{n.T-(x+h)} f_z(z). \left(x+h+z+t_f+t_{UD} \right) dzdhdx \end{array} \tag{33}$$

$$EL_{10}(T,M) = \sum_{n=1}^{M} EL_{10n}(T,n) \tag{34}$$

Scenario 11 (any M) – preventive replacement at $M.T$, with the system in good state:

$$P_{11}(T,M) = (1-w)^{M-1}.R_x(M.T) \tag{35}$$

$$EC_{11}(T,M) = \left(M.c_v + c_p \right).P_{11}(T,M) \tag{36}$$

$$EL_{11}(T,M) = \left(M.T + t_p\right).P_{11}(T,M) \quad (37)$$

Scenario 12 (any *M*) – preventive replacement at *M.T*, with the system in defective state:

$$P_{12}(T,M) = \sum_{n=1}^{M} \left[\begin{array}{l} (1-w)^{n-1}.{q_1}^{M-n}. \\ \int_{(n-1).T}^{n.T} f_x(x).R_h(M.T - x)\,dx \end{array} \right] \quad (38)$$

$$EC_{12}(T,M) = (M.c_v + c_d).P_{12}(T,M) \quad (39)$$

$$EL_{12}(T,M) = (M.T + t_d).P_{12}(T,M) \quad (40)$$

4.2.2 Joining the Scenarios

The expected value of a measure of interest *EQ* for a renewal cycle is equal to the sum of the contributions of each scenario, therefore:

- If $M = 1$:

$$EQ(T,M) = \sum_{j \in \{6,10,11,12\}} EQ_j(T,M) \quad (41)$$

- If $M = 2$:

$$EQ(T,M) = \sum_{j \in \{1,2,4,5,6,8,9,10,11,12\}} EQ_j(T,M) \quad (42)$$

- If $M \geq 3$:

$$EQ(T,M) = \sum_{j=1}^{12} EQ_j(T,M) \quad (43)$$

The measure *EQ* can be the cost (*EC*) or the length of the renewal cycle (*EL*).

4.2.3 Calculating the Cost Rate (Cost Criterion)

According to the reward renewal theorem (Tijms 1994), the expected cost per unit of time in the long run can be calculated according to eq. (44):

$$C_\infty(T,M) = \frac{EC(T,M)}{EL(T,M)} \quad (44)$$

4.2.4 Calculating the Rate of Unmet Demands (Risk Criterion)

The probability that a renewal cycle ends in an unmet demand can be calculated as the sum of the probabilities of occurrence of the scenarios that end in this way, therefore:

- If $M = 1$:

$$\rho(T, M) = \sum_{n=1}^{M} P_{10}(T, n) \tag{45}$$

- If $M = 2$:

$$\rho(T, M) = \sum_{n=2}^{M} P_8(T, n) + \sum_{n=2}^{M} P_9(T, n) + \sum_{n=1}^{M} P_{10}(T, n) \tag{46}$$

- If $M \geq 3$:

$$\rho(T, M) = \sum_{n=3}^{M} P_7(T, n) + \sum_{n=2}^{M} P_8(T, n) + \sum_{n=2}^{M} P_9(T, n) + \sum_{n=1}^{M} P_{10}(T, n) \tag{47}$$

As the renewal cycles are statistically independent of each other, then, in agreement with the observations of Scarf et al. (2009), it is possible to state that the time intervals between unmet demands are approximately exponentially distributed with the rate λ calculated according to eq. (48).

$$\lambda(T, M) = \frac{\rho(T, M)}{EL(T, M)} \tag{48}$$

Once estimated the expected losses for a situation of unmet demand, the rate λ (T, M) can be used to calculate the expected rate of loss for non-financial dimensions (human, environmental, etc.). The rate λ (T, M) can then be understood as a measure of risk (Aven 2012).

5 The Multicriteria Model

In several contexts, maintenance planning addresses several objectives, which characterize it as a multicriteria decision problem (Jiang and Ji 2002). For example, the objectives of maintenance policies for shut-off valves are to minimize the expected cost rate in the long run and to minimize the impacts resulting from unmet demands. These two objectives characterize well problems involving the maintenance of protection systems. Multicriteria models then present an important contribution to decision-making in these cases (De Almeida et al. 2015b).

In this work, the multi-attribute value theory (MAVT) is considered for the construction of the multicriteria model. MAVT is a method of aggregation with a single synthesis criterion, based on compensatory logic, and suitable for situations where there is complete knowledge about the state of the nature of the problem, so that the consequences of possible decision alternatives are deterministic (Keeney and Raiffa 1976). Although these conditions are difficult to verify in maintenance planning problems, due to the difficulty of accurately estimating the parameters needed for mathematical model applications, this method can provide important insights into the process and has been used in other studies, as verified by De Almeida et al. (2015b).

For the construction of the multicriteria model, it is important to follow a structured protocol to provide a good understanding of the faced problem, as well as to elicit and evaluate the relations of preference of the decision maker (DM). A framework for the construction of the multicriteria model for decision support in the context addressed is presented in Fig. 3.

The first steps of the framework aim to provide a good understanding of the problem, with a characterization of the DM and the system under study, and the identification of possible maintenance strategies. Next, the DM should assess the risk dimensions associated with the problem, which should be considered in the multicriteria model construction.

The mathematical modelling of the maintenance policies, in turn, allows the calculation of the performance of the alternatives in the considered attributes, and for the effective use of the developed models it is essential to analyse the input parameters, particularly the parameters for which the model presents greater sensitivity. In the absence of objective historical data or to complement them, elicitation procedures can be used to estimate, from the experts' knowledge, the input parameters for the mathematical models (Berger 1985).

The evaluation of the alternatives of action is done by assessing their consequences (Keeney and Raiffa 1976); thus, the next step is the delimitation of the space of consequences (CS) to be considered for eliciting the preferences of the DM. The procedure was elaborated so that the resulting CS represents a space of alternatives that are limited to non-dominated alternatives and that respect the thresholds of acceptable risk. In this way, the analysis is restricted only to the alternatives that in fact interest the DM, and the model shows greater sensitivity, being able to better differentiate between the best and worst alternatives.

With the problem duly contextualized and the initial conditions for the analysis established, the modelling of the preferences of the DM using MAVT can be started. This stage consists of eliciting the multi-attribute value function, which will be used to evaluate the performance of maintenance policies.

From Keeney and Raiffa (1976) and De Almeida et al. (2015a), it is concluded that the process of eliciting the multi-attribute value function can be divided into five main steps: introduction of ideas and terminology; elicitation of one-dimensional value functions (intracriterion evaluation); verification of the relations

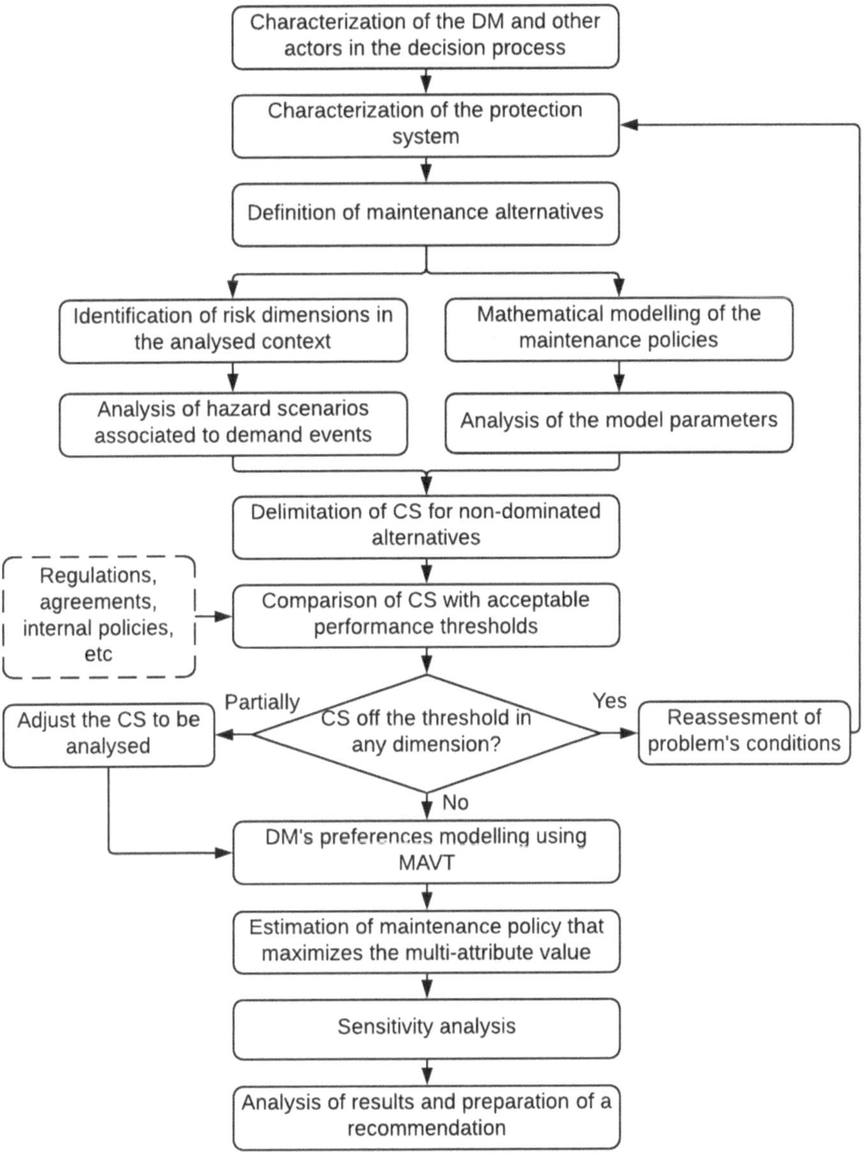

Fig. 3 Framework for the construction of the multicriteria model using MAVT. *CS* Consequences Space

of mutual preferential independence; elicitation of the multi-attribute value function (intercriteria evaluation); and finally, tests for consistency checking.

The multi-attribute value function associates a real number with each point in the CS and can be obtained from the combination of one-dimensional value functions, which evaluate the consequences by considering only one criterion at a time (Keeney and Raiffa 1976). One-dimensional value functions result in a scale transformation (usually on a scale of 0 to 1) and indicate the value gains (or losses) that can be obtained with an improvement (or worsening) in the performance of the criterion according to the initial reference point. Linear functions indicate that marginal gains are constant and are widely used in practical applications, where they are often good approximations (De Almeida et al. 2015a). Belton and Stewart (2002) present methods for eliciting one-dimensional value functions.

The intercriteria evaluation, in turn, can begin with the verification of the condition of mutual preferential independence between the criteria. The mutual preference independence between two criteria Y and Z occurs if and only if the conditional preference in the space of Y (intracriterion evaluation) does not depend on the level of performance in criterion Z, and vice versa. Once this condition is verified, the multi-attribute value function can be expressed in the additive form, according to eq. (49) (De Almeida et al. 2015a).

$$V(A) = \sum_{j=1}^{nc} k_j . V_j \left(A_j\right) \tag{49}$$

where:

$$\sum_{j=1}^{nc} k_j = 1$$

The scale constants (k_j) can be calculated from trade-offs between the criteria. This procedure is presented in detail by Keeney and Raiffa (1976).

The maintenance policy that maximizes the multi-attribute value presents the best compromise relationship between the considered criteria. A sensitivity analysis is then required to assess the impact of possible inconsistencies in the process of elicitation of the DM's preferences in order to evaluate whether the recommendation obtained is robust to possible disturbances or whether extra verifications are required to ensure greater reliability of the results (De Almeida et al. 2015a). Once the necessary checks and a critical analysis of the results have been made, the recommendation for the DM is drawn up.

6 Numerical Application

For the numerical application, the context of shut-off valves used in a lifting station of a sewage collection and transportation system was considered. The maintenance of such a system not only has financial but also environmental impacts, so a multicriteria approach can better address the multidimensionality of the problem of maintenance policies evaluation. The criteria of cost (C_∞ (T, M)) and risk (λ (T, M)) were considered in the analysis.

Simulated data were used to illustrate the use of the model and to obtain some results for discussion.

6.1 Specifying the Model Input Parameters

For the definition of the parameters, realistic values were considered for the specific example of a shut-off valve operated by a Brazilian sanitation company.

As is common in mechanical equipment, the system's deterioration and failure process can be modelled considering the delay-time concept; that is, before a failure occurs, the system presents a defective state that can be identified (Christer 1999). In an investigation of the factors that affect the performance of the system, it was verified that the variation in the quality of the maintenance service or in the quality of the material used could lead to early failures. A simplification of the presented model is that the protection system is a non-repairable and single-component system, so to model this aspect of maintenance quality, it is considered that the components used in the replacements come from a heterogeneous population composed of weak items and strong items.

Weibull distributions (with a shape parameter β and scale parameter η) were used to characterize the probability distributions of X for weak items ($\beta_1 = 1.5$, $\eta_1 = 1$ years) and strong items ($\beta_2 = 2.5$, $\eta_2 = 4$ years) and the probability distribution of H ($\beta_3 = 1$, $\eta_3 = 0.25$ years). The resulting probability distribution of X can then be calculated as a mixture of the distributions of the weak items and the strong items, as indicated in eq. (50) (Scarf et al. 2009). The mixing parameter was estimated as $p = 0.10$.

$$f_x(x) = p.f_{x1}(x) + (1 - p).f_{x2}(x) \tag{50}$$

Regarding the inspection quality parameters, a low probability of false positives during the inspection was verified, whereas the probabilities of the two types of false negatives were more significant, with the probability of false negative type 1 being greater compared to the probability of false negative type 2. The following values were estimated for the inspection quality parameters: $w = 0.05$, $q_1 = 0.30$ and $q_2 = 0.10$.

The costs were established taking the cost of preventive replacement of a component when it is in good state as reference, so that $c_p = 1$ un. (the quantity is not specified in monetary units). The inequality ratios $c_f > c_d > c_p$ and $t_f > t_d > t_p$ are appropriate because it is reasonable to consider that the cost and time for system recovery tend to be larger the worse their condition is.

It is worth noting that in the case of protection systems, the higher penalty is associated with an unmet demand (UD) event and not necessarily with the replacement of a failed component because in this case, there is no interruption

Table 2 Specifying the model input parameters

Costs	Times	Demands
$c_v = 0.04$ un.	$t_v = 0$ (negligible)	$\mu = 2$ demands / year
$c_p = 1$ un.	$t_p = 3$ h	$C_{UD} = 30$ um
$c_d = 1.5$ un.	$t_d = 6$ h	$t_{ud} = 72$ h
$c_f = 3$ un.	$t_f = 12$ h	

of the main production process due to the occurrence of such failure. The values of the other parameters are shown in Table 2.

Finally, due to limitations in the availability of resources, the following constraints are added to the problem: $T \geq 1$ month, $M.T \geq 6$ months, i.e. the time interval between two inspections cannot be less than one month, and the scheduled preventive replacements can only occur with at least 6 months of component operation. Acceptable risk thresholds are not considered, so that the space of alternatives to be analysed covers the whole set of non-dominated policies.

6.2 *Results*

The results obtained are presented in three stages: first, an analysis of the impact of the different aspects of maintenance quality on the recommendation and performance of the optimal maintenance policy is made, considering the two criteria separately. Then, a multicriteria additive aggregation model, based on MAVT, is applied considering different combinations of scale constants, in order to verify how the maintenance policy recommendation varies. Finally, certain investment scenarios to improve the maintenance quality are evaluated considering a multicriteria perspective.

6.2.1 Assessing the Criteria Separately

To minimize the cost rate, it is necessary to find a balance between the cost of preventive maintenance actions and the cost due to failures and unmet demands. On the other hand, the minimization of the rate of unmet demands depends basically on the constraints of the problem because the more frequent the maintenance actions are carried out, the lower the rate tends to be.

For the case considered (base case), the policy that minimizes the cost criterion is $T = 0.163$ year and $M = 17$, which results in $C_\infty = 1.243$ un./year and $\lambda = 8206.10^{-3}$ UD/year. On the other hand, the policy that minimizes the risk criterion is $T = 0.083$ year (minimum T) and $M = 6$ (minimum M given minimum T), which results in $C_\infty = 2.852$ un./year and $\lambda = 1080.10$–3 UD/year. From these results, it is possible to delimit the CS regarding the non-dominated policies.

In Table 3, the results of an analysis of the model's sensitivity to the variation in the maintenance quality parameters are presented. These results aim to evaluate

Table 3 Optimum maintenance policies (for the criteria cost and risk) for different values of the maintenance quality parameters

	Quality parameters				Minimizing cost			
Case	p	w	q_1	q_2	M	T	C_∞*	λ
1	0.1	0.05	0.3	0.1	17	0.163	1.243	8.206
2	0.1	0.05	0.3	0	16	0.173	1.201	7.521
3	0.1	0.05	0	0.1	16	0.183	1.116	6.42
4	0.1	0	0.3	0.1	23	0.123	1.086	6.068
5	0	0.05	0.3	0.1	10	0.213	1.071	6.153
6	0.1	0.05	0.3	0.2	18	0.153	1.288	8.841
7	0.1	0.05	0.4	0.1	18	0.153	1.29	8.599
8	0.1	0.1	0.3	0.1	13	0.203	1.38	10
9	0.2	0.05	0.3	0.1	22	0.143	1.396	9.333
10	0.1	0	1	0.1	17	0.113	1.503	13
11	0.1	0	0	0	22	0.143	0.949	4.082
12	0	0	0	0	17	0.163	0.866	3.577
	Quality parameters				Minimizing risk			
Case	p	w	q_1	q_2	M	T	C_∞	λ*
1	0.1	0.05	0.3	0.1	6	0.083	2.852	1.08
2	0.1	0.05	0.3	0	6	0.083	2.847	0.913
3	0.1	0.05	0	0.1	6	0.083	2.835	0.645
4	0.1	0	0.3	0.1	6	0.083	2.588	1.1
5	0	0.05	0.3	0.1	6	0.083	2.757	0.142
6	0.1	0.05	0.3	0.2	6	0.083	2.857	1.28
7	0.1	0.05	0.4	0.1	6	0.083	2.859	1.27
8	0.1	0.1	0.3	0.1	6	0.083	3.146	1.07
9	0.2	0.05	0.3	0.1	6	0.083	2.948	2.04
10	0.1	0	1	0.1	6	0.083	2.661	2.935
11	0.1	0	0	0	6	0.083	2.568	0.544
12	0	0	0	0	6	0.083	2.491	0.081

T in years, C_∞ in un./year and λ in 10^{-3} UD/year
*An approximate optimization algorithm was used to find approximately optimal maintenance policies, and because of this the last digit of the values indicated for T is 3 in all case considered

the impact of these parameters on the definition and performance of the optimal maintenance policies, defined based on each criterion separately. The base case is highlighted in row 1 of the table, and the quality parameters are then varied from there.

It is observed that the optimal maintenance policy varies depending on the case when considering the cost criterion, whereas when considering the risk criterion, the recommendation remains the same: to carry out the maintenance actions as frequently as possible, with the maintenance quality having an impact on policy performance.

When only the cost criterion is considered, certain tendencies are observed: the greater the probabilities of false negative type 1 (q_1 – see cases 1, 2 and 6) and false negative type 2 (q_2 – see cases 1, 3 and 7), more frequent inspections (T minor) and preventive replacements ($M.T$ minor) should be in order to increase the probability of detection of a defect or failure before the occurrence of an unmet demand and compensate for possible negative effects on system availability. On the other hand, the greater the probability of false positives (w – see cases 1, 4 and 8), the less frequent the inspections (T major) should be to reduce the probability of an early replacement; however, less frequent inspections increase the vulnerability of the system, an effect that is compensated by the reduction in the time interval between scheduled preventive replacements ($M.T$ minor). Finally, with respect to the mixing parameter, it is observed that the greater the proportion of weak items (p – see cases 1, 5 and 9), the more frequent the inspections (T minor) should be to increase the probability of detection of weak items before the occurrence of a failure or an unmet demand. Interestingly, for higher p values, preventive replacement was recommended later (greater $M.T$) because the increased inspection frequency compensated for the negative effects of imperfect inspection on system availability.

Case 10 simulates a situation where the maintenance policy does not guide the preventive replacement in the case of defect identification. Because the defective state must be ignored in this situation, it was considered $q_1 = 1$ (as if the defect was not seen), and with respect to w, it was assumed that the probability of a false positive that indicated that the system had failed is zero. The cost resulting from the optimal configuration of this policy was 20% greater than that obtained for the optimal policy for the base case, which indicates that it was advantageous to adopt a policy that also indicated preventive replacement in case of a defect, even considering the possibility false positives (see cases 1, 4, 8 and 10). The same conclusion was obtained when considering the risk criterion.

When considering only the risk criterion, the following trends were observed: the higher the probability of a false positive (w), the lower the optimal risk obtained as a consequence of the increase of the early replacements. On the other hand, the greater the probability of false negatives type 1 and type 2 (q_1 and q_2), and the greater the proportion of weak items in the population of components (p), the greater the risk obtained. The greater impact was due to variations in the value of p.

Cases 11 and 12 show the mistakes that could be made when such aspects of imperfect maintenance were observed in practice but ignored in the mathematical modelling of maintenance policy. Interestingly, the optimal policies indicated for case 12 were the same as those obtained for the base case, but this result should not be expected. However, by ignoring the effect of imperfect maintenance, cost and risk assessments could be quite misleading, with negative impacts on maintenance decision-making.

The divergences between the optimal maintenance policies obtained for each criterion confirmed the need for a multicriteria model for a better evaluation of maintenance policies.

6.2.2 Using a Multicriteria Model to Define the Maintenance Policy

A multicriteria model based on MAVT was used to define the maintenance policy that presents the best compromise relationship between the criteria cost and risk for different scenarios. Simulated data were used to evaluate how the recommendations vary according to different preference structures.

To initiate the elicitation of the DM's preferences, it was necessary to delimit the CS referring to the set of alternatives that would be analysed in the decision problem. As indicated earlier, the alternatives that actually matter have the following properties: they are non-dominated and met the limits of acceptable performance in all dimensions of analysis. In the present application, the set of alternatives analysed is continuous, so the problem of maximizing the multi-attribute value took the form of a classical optimization problem.

As there were no acceptable risk constraints, the CS considered in this application refers to the set of non-dominated alternatives and was delimited from the optimization of each criterion separately: let (C_∞*, λ^0) be the performance of the policy that minimizes the cost (policy 1) and ($C_\infty{}^0$, λ*) the performance of the policy that minimizes the risk (policy 2), then any alternative with performance ($C_\infty{}^x$, λ^x) such that $C_\infty{}^x \geq C_\infty$* (because no viable alternative had a cost lower than C_∞*) and $\lambda^x > \lambda^0$ is dominated by policy 1, so the latter does not matter to the DM. The same logic could be applied considering the risk criterion as a reference; then, it was concluded that the CS for the cost and risk criteria could be represented by the intervals [C_∞*, $C_\infty{}^0$] and [λ*, λ^0].

Linear value functions were considered for the intracriterion evaluations, taking the scale [0,1] as a reference. A linear value function indicates that the marginal gain related to the considered criterion is constant. The CS limits (presented in the previous subsection) and the one-dimensional value functions considered are shown in Table 4.

For the multi-attribute value function, it was considered that there is preferential independence between the criteria, so that this function could be represented in the additive form, according to eq. (51). Because the cost and risk are functions of the maintenance policy decision variables (T and M), the representation presented in eq. (52) is also adequate:

$$V\left(C_\infty\left(T,M\right),\lambda\left(T,M\right)\right)=k_c.V_c\left(C_\infty\left(T,M\right)\right)+k_r.V_r\left(\lambda\left(T,M\right)\right) \tag{51}$$

Table 4 CS limits and one-dimensional value functions for the multicriteria model

Criterion	Lower	Upper	Value functions
Cost ($C_\infty(T,M)$)	1.243	2.852	$V_c(C_\infty(T,M)) = 1.773 - (0.622).C_\infty(T,M)$
Risk ($\lambda(T,M)$)	$(1.080)0.10^{-3}$	$(8.206)0.10^{-3}$	$V_r(\lambda(T,M)) = 1.152 - (140.331).\lambda(T,M)$

C_∞ in un. /year and λ in UD/year

Table 5 Results of multicriteria evaluation

Case	k_c	k_r	M	T	$V^*(T,M)$	$C_\infty(T,M)$	$\lambda(T,M)$
1	0.1	0.9	8	0.083	0.902	2.39	1.295
2	0.2	0.8	16	0.083	0.857	1.734	1.813
3	0.3	0.7	20	0.083	0.840	1.619	1.999
4	0.4	0.6	24	0.083	0.831	1.552	2.174
5	0.5	0.5	28	0.083	0.828	1.512	2.341
6	0.6	0.4	33	0.083	0.829	1.484	2.535
7	0.7	0.3	30	0.093	0.842	1.405	3.163
8	0.8	0.2	23	0.113	0.872	1.319	4.306
9	0.9	0.1	20	0.133	0.921	1.267	5.754

T in years, C_∞ in un./year and λ in 10^{-3} UD/year
*An approximate maximization algorithm was used to determine the maintenance policies with a maximum value, and because of this the last digit of the indicated values for T is 3 in all cases considered

$$V(T, M) = k_c.V_c(T, M) + k_r.V_r(T, M) \tag{52}$$

The scale constants k_c and k_r are calculated from trade-offs between the criteria and indicate how much the loss of performance in one criterion can be compensated for by the improvement in another criterion. Different combinations of the scale constants k_c and k_r were considered to observe how the obtained recommendations vary. The results are shown in Table 5.

In Table 5, it is observed that in all cases, the recommended value of T was significantly lower compared to the policy that minimizes the cost criterion (see Table 3): even in the case where $k_c = 0.9$, the value of T recommended was approximately 20% lower, whereas for $k_r \geq 0.4$, the lowest possible value ($T = 0.083$ year) was recommended. These results demonstrated the strong impact of considering the risk criterion in the decision, which tends to result in the recommendation of significantly more conservative policies. This phenomenon could also be observed in the reduction in the value of M indicated when increasing k_r in cases 1 to 6: such variation in this decision variable occurred as a consequence of the restriction applied to T.

The results presented in Table 5 demonstrate the importance of implementing a structured protocol to elicit the DM's preferences because the policy that presented the best compromise relationship between the criteria could vary significantly depending on the combination of the scale constants (compare cases 1 and 9, for example).

Once the DM's preferences were mapped, the scale constants were calculated, and an initial compromise solution was obtained, it is important to analyse the sensitivity of the model to verify the robustness of the recommended solution in the face of possible inconsistencies that may occur in the elicitation process. One possibility for this analysis is to vary the scale constants and evaluate how the results respond to such a disturbance. In the example, it is observed that for k_r between 0.5

and 0.8 (cases 2 to 5), the recommended maintenance policies differ little, with small variations in the value of M, which indicates that in this range of values of the scale constants, the effectiveness of the recommended solution was little affected by inconsistencies that result in small differences in their actual values.

6.2.3 Evaluating the Investment in Higher Quality Maintenance

The maintenance quality could be improved in different ways: by training maintenance teams, by adopting more effective technologies, among other examples, and improvement of the maintenance quality usually requires financial investment. A multicriteria model could also be used to evaluate possible investment scenarios in maintenance quality, indicating which strategy was most appropriate according to the DM's preferences. In this application, a model based on MAVT was used.

For the construction of the multicriteria model, it was initially necessary to list the possible investment scenarios, indicating the improvement in the maintenance quality parameters and the associated cost increase. One way to accomplish the latter is to consider that such an investment entails increasing the cost associated with the maintenance actions, for example: an improvement in the inspection quality parameters (reduction in w, q_1, q_2) could be achieved with an investment that resulted in an increase in unit cost of an inspections (c_v).

Once the possible investment scenarios had been identified, it is necessary to delimit the CS associated with each of them, according to the procedure presented previously. If, for a certain scenario, the CS referring to the non-dominated alternatives did not comply with the limits of acceptable risk, this scenario should not be considered. The CS to be considered for the general analysis should then be defined in a way that includes the CSs related to the scenarios addressed.

The investment scenarios considered for this application are presented in Table 6. The original scenario, which is maintained when no investment is made, is highlighted in the first line, and the other scenarios were related to investments in the inspection quality (scenario 2), replacement quality (scenario 3) or both (scenario 4).

For each scenario, the cost and risk criteria were optimized separately, as presented in Table 7, where the values considered to delimit the CS for the general analysis are highlighted.

It is observed that for the cost criterion, only the scenario of investment in the inspection quality (scenario 2) was more advantageous compared to the initial

Table 6 Investment scenarios. Costs in un

Scenario	p	w	q_1	q_2	c_v	c_p	c_d	c_f
1 (base)	0.1	0.05	0.3	0.1	0.04	1	1.5	3
2	0.1	0.01	0.1	0	0.08	1	1.5	3
3	0.03	0.05	0.3	0.1	0.04	1.2	1.8	3.6
4	0.03	0.01	0.1	0	0.08	1.2	1.8	3.6

Table 7 Optimum cost and risk for each investment scenario

Scenario	M	T	C_∞*(T,M)	$\lambda(T,M)$	M	T	$C_\infty(T,M)$	λ*(T,M)
1 (base)	17	0.163	1.243	**8.206**	6	0.083	2.852	1.08
2	13	0.203	**1.238**	7.428	6	0.083	3.102	0.655
3	13	0.193	1.268	7.610	6	0.083	3.244	0.423
4	11	0.233	1.26	7.300	6	0.083	**3.461**	**0.262**

T in years, C_∞ in un./year and λ in 10^{-3} UD/year
*An approximate optimization algorithm was used to determine approximately optimal maintenance policies, and because of this the last digit of the values indicated for T is 3 in all cases considered

Table 8 CS limits and one-dimensional value functions for the multicriteria model

Criterion	Lower	Upper	Value functions
Cost ($C_\infty(T,M)$)	1.238	3.461	$V_c(C_\infty(T,M)) = 1.557 - (0.450).C_\infty(T,M)$
Risk ($\lambda(T,M)$)	$(0.262)0.10^{-3}$	$(8.206)0.10^{-3}$	$V_r(\lambda(T,M)) = 1.033 - (125.878).\lambda(T,M)$

C_∞ in un./year and λ in UD/year

Table 9 Results of the multicriteria evaluation of investment scenarios

Case	k_c	k_r	Best invest. Scenario	M	T	$V(T,M)$	$C_\infty(T,M)$	$\lambda(T,M)$
1	0.2	0.8	4	16	0.083	0.893	2.03	0.621
2	0.5	0.5	4	19	0.103	0.847	1.611	1.368
3	0.8	0.2	4	15	0.153	0.887	1.344	3.249

T in years, C_∞ in un./year and λ in 10^{-3} UD/year
*An approximate maximization algorithm was used to determine the maintenance policies with a maximum value, and because of this approach the last digit of the indicated values for T is 3 in all cases considered

scenario, but with a small difference. For the risk criterion, as expected, the higher the effect of the investment, the better the result. No dominance relationships were observed between the scenarios, so it is appropriate to consider all of them in the multicriteria evaluation.

Linear value functions were considered for the intracriterion evaluations, taking the scale [0,1] as a reference. The CS limits and the one-dimensional value functions considered are presented in Table 8.

For the multi-attribute value function, it was considered that there exists preferential independence between the criteria, so that this function could be represented in the additive form. Different combinations of the scale constants k_c and k_r, representing different preference structures, were considered to evaluate how the obtained recommendation varies. The results are shown in Table 9.

In all cases listed in Table 9, the investment scenario 4 obtains the best compromise relationship between the criteria, which again demonstrates the impact that the consideration of the risk criterion has on the recommendations. Although in all cases there is agreement on the recommendation of the investment to improve the inspection and replacement quality, the recommended maintenance policies are significantly different, which demonstrates the importance of the execution of a structured protocol for elicitation of the DM's preferences in order to avoid wrong decisions.

7 Conclusion

This chapter presents some contributions to the process of evaluating maintenance policies for protection systems, which were motivated by a case study concerning a real system.

First, a mathematical model is proposed for a hybrid inspection and preventive replacement policy applied to a protection system, which considers the possibility of errors in the identification of the state of the system during an inspection (false positives and false negatives). The mathematical model is innovative and developed based on the delay-time concept but with a different approach than that used in previous studies presented in the literature, which allows us to more accurately estimate the cost and risk of unmet demands as a function of the decision variables of the maintenance policy.

In many contexts, including the context that motivated the development of this work, maintenance planning addresses multiple and sometimes conflicting objectives; thus, a multicriteria approach based on MAVT is proposed to more appropriately treat the multidimensionality inherent to the problem.

With the numerical application, it was possible to observe the suitability and importance of the proposed approach to the presented problem. From a broad analysis of the results obtained with the application of the model, the DM can obtain certain answers to important questions that arise during the decision process, for example: is it worth recommending preventive replacement when the inspection indicates a defective state? What are the effects of the different aspects of maintenance quality on the indication of the optimal maintenance policy for different criteria? Is it worth investing in higher quality maintenance? If so, how should this investment be performed? When obtaining answers to questions such as this, the DM has important information that can result in better decisions.

Finally, the results demonstrate the importance of executing a structured protocol for the construction of the multicriteria model because the recommendations obtained can vary significantly depending on the structure of the preferences of the DM.

Acknowledgements The work of the authors has been supported by CAPES (Coordination for the Improvement of Higher Education Personnel), CNPQ (Brazilian National Council) and FACEPE (Foundation for Science and Technology of the State of Pernambuco).

References

Alberti AR, Cavalcante CAV, Scarf P, Silva ALO (2018) Modelling inspection and replacement quality for a protection system. Rel Eng Syst Saf 277:145–153

Ascher H, Feingold H (1984) Repairable systems reliability. Marcel Dekker, New York

Aven T (2012) The risk concept—Historical and recent development trends. Rel Eng Syst Saf 99:33–44

Baah K, Dubey B, Harvey R, Mcbean E (2015) A risk-based approach to sanitary sewer pipe asset management. Sci Total Environ 505:1011–1017

Belton V, Stewart TJ (2002) Multiple criteria decision analysis: An integrated approach. Kluwer Academic Publishers, Dordrecht

Berger JO (1985) Statistical decision theory and Bayesian analysis. Springer-Verlag, New York

Berrade MD, Cavalcante CAV, Scarf PA (2012) Maintenance scheduling of a protection system subject to imperfect inspection and replacement. Eu J Operation Res 218:716–725

Berrade MD, Scarf PA, Cavalcante CAV (2015) Some insights into the effect of maintenance quality for a protection system. IEEE Trans on Reliab 64(2):661–672

Cavalcante CAV, Scarf PA, Almeida AT (2011) A study of a two-phase inspection policy for a preparedness system with a defective state and heterogeneous lifetime. Rel Eng Syst Saf 96:627–635

Christer AH (1999) Developments in delay time analysis for modelling plant maintenance. J Oper Res Soc 50:1120–1137

DE Almeida AT, Cavalcante CAV, Alencar MH, Ferreira RJP, DE Almeida-Filho AT, Garcez TV (2015a) Multicriteria and multiobjective models for risk, reliability and maintenance decision analysis. Springer, New York

DE Almeida AT, Ferreira RJP, Cavalcante CAV (2015b) A review of the use of multicriteria and multi-objective models in maintenance and reliability. IMA J Manag Math 00:1–23

Diogo AF, Barros LT, Santos J, Temido JS (2018) An effective and comprehensive model for optimal rehabilitation of separate sanity sewer systems. Sci Total Environ 612:1042–1057

Grigg NS (2012) Water, wastewater, and Stormwater infrastructure management. CRC Press, Boca Raton

Humes D, Stolberg CG (eds) (2006) Submersible sewage pumping systems (SWPA) handbook. Submersible wastewater pump association (SWPA), Sheridan Road

Jia X, Christer AH (2002) A periodic testing model for a preparedness system with a defective state. IMA J Manag Math 13:39–49

Jiang R, Ji P (2002) Age replacement policy: a multi-attribute value model. Rel Eng Syst Saf 76:311–318

Keeney RL, Raiffa H (1976) Decisions with multiple objectives: Preferences and value Tradeoffs. Cambridge University Press, Cambridge & New York

Scarf PA, Cavalcante CAV (2012) Modelling quality in replacement and inspection maintenance. Int J Produc Economic 135:372–381

Scarf PA, Cavalcante CAV, Dwight RA, Gordon P (2009) An age-based inspection and replacement policy. IEEE Trans on Reliab 58(4):641–648

Tijms HC (1994) Stochastic models: An algorithmic approach. John Wiley & Sons, New Jersey

Vaurio JK (1999) Availability and cost functions for periodically inspected preventively maintained units. Rel Eng and Syst Saf 63:133–140

A Multicriteria Model to Determine Maintenance Policy for a Protection System Subject to Imperfect Maintenance

Alexandre Ramalho Alberti and Cristiano Alexandre Virgínio Cavalcante

1 Introduction

Protection systems usually remain inactive during normal operating periods of the main production system, and are required to function when facing specific demands, such as emergency events. Consequently, such systems have hidden failures that can only be detected through inspections or at a demand event. Alarms, isolation valves, fire protection systems and emergency brakes are examples of such systems (Vaurio 1999; Jia and Christer 2002; Cavalcante et al. 2011). It is generally observed that the occurrence of unmet demands due to an unavailable protection system can have very negative effects that potentially cross multiple dimensions (financial, human, environmental, etc.), which justifies the concern on appropriate maintenance planning for this type of system.

Usually, the maintenance planning process can be divided into three main steps, which in brief consist of obtaining answers to the following questions. (1) Which maintenance actions are appropriate for the system under study? (2) How often should these actions be carried out? (3) What resources should be used? These questions are usually answered in this order and hierarchically (De Almeida et al. 2015a). However, the impact of the quality of resources involved in the performance of the maintenance policy is sometimes overlooked in this process (Alberti et al. 2018).

A. R. Alberti (✉)
RANDOM – Research Group on Risk and Decision Analysis in Operations and Maintenance, Universidade Federal de Pernambuco, Recife, Brazil
e-mail: a.r.alberti@random.org.br

C. A. V. Cavalcante
INSID – National Institute of Information and Decision Systems, Universidade Federal de Pernambuco, Recife, Brazil
e-mail: c.a.v.cavalcante@random.org.br

A. T. de Almeida et al. (eds.), *Multicriteria and Optimization Models for Risk, Reliability, and Maintenance Decision Analysis*, International Series in Operations Research & Management Science 321, https://doi.org/10.1007/978-3-030-89647-8_10

To incorporate this element into the maintenance decision-making process, some studies have presented mathematical models that allow one to consider certain aspects of maintenance quality. For example, Scarf et al. (2009) model the component replacement quality by considering that it comes from a heterogeneous population composed of weak items, which have low reliability and are susceptible to early failures, and strong items. The source of this heterogeneity may be the variation in the quality of the components' manufacturing or recycling, variations in the quality of the installation service, or both. Berrade et al. (2012, 2015) and Alberti et al. (2018) share the same notion for the models applied to protection systems. To model the inspection quality, some studies consider the possibility of misclassification in the identification of the state of a component (false positives and false negatives). Okumura et al. (1996), Berrade et al. (2012, 2015) and Alberti et al. (2018) consider that the inspections present a constant probability of misclassification. Conversely, Driessen et al. (2017) consider that this probability can be variable. Other studies consider the possibility that inspections themselves may induce defects (Scarf and Cavalcante 2012; Alberti et al. 2018) or failures (Flage 2014) in the system. Mathematical models seek to represent the reality in an approximate way and are not able to involve all its aspects, but they can be very useful tools to guide the maintenance decision-making process.

In this chapter, we present a mathematical model for a maintenance policy applied to a protection system that incorporates an important aspect of inspection quality: the probability of defect induction. The delay-time concept (Christer 1999) is used to model the deterioration and failure process of a single component protection system. It is an innovative model that allows to more accurately estimate the cost resulting from the adoption of a maintenance policy, as well as the rate of occurrence of unmet demands, which can be understood as a risk measure.

We verified that the model is very sensitive to the probability of defect induction at an inspection. A small variation in this parameter can result in substantial variations in the performance of the maintenance policy, as well as in the recommendation of the best policy to be adopted. In addition, its estimation is a difficult process and involves a certain degree of uncertainty. Therefore, in order to ensure the appropriate treatment of the problem under these conditions, this parameter is modelled through a probability distribution.

A multicriteria model for evaluating maintenance policies is then presented. It considers two criteria: the expected cost rate (cost) and the rate of unmet demands (risk) that result from the adoption of a maintenance policy in the long run. Depending on the scenario, the losses in non-financial dimensions (human, environmental, etc.) are proportional to the rate of unmet demands. Thus, the multicriteria model allows for an appropriate treatment of the multidimensionality characteristic of the problem. Multi-attribute utility theory (MAUT) was considered for the model development since it is suitable for scenarios where there is uncertainty about the state of nature, and the parameters of the problem can be described by probability distributions (Keeney and Raiffa 1976), as is the case of the presented problem.

This chapter is organized as follows. The next section briefly presents the problem statement, and the following section presents the notations used throughout this text. In Sect. 4, the mathematical model for the maintenance policy applied to a protection system is developed taking into consideration the previously mentioned aspect of inspection quality, and in Sect. 5, a framework for the construction of the multi-criteria model and obtaining results is presented. In Sect. 6, a numerical application is presented using simulated data, and finally, in the last section, some conclusions are pointed out.

2 Problem Statement

The level of availability of a protection system indicates how well it is able to meet demand events. In this sense, in order to ensure a satisfactory level of availability for protection systems, several papers propose the adoption of periodic inspection policies (Jia and Christer 2002).

Inspections have the objective of obtaining information about the system's state without affecting its condition. However, as indicated previously, an inspection can induce defects in the system, thereby negatively affecting its reliability. Therefore, some questions arise: What is the impact of the inspection quality on the maintenance policy's performance? How can inspection quality influence the definition of the most appropriate policy? Finally, how can we properly approach the multidimensionality of the problem in some application contexts? This chapter addresses these questions.

Regarding equipment maintenance, a set of maintenance procedures does not necessarily guarantee any improvement in performance. Oftentimes, contrary to expectations, it may even have a negative impact that can reduce a system initial performance or increase the chances of failure. In this sense, it is of vital importance to observe the main aspects that can result in maintenance work causing performance degradation. With respect to inspections in particular, examples have been presented in the literature demonstrating how inspecting equipment can cause damages to the system (Scarf and Cavalcante 2012; Flage 2014; Alberti et al. 2018). Alberti et al. (2018) point out that this situation is not always easy to observe, and even if the probability of defect induction is low, it can have a significant impact on the system performance. This demonstrates the importance of the methods for investigating trends that affect system performance, such as snapshot modelling (Christer and Whitelaw 1983) and the use of models that consider this aspect of maintenance quality.

3 Notation

The notation presented is used throughout this chapter.

- Decision Variables:

  M – maximum number of inspections until the preventive replacement.
  T – time interval between two consecutive inspections.

- Notation for the development of the maintenance policy's mathematical model:

  X – time until defect arrival – non-negative random variable.
  H – sojourn time in the defective state (delay-time) – non-negative random variable.
  Z – time between the occurrences of two consecutive demands – non-negative random variable.
  f_x, F_x, R_x – for X, functions of the probability density, cumulative probability distribution and reliability, respectively.
  f_h, F_h, R_h – for H, functions of the probability density, cumulative probability distribution and reliability, respectively.
  f_z, F_z, R_z – for Z, functions of the probability density, cumulative probability distribution and reliability, respectively.
  μ – rate of demands.
  p – probability of defect induction at the inspection.
  f_p – probability density function of the distribution of p.
  c_v – cost of an inspection.
  c_p, c_d, c_f – replacement costs when the component is good, defective and failed, respectively.
  CUD – cost of an unmet demand.
  t_v – time required to perform an inspection.
  t_p, t_d, t_f – time for performing the replacement when the component is good, defective and failed, respectively.
  t_{ud} – time to normalize the system when an unmet demand occurs.
  EC, EL – expected cost and expected length of a renewal cycle, respectively.
  ρ – probability of a renewal cycle ending in an unmet demand.
  C_∞ – expected cost per unit of time in the long run.
  λ – rate of unmet demands.

- Notation for multicriteria model development:

  A – alternative/solution – in this context, a maintenance policy is defined by the combination of decision variables T and M.
  A_j – performance of alternative A for criterion j.
  U, EU – respectively, multi-attribute utility and expected multi-attribute utility.
  U_j, EU_j – respectively, one-dimensional utility for criterion j and expected one-dimensional utility for criterion j.
  k_j – scale constant for criterion j.
  c, r – indexes for the cost (cost rate) and risk (rate of unmet demands) criteria, respectively.

4 The Proposed Mathematical Model

To support the process of evaluating maintenance policies in contexts such as the one presented, a delay-time model was developed for a protection system subject to the demands that occur according to an homogenous Poisson process.

A single component protection system is considered, which is composed of a component and a socket that together play an operational function (Ascher and Feingold 1984). The component deteriorates over time, and the operating status of the system depends on its state. If the component is in a good or defective state, the system is operational, and if the component is failed, the system is unable to fulfil its function. With the replacement of the component, the system as a whole is renewed.

A mixed inspection and preventive replacement policy (MT policy) is proposed, similar to the one proposed by Vaurio (1999), which guides the performance of M inspections with a time interval T between the beginning of two consecutive inspections. When an inspection indicates that the component is defective or failed, it is recommended to replace it with a new unit, and at the M-th inspection the component is replaced regardless of its state. It is a flexible policy format, and its special cases are the pure inspection policy (policy PI, $M = \infty$) (inspections will occur until a defect or failure occurs or a demand identifies a failure without prior planning to replace the component) and the policy of pure preventive replacement or age-based replacement (policy ABR, $M = 1$) (at the first inspection, the component is replaced regardless of its state).

Finally, to model the quality of the inspection, it is considered that there is the probability of defect induction when performing an inspection when the component is in a good state.

4.1 *Model Assumptions*

The following assumptions are considered for model development:

1. The component replacement is performed when an inspection indicates a defective or failed state, when an unmet demand occurs, or at the M-th inspection, whichever occurs first.
2. At the end of the renewal cycle, an inspection is also carried out to check the component's state and record the information.
3. The distributions F_x, F_h and F_z are known and statistically independent.
4. At an inspection performed when the component is good, there is a probability p of inducing the defective state.
5. There are no misclassification errors regarding the identification of the component's state.
6. The time and costs of maintenance actions are constant and known.
7. The time spent on maintenance actions is not configured as downtime for the protection system. Here, it is considered that during the maintenance actions,

there is an interruption in the process of demands' arrival, or that demands are met through alternative solutions.
8. Demands occur according to a homogeneous Poisson process with a known rate. Hence, f_z is an exponential distribution with the characteristic parameter μ.
9. $t_v << T$.

A discussion on assumptions is required. Assumption 7 is particularly true for examples where, in order to maintain the protection system, it is necessary to interrupt the main production system operation, as it is the case of emergency brakes and isolation valves (Alberti et al. 2018). Moreover, this is a good approximation when t_p, t_d, t_f and $t_{cud} << \mathrm{T}$.

The mathematical models for the protection systems previously presented in the literature are based on the characterization of the system's expected renewal cycle according to the decision variables of the maintenance policy, and a penalty for the downtime incurred with a cost rate proportional to the rate of demands and the cost of an unmet demand (Vaurio 1999; Cavalcante et al. 2011; Berrade et al. 2012, 2015; Alberti et al. 2018). Different from these models, the model presented in this chapter assumes that the component is replaced in the case of unmet demand, which is a more realistic scenario and allows for estimating the rate of unmet demands with greater accuracy. This assumption can be incorporated into an analytical model once the assumptions 7 and 8 are considered.

4.2 Model Development

The model was developed from the enumeration of all possible renewal cycle scenarios that may occur under the presented conditions (represented graphically in Fig. 1). It is an exhaustive and mutually exclusive set of scenarios such that the sum of the probabilities of occurrence of these scenarios must be equal to 1 regardless of the decision variables' values.

For each scenario, we present the calculation of its probability of occurrence ($P_{scenario}$) and, from this, the expressions to calculate the expected values for a renewal cycle of the following measures: cost ($EC_{scenario}$) and length of the renewal cycle ($EL_{scenario}$). After the scenarios are detailed, the expressions are grouped in order to model the performance of the maintenance policy as a function of the decision variables.

4.2.1 Mathematical Development of the Scenarios

Scenario 1 – the defect arrives naturally at the *i*-th interval between inspections and is identified at the subsequent inspection:

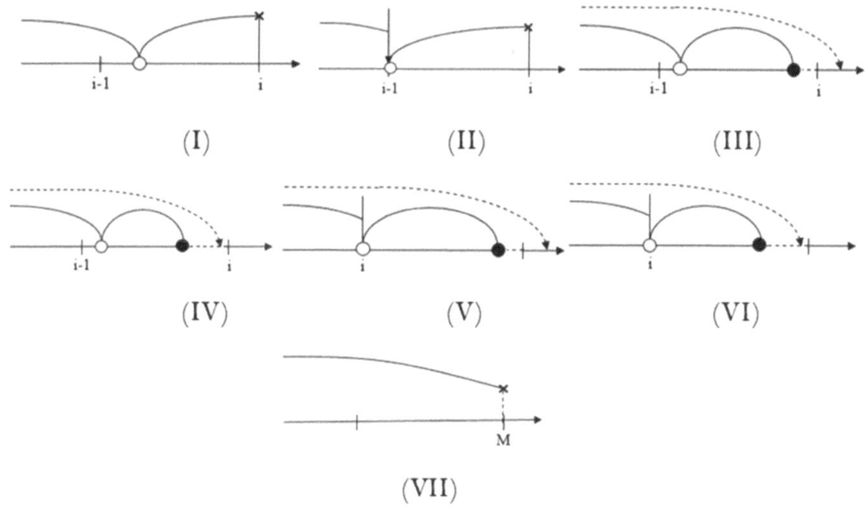

Fig. 1 Graphical representation of the possible scenarios of renewal cycles. ∘ defect arrival, • failure, and — demand arrival process

$$P_1(T, i) = (1 - p)^{i-1}. \int_{(i-1).T}^{i.T} f_x(x).R_h(i.T - x)\, dx \tag{1}$$

$$EC_1(T, M) = \sum_{i=1}^{M} [(i.c_v + c_d).P_1(T, i)] \tag{2}$$

$$EL_1(T, M) = \sum_{i=1}^{M} [(i.T + t_d).P_1(T, i)] \tag{3}$$

Scenario 2 – the defect is induced at the i-th inspection and is identified at the subsequent inspection:

$$P_2(T, i) = (1 - p)^{i-1}.p.R_x(i.T).R_h(T) \tag{4}$$

$$EC_2(T, M) = \sum_{i=1}^{M-1} [((i + 1).c_v + c_d).P_2(T, i)] \tag{5}$$

$$EL_2(T, M) = \sum_{i=1}^{M-1} [((i + 1).T + t_d).P_2(T, i)] \tag{6}$$

Scenario 3 – the defect arrives naturally at the i-th interval between inspections. The failure occurs and is identified at the subsequent inspection before the occurrence of a demand:

$$P_3(T,i) = (1-p)^{i-1}. \int_{(i-1)T}^{i.T} f_x(x). \int_0^{i.T-x} f_h(h).R_z[i.T-(x+h)]\,dhdx \tag{7}$$

$$EC_3(T,M) = \sum_{i=1}^{M} \left[\left(i.c_v + c_f\right).P_3(T,i)\right] \tag{8}$$

$$EL_3(T,M) = \sum_{i=1}^{M} \left[\left(i.T + t_f\right).P_3(T,i)\right] \tag{9}$$

Scenario 4 – the defect arrives naturally at the i-th interval between inspections. The failure occurs and is identified at the occurrence of a demand:

$$P_4(T,i) = (1-p)^{i-1}. \int_{(i-1)T}^{i.T} f_x(x). \int_0^{i.T-x} f_h(h).F_z[i.T-(x+h)]\,dhdx \tag{10}$$

$$EC_4(T,M) = \sum_{i=1}^{M} \left[\left(i.c_v + c_f + C_{UD}\right).P_4(T,i)\right] \tag{11}$$

$$EL_4(T,M) = \left(M.T + t_f\right).P_4(T,M) \tag{12}$$

Scenario 5 – the defect is induced at the i-th inspection. The failure occurs and is identified at the subsequent inspection before a demand occurs:

$$P_5(T,i) = (1-p)^{i-1}.p.R_x(i.T). \int_0^T f_h(h).R_z(T-h)\,dh \tag{13}$$

$$EC_5(T,M) = \sum_{i=1}^{M-1} \left[\left((i+1).c_v + c_f\right).P_5(T,i)\right] \tag{14}$$

$$EL_5(T,M) = \sum_{i=1}^{M-1} \left[\left((i+1).T + t_f\right).P_5(T,i)\right] \tag{15}$$

Scenario 6 – the defect is induced at the i-th inspection. The failure occurs and is identified at the occurrence of a demand:

$$P_6(T, i) = (1-p)^{i-1}.p.R_x(i.T).\int_0^T f_h(h).F_z(T-h)\,dh \tag{16}$$

$$EC_6(T, M) = \sum_{i=1}^{M-1}\left[\left((i+1).c_v + c_f + C_{UD}\right).P_6(T, i)\right] \tag{17}$$

$$EL_6(T, M) = \sum_{i=1}^{M-1}\left[\begin{array}{l}(1-p)^{i-1}.p.R_x(i.T). \\ \int_0^T f_h(h).\int_0^{T-h} f_z(z).\left(i.T + h + z + t_f + t_{ud}\right)dzdh\end{array}\right] \tag{18}$$

Scenario 7 – the component is preventively replaced at $M.T$:

$$P_7(T, M) = (1-p)^{M-1}.R_x(M.T) \tag{19}$$

$$EC_7(T, M) = \left(c_p + M.c_v\right).P_7(T, M) \tag{20}$$

$$EL_7(T, M) = \left(M.T + t_p\right).P_7(T, M) \tag{21}$$

4.2.2 Joining the Scenarios

The expected value of a measure of interest EQ for a renewal cycle is equal to the sum of the contributions of each scenario:

- If $M = 1$:

$$EQ(T, M) = EQ_1(T, M) + EQ_3(T, M) + EQ_4(T, M) + EQ_7(T, M) \tag{22}$$

- If $M > 1$:

$$EQ(T, M) = \sum_{j=1}^{7} EQ_j(T, M) \tag{23}$$

The EQ measure can be the cost (EC) or the length of the renewal cycle (EL).

4.2.3 Calculating the Cost Rate (Cost Criterion)

According to the reward renewal theorem (Tijms 1994), the expected cost per unit of time in the long run can be calculated according to Eq. (24).

$$C_{\infty}(T,M) = \frac{EC(T,M)}{EL(T,M)} \tag{24}$$

4.2.4 Calculating the Rate of Unmet Demands (Risk Criterion)

The probability that a renewal cycle ends in an unmet demand can be calculated as the sum of the probabilities of occurrence of the scenarios that end in this way:

- If M = 1:

$$\rho(T,M) = \sum_{i=1}^{M} P_4(T,i) \tag{25}$$

- If $M > 1$:

$$\rho(T,M) = \sum_{i=1}^{M} P_4(T,i) + \sum_{i=1}^{M-1} P_6(T,i) \tag{26}$$

As the renewal cycles are statistically independent of each other, then, in agreement with the observations of Scarf et al. (2009), it is possible to state that the time intervals between unmet demands are approximately exponentially distributed with the rate λ calculated according to Eq. (27):

$$\lambda(T,M) = \frac{\rho(T,M)}{EL(T,M)} \tag{27}$$

In some contexts, the occurrence of an unmet demand may incur risks to the health of workers and the environment. For example, the occurrence of a demand not met by an emergency brake of handling devices for suspended cargo may lead to human losses. In another example, the occurrence of a demand not met by a containment system for chemical leaks can lead to serious environmental losses. Once the expected losses for an unmet demand are estimated, the rate $\lambda(T,M)$ can be used to calculate the expected losses per unit of time in the long run for non-financial dimensions. The rate $\lambda(T,M)$ can be understood as a measure of risk (Aven 2012).

5 The Multicriteria Model

Multicriteria models present a major contribution to decision-making in cases involving multiple, sometimes conflicting, objectives that cannot be transformed into a single evaluation metric. This situation is often verified in the context of service production systems, and in contexts where the operation of the system involves risks to human life and the environment (De Almeida et al. 2015b; Brito and De Almeida 2009).

In the previous section, a mathematical model is presented for a maintenance policy applied to a protection system subject to defect induction in inspections. It is expected that the higher the cost of an unmet demand (*CUD*), the more conservative the optimal maintenance policy defined based on the cost criterion ($C_{\infty}(T,M)$), thus reducing the risk of unmet demands ($\lambda(T,M)$). However, the *CUD* value may not be well representative of the consequences of unmet demand events, which may have impacts in multiple dimensions. In this case, a multicriteria model can help to evaluate the compromise relationships between different criteria considered in the analysis.

For the development of the multicriteria model, it is necessary to comply with the protocol for elicitation and evaluation of the decision maker's preferences (DM's preferences) regarding the problem under analysis. In this work, the multi-attribute utility theory (MAUT) is considered for the model development. MAUT has wide applicability in maintenance problems, especially when the DM presents compensatory rationality between the criteria. MAUT presents an appropriate axiomatic structure to deal with uncertain situations, such as those treated in this work (De Almeida et al. 2015a).

A framework for constructing the multicriteria model for decision support in the evaluation of maintenance policies for protection systems subject to imperfect inspection is presented in Fig. 2.

The first step of the framework is the characterization of the DM (or group of DMs) and the other actors in the decision-making process. In the industrial context, the DM may be a maintenance manager, a production manager or even a unit leader. The DM exerts influence over risk management since its preference relationships are considered for the construction of the multicriteria model (Brito and De Almeida 2009). Professionals in specific areas, such as environmental management and occupational safety, can act as specialists that provide important information and assist in the decision-making process.

The second step is an assessment of the operational and failure aspects of the protection system. A complete description of the system under study, including the details of its context of use, the situations that characterize a demand, and its process of deterioration and failure, is an important input for the process of defining the maintenance strategies and alternative strategies for risk management. The third step consists of the definition of the alternatives to maintenance, which will be evaluated by the multicriteria model. In this problem, a continuous set of alternatives is considered, since the alternatives are any policy obtained from the combination of *M* and *T*.

Next, the DM should assess the risk dimensions associated with the problem, which should be considered in the multicriteria model. One way to obtain the appropriate criteria for a problem involving multiple objectives is to elaborate a hierarchy of objectives (Keeney and Raiffa 1976). Using this method, the hierarchy shown in Fig. 2 was obtained. Here, the rate of unmet demands is considered directly, but this measure can be used to calculate the expected loss rates in

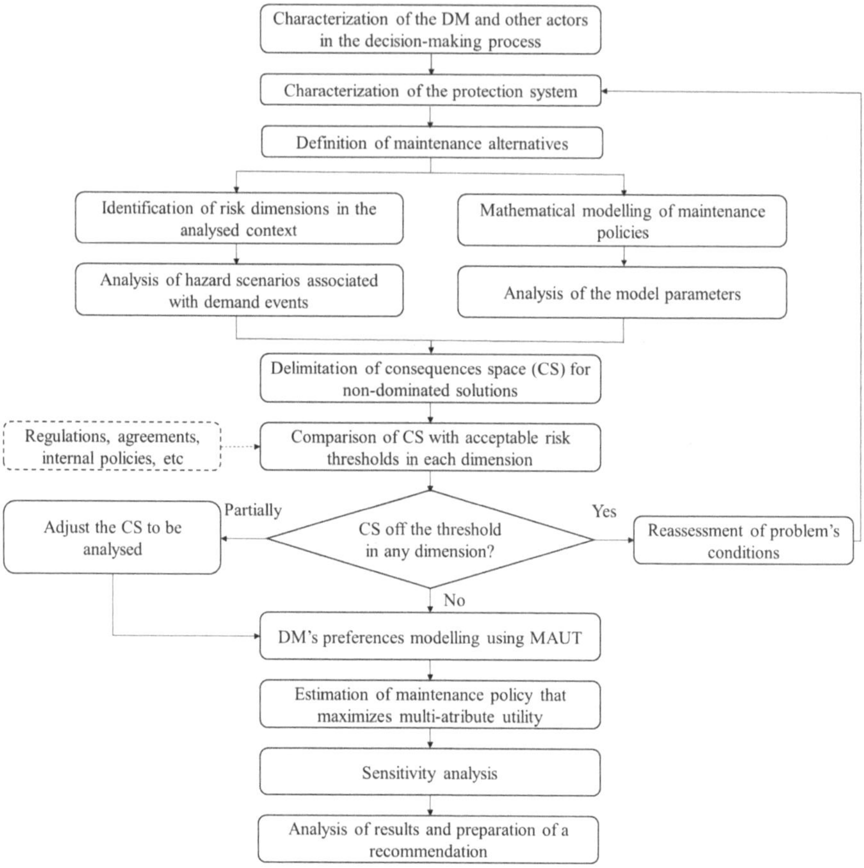

Fig. 2 Framework for the construction of the multicriteria model using MAUT. (Note: ALARP (as low as reasonably possible) is a concept that refers to the acceptable risk range for a given dimension of analysis (De Almeida et al. 2015a))

non-financial dimensions (human, environmental, etc.), which can be treated as criteria for decision. In this case, an analysis of the hazard scenarios associated with demand events is required. Different from cost, which is also affected by the frequency of maintenance actions, loss rates in these dimensions are generally directly proportional to the rate of unmet demands. In this sense, it is observed that while the losses related to the rate of unmet demands are minimized by an intense maintenance action (a large frequency of inspections, which corresponds to short intervals of T), the cost has (as a minimization strategy) less intense actions, less frequent inspections, and longer T intervals. It should be noted that conducting inspections that frequently make the main system unavailable would result in prohibitive operating costs.

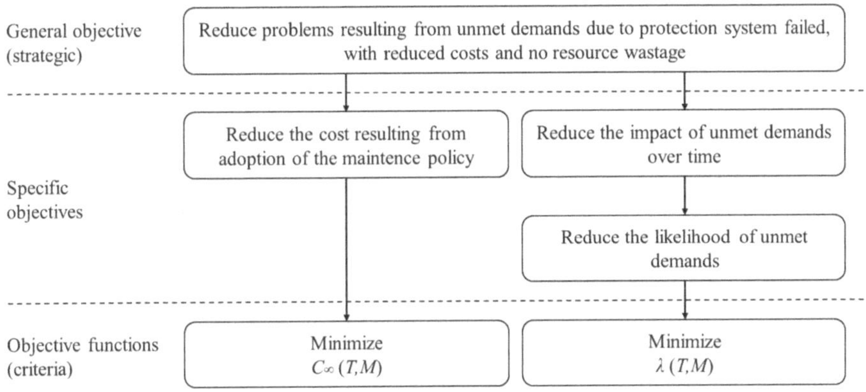

Fig. 3 Hierarchy of objectives of the problem

In the presented model, one of the considered assumptions is that during maintenance actions, the process of the arrival of demands is interrupted or attended by alternative solutions. In some scenarios, this disruption may be the result of the partial or total unavailability of the main productive system (Alberti et al. 2018). In this case, the level of unavailability of the main productive system can be incorporated into the problem as a constraint or as an additional criterion (Fig. 3).

To effectively use the mathematical model of the maintenance policy, it is important to analyse the input parameters, especially the parameters for which the model has the highest sensitivity. In the absence of objective historical data or to complement them, elicitation procedures can be used to estimate the a priori probability distributions of uncertain parameters based on the experts' knowledge (Berger 1985).

The next stage consists of the delimitation of the consequence space (CS) to be considered for the evaluation of the DM's preference relations since the evaluation of the alternative actions is performed by evaluating its consequences (Keeney and Raiffa 1976). It is desired that the analysed space of alternatives presents the following characteristics: the alternatives are not statistically dominated and respect the thresholds of acceptable risk. In this way, the analysis is restricted only to the alternatives that interest the DM. If these characteristics cannot be achieved, it is necessary to reassess the conditions of the problem, such as the operational constraints, the system's design, etc.

Once the problem has been contextualized and the initial conditions of analysis have been established, the CS is considered to model the DM's preferences using MAUT. This stage consists of eliciting the multi-attribute utility function, which will be used to evaluate the performance of maintenance policies.

To elicit the multi-attribute utility function, five main steps are recommended (Keeney and Raiffa 1976): the introduction of ideas and terminology, the verification of relevant preferential independence relationships, the elicitation of one-dimensional utility functions (intracriterion evaluation), the elicitation of scale constants for the composition of the multi-attribute utility function (intercriteria evaluation), and finally, the consistency tests.

The multi-attribute utility function aggregates the utility functions obtained for each criterion individually. For the elicitation of the one-dimensional utility functions, the consequence space obtained in the previous step is observed. Then, for each dimension of analysis, a utility equal to 1 (maximum) is assigned to the most desired end of the CS, and a utility equal to 0 (minimum) to the opposite end. In addition to converting to a normalized reference scale, the utility function allows for a suitable treatment of the uncertainty conditions of the problem through the mathematical modelling of the DM's risk behaviour (prone, neutral or averse) (Keeney and Raiffa 1976). Keeney and Raiffa (1976) present a structured procedure for the elicitation of one-dimensional utility functions, which consists of the verification of some intermediate utility points, and later an adjustment using a function obtained from such points.

The intercriteria evaluation, in turn, can be performed by verifying the conditions of preferential independence and trade-offs between criteria. Once the DM's preference relations have shown utility independence and additive independence between the criteria, the multi-attribute utility function can be expressed in the additive form according to Eq. (28) (Keeney and Raiffa 1976).

$$U(A) = \sum_{j=1}^{n} k_j . U_j\left(A_j\right) \tag{28}$$

Where

$$\sum_{j=1}^{n} k_j = 1.$$

The maintenance policy that maximizes the multi-attribute utility presents the best compromise relationship among the considered criteria. When a continuous set of alternatives is considered, it becomes a classical optimization problem. A sensitivity analysis is needed to assess the impact that possible inconsistencies in the elicitation process may have on the recommended solution. If the model is sensitive to perturbations in the input parameters, extra verifications may be required to guarantee greater reliability of the results (De Almeida et al. 2015a). From then on, through a critical analysis of the results, the recommendation can be elaborated for the DM.

6 Numerical Application

For the considered application, the protection system acts as a safety system and prevents the occurrence of accidental scenarios in emergencies (demands). The maintenance of this system has an impact not only financially but also on other dimensions. Therefore, a multicriteria approach is required to give the best treatment to the multidimensionality inherent to the problem of the maintenance policy's definition. The cost ($C_{\infty}(T,M)$) and risk ($\lambda(T,M)$) criteria are considered in the analysis.

The numerical application is presented to illustrate the model usage and obtain some results for discussion. Simulated data were used for this analysis.

6.1 Specifying the Model Input Parameters

System deterioration and failure can be modelled considering the delay-time concept. That is, before a failure occurs, the system presents a defective state that can be identified in an inspection (Christer 1999). In an investigation of the factors that impact the system's performance, it was verified that there is the possibility of defect induction in an inspection, but the probability that this happens is uncertain. For the parameters' definition, the study of maintenance policies for isolation valves in a water distribution system was taken as a realistic scenario. For this particular system, some approximations were made.

Weibull distributions (with shape parameter β and scale parameter η) were used to characterize the distributions of X ($\beta_1 = 2.1$, $\eta_1 = 4$ years) and H ($\beta_2 = 1.5$, $\eta_2 = 0.3$ year). The possibility of defect induction in the case of system activation for meeting a demand is part of its operational characteristics. It has an effect on the distribution of X, which tends to be more dispersed, and consequently results in a reduced shape parameter. Defect induction in the inspections can be considered as an external factor, and the impact on the system's performance is higher or lower according to the frequency of inspection actions.

The probability of defect induction in an inspection (p) is an uncertain parameter, but it is estimated that it should have some value between 0.03 and 0.07. A triangular distribution is considered to model the probability distribution of the possible values that this parameter can assume. This distribution is usually used when the true distribution of a given parameter is not known, and it yields good results (Fishman 1995). In this case, the analyst only needs to estimate the extreme values and the most likely value of the parameter to establish the distribution. For this application, it is estimated that the most likely value of p is 0.05.

The costs were established by taking the costs of a component's preventive replacement when it is in good state, which means that $c_p = 1$ un. (the amount is not specified in monetary units but may be converted by the combination of the amount incurred in performing a preventive replacement). The inequality relations $c_f > c_d > c_p$ and $t_f > t_d > t_p$ are appropriate, since it is valid to consider that the cost and time for the system recovering tend to be larger the worse their condition. It is worth noting that in the case of protection systems, the higher penalty is associated with an unmet demand (UD) event and not necessarily the substitution of a failed component, since in this case there is no interruption of a main production process

Table 1 Specifying the model input parameters

Costs	Times	Demands
$c_v = 0.04$ un.	$t_v = 0$ (negligible)	$\mu = 3$ demands/year
$c_p = 1$ un.	$t_p = 3$ h	$CUD = 25$ un.
$c_d = 1.2$ un.	$t_d = 6$ h	$t_{ud} = 72$ h
$c_f = 2$ un.	$t_f = 12$ h	

due to the occurrence of the failure. The values of the other parameters are shown in Table 1.

Due to limitations in the availability of resources, the following constraints are added to the problem: $T \geq 1$ month and $M.T \geq 6$ months, i.e. intervals between inspections cannot be less than 1 month and scheduled preventive replacements can only occur after at least 6 months of component operation.

6.2 *Results and Discussion*

The results' analysis is presented in five steps. First, an analysis of the model's sensitivity to variations of p is made in order to verify if there is a need to develop a multicriteria model that incorporates this aspect of uncertainty. Then, a procedure for delimiting the CS for the analysis is presented. The third and fourth parts are dedicated to the evaluation of the intracriterion and intercriteria preferences. Finally, the sensitivity analysis of the multicriteria model is presented.

6.2.1 Analysis of the Model Sensitivity to Variations in *P*

The first step of the analysis consists of an evaluation of the sensitivity of the mathematical model presented in Sect. 4 to variations in p (uncertain parameter). This analysis allows us to verify if, with the variation in the value of p, there are significant variations in the recommendation of the maintenance policy when considering a certain criterion. The analysis also allows us to verify if there is any conflict in the recommendation when different criteria are considered. Depending on the obtained results, it may not be necessary to develop the framework presented in Fig. 2 for the construction of the multicriteria model.

In Table 2, the maintenance policies (M and T) that optimize the cost and risk criteria separately are presented for different values of p.

As seen in Table 2, for the analysed case, the optimal policies are always the special cases: policy ABR or policy PI. For the cost criterion, it can be observed that, within the range of possible values of p, the recommendation can vary from one case to another, which demonstrates the impact that a sensitive increase in this parameter may have on the performance of an inspection policy in regard to this criterion. However, for the risk criterion, it is not observed, and an ABR policy with a minimum T is indicated in all cases. Because this attribute presents a lower

Table 2 Analysis of the model sensitivity with respect to p. T in years, C_∞(T,M) in un./years and $\lambda(T,M)$ in 10^{-2} UD/year

Minimizing C_∞	p	M	T	C_∞*	λ	Minimizing λ	M	T	C_∞	λ*
	0.03	∞	0.105	0.957	0.795		1	0.5	2.086	0.221
	0.04	∞	0.106	1.125	1		1	0.5	2.086	0.221
	0.05	∞	0.107	1.296	1.3		1	0.5	2.086	0.221
	0.06	1	1.048	1.394	1.6		1	0.5	2.086	0.221
	0.07	1	1.048	1.394	1.6		1	0.5	2.086	0.221

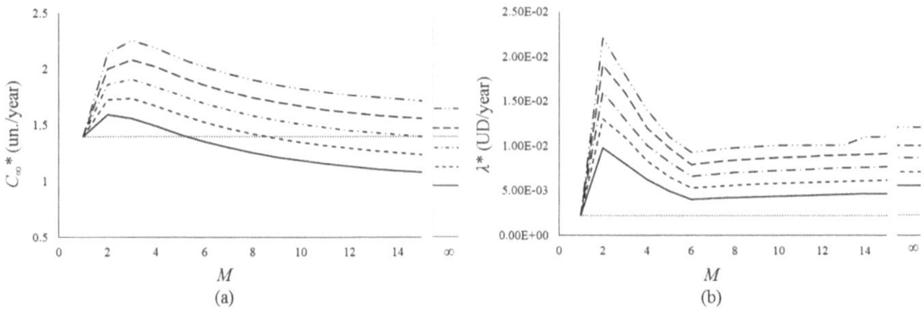

Fig. 4 Optimum values of $C_\infty(T,M)$ and $\lambda(T,M)$ as a function of M and p: $p = 0.03$ (), $p = 0.04$ (—), $p = 0.05$ (_ . _), $p = 0.06$ (_ _) and $p = 0.07$ (_ .. _). The dotted horizontal line (...) is a baseline referring to the optimal ABR policy performance, which is not influenced by the quality of the inspection

value when practising frequent replacements, it eliminates from the scope of actions the inspection that provides an increase in the rate of unmet demands due to defect induction that can happen regardless of the intensity with which this occurs. It is also observed that even in cases where a policy of the same format is recommended to optimize both criteria, there is no agreement with the value of T, which demonstrates the need for a multicriteria approach.

In addition to Table 2, two graphs are presented in Fig. 4 in which it is possible to observe the impact of the variation in p on the performance of the maintenance policies that consider inspections. In the graphs of Fig. 4, the optimal values of $C_\infty(T,M)$ and $\lambda(T,M)$ are presented for different fixed values of M and p.

It is observed that for the cost criterion (Fig. 4a), the optimal value of M is strongly influenced by the variation in the inspection quality, and for the policies with $M > 1$, a variation of 0.01 in the value of p has a significant impact on the performance of the maintenance policy. For the risk criterion (Fig. 4b), the ABR policy presents the best performance regardless of the value of p, and for policies with $M > 1$, the change in the value of p has an impact given that the optimum value of $\lambda(T,M)$ changes according to the variation in p.

In this preliminary analysis, conflicts between the criteria are observed, which makes it necessary to take a multicriteria approach that is capable of incorporating uncertainty about the state of nature in the process of maintenance policy evaluations for the protection system under study.

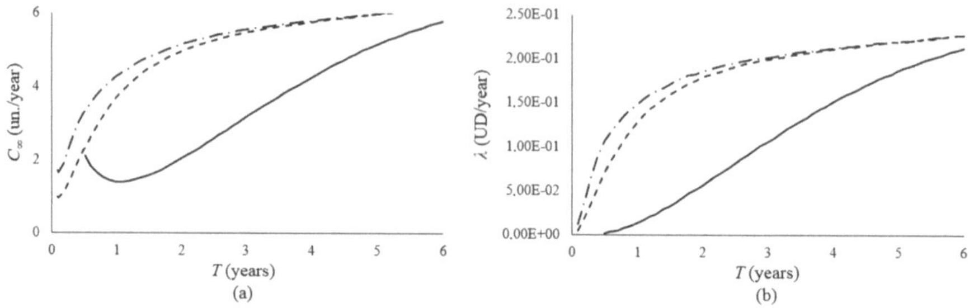

Fig. 5 Graphs of $C_\infty(T,M)$ and $\lambda(T,M)$ as a function of T for different values of M and p: $M = 1$ and any p (), $M = \infty$ and $p = 0.03$ (—), and $M = \infty$ and $p = 0.07$ (_ . _)

6.2.2 Delimitation of the CS to Be Analysed

To optimize the cost criterion, it is necessary to find a balance between the costs of preventive maintenance actions and the costs due to failures and unmet demands. Regarding the risk of unmet demands, this tends to be lower when the frequency of maintenance actions is greater, and so the optimization of this criterion basically depends on the constraints of the problem. These trends can be observed in the graphs shown in Fig. 5.

As seen in Fig. 5, the values of $C_\infty(T,M)$ and $\lambda(T,M)$ tend to increase together, except for a range relative to small values of T where the Pareto frontier is delimited, that is, the set of non-dominated solutions. The delimitation of the consequence space considering only the non-dominated solutions can help to avoid inconsistencies in the process of eliciting DM's preference relations.

For the delimitation of the CS for non-dominated solutions, the following procedure is proposed. First, each criterion is optimized separately by considering the scenario with the lowest possible value of p (best scenario) and then the scenario with the highest value (worst scenario). Then, the optimal policies obtained in the previous step for the best scenario are applied to the worst case scenario. The opposite is not necessary since the obtained results are expected to be within the range of values that are obtained. Finally, the smaller and larger values of $C_\infty(T,M)$ and $\lambda(T,M)$ obtained in the previous steps are taken as the limits of the CS. Solutions that present values outside this range (for any value of p) are dominated solutions, and so they do not need to be considered in the decision process.

The results that were obtained for the present application are presented in Table 3.

For a scenario with $p = 0.03$, policies 1 and 2 delimit the Pareto frontier since any alternative with performance outside this limit can be considered a dominated alternative. Assuming that a policy W has higher cost than that resulting from policy 2, and considering that it carries a risk greater than or equal to that resulting from policy 2, then it can be considered that policy W is dominated by policy 2. The same logic can be applied to policy 1. Likewise, policies 3 and 4 delimit the Pareto frontier for a scenario with $p = 0.07$.

Table 3 Summary of the procedure for CS delimitation

Description	p	M	T	C_∞	λ
Policy 1 (minimizes $C_\infty(T,M)$ for p = 0.03)	0.03	∞	0.105	**0.957**	0.795
Policy 2 (minimizes $\lambda(T,M)$ for p = 0.03)	0.03	1	0.500	**2.086**	**0.221**
Policy 3 (minimizes $C_\infty(T,M)$ for p = 0.07)	0.07	∞	1.048	1.394	1.600
Policy 4 (minimizes $\lambda(T,M)$ for p = 0.07)	0.07	1	0.500	**2.086**	**0.221**
Policy 1	0.07	∞	0.105	1.643	**1.700**
Policy 2	0.07	1	0.500	**2.086**	**0.221**
CS delimitation:	Criterion	C_∞	λ		
		Min	**0.957**	**0.221**	
		Max	**2.086**	**1.700**	

T in years, $C_\infty(T,M)$ in un./year and $\lambda(T,M)$ in 10^{-2} UD/year

Considering a scenario with uncertain p that ranges from 0.03 to 0.07, we have that policies 1 and 2 present the best values for each criteria separately for the best scenario ($p = 0.03$). When applying these policies in the worst scenario ($p = 0.07$), one may or may not have a performance range that fits within the range delimited by policies 3 and 4 in this scenario. In this second case, it is important to consider this extension of the CS range to incorporate the possibility of the best scenario for each criterion, which is its individual optimization with $p = 0.03$. On the other hand, when applying policies 3 and 4 for the best scenario, the results will necessarily fall within the range of the values delimited according to the presented procedure.

For this application, acceptable risk limits were not considered, and so it is not necessary to adjust the obtained CS for the next steps of the analysis.

6.2.3 Intracriterion Evaluation

Three different hypothetical utility functions are considered for each criterion by modelling three different behaviours that the DM may have in relation to their risk behaviour: neutrality, aversion or propensity. Note that this concept of risk is different from the risk criterion that is related to the rate of unmet demands. Here, risk refers to the DM's behaviour with regard to the uncertainty inherent in the problem.

The functions considered for the analysis are presented in Table 4, and the graphs of the utility according to the level of performance of each criterion are presented in Fig. 6.

The utility functions presented represent a pronounced behaviour of risk aversion or propensity (with the exception of the linear function), as can be observed in Fig. 6. This marked behaviour was considered for illustration purposes.

Before proceeding to the intercriteria evaluation, it is interesting to evaluate which maintenance policies maximize the expected utility for each criterion separately. The calculation of the expected utility can be performed according to Eq. (29).

Table 4 Investment scenarios. Costs in un

Cost criterion ($C_\infty(T,M)$)	Risk criterion ($\lambda(T,M)$)
DM risk neutral: $U_c(C_\infty) = -(0.866).\ C_\infty + 1.848$	DM risk neutral: $U_r(\lambda) = -(6.946).10.\ \lambda + 1.154$
DM risk averse: $U_c(C_\infty) = 1 - (7.590).10^{-4}.\ \exp[(3.444).\ C_\infty]$	DM risk averse: $U_r(\lambda) = 1 - (1.171).10^{-2}.\ \exp[(2.677).10^2.\ \lambda]$
DM risk prone: $U_c(C_\infty) = (5.062).10.\ \exp[1 - (4.101).\ C_\infty]$	DM risk prone: $U_r(\lambda) = (2.094).\ \exp[-(3.340).10^2.\ \lambda]$

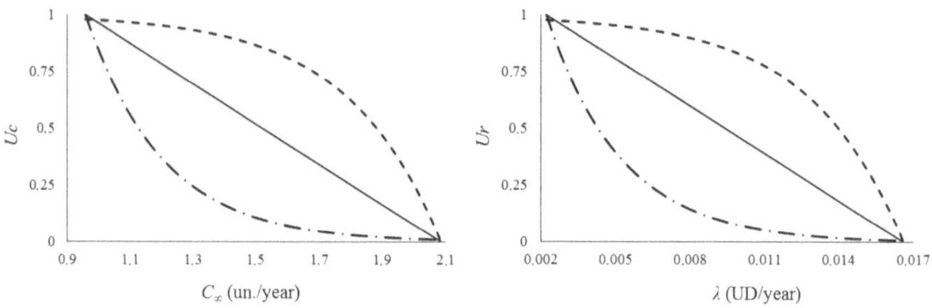

Fig. 6 Graphical representation of the utility functions for $C_\infty(T,M)$ and $\lambda(T,M)$ for different behaviours with regard to the risk: neutrality (), aversion (—) and propensity (_. _)

$$EU_j(A) = \int_{p_{\min}}^{p_{\max}} U_j\left[A_j(p)\right].f_p(p)dp \tag{29}$$

For the present application, maintenance policies that maximize the expected utility for the cost and risk criteria, respectively, are as follows: $M = \infty$ and $T = 0.105$ year (PI policy), and $M = 1$ and $T = 0.5$ year (ABR policy). This second result is expected, considering the results obtained in the first stage of the analysis. The policies that are indicated were the same regardless of the behaviour assumed by the DM with regards to the risk, but it is a particularity observed in this application, not an expected result. The observed conflict indicates the need for a multicriteria approach to evaluate the compromise relationship between the two criteria.

6.2.4 Intercriteria Evaluation

For the intercriteria evaluation, different combinations of the utility functions presented in the previous subsection are considered in order to observe how the recommendation of the maintenance policy varies. For this evaluation, it is considered that the conditions of utility independence and additive independence are verified. Furthermore, $k_c = k_r = 0.5$ is adopted, which in practice means that the DM

Table 5 Results of the multicriteria evaluation. T in years

	DM behaviour with regard to the criteria		Policy that maximizes the expected multi-attribute utility					
Case	Cost (C_∞)	Risk (λ)	M	T	EU	M	T	EU
1	Neutral	Neutral	1	0.68	0.593	∞	**0.083**	**0.609**
2	Averse	Averse	1	0.76	0.887	∞	**0.083**	**0.896**
3	Prone	Prone	**1**	**0.50**	**0.505**	∞	0.083	0.19
4	Averse	Prone	**1**	**0.58**	**0.622**	∞	0.083	0.521
5	Prone	Averse	1	0.74	0.509	∞	**0.083**	**0.563**

is indifferent between two hypothetical alternatives with the performance defined by the combination of the extreme CS points, namely, [C_∞* = 0.957 un./year; λ = (1.700)0.10–2 UD/year] and [C_∞ = 2.086 un./year; λ* = (2.210)0.10–3 UD/year]. Thus, the multi-attribute utility of a maintenance policy A with a probability of defect induction p is calculated according to Eq. (30), and the expected utility, considering the probability distribution of p is calculated according to Eq. (31).

$$U(A, p) = (0.5).U_c(A_c, p) + (0.5).U_r(A_r, p) \tag{30}$$

$$EU(A) = \int_{p_{\min}}^{p_{\max}} U(A, p).f_p(p)dp \tag{31}$$

The obtained results are presented in Table 5. For comparison purposes, the policies PI and ABR are presented with the maximum expected utility, and the policy to be recommended is highlighted in bold. In the evaluated cases, no mixed policy (M finite and greater than 1) had an expected utility higher than the recommended policy indicated in Table 5.

In the case of the policy PI, in all cases, the lowest possible value of T (equal to the restriction) was recommended. This is due to the consideration of the risk criterion in the analysis, which results in the indication of more conservative policies. The same is not true for the ABR policy (which is unaffected by the quality of the inspection), where the recommended value of T varies according to the case and is generally less conservative than the policy that maximizes the expected utility of the risk. It is observed that, depending on the case, the recommendation varies between one policy format and another.

Some results are different from what is expected a priori. In the case of a risk-averse DM, it is expected that a policy ABR, which has a certain performance that is not variable according to the value of p, is preferable. Meanwhile, it is expected that for the risk-prone DM, a policy whose performance is influenced by the value of p is preferable since it has the potential to have good cost results. What is observed in Table 5, however, is just the opposite.

Table 6 Sensitivity analysis for the case of a risk neutral DM

case	k_c	k_r	M	T	EU	M	T	EU
1	0.5	0.5	1	0.68	0.593	∞	**0.083**	**0.609**
2	0.6	0.4	1	0.74	0.561	∞	**0.083**	**0.62**
3	0.4	0.6	**1**	**0.62**	**0.641**	∞	0.083	0.599
4	0.45	0.55	**1**	**0.65**	**0.615**	∞	0.083	0.604
5	0.475	0.525	1	0.66	0.603	∞	**0.083**	**0.607**

This result is due to the mathematical properties of the one-dimensional utility functions that are considered, which represent the accentuated behaviour of risk aversion and propensity. The utility functions for risk aversion have a sharper utility reduction rate close to the upper end of the CS. Meanwhile, the utility functions for risk propensity show a marked reduction rate along the lower end of the CS, which results in significantly lower utility for the larger range of the CS. As a result, the expected maximum utilities in the cases presented in lines 3, 4 and 5 are significantly lower than that obtained in the case of line 2.

In addition to this, it is observed that the rate of unmet demands (risk criterion) presents a significant rate of increase with the increase of T (growth more accentuated than to cost), especially for the case of the policy PI. This has resulted in the recommendation of policy ABR for cases where the DM is risk prone with respect to this criterion (lines 3 and 4).

The results obtained in this stage indicate that a priori expectations are not always confirmed through the mathematical modelling of the DM's preference structure, which demonstrates the need to comply with a structured protocol for the construction of the multicriteria model.

6.2.5 Sensitivity Analysis

The case presented in line 1 of Table 5 is considered for a sensitivity analysis. The sensitivity analysis that is presented consists of checking the maintenance policies that are recommended when variations in the scale constants are considered in the multicriteria model. The obtained results are presented in Table 6. Again, the results for the policies PI and ABR are presented, and the recommended policy is highlighted in bold.

As shown in Table 6, for the case of an increase in the value of k_c (and consequently a reduction of k_r), the recommendation of policy PI with the minimum T remains stable (lines 1 and 2). For the case of an increase in the value of k_r, there is a change in the recommendation of the maintenance policy, which indicates policy ABR with T varying according to the case (lines 1 and 3 to 5). In the case presented in line 5, the difference between the expected maximum utilities obtained for the PI and ABR policies is small, indicating that this is a transition region, which is more sensitive to inconsistencies in the DM's preferences elicitation process. It can be said that, for this range of scale constants, the DM is practically indifferent between the two maintenance policies, PI and ABR. It is observed that, in practice, the operationalization of these policies demand quite different resources. One is inspection intensive and it impacts the frequent actions of maintainers for state identification. The other requires state-independent component replacement schedules, resulting in heavier investments in assets with less need for staff.

Consistency tests can be performed to more precisely identify for which ranges of values that the scale constants tend to. If k_c tends to have a value equal to or greater than 0.5, the obtained recommendation is consistent, and otherwise new validations may be necessary in order to give greater security to the results.

In a practical application, the next step is the preparation of the maintenance policy recommendation so that the DM can decide about its implementation.

7 Conclusions

This chapter presents some contributions to the process of evaluating maintenance policies for protection systems. First, a model is proposed for a mixed inspection and preventive replacement policy, which considers an important aspect of inspection quality: the probability of defect induction. This is an innovative model that allows us to more accurately estimate the cost and risk of unmet demands resulting from the adoption of a maintenance policy. Given the generic format of the considered policy, with the presented model, it is possible to evaluate the performance of the classic maintenance policies: the pure inspection policy and the age-based replacement policy.

Considering that cost measures are not always representative of the consequences associated with the adoption of a maintenance policy for a protection system, a multicriteria approach based on MAUT is proposed, which allows the handling of the multidimensionality of the problem and the uncertainty conditions in a more appropriate way.

Through the numerical application that is presented, it is possible to observe the adequacy and importance of the proposed approach to the presented problem. The results demonstrate the importance of properly executing a structured protocol for the construction of the multicriteria model since the recommendation of a maintenance policy can vary significantly depending on the DM's preference structure, and the expected a priori results may not be confirmed by the mathematical model.

Acknowledgements The work of the authors has been supported by CAPES (Coordination for the Improvement of Higher Education Personnel), CNPQ (Brazilian National Council) and FACEPE (Foundation for Science and Technology of the State of Pernambuco).

References

Alberti AR, Cavalcante CAV, Scarf P, Silva ALO (2018) Modelling inspection and replacement quality for a protection system. Rel Eng Syst Saf 277:145–153

Ascher H, Feingold H (1984) Repairable systems reliability. Marcel Dekker, New York

Aven T (2012) The risk concept—Historical and recent development trends. Rel Eng Syst Saf 99:33–44

Berger JO (1985) Statistical decision theory and Bayesian analysis. Springer-Verlag, New York

Berrade MD, Cavalcante CAV, Scarf PA (2012) Maintenance scheduling of a protection system subject to imperfect inspection and replacement. European Journal of Operational Research 218:716–725

Berrade MD, Scarf PA, Cavalcante CAV (2015) Some insights into the effect of maintenance quality for a protection system. IEEE Trans on Reliab 64(2):661–672

Brito AJ, DE Almeida AT (2009) Multi-attribute risk assessment for risk ranking of natural gas pipelines. Rel Eng Syst Saf 94: 187–198.

Cavalcante CAV, Scarf PA, Almeida AT (2011) A study of a two-phase inspection policy for a preparedness system with a defective state and heterogeneous lifetime. Rel Eng Syst Saf 96:627–635

Christer AH (1999) Developments in delay time analysis for modelling plant maintenance. The Journal of the Operational Research Society 50:1120–1137

Christer AH, Whitelaw J (1983) An operational research approach to breakdown maintenance: Problem recognition. The J Oper Res Soc 34(11):1041–1052

DE Almeida AT, Cavalcante CAV, Alencar MH, Ferreira RJP, DE Almeida-Filho AT, Garcez TV (2015a) Multicriteria and multiobjective models for risk. Reliability and Maintenance Decision Analysis, Springer, New York

DE Almeida AT, Ferreira RJP, Cavalcante CAV (2015b) A review of the use of multicriteria and multi-objective models in maintenance and reliability. IMA Journal of Management Mathematics 00:1–23

Driessen JPC, Peng H, Van Houtum GJ (2017) Maintenance optimization under non-constant probabilities of imperfect inspections. Rel Eng Syst Saf 165:115–123

Fishman GS (1995) Monte Carlo: Concepts, algorithms and application. Springer, New York

Flage R (2014) A delay time model with imperfect and failure-inducing inspections. Rel Eng Syst Saf 124:1–12

Jia X, Christer AH (2002) A periodic testing model for a preparedness system with a defective state. IMA Journal of Management Mathematics 13:39–49

Keeney RL, Raiffa H (1976) Decisions with multiple objectives: Preferences and value Tradeoffs. Cambridge University Press, Cambridge & New York

Okumura S, Jardine AKS, Yamashina H (1996) An inspection policy for a deteriorating single-unit system characterized by a delay-time model. International Journal of Production Research 34:2441–2460

Scarf PA, Cavalcante CAV (2012) Modelling quality in replacement and inspection maintenance. International Journal of Production Economics 135:372–381

Scarf PA, Cavalcante CAV, Dwight RA, Gordon P (2009) An age-based inspection and replacement policy. IEEE Trans on Reliab 58(4):641–648

Tijms HC (1994) Stochastic models: An algorithmic approach. John Wiley & Sons, New Jersey

Vaurio JK (1999) Availability and cost functions for periodically inspected preventively maintained units. Rel Eng Syst Saf 63:133–140

Multi-criteria Decision Model to Support the Maintenance Policy for Circuit Breakers in an Electrical Substation

Cristiano Alexandre Virgínio Cavalcante, Rodrigo Sampaio Lopes, Marcelo Hazin Alencar, Adiel Teixeira de Almeida, Ana Paula Cabral Seixas Costa, Heldemarcio Leite Ferreira, and Rodrigo José Pires Ferreira

1 Introduction

Maintenance decisions regarding protection systems must ensure that the protective devices are reliable and that they can securely operate at the performance level required to exercise their protective function. The equipment responsible for protection systems in substations and power lines requires special attention from the maintenance management to ensure the stability and security of the substation and network equipment components.

Allan (1988) highlights two main functions of the protection system in a power system: to isolate a fault that occurs in the system so that it does not propagate and to protect critical system components. In this way, the readiness of protection systems is critical for the safe operation of power systems and depends on the correct planning of the maintenance of the circuit breakers that comprise the protection system.

C. A. V. Cavalcante (✉)· R. S. Lopes
RANDOM – Research Group on Risk and Decision Analysis in Operations and Maintenance, Universidade Federal de Pernambuco, Departamento de Engenharia de Produção, Recife, PE, Brazil
e-mail: c.a.v.cavalcante@random.org.br; r.s.lopes@random.org.br

M. H. Alencar · A. T. de Almeida · A. P. C. S. Costa · R. J. P. Ferreira
INSID – National Institute of Information and Decision Systems, Universidade Federal de Pernambuco, Departamento de Engenharia de Produção, Recife, PE, Brazil
e-mail: marceloalencar@cdsid.org.br; almeida@cdsid.org.br; apcabral@cdsid.org.br; rodrigo@cdsid.org.br

H. L. Ferreira
IFPE – Instituto Federal de Pernambuco, Universidade Federal de Pernambuco, Departamento de Engenharia de Produção, Recife, PE, Brazil
e-mail: heldemarcioferreira@recife.ifpe.edu.br

A. T. de Almeida et al. (eds.), *Multicriteria and Optimization Models for Risk, Reliability, and Maintenance Decision Analysis*, International Series in Operations Research & Management Science 321, https://doi.org/10.1007/978-3-030-89647-8_11

Meeuwsen and Kling (1997) state that electrical utilities have shown a growing interest in management and maintenance practices, including actions directed at circuit breakers and protection systems. These authors list reasons for this growing interest, such as the increasing number of pieces of equipment and the increasing age of the equipment, together with the increased need for the maintenance of older equipment.

Mijailovic (2003) states that many electrical utilities plan preventive maintenance activities based on their own experience or on the recommendations of manufacturers, which are general and may not cover the specifics and peculiarities of a particular power utility.

Another factor that should be considered when planning maintenance activities is that circuit breakers operate within tolerance limits; for example, when a disturbance is detected on a network, the circuit breaker should be triggered to protect the equipment that is the most expensive and critical to the system function (Vilaithong et al. 2007).

Hussain et al. (2015) note that most failures in circuit breakers and high-voltage transformers are of mechanical origin. Mechanical malfunctions caused by mechanical wear can be detected through scheduled maintenance.

Meeuwsen and Kling (1997) argue that reducing the maintenance frequency in protection systems can reduce costs in the short term, but it can have an adverse effect on system security in the long term.

Meeuwsen et al. (1996) emphasize that determining the positive or negative influence on the frequency of performing preventive maintenance on circuit breakers and protection systems is a difficult task, and by performing preventive maintenance, the probability of a circuit breaker operation failure can be reduced if the preventive maintenance frequency is well programmed.

Maintenance managers in substations must consider cost reductions while simultaneously providing the desired level of customer service. Hence, the managers must establish periodic maintenance schedules for circuit breakers to reduce system failures and mitigate the associated risks.

Circuit breakers are essential to power supply systems as part of their protection systems. They are subject to hidden failures and must be operational when needed because when they are not operational, they do not fulfill their protective function; thus, their failure could trigger other system failures with serious consequences associated with power supply interruption.

This work proposes a model that evaluates the frequency of predictive and preventive maintenance in protection systems, specifically that of substation circuit breakers, to minimize the maintenance costs and the risk of the circuit breaker being in demand but not operational. This problem is important considering the possible effects associated with circuit breaker failure. The choice of variables determines the schedule of verification inspections (failure-finding inspections), in addition to the renovation actions (replacement). Thus, a multi-criteria decision support model is proposed to support the manager in maintenance planning, which is composed of inspection times and protection system renewals. The model is tested in a power utility substation in Brazil.

This chapter is organized as follows. After the introduction, Sect. 2 presents an important example of a strategic problem in the context of power supply process, and then Sect. 3 presents a background of work related to circuit breaker maintenance and multi-criteria decisions. Section 4 describes a framework for building the multi-criteria model. In Sect. 5, we propose a multi-criteria model. In Sect. 5, a case study implementation of the model is described, along with a discussion of the results, and in Sect. 6, we present some conclusions.

2 A Strategical Problem in Supply Electricity Using Additive-Veto Model

Companies that produce and supply electricity follow the requirements of agencies and regulators regarding the acquisition and maintenance of equipment to ensure a quality standard of the service offered. However, when failures and interruptions are observed in the provision of services, they are easily detected by customers, since services are generally produced and delivered to customers at the same instant of time, being directly affected by failures and/or interruptions. In this context, de Lima et al. (2016) propose a multi-criteria model based on the additive-veto model to prioritize locations where voltage regulators will be installed in an electricity distribution network of 15 kV. The proposed model considers technical, regulatory, economic, and social aspects related to consequence assessment of the installation of this equipment and operational performance, considering the decision-maker's preferences for the allocation of this equipment in these available locations to supply as many consumer units as possible. The proposed model contributes to the context under study since the distribution network sizing is the central aspect that signals the potential difficulties in maintaining the quality of service offered to consumers. Maintaining quality is directly associated with investments, restricted by the availability of resources provided by the organization. Investment planning suffers interference from multiple objectives that are aligned with the company's strategic issues. Thus, different geographic regions are analyzed by criteria, representing the objectives of the organization. De Lima et al. (2016) define as criteria sector regulation, size of the affected consumer unit, company image, and revenues. The application of the proposed model supports decisions related to the strategic and financial planning in the electric sector to prioritize the areas that most need voltage regulators, to prevent interruptions and minimize losses in the power supply process. The multi-criteria modeling application seeks to prioritize localities to receive regulators, ensuring the aggregation of consequences and the treatment of uncertainties. These procedures aim to evaluate the set of alternatives contemplating aspects inherent to the real world. Seven criteria are defined based on two voltage measurement indicators in the power supply process. Two criteria are defined based on regulatory aspects; one criterion is based on requirements for company image; two criteria consider the number of low voltage consumer units in a given region to identify potential operational problems. The last two criteria relate to the business vision of the organization. The territorial area considered in the study is subdivided into 8 regions, containing 12 power feeders. Alternative evaluation regarding the location of voltage regulators depends on the value function for each criterion. The consequences in this modeling are presented deterministically. The additive-veto model represents the performance compensation of alternatives against the set of criteria. This model is used due to the need to consider alternative veto in potential compensation situations of poor performance in one criterion, for excellent performance in another criterion. The decision-maker may not be willing to choose a location where performance in a given criterion is below a certain level. Additionally, the article presents a decision support system (DSS) that incorporates the concepts of the proposed model, facilitating the graphical information visualization and, consequently, a better understanding and interaction of the decision-maker. Elicitation of the veto threshold for each criterion and the elicitation of scale constants are resources available on DSS, being essential steps of the proposed multi-criteria decision model. The comparison of the ordering obtained with the additive-veto model and the additive model without veto is verified in the article, and the results obtained differ between the models used. The robustness of the results obtained is also verified concerning the data inputs and parameters adopted in the decision model. Finally, the application of the additive-veto model for the allocation of voltage regulators contributes to the decision support process based not only on financial aspects but also on strategic issues, essential points for profit generation, and the provision of quality services on electric sector.

3 Related Works on Circuit Breaker Maintenance

To propose a suitable maintenance policy for a circuit breaker, it is interesting to understand the many existing maintenance policies in the literature focused in circuit breakers. Thus, papers related to maintenance decisions in circuit breakers are presented, and they address the decision problem and methods used for decision support. The brief descriptions provided aim to show the diversity of the different models and to provide insights into the main aspects relevant to the process of developing maintenance policies. Thus, the different works presented are not exhaustive.

Jian and Tianyuan (2015) propose an optimal time maintenance schedule for circuit breakers via an algorithm using least squares support vector machines (LS-SVM), aiming to increase the reliability of the substation and reduce maintenance costs. This algorithm is used in a power supply company, and an optimal result is presented that helps make decisions about performing different types of long-term maintenance. The algorithm can be used under conditions of different failures and maintenance management structures.

Mingliang et al. (2015) present a method based on empirical mode decomposition (EMD) entropy to extract parameter characteristic of the vibration signal of HV (high voltage) of circuit breakers. The authors argue that the proposed method is convenient and can be applied directly to diagnose a fault.

Boudreau and Poirier (2014) present a methodology for analyzing the end-of-life of electric power equipment, investigating the effect of equipment aging. Parametric and nonparametric statistical methods are used to provide a risk assessment based on the average residual life, reliability, and life expectancy of components as a function of the time since installation or manufacture. This methodology was applied to an oil circuit breaker.

Lopez-Roldan et al. (2014) [10] describe methods for monitoring and diagnosing the performance of circuit breakers based on the nonintrusive evaluation of key parameters, such as operating hours, restrike features, reignition, and detection, to support maintenance planning.

Javadi (2010) proposed a combination of bus sectionalizing and an LC resonant circuit as a solution to limit the fault current level of power grid substations. An application of this method in the Iranian power system network was presented. Case studies performed at some substations showed that the method can be successfully used to reduce the current defect from 28 to 18%, and an economic study showed that the method could be used at high-voltage substations containing more than 12 breakers.

Mijailovic (2003) employs a probabilistic method for calculating the operational cost during a planning period for the evaluation of substation components availability in connection with planned maintenance activities. The proposed method analyzes fault repair costs, maintenance costs, and the availability of substation components. The application of the method was demonstrated on a circuit breaker.

Meeuwsen and Kling (1997) demonstrate that the effects of preventive maintenance on circuit breakers and protection systems in substations can reflect the reliability of the electricity supply. The optimal frequency of preventive maintenance is evaluated according to the topology of the substation.

Based on these descriptions, it can be verified that the development of methods for decision-making regarding circuit breaker maintenance can greatly affect electrical systems, making this development highly interesting for research and applications. With the increasing demand for energy and more efficient systems, precise decision-making in the maintenance of these systems is increasingly essential. The model that we propose in this chapter is different from that in previous studies in the following perspectives:

- This model describes the decision problem in determining the frequency of verification inspection and renewal actions for the circuit breaker.
- It establishes the consequences of actions, evaluating their risk and costs.
- It proposes a multi-criteria model for the decision-making process, considering the preference of the decision-maker about the risk and cost criteria and describing the veto for alternatives.

Thus, a multi-criteria model to support maintenance planning, which considers not only different aspects involved in the problem but also the decision-maker's preferences, appears useful for the manager to determine an appropriate planning strategy that incorporates the characteristics of the circuit breaker and the substation and that reflects the priority actions, contextual issues, and preferences.

4 Multi-Criteria Decision and Additive-Veto Model

Maintenance decisions involving the provision of services such as power supply invite us to consider and balance multiple criteria. These criteria are essential because of the important effects described both quantitatively, for example, cost, reliability, and operating availability, and qualitatively, such as affected consumers, social aspects, and safety perception. Therefore, there are many alternatives for performing maintenance actions, considering some of these essential aspects. Occasionally, many stakeholders are involved in the decision-making process, for example, system managers, customers, and government regulatory agencies. These elements complicate the decisions regarding maintenance in electrical systems.

Vincke (1992) states that for complex decisions, modeling and structuring of the problem using a multi-criteria method is suitable to identify and evaluate criteria and alternatives consistent with the decision.

Multi-criteria decision-making (MCDM) has gained notoriety, as evidenced by the number of applications and papers in the literature (Wallenius et al. 2008; Zavadskas et al. 2014; Kadziński et al. 2016; Cavalcante and Lopes 2014). The MCDM methods are intended to support the decision-maker (DM) for different types of problems, such as selecting the best alternative from a set of alternatives, defining a ranking of alternatives, or sorting a set of alternatives into pre-defined classes. In all of these problems, the decision-maker plays an essential role, and preferences are considered in the model.

The increasing number of MCDM applications for different issues in the context of maintenance decisions highlights the contribution level that these methods provide to support maintenance managers. One of their advantages is related to the aggregation of complex information and the ability to address aspects that are often in conflict with the decisions regarding maintenance actions (de Almeida et al. 2015a, b; Alencar and Almeida 2015).

For some decisions, the DM would not accept an alternative that pays off a criterion whose performance is below a certain level. To represent this situation, the veto function is evaluated to avoid the selection of this alternative or to correct the final order of the alternative. The veto concept has been introduced in compensatory methods such as additive models (De Almeida 2013).

The concept of veto over MCDM problems is found in some outranking methods, which are not compensatory methods, such as the family of ELECTRE methods (Roy 1996).

De Almeida (2013) proposed an additive-veto model considering veto alternatives for the ranking problematic in which the overall value for an alternative is given by Eq. (1):

$$v(a) = r(a)\sum_{i=1}^{n} k_i v_i(a) \quad (1)$$

where

$$z_i = \begin{Bmatrix} 0.if.v_i(a) \leq l_i \\ 1.if.v_i(a) \geq u_i \\ \frac{v_i(a)-l_i}{u_i-l_i}.if.l_i < v_i(a) < u_i \end{Bmatrix} \quad (2)$$

For the veto function z_i, which is defined for each criterion i, l_i is the lower threshold and u_i is the upper threshold for the performances on criterion i. For a criterion to be maximized, the lower limit penalizes the overall value of a given alternative if the performance of that alternative is under l_i. The upper limit establishes that alternatives with performance bigger than u_i are save from the veto mechanism. Finally, alternatives with performances inside the range $[l_i, u_i]$, in some way, are affected by the veto mechanism that penalize the overall value of those alternatives.

The weighed veto function $r_i(a)$ for each criterion is given by Eq. (3):

$$r_i(a) = z_i(a)k_i \quad (3)$$

Naturally the veto functions for all criteria is obtained by the sum of $r_i(a)$ for all i.

where $z_i(a)$ is given by Eq. (2) and $v_i(a)$ is the value of the performance of alternative (a) in criterion i, k_i is the scale constant for criteria i, and n is the number of criteria. In this way, the final expression for $v(a)$ is given by Eq. (4):

$$v(a) = r(a)\sum_{i=1}^{n} k_i v_i(a) \quad (4)$$

5 Multi-criteria Model Proposed for Maintenance Planning for Circuit Breakers in an Electrical Substation

The construction of decision support models for maintenance management considering the characteristics of the analyzed system is essential because the model reflects the system. In the case of circuit breakers, once they operate only under demand of protection, failures are hidden and can be dangerous for the substations, since they may result in serious consequences, beyond those related with costs. Therefore, multidimensional perspectives are recommended.

We are interested in determining the maintenance frequency of circuit breakers and measuring the consequences of these maintenance actions as a function of the cost and associated risk. Note that the criteria of cost and risk are in conflict, i.e., no alternative exists that simultaneously results in minimum cost and minimum risk.

Therefore, a multi-criteria model is proposed to analyze the maintenance decision problem in circuit breakers, searching for a more balanced decision and considering the DM preference. Note that the DM preference based on cost and risk criteria may differ depending on the substation being analyzed. For example, for a substation at a port, airport, or industrial complex, the risk criterion may have a greater importance than that for a substation serving residential customers. Thus, no single general solution exists for all substations.

5.1 Risk Criterion

For this criterion, the risk is defined as the probability of the circuit breaker that is in demand but unavailable due to a hidden failure that prevents its protective function for various equipment in the substation and transmission lines against damage.

The unavailability of the system protection with hidden failures can be measured using the concept of mean fractional dead time (MFDT), which measures the proportion of time during which protection systems are not available, as discussed in the literature (Rausand and Vatn 1998; Aven 1986).

The MFDT concept can be used for circuit breakers, whereby the unavailability of the circuit breaker is influenced by the frequency at which maintenance is performed.

For the specific policy, the maintenance actions are of two kinds: predictive and preventive maintenance. The frequency of inspections (*T*) is defined in order to reduce the risk of the hidden failures of a circuit break, in addition to the renewal action (preventive maintenance) of the system after *N* inspection, at time (*NT*).

According to the literature (Vaurio 1997; Ahmadi et al. 2012; Ahmadi and Kumar 2011; Ahmadi et al. 2010), it is possible to describe a function that represents the MFDT relative to the variables *T* (verification/inspection frequency) and *N* (cycles until renewal), as described by Eq. (5).

$$\mathrm{MFDT}\,(T,N)=\frac{1}{T}\int_{(N-1)T}^{NT}1-\exp\left[\left(\frac{(N-1)\,T}{\alpha}\right)^{\beta}-\left(\frac{(N-1)\,T+t}{\alpha}\right)^{\beta}\right]dt \tag{5}$$

To describe the risk criterion, λ is the rate at which the circuit breaker is triggered over time, and it is assumed that the demand by the system is independent of the occurrence of failure. Thus, it is possible to describe the risk criterion using Eq. (6).

$$R\,(T,N)=\lambda\mathrm{MFDT}\,(T,N) \tag{6}$$

The rate λ can be obtained from circuit breaker counters implemented to provide this information. Combining the values for *T* and *N*, the value of the criterion risk can be determined.

However, as the risk function is only a product of MFDT(T, N) by a constant λ, we will consider the risk as corresponding of MFDT(T, N).

5.2 Cost Criterion

Cost is a criterion most commonly used in the definition of frequency of maintenance actions (De Almeida et al. 2015a, b). Maintenance managers are interested in understanding how maintenance actions affect costs, including the costs of preventing failure and repairing the failures that occur.

A suitable maintenance policy is critical for protection systems. The goal of these devices is to protect the system in abnormal situations that threaten the integrity of the system or mitigate serious consequences of the principal system. Thus, the preparedness of protection systems is crucial. For circuit breakers, when a protection demand from the main system is not met because of a previous failure of the protection system, the repercussion in terms of cost is tremendous and may be subject to penalties from regulatory agencies.

Previous studies (Ahmadi et al. 2012; Ahmadi and Kumar 2011) were used to construct this criterion. To model the cost criterion, it is necessary to make some assumptions related to the circuit breaker, as follows:

1. Failures are hidden and are detected by a verification inspection, or in cases in which a demand occurs (interrupted when protection is required, it is already in the failed state). A failure of the protection system may reflect more severe damage to equipment in the substation and transmission line.
2. The inspection and repair actions prevent the system to operate because the power system (substation) must be shut down and de-energized for maintenance.
3. Failures are detected by a verification inspection.
4. The circuit breaker is functional after an inspection or repair.

Our goal is to determine the optimal (T) interval between verification inspections in circuit breakers and the number of inspections until the replacement (N). This means that at NT, an overall action for the circuit breaker is planned.

The verification (inspections) that have an interval T include tests on contact resistance, insulation resistance, and dielectric strength of the insulating oil and rigorous visual inspection.

The renewal actions that occur after N cycles of verification inspections include cleanup actions, gasket replacement, anti-rust treatment of the tank, replacement of conduit box gaskets, and painting. Following these actions, final testing (of the contact resistance, insulation resistance, and dielectric strength of insulating oil) is performed. Then, the system is normalized, and operation is re-established by starting a new cycle. The age of the system can be described by the renewal frequency and the interval between these renewals is described by $T_N = TN$.

The costs of maintenance actions considered to model the cost criteria are as follows:

c_r Cost of repairing a possible failure found in the verification inspection of the circuit breaker.

c_i Cost of the verification inspection at the circuit breaker, including the testing costs of contact resistance, insulation resistance, and dielectric strength of the insulating oil.

c_{re} Cost of the renewal action on the circuit breaker, including costs related to disassembly, verification, contact replacement and adjustment, gasket replacement, oil draining, and oil reconditioning.

c_{tp} Cost associated with not operating the power substation during the time of inspection and repair when the substation is turned off and de-energized. These costs are managed by a contract between the utility company and the customers, reflecting the downtime cost.

c_a Cost associated with undesirable consequences of the circuit breaker failure, such as the cost of the devices affected by the failure in the protection system or the occurrence of multiple failures, causing difficulty in restoring the substation functionality.

To determine the function that describes the cost, the time to perform the verification inspection t_i and the time to perform the repair of failures t_r must be considered.

When the circuit breaker is minimally repaired following the verification inspection and repair, it returns to a functional state, although it is not as good as new. It is also assumed that the occurrence of faults in the breaker follows a non-homogeneous Poisson process. Thus, the conditional probability of failure during the Nth inspection cycle is given by $F_N(t)$, which describes the probability that a circuit breaker tested at T_1 can survive until T_2 after inspection. $F_N(t)$ can be described using Eq. (7):

$$F_N(t) = 1 - \exp\left[\left(\frac{(N-1)\,T}{\alpha}\right)^{\beta} - \left(\frac{(N-1)\,T + t}{\alpha}\right)^{\beta}\right] \tag{7}$$

where β and α are the parameter of shape and scale, respectively, and t is the time until the Nth inspection cycle. Given the assumed, the cost can be described using Eq. (8), which represents the cost criterion.

$$C\,(T,N) = \frac{c_i}{T} + \frac{c_r \sum_{N=1}^{T} F_N(T)}{NT} + \frac{c_{tp} t_i}{T} + \frac{c_{ip} t_r \sum_{N=1}^{T} F_N(T)}{NT} + \frac{c_a \lambda T \sum_{N=1}^{T} \mathrm{MFDT}(T,N)}{NT} + \frac{c_{re}}{NT} \tag{8}$$

Equation 8 can be used to determine the cost related with the decision variables T and N.

5.2.1 Veto Definition and Additive Aggregation

We are interested in defining the veto and additive aggregation functions, which require defining the upper and lower limits of the veto function for each criterion. The decision variables of the problem are T and N. Here, we consider these vectors (T, N) as alternative action, where (T) indicates the interval between verification inspections and (N) indicates the number of cycles. For interval T, it is possible to consider any unity of time, i.e., hours, days, months, or years, and for the number of cycles N, the alternatives are integers.

For each combination of T and N, the results can be obtained for the cost criterion using Eq. (8), and for the risk criterion using Eq. (6), thus we can determine the space of consequence.

After determining the space of consequence, the value function that represents the DM preference for each criterion can be found.

The veto is the minimum acceptable performance value for the DM to consider the alternatives of T and N for each criterion, $C(T,N)$ and $R(T,N)$. The DM can accept alternatives with a performance above this threshold according to the additive model. For example, the DM can set a veto value for the risk criterion that only considers alternatives resulting in a risk value below the acceptable risk limit.

After defining the parameters of the veto function and the constant scales, the ranking function can be established using Eq. (1).

5.2.2 Decision-Making Process Using the Proposed Model

The decision-making process for the maintenance model for circuit breakers formulated in this study is illustrated in Fig. 1, including the main steps of the proposed approach. Step 1 requires specifying the maintenance decision problem in the circuit breakers; specifying the characteristics of the circuit breaker; establishing the DM for the analysis; understanding the failure mechanism of the circuit breaker, the fault data, their interaction, and the effect of the failure on the power system; and studying the maintenance policies adopted by the company analyzed.

Step 2 – Establish the criteria and space of consequence. Using the data, the model can calculate the criteria, first determining alternative times between inspections T and the number of cycles N. In accordance with the energy supplier priorities, convenient alternatives for maintenance planning can be considered, using Eq. (8) to calculate the cost and Eq. (6) to calculate the risk. Then, the space of consequences can be established.

Step 3 – Model the DM preference by establishing a preference function to transform the cost and risk measures into a preference measure. After establishing the preference function, it is possible to determine the scales constant, and then the veto function can be established defining the thresholds of veto l and u for each criterion.

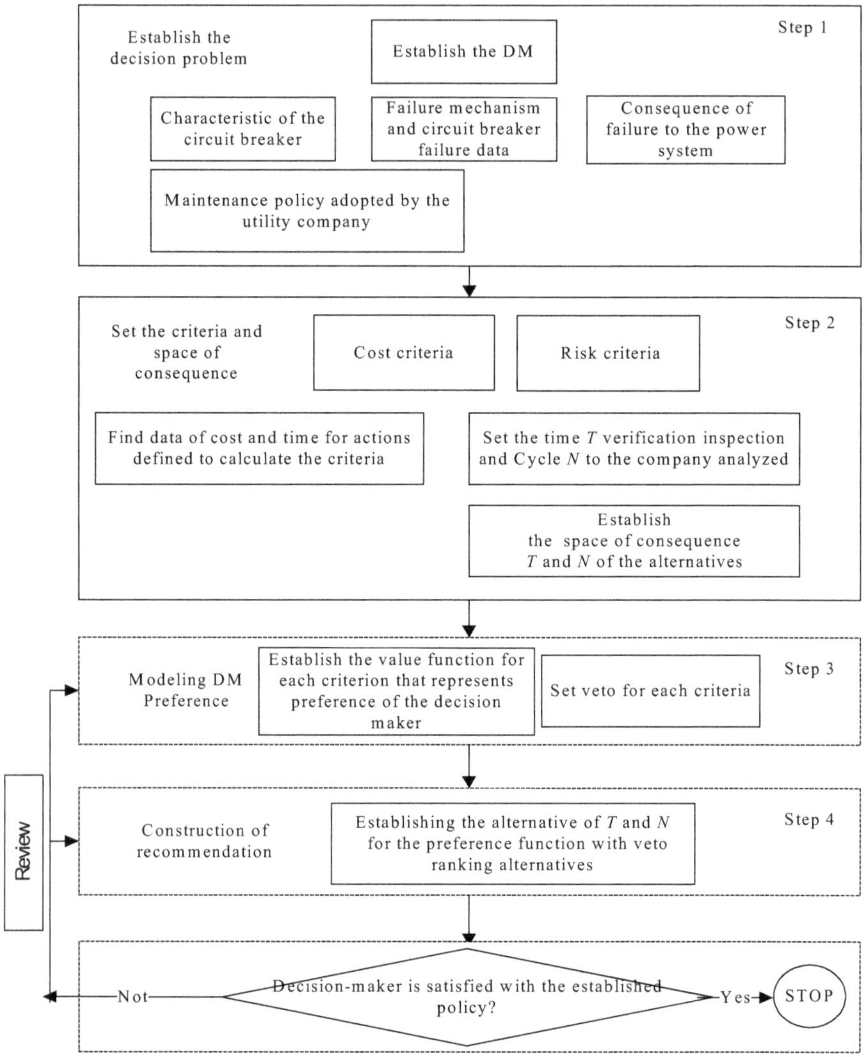

Fig. 1 Decision-making process

In the fourth step, using the results obtained in the previous steps, the best inspection policy can be found in accordance with the DM preference using Eq. 1. If the DM is satisfied with the recommendation, then the process ends; otherwise, it is necessary to review the elements in steps 3 and 4.

6 Case Study Implementation and Discussion of the Results

The situation structured mimics a real company that serves 200 cities in Brazil and more than three million customers. It operates in the generation, transmission, cogeneration, trade, and distribution of energy. The maintenance management of this company is responsible for maintaining 140 substations. To implement the proposed model, a circuit breaker in a pilot substation was studied. This substation is responsible for an industrial complex and is considered critical by the company to serve industrial customers; therefore, this substation focuses on maintenance planning.

A 230 kV oil circuit breaker was analyzed using parameters that support the minimum symmetric and asymmetric short circuit faults related to the nominal short-circuit current of 40 kA and a peak value of the rated circuit that can withstand a current of 104 kA (with an asymmetry factor of 2.6). The DM of this process is responsible for managing the maintenance of the pilot substation studied.

The procedures for the planned maintenance of substations in the studied company are classified into two possible types, each with specific characteristics related to the actions performed:

- Planned maintenance type I – maintenance that results in operational downtime but does not require disassembly of all or part of the equipment or facility; defined for fixed periods T.
- Planned maintenance type II – produces operational unavailability, with full or partial disassembly of the equipment or installation; defined for fixed periods NT.

When analyzing the circuit breaker based on the company data, the most common causes of circuit breaker failures were related to wear due to the age of the contacts, including the "deregulation" of mechanical poles and the decreased insulation resistance. This finding agrees with other studies related to circuit breakers, such as those by Hussain et al. (2015) and Boudreau and Poirier (2014).

Thus, knowledge of the causes and effects of failures is important. The circuit breaker monitoring parameters used by the studied company are as follows:

- Contact wear. The circuit breaker manufacturer provides a maintenance curve relating the number of closing/opening operations ("close-to-open") to the levels of current disruptions. The function of this curve is to analyze the wear of the circuit breaker contacts.
- Total number of operations. Incremental counters for closing/opening operations ("close-to-open") are used to send this information to the system historic data.
- Mechanical operating time. The mechanical circuit breaker operation time can be obtained using an oscillograph. Deviations in this value can indicate problems in the operating mechanism.
- Electrical operating time. Similar to the mechanical operation time, this time measures the interval between the trip command or closing command and the normalization of the circuit breaker current measurements. An increasing trend of this parameter over time may indicate a failure of the contacts.

Table 1 Parameters

Parameters	Value
$c_i c_i$ cost of verification inspection at the circuit breaker	3 m u
$c_r c_r$ cost to repair a possible failure found in the verification inspection	10 m u
$c_{re} c_{re}$ cost of renewal action on the circuit breaker	120 m u
$c_{tp} c_{tp}$ cost associated with no power supply	5 m u (loss of profit)/day
$c_a c_a$ cost associated with undesirable consequences	3000 m u
$t_i t_i$time to perform the verification inspection	7.2 h (type A)
$t_r t_r$time to perform the repair when a failure is found	12 h (type B)
Shape parameter	2.4
Scale parameter (hours)	7200
λ rate at which the demand occurs in the circuit breaker per unit time	8 (1/year)

- Time to inactivity. By monitoring the activity of the number of operations, the number of days that the circuit breaker has not tripped can be calculated.

By analyzing these parameters and the procedures adopted by the company studied, the maintenance practices that were used can be understood and data can be obtained in order to be possible to implement the proposed model.

6.1 Establish the Criteria and the Space of Consequences

To establish the space of consequence, the time and costs involved were collected and are shown in Table 1. The costs and time were rescaled to maintain the confidentiality of the company where the data were obtained, and these values are measured in a (defined) monetary units (m.u.) and days, respectively.

The alternatives are the time values ranging from 1 month to 2 years, in multiples of 30 days. That is, the number of alternatives for T ranges from T [30, 60, 90, ... 720] and for the number of cycles for the overall, N, ranges from 1 to 50, $TN = 7300$, that corresponds to 20 years, the maximum. In this way, we didn't consider the value of N, where the product TN is bigger than 7300. As examples of alternatives, let's say alternative *A1* corresponding to $T = 30$ and $N = 1$ (T = 30, N = 1), alternative *A2* to $T = 30$ and $N = 2$ (T = 30, N = 2), and so on until alternative that corresponds to $T = 720$ and $N = 10$. There are 1200 alternatives and 2 criteria.

After determining the space of consequence, each alternative and its consequences can be evaluated for each criterion.

6.2 Modeling Preference

Once the matrix of consequence is established, a study of the value functions for each criterion must be performed, considering the full range of alternatives. For this study, we considered a linear value function, i.e., for a criterion that must be minimized, a smaller value of the alternative in this criterion indicates a greater preference for this alternative.

Thus, it is necessary to perform a scale transformation and to determine the value function for each criterion. The criterion cost per time determines the extremes, i.e., the alternative of lower cost $C(T^*, N^*)$ and the alternative of higher cost $C(T^0, N^0)$. These alternatives, in terms of preference scale, are given values of 1 (low cost) and 0 (high cost), respectively. Hence, $C(T^*, N^*)$ is the most highly preferred alternative, and $C(T^0, N^0)$ is the least preferred alternative. All other alternatives should receive intermediate values according to the function that presents the cost values. Equation 9 describes the value function for the cost.

$$vC_i\left(T_i, N_i\right) = \frac{C\left(T^0, N^0\right) - C_i\left(T_i, N_i\right)}{C\left(T^0, N^0\right) - C\left(T^*, N^*\right)} \tag{9}$$

The risk criterion also determines the extremes, i.e., the alternative of lower risk $R(T^*, N^*)$ and that of the highest risk $R(T^0, N^0)$. The lower risk value function is equal to 1, and the greater risk value function is equal to 0. Equation 10 describes the value function for the risk.

$$vR_i\left(T_i, N_i\right) = \frac{R\left(T^0, N^0\right) - R_i\left(T_i, N_i\right)}{R\left(T^0, N^0\right) - R\left(T^*, N^*\right)} \tag{10}$$

Thus, the DM must realize the importance of the criterion, not only of the criterion itself but also the changes in the criteria of the alternatives under evaluation.

The additive value function is determined in the aggregate by adding the cost and risk value function. This function calculates the value for each criterion, and then, the resulting values of the cost and risk value function are added, thus determining the scale constant of each criterion according to their relative importance.

The additive value function should fulfill the requirement of independence, meaning that a trade-off relationship between the two criteria cannot depend on any other criteria, i.e., the ratio between cost and risk cannot depend on other criteria, which is the case for the decision problem under analysis.

A scale constant will establish the DM's preference according to the criteria, where k1 is the scale constant for risk and k2 is the scale constant for the cost. By definition, $\sum_{I=1}^{n} k_i(v) = 1$. Because there are two criteria, only one scale constant needs to be defined because the other is $1 - k$.

The upper veto threshold *(u)* is the minimum performance value of the risk and cost acceptable to the DM for any alternative of T and N. If there is an alternative to

Table 2 Input parameters

Parameters	Risk	Cost
k	0.8	0.20
l	0.07	0.5
u	0.000001	0.15
k	0.8	0.20

Table 3 Rankings of alternatives

Alternative	TN (months)	v(a)	Rank	Risk	Cost
(T = 90, N = 13)	39	0.96972	1	0.00108	0.156
(T = 60, N = 27)	54	0.96957	2	0.00119	0.1528
(T = 60, N = 26)	52	0.96952	3	0.00113	0.1555
(T = 60, N = 25)	50	0.96879	4	0.00107	0.1583
(T = 120, N = 9)	36	0.96852	5	0.00124	0.1526
(T = 60, N = 24)	48	0.96804	6	0.001	0.1615
(T = 90, N = 12)	36	0.96691	7	0.00096	0.1642
(T = 60, N = 23)	46	0.96661	8	0.00094	0.1649
(T = 60, N = 22)	44	0.96516	9	0.00089	0.1687
(T = 90, N = 14)	42	0.96445	10	0.00121	0.149

this performance (or above), the DM is willing to accept it. The DM is asked what values of risk and cost are acceptable to consider the decision alternatives.

The lower veto threshold *(l)* is the maximum value for the criteria of cost and risk that the DM is certain to reject, independent of their performance on other criteria. The DM is asked which risk value with an inferior performance he/she will reject, independent of their cost performance. When the cost value is above this limit, he/she rejects the alternative.

After obtaining the answers to the questions from the DM of the studied pilot substation, the scale constants (k_R and k_C), upper veto threshold (u), and lower veto threshold (l) for each criterion were determined, as shown in Table 2.

Having defined the upper and lower limits for each criterion, the veto function can be established, as described in Eq. (2).

6.3 Construction of the Recommendation

After establishing the value function, the scale constant, and the veto, the alternatives can be ranked and the order of the alternatives can be defined using Eq. (1). Table 3 shows the first ten rankings of alternatives.

As a result, alternative (T = 90, N = 13) has the highest ranking, indicating that verification inspections of type I should be performed every 90 days, and after every 13 inspection cycles, maintenance type II should be performed. Implementing this policy results in an expected cost of 0.156 with a risk of 0.00108.

Observe that despite the value of the scale constant to risk (k_R) is 0.8, the alternative (T = 60, N = 22) with the best performance in the risk criterion was positioned in the ninth position. This may be explained by the fact that the increment of improvement on this criterion, resulting from an increasing in the frequency of inspection, is not sufficient to compensate the cost necessary to implement this alternative.

It is worth mentioning that the model provides some very different options of maintenance policies. Notice that related with maintenance type I, there are alternatives in which this kind of maintenance happens more often, for example, for the case where the overall happens with 36 months, i.e. (T = 120, N = 9) and (T = 90, N = 12), the last policy means that inspection is much more frequent; it happens every 90 days. On the other hand, in the former policy (T = 120, N = 9), it only happens every 120 days.

Another important aspect is the fact that when the interval of the overall is too long, for example, every 54 months (T = 60, N = 27), notice that maintenance type I has to be intensified (every 60 days), in order to be possible to postpone the maintenance type II, without incurring in great increasing in risk.

The next step is to check whether the DM is satisfied with the recommendation of the resulting policy.

6.4 Verification

The verification step by the DM is important because the DM's commitment to implementing the steps recommended by a model depends on how this model can represent the reality of the DM for the problem addressed. If the model does not properly represent the DM preference, then steps 3 and 4 are reviewed.

To perform this verification, variations in the preference parameters and the vetoes previously elicited as well as variations in the performance are presented to the DM. As an example of this process, the original scale constants were changed; in this way, we observe the results for $k_C = 0.8$ and $k_R = 0.2$. Table 4 shows the effect of this variation.

By varying the scale constant k, it can be seen how the DM preference is related to determining the best alternative for the model. When the scale constant for the cost increases, the best alternative has a higher risk and a smaller maintenance cost.

It is possible to observe that when the cost is prioritized by the decision-maker, in general it results in a postponement of the maintenance type II. It is also important to observe that the maintenance strategy with smallest cost is in the position number 10. This gives to us a good idea how important it is to consider the multiple criteria perspectives. Like in the first analyses presented in Table 3, the trade-off between the performance of the two criteria involved in the problem that each alternative is capable to provide is more important than its isolated performance in each criterion. This becomes even more evident when the veto function corrects some possible imbalance between the performances of the two criteria.

Table 4 Effect of this variation

Alternative	TN (months)	v(a)	Rank	Risk	Cost
(T = 120, N = 21)	84	0.97235	1	0.00436	0.0985
(T = 120, N = 20)	80	0.97224	2	0.00406	0.1000
(T = 120, N = 19)	76	0.9722	3	0.00456	0.1017
(T = 120, N = 18)	72	0.97211	4	0.00377	0.1038
(T = 150, N = 15)	75	0.97199	5	0.00412	0.0974
(T = 150, N = 14)	70	0.97196	6	0.00349	0.0999
(T = 90, N = 17)	51	0.97178	7	0.00321	0.1334
(T = 150, N = 13)	65	0.97175	8	0.00369	0.1030
(T = 120, N = 22)	88	0.97167	9	0.00467	0.0972
(T = 150, N = 16)	80	0.97158	10	0.00501	0.0954

Notice that for the second analysis, the best alternative is (T = 120, N = 21); this alternative doesn't appear as a good option for the decision-maker in the first analysis. The same occurs when considering the best alternative for the first analysis (T = 90, N = 13); notice that it is not presented as a good option in the second analysis.

7 Conclusion

Implementing an effective maintenance policy for the circuit breakers of a substation is a concern for all power utilities to ensure the efficient operation and to reduce the damage of the substation when a failure occurs. Due to the various factors affecting this decision, it is not easy for the DM to select maintenance actions on circuit breakers. This chapter presents a model of multi-criteria decision-making to determine the maintenance of circuit breakers, considering the criteria of cost and risk. Using a multi-criteria model with a veto, the preference of the DM can be represented, and the order of maintenance actions can be corrected, by penalties defined by the veto function, integrated with the value function.

The decision-making process using the proposed model and the steps for its implementation are presented. The resulting recommendation of the model allows the DM to set up the maintenance plan for the circuit breakers, not only the frequency of the (T) verification/inspections, as well as the number of cycles (N) until the renewal.

The most important aspect is that the model provides a set of the best alternatives that are aligned with the preference of the decision-maker. It could be observed that the overall values of these alternatives are very close, indicating that any one alternative from this set represents a good alternative. This gives an important flexibility for the decision-maker that can choose the alternative that best fits in any pre-existing package plan, taking advantage for resources that are already planned to be used with that periodicity.

Acknowledgments The work of the authors has been supported by CAPES (Coordination for the Improvement of Higher Education Personnel), CNPQ (Brazilian National Council), and FACEPE (Foundation for Science and Technology of the State of Pernambuco).

References

Ahmadi A, Kumar U (2011) Cost based risk analysis to identify inspection and restoration intervals of hidden failures subject to aging. IEEE Trans Reliab 60(1):197–209. https://doi.org/10.1109/TR.2011.2104530

Ahmadi A, Kumar U, Ghodrati B (2010) Risk based maintenance decision for periodically tested repairable components subject to hidden failure. In: 2nd International Conference on Reliability, Safety and Hazard (ICRESH), pp 197–204

Ahmadi A, Block JM, Kumar U (2012) Risk based maintenance deferral for components subject to hidden failure. In: Reliability and Maintainability Symposium (RAMS)

Alencar MH, De Almeida AT (2015) A multicriteria decision model for assessment of failure consequences in the RCM approach. Math Probl Eng (Print) 1–10. https://doi.org/10.1155/2015/729865

Allan RN (1988) Effects of protection systems operation and failures in composite system reliability evaluation. Int J Electr Power Energy Syst 10(3):180–189. https://doi.org/10.1016/0142-0615(88)90034-8

Aven T (1986) Formulae for the average unavailability (MFDT) of a coherent system with periodically tested components. Microelectron Reliab 26(2):283–288. https://doi.org/10.1016/0026-2714(86)90726-2

Boudreau JF, Poirier S (2014) End-of-life assessment of electric power equipment allowing for non-constant hazard rate – application to circuit breakers. Int J Electr Power Energy Syst 62:556–561. https://doi.org/10.1016/j.ijepes.2014.05.016

Cavalcante AVC, Lopes RS (2014) Multi-criteria model to support the definition of opportunistic maintenance policy: a study in a cogeneration system. Energy 80:32–40. https://doi.org/10.1016/j.energy.2014.11.039

De Almeida AT (2013) Additive-veto models for choice and ranking multicriteria decision problems. Asia-Pac J Oper Res 30:1350026-1–1350026-20. https://doi.org/10.1142/S0217595913500267

De Almeida AT, Cavalcante CAV, Alencar MH, Ferreira RJP, De Almeida-Filho AT, Garcez TV (2015a) Multicriteria and multi-objective models for risk, reliability and maintenance decision analysis, International series in operations research & management science, vol 231. Springer, New York

De Almeida AT, Ferreira RJP, Cavalcante CAV (2015b) A review of the use of multicriteria and multi-objective models in maintenance and reliability. IMA J Manage Math (Print) 26:249–271. https://doi.org/10.1093/imaman/dpv010

De Lima MAX, Clemente TRN, De Almeida AT (2016) Prioritization for allocation of voltage regulators in electricity distribution systems by using a multicriteria approach based on additive-veto model. Int J Electr Power Energy Syst 77:1–8

Hussain A, Lee SJ, Choi MS, Brikci F (2015) An expert system for acoustic diagnosis of power circuit breakers and on-load tap changers. Expert Syst Appl 42(24):9426–9433. https://doi.org/10.1016/j.eswa.2015.07.079

Javadi H (2010) Fault current limiter using a series impedance combined with bus sectionalizing circuit breaker. Int J Electr Power Energy Syst 33(3):731–736. https://doi.org/10.1016/j.ijepes.2010.11.023

Jian L, Tianyuan T (2015) LS-SVM based substation circuit breakers maintenance scheduling optimization. Int J Electr Power Energy Syst 64:1251–1258. https://doi.org/10.1016/j.ijepes.2014.09.013

Kadziński M, Słowiński R, Greco S (2016) Robustness analysis for decision under uncertainty with rule-based preference model. Inf Sci 328:321–339. https://doi.org/10.1016/j.ins.2015.07.062

Lopez-Roldan J, Pater R, Poirier S, Birtwhistle D, Tang T, Doche R, Blundell M (2014) Development of non-intrusive monitoring for reactive switching of high voltage circuit breaker. Int J Electr Power Energy Syst 61:P219–P228. https://doi.org/10.1016/j.ijepes.2014.03.048

Meeuwsen JJ, Kling WL (1997) Effects of preventive maintenance on circuit breakers and protection systems upon substation reliability. Electr Power Syst Res 40(I3):181–188. https://doi.org/10.1016/S0378-7796(96)01154-6

Meeuwsen JJ, Kling WL, Ploem WAGA (1996) The influence of protection system failures and preventive maintenance on protection systems in distribution systems. IEEE/PES 1996 Winter Meeting, paper no. 96 WM 065–3 PWRD

Mijailovic V (2003) Probabilistic method for planning of maintenance activities of substation components. Electr Power Syst Res 64(1):53–58. https://doi.org/10.1016/S0378-7796(02)00148-7

Mingliang L, Keqi W, Laijun S, Jianju Z (2015) Applying empirical mode decomposition (EMD) and entropy to diagnose circuit breaker faults. Optik 126(20). https://doi.org/10.1016/j.ijleo.2015.05.145

Rausand M, Vatn J (1998) Reliability modelling of surface controlled subsurface safety valves. Reliab Eng Syst Saf 61(1–2):159–166. https://doi.org/10.1016/S0951-8320(97)00066-5

Roy B (1996) Multicriteria methodology for decision analysis. Kluwer, Dordrecht

Vaurio JK (1997) On time-dependent availability and maintenance optimization of standby units under various maintenance policies. Reliab Eng Syst Saf 56(1):79–89. https://doi.org/10.1016/S0951-8320(96)00132-9

Vilaithong R, Tenbohlen S, Stirl T (2007) On-line tap changer diagnosis based on acoustic technique Mat Post, pp 15–16. http://www.uni-stuttgart.de/ieh/forschung/veroeffentlichungen/MATPOST2007.pdf

Vincke P (1992) Multicriteria decision-aid. Wiley, London

Wallenius J, Dyer JS, Fishburn PC, Steuer RE, Zionts S, Deb K (2008) Multiple criteria decision making, multiattribute utility theory: recent accomplishments and what lies ahead. Manag Sci 54(7):1336–1349

Zavadskas EK, Vilutienė T, Turskis Z, Šaparauskas J (2014) Multi-criteria analysis of projects' performance in construction. Arch Civil Mech Eng 14(1):114–121. https://doi.org/10.1016/j.acme.2013.07.006

Part IV
Optimization and Multiobjective Models for Reliability and Safety Models for Systems

A Bayesian Model for Monitoring and Generating Alarms for Deteriorating Systems Working Under Varying Operating Conditions

Ramin Moghaddass, Zachary Bohl, Raul Billini, and Shihab Asfour

1 Introduction

Mechanical systems often operate under various levels of stress, load, and environmental factors during their lifetimes, which can significantly influence the rate of damage and failure for their components over time. For example, wind turbines are exposed to variable loading conditions that may be extracted from their supervisory control and data acquisition (SCADA) systems (Vera-Tudela and Khn, 2017). Depending on the conditions, load, and stress levels the turbines operate in, their age and degradation pattern may vary. The control of wind turbines, which is a complex and interdisciplinary subject, is highly dependent on the wind turbine operational regions (Novaes Menezes et al., 2018). Thus, it is critical to monitor the history of operating conditions for mechanical systems, such as wind turbines, for better real-time diagnostics and prognostics, particularly for better remaining useful life (RUL) estimation. From small equipment to large machinery, accurate modeling and detection of damage levels and forecasting of breakdown time can result in significant cost savings, better maintenance scheduling, and more uninterrupted service periods. Cumulative fatigue models have been widely used for both degradation diagnosis and prognosis of mechanical components and systems working under varying operating conditions. Although many cumulative damage models have been developed in the literature, none of them enjoys universal acceptance, and the applicability of each model varies from case to case. Each

R. Moghaddass (✉) · Z. Bohl · S. Asfour
University of Miami, Coral Gables, FL, USA
e-mail: ramin@miami.edu; z.bohl@umiami.edu; sasfour@miami.edu

R. Billini
World Fuel Services, Miami, FL, USA
e-mail: rbillini@wfscorp.edu

A. T. de Almeida et al. (eds.), *Multicriteria and Optimization Models for Risk, Reliability, and Maintenance Decision Analysis*, International Series in Operations Research & Management Science 321, https://doi.org/10.1007/978-3-030-89647-8_12

damage model can only account for a few phenomenological factors, such as load dependence, multiple damage stages, nonlinear damage evolution, load sequence, interaction effects, overload effects, or small amplitude cycles below the fatigue limit. It is widely known that none of the existing damage models can encompass all of these factors (Fatemi and Yang, 1998). Considering all the limitations, more efforts in the study of cumulative fatigue damage are needed in order to provide design engineers and maintenance decision makers with a general and reliable fatigue damage analysis and life prediction model (Fatemi and Yang, 1998).

Some of the most commonly used models for cumulative damage modeling are the Palmgren–Miner's rule and its extensions. Miner's rule is the epitome of the linear damage accumulation approach and receives extensive usage in engineering due to its simplicity (Zuo et al., 2015). This model is very simple and highly interpretable due to its physics-based nature. In this method, increments of damage, expressed as fractions of a lifetime at particular stress levels, are linearly added together to express the total damage and the overall lifetime (Christensen, 2008). Based on this rule, at a constant stress level, if a component is cyclically loaded for a known number of cycles, each cycle becomes a small piece of the total life of the component (Stillinger et al., 2012). Based on the Palmgren–Miner's rule, the overall damage C can be calculated in terms of the number of cycles applied at a given stress range l (denoted by n_l) divided by the corresponding number of cycles to failure at the same stress level l (denoted by N_l), as

$$n_1/N_1 + n_2/N_2 + \cdots + n_L/N_L = C,$$

where L is the number of stress levels. Based on this model, the failure occurs when the summation of the damage increment at the intervening stress ranges reaches a pre-defined threshold (Blason et al., 2016). Experience has shown that many components exposed to varying loads fail in a manner that is consistent with the Palmgren–Miner's rule (Stephens et al., 2000). The model predictions based on this framework coincide well with engineering hypotheses and experimental results, even though further verification and validation are required (Sun et al., 2014). In spite of the fact that Miner's rule is suggested many decades ago, it is still regarded by many scientists and physicists as a suitable model to quantify cumulative fatigue damage (Suhir et al., 2017). Although many linear and nonlinear damage models have been developed over the years for many applications, the Palmgren linear damage rule is frequently used because of its simplicity and the experimental fact that other complex damage theories do not always hold (Stillinger et al., 2012).

Despite its popularity, Miner's rule has some important limitations that made its original version and some of its extensions difficult to use and hard to justify in practice. First, the original Miner's rule model and many of its extensions, such as the nonlinear cumulative damage rule, the damage curve approach (DCA), and approaches based on crack growth, are deterministic in nature. Thus, they lack the probabilistic nature that is required for the proper analysis of many fatigue failures (ReliaSoft, 2007). Also, given the deterministic approach of Miner's rule, it does not account for the statistical dispersion of cumulative damage (Sun et al., 2014).

The second limitation of Miner's rule is that it considers a fixed deterministic failure threshold for all systems/components of the same type; however, many experimental results show that for different components, the individual failure thresholds are different and usually vary between 0.5 and 2.5 (Sun, 1994). Another limitation is that almost all work in this domain has assumed that the model parameters (such as the life expectancy at each stress level and failure thresholds) are fully known, and no details are available on how these parameters should be estimated/updated from historical data (with possible missing points). Many extensions of Miner's rule are reported in the literature to address some of the above limitations. For example, the work of Zuo et al. (2015) accounts for low-to-high load sequence and high-to-low load sequence, and in Suhir et al. (2017), the classic Palmgren–Miner's rule is extended for the case of random loading. Sun (1994) developed a revised Miner's rule that eliminates the drawback in the original model by distinguishing between a component population and an individual in that population and taking into account the fact that the damage accumulation prior to failure for different individuals in a population is not a constant but a random variable following a probability distribution. In Paolino and Cavatorta (2014), a stochastic version of Miner's rule was implemented to include random stress thresholds, but no investigation was carried out on the stochastic behavior of the remaining useful life estimation or other prognosis purposes. In spite of its major shortcomings, Miner's rule is still commonly used for damage modeling and accumulation (Fatemi and Yang, 1998; Liang and Chen, 2016). In fact, it is by far the most well-known and used damage summation law (Ciavarella et al., 2017).

In addition to changes in the structure of the basic Miner's rule to address its limitations, relatively few articles have used it for purposes other than damage modeling and tracking, such as remaining useful life estimation and prognosis. For example, in Gu et al. (2007), Miner's rule was used to predict the life consumed and remaining life for electronics under vibration loading. Also, in Gu et al. (2009), a health monitoring and prognostics methodology based on the deterministic Miner's rule approach was discussed for assessing the reliability of a group of electronic components mounted on a printed circuit board by using strain gauges and an accelerometer to monitor the life-cycle vibration loads. It is clear that the remaining life estimation using the basic Miner's rule cannot accommodate for uncertainty and gives only a point estimate for the prediction. Thus, RUL estimation needs to be studied using stochastic frameworks for damage modeling. One potential use of estimated RUL is in decision making with respect to maintenance. For example, for systems such as wind turbines with relatively long service lives (20–40 years), a large number of wind turbines may reach the end of their service lives at the same point in time (Ortegon et al., 2013). Thus, it is important for system operators to decide when to initiate and prepare maintenance setup activities before the actual failure occurs.

The chapter is mainly motivated by the need to develop reliability-based mathematical models that can efficiently generate real-time actionable insights and decision-making intelligence for systems working under dynamic operating and environmental conditions, which cover the vast majority of condition-monitored systems, particularly in the wind and power industries. In this chapter, we propose a new stochastic version of the Miner's rule using a hierarchical Bayesian framework that more realistically accounts for uncertainty and nonlinearity of damage. We also provide an efficient approach for parameter estimation, remaining useful life prediction, and a warning generation policy that determines the optimal time to issue a warning/alarm. The results of the warning generation policy can be transformed into an alarm system that gives audible, visual, or other forms of signals to indicate a potential need or hazardous condition (Deb and Claudio, 2015). With the proposed Bayesian hierarchical modeling approach, the parameter and hyperparameters have a reasonable role in the framework to control the complexity of the model in an interpretable manner.

Our main contributions in this chapter are summarized below. Unlike the deterministic Miner's rule, our model is stochastic and accommodates for the uncertainty in both model parameters and model structure. The model does not assume that model parameters are known a priori and instead trains the model's parameters with past data. The framework contains a full reliability pipeline, including model formulation, parameter estimation, diagnosis and prognosis, and decision making. Note that with the proposed flexible formulation, the classic Miner's rule and some of its basic extensions can be considered as special cases of our model. Another new aspect of our model is that it can incorporate all covariates and stress factors in the model and also can determine their importance in terms of affecting the damage process.

Our work is different from the recently published work of Liu et al. (2017), which employed hierarchical Bayesian structure only for fatigue curves estimation, and the work of Suhir et al. (2017), which extended the deterministic Palmgren–Miner's rule to accommodate random loading, for the following three reasons: (i) Our model has a hierarchical Bayesian structure that simultaneously imposes uncertainty on the number of cycles to failure at each stress and system damage level and at the failure threshold; (ii) our model considers a full reliability pipeline, including degradation modeling, parameter estimation with data, damage tracking, remaining useful life prediction, and decision making for generating alarms; and (iii) our model is covariate-based, that is, the effect of internal and external covariates can be taken into account in the damage modeling. In addition, the proposed model allows for both structural priors and noninformative priors and clearly defines the role of the model's hyperparameters in the control of model complexity and model interpretation. Also, we do not claim that our model works on all application areas since the effectiveness of any damage model highly depends on the application and the characteristics of the degradation process.

The remainder of the chapter is organized as follows. Section 2 presents the developed framework and the interpretation of parameters and hyperparameters. In Sect. 3, a model training framework is developed. Section 4 discusses how to predict remaining useful life and the uncertainty associated with it. A dynamic policy is introduced in Sect. 5 that can use the results of our model to determine (in real time) the optimal time to issue a warning to the operator based on the estimated damage level of the deteriorating system. Finally, in Sects. 6 and 7, we demonstrate the application and correctness of our model with comprehensive numerical experiments using simulation-based and real data.

2 The Model

We list below the main notation used throughout the chapter below:

$^{(i)}$: The index for the ith system (or the ith life trajectory sample)
$T^{(i)}$: The actual age (failure point) of the ith system in the training data
L: The number of distinct load/stress levels the system operates over its life cycle
P: The number of covariates or attributes
$\boldsymbol{x}^{(i)} = [x_1^{(i)}, \ldots, x_P^{(i)}]$: The covariate vector for the ith system
$o_t^{(i)}$: The binary operating status of the ith system at time t (1 means failure)
N_l: The expected life of the system (number of cycles) if operated at the load level l
$n_{l,t}^{(i)}$: The total number of cycles up to time t that system i spent on the load level l
$\alpha_l^{(i)}$: The overall cycle ratio for the ith system at the lth stress level
$c_t^{(i)}$: The damage index of the ith system at time t
W: The damage capacity of the system

Without loss of generality, we use the term damage, fatigue, degradation, and deterioration interchangeably.

2.1 A Deterministic Linear Damage Framework

Cumulative damage models have been widely used for various types of mechanical components to estimate the actual damage and the number of cycles that the system will last over time. The linear cumulative damage rule was first proposed by Palmgren (1924) for predicting ball bearing life. Then, it was independently adopted by Miner (1945) as a tool to calculate fatigue of aircraft components. The main assumption of this model is that mechanical components operate under L different finite stress levels $\{1, \ldots, L\}$ with stress factor $\{S_1, \ldots, S_L\}$. By definition, the larger the stress level, the shorter the expected lifetime for the component under study. In most original articles on Miner's rule, a simple idea was introduced: If a component is cyclically loaded at the stress level l, then it would cause fatigue failure in N_l cycles. Thus, each cycle would exhaust one unit in N_l of the life of the component. Under the assumption that the sum of individual damages equals C, the following equation represents the original Miner's rule:

$$n_1/N_1 + n_2/N_2 + \cdots + n_L/N_L = C, \tag{1}$$

where n_l is the number of cycles at the lth stress level (n_l is the number of cycles accumulated at stress level S_l), N_l is the number of cycles to failure at the lth stress level, n_l/N_l is the damage ratio at the lth stress level, and C is the overall damage index. This model is fully deterministic and does not include a component to account for random variation either in model parameters or in its structure. For this reason, Miner's rule is usually defined in terms of a critical value for C, which is experimentally found to be between 0.7 and 2.2 (Akbarzadeh and Khonsari, 2016). Note that the damage fraction C at any stress level is linearly proportional to the ratio of the number of cycles of operation (n_l) to the total number of cycles that would produce failure at that stress level l (N_l). This model simply states that if a part is cyclically loaded at a constant stress level for a given number of cycles, then each cycle becomes a small part of the total fatigue life (Stillinger et al., 2011). Many experiments reported in the literature verified that components exposed to varying loads fail in a manner that is consistent with the Palmgren–Miner algorithm. Despite its success, this model has some shortcomings as discussed before. Some of these shortcomings are addressed in a structured manner in this chapter. Our main focus is to relax its deterministic structure and make it more useful for generating diagnostics, prognostics, and decision-making insights.

2.2 A Stochastic Damage Model with a Bayesian Hierarchical Structure

The proposed model has a hierarchical structure that allows it to account for uncertainty and other important features associated with the degradation. First, the damage control index in our model can change based on the observable covariate vector, which includes factors that may change the stochastic behavior of the damage over time. Examples of covariates are location, environmental factors (e.g., temperature and humidity), or any other fixed or time-dependent variables. For simplicity, we only include fixed covariates in this work; however, results can be easily extended to include time-dependent covariates. If time-dependent covariates were to be included in the model, then $\boldsymbol{x}^{(i)}$ would be replaced by $\boldsymbol{x}^{(i)}{}_t$ in Eq. (2) (this is beyond the scope of this chapter). Based on our model, the damage control index at time t for system i, which is the sum of individual damages at each stress level up to time t, can be computed as

$$c_t^{(i)} = \left(\frac{n_{1,t}^{(i)}}{N_1} + \frac{n_{2,t}^{(i)}}{N_2} + \ldots + \frac{n_{L,t}^{(i)}}{N_L} \right) \exp(\boldsymbol{\beta} \boldsymbol{x}^{(i)}), \tag{2}$$

where $\boldsymbol{\beta} = [\beta_1, \ldots, \beta_p]$ is the regression coefficient vector associated with P covariates, $\boldsymbol{x}^{(i)} = [x_1^{(i)}, \ldots, x_P^{(i)}]$ is the covariate vector for the ith system, $N_1, N_2, \ldots, N_L$ are the age of the system at each stress level if the system is operated only at that stress level, and $n_{l,t}^{(i)}$ is the total number of cycles up to time t that system i spent on load level l. Based on this formulation, the damage monitoring system only needs to track the number of cycles spent at each load level $\{1, \ldots, L\}$ over time and the values of the covariates. The system is assumed to fail if its control index exceeds the threshold $\gamma^{(i)}$ as

$$\textbf{System Failure Rule}: \quad \text{If } \left\{c_t^{(i)} > \gamma^{(i)}\right\} \rightarrow \left\{o_t^{(i)} = 1 \ \& \ T^{(i)} = t\right\}. \tag{3}$$

Based on the above equation, the effective age of the system ($T^{(i)}$) is t, and the health/operating status at time t is failure (i.e., $o_t^{(i)} = 1$). Unlike many traditional damage models, the failure threshold $\gamma^{(i)}$ is not fixed and may vary for different systems of the same type. Also, due to the stochastic nature of $N_1, \ldots, N_L$, the damage index is also stochastic. By defining structural priors, we accommodate uncertainty and stochastic modeling into our framework as described in the next subsection.

The Full Bayesian Hierarchical Model for Damage Progression

The model described in the previous subsection was the basic model that does not define the relationship between model variables and does not account for the accommodation of uncertainty. The hierarchical form described here will explain the structure of the model and will assist in the interpretation of model parameters. All parameters and hyperparameters in our model have a meaning and can be interpreted, thus our model is not a black-box model. The model is designed so that important parameters, which are often assumed to be known in the literature, can be trained/estimated by past data. This is a benefit over models that simply assume that many of the model parameters are known a priori and do not change with time. The variables and their stochastic characteristics considered in this chapter are explained below:

I. *System Damage Index (W)*: This parameter reflects the maximum bearable damage in the system, which is sometimes referred to as a critical damage. This parameter is often assumed to be proportional to the stress level and life expectancy at each stress level. For a fixed stress level, the larger the value of W, the larger the overall life and the life expectancy at each stress level. It is assumed in this chapter that parameter W is not known a priori and should be estimated with data. We let W follow a Gaussian prior as

$$W \sim \mathcal{N}(0, \sigma_w), \tag{4}$$

where σ_w is a hyperparameter that controls the level of information available for W. In the presence of more specific knowledge on the distribution of W, one may use a predefined prior distribution. For an informative prior, a normal distribution with mean zero and a large variance can be used. Various noninformative prior distributions for σ_w can be used as well, including inverse-gamma and distributions that depend on the data-level variance. In this chapter, we treat σ_w as a hyperparameter and tune it with cross validation. In our numerical experiments, we let $\sigma_w = 10^5$ to have a super-weak prior, since we assume that no information is available on the range of W. We should point out that any informative prior for W should be selected very carefully given the fact that W is a positive variable. Our super-weak prior essentially had no computational effect on this variable and did not lead to the negative domain for this variable. If some information is available with regard to W, the users of our model should employ a more appropriate prior to state the knowledge about this parameter and its positive domain. We should point out that parameter W is fixed for each system and does not change throughout the life of the system.

II. *Stress Level Life Expectancy (N_l)*: This parameter is defined for all stress levels ($l \in \{1, \ldots, L\}$) and reflects the life expectancy of the system at the lth stress level (number of cycles to failure at the lth stress). In the conventional Miner's rule model, the critical damage is the same across all the stress levels and is represented by a one-parameter equation $W = N_l S_l$. In this chapter, we allow for some level of uncertainty in this relationship at each stress level as

$$N_l \sim \mathcal{N}\left(\frac{W}{S_l}, \sigma_N\right), \quad l \in \{1, \ldots, L\}, \tag{5}$$

where σ_N controls the uncertainty in the deterministic relation between W, N_l, and S_l and the deviation from a linear model. Note that the larger the value of σ_N, the greater the deviation from the zero-noise linear relationship between W, N_l, and S_l.It is important to remind that the value of the mean is assumed to be much larger than the standard deviation and thus, we expect not to observe any negative values for N_l. We should point out that we can easily replace $\frac{W}{S_l}$ in the mean of Eq. (5) by other extensions of the Palmgren–Miner's rule, such as the two-parameter Basquin relation given in (Basquin, 1910) (that is $N_l = \frac{W}{S_l^{\alpha_1}}$) and the three-parameter formula given in (Liu et al., 2017) (that is $N_l = \frac{W}{(S_l - \alpha_2)^{\alpha_1}}$), where α_1 and α_2 are new parameters. In this chapter, it is assumed that there is no prior information for σ_N, and we let the model decide whether an estimate best fits the data. The nonnegative parameter σ_N follows the noninformative prior as $\log(\sigma_N) \sim \mathcal{U}(-a, a)$, where the hyperparameter a is set to 100 in this chapter to impose a very weak prior. Note that we also tried the noninformative inverse-gamma(ϵ, ϵ) for $p(\sigma_N^2)$ in our numerical experiments, and the results were very similar. It should be pointed out that since W is fixed and S_ls are monotonically increasing, then we expect the mean in Eq. (5) to be non-increasing over l. Also, since we expect σ_n to be much lower than the mean, then the non-increasing trend of N_l should hold true. During the estimation phase, we will define an additional step to make sure the non-increasing trend of N_l holds true.

III. *Regression Coefficients ($\boldsymbol{\beta} = [\beta_1, \ldots \beta_P]$)*: This vector determines the importance of covariates $\{1,..,P\}$ in the model; the larger the β_p, the more important covariate p becomes. In order to have a sparse model with less variables but that imposes regularization at the same time, these variables are centered at zero with covariance $\sigma_{\boldsymbol{\beta}}$, which is a hyperparameter that is set to 0.01 in this chapter. Thus, we have

$$\beta_p \sim \mathcal{N}(0, \sigma_{\boldsymbol{\beta}}), \ p \in \{1, \ldots, P\}. \tag{6}$$

In other words, we regularize parameter β_p by imposing the Gaussian prior (which can be interpreted as a L2 regularization term) on regression coefficients, where $\sigma_{\boldsymbol{\beta}}$ is a strictly positive scalar. By imposing Eq. (6), we expect to remove unrelated covariates from the damage process. Note that as $\sigma_{\boldsymbol{\beta}}$ decreases, more coefficients are set to zero (i.e., more covariates are removed), and more shrinkage is employed among the remaining covariates.

IV. *Failure Threshold* ($\gamma^{(i)}$): This parameter shows the failure threshold for the ith system and follows a normal distribution centered at γ_0 as

$$\gamma^{(i)} \sim \mathcal{N}(\gamma_0, \sigma_\gamma). \tag{7}$$

We have assumed throughout the chapter that γ_0 and σ_γ are known hyperparameters defined by the users. Here, hyperparameter σ_γ controls the uncertainty of the failure threshold, which is not considered in many models (e.g., Miner's rule-based model) in the literature where failure threshold was assumed to be fixed over time (that is where $\sigma_\gamma = 0$). The larger the σ_γ, the less centered are the thresholds for different systems. We can also define a noninformative prior, such as inverse-gamma for σ_γ and let the model fit its best value with data. With such a definition, we do not simply expect all systems to fail at the same threshold values. Thus, the model accounts for uncertainty and the stochastic nature of uncertainty. For better numerical stability, we let $\gamma_0 = 1$ throughout the chapter so that the failure thresholds center around 1.

V. *System's Overall Health/Operating Status* ($o_t^{(i)}$): This is a binary variable indicating whether or not the system is operating. Based on the failure definition in Eq. (3), we can write the relationship between $o_t^{(i)}$ and $c_t^{(i)}$ as follows:

$$p\left(o_t^{(i)} | c_t^{(i)}, \gamma^{(i)}\right) = \begin{cases} 1, & \text{if } c_t^{(i)} \geq \gamma^{(i)} \\ 0, & \text{otherwise} \end{cases}. \tag{8}$$

For numerical stability and better interpretability of the hierarchical model, we impose the following two constraints in our numerical experiments:

(i) $0.7 \leq \gamma^{(i)} \leq 2.2$ for the ith system, and
(ii) $N_1 \geq N_2 \geq \ldots \geq N_L \geq 0$, that is, the age of the system at each stress level should be greater than higher stress levels.

With the above constraints, we can make sure that (i) the failure threshold values are within the ranges already reported in the literature and (ii) the higher the stress level, the lower is the life expectancy. We should point out that depending on the application, we may need to adjust the range of $\gamma^{(i)}$, especially due to the effect of covariates (i.e., $\exp(\boldsymbol{\beta x}^{(i)})$). The hierarchical model can now be explained in a generating form as follows:

A. Generative Structure of the Damage Model: After the overall damage index W is generated from Eq. (4) and the variance σ_N is generated from its uninformative prior, they together generate $N_1, \cdots, N_L$ (see Eq. (5)). Then, the regression coefficients $\beta_1, \ldots, \beta_P$ are generated from the Gaussian distribution, which is centered at zero (see Eq. (6)).

B. Generative Structure of the Individual Damage Progression: Over time, the number of cycles spent at each stress level is recorded. Any time t, the operation data collected as $n_{1,t}^{(i)}, \ldots, n_{L,t}^{(i)}$ with known covariate vector $\boldsymbol{x}^{(i)}$ are used to generate the damage control index $c_t^{(i)}$. Then the failure threshold $\gamma^{(i)}$ is generated from $\mathcal{N}(\gamma_0, \sigma_\gamma)$. If $c_t^{(i)} > \gamma^{(i)}$, then the system fails, otherwise it continues operating until the next cycle.

According to this model, the structure of the hierarchical model contains four levels of variables that are related to (i) degradation factors, which include variables $W, \sigma_N, N_1, \ldots, N_L$ and hyperparameters $\{a, \sigma_w\}$; (ii) the failure threshold, which includes variable $\gamma^{(i)}$, and hyperparameters $\{\gamma_0, \sigma_\gamma\}$; (iii) the control index, which includes damage index $c_t^{(i)}$, variables $\boldsymbol{\beta}$, and hyperparameter $\sigma_{\boldsymbol{\beta}}$; and (iv) the health/operating status, which includes the observable variable $o_t^{(i)}$. The set of unknown parameters in the model (denoted by $\boldsymbol{\theta}$) is

$$\boldsymbol{\theta} = \left\{W, \sigma_N, N_1, \ldots, N_L, \beta_1, \ldots, \beta_P, \gamma^{(1)}, \ldots, \gamma^{(m)}\right\},$$

and the set of model's hyperparameters (denoted by $\boldsymbol{\vartheta}$) is $\boldsymbol{\vartheta} = \{\sigma_W, a, \gamma_0, \sigma_\gamma, \sigma_{\boldsymbol{\beta}}\}$. The graphical model illustrated in Fig. 1 represents the dependencies and causal relationships among model parameters. Although it is assumed that the model hyperparameters are tuned in with cross-validation and/or prior knowledge, one may

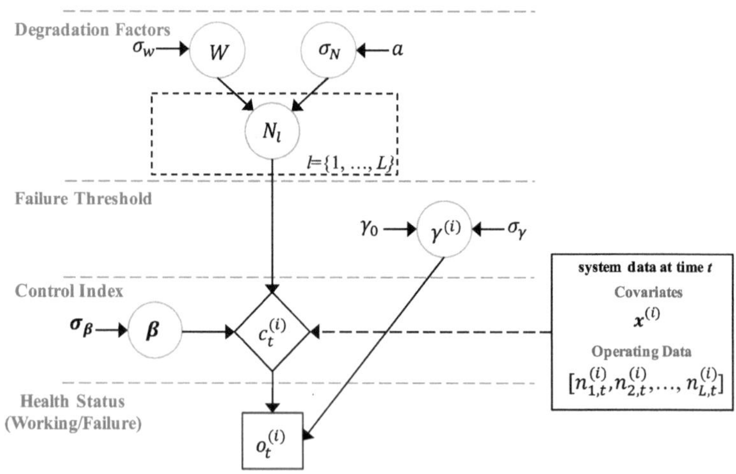

Fig. 1 Directed acyclic graph for the hierarchical damage model. Circles indicate stochastic nodes, rectangles indicate observable factors, and the rhombus indicates the dynamic damage index. The hyperparameters are denoted by $\{a, \sigma_w, \gamma_0, \sigma_\gamma, \sigma_\beta\}$. The box on the right side of the plot represents the covariate vector and collected operating data for system i over time

use other informative or noninformative priors for them and let the hierarchical model find their best values. Also, we point out that the users of the model may choose their preferred distributions for the informative priors used for model parameters. We will show in the remainder of this chapter how to train this model using past data and then how to employ the trained model for real-time diagnosis and prognosis.

3 Model Training with Data

In this section, it will be shown how to use historical data to estimate the parameters of the proposed model, which are $\boldsymbol{\theta} = \{W, \sigma_N, N_1, \ldots, N_L, \beta_1, \ldots, \beta_P, \gamma^{(1)}, \ldots, \gamma^{(m)}\}$, assuming that the set of hyperparameters that control model complexity, denoted by $\boldsymbol{\vartheta} = \{\sigma_W, a, \gamma_0, \sigma_\gamma, \sigma_{\boldsymbol{\beta}}\}$ is known. Suppose that historical data for M independent life trajectories/systems with failure time $T^{(1)}, \ldots, T^{(M)}$ are available. Thus, the historical data, denoted by $\mathcal{D} = [\mathbf{X}, \mathbf{O}, \boldsymbol{T}]$, include a set of known covariates, $\mathbf{X} = [\boldsymbol{x}^{(i)}]$, failure status $\mathbf{O} = [o_t^{(i)}]$, and $\boldsymbol{T} = [T^{(i)}]$, where $i \in \{1, \ldots, M\}$ and $t \in \{1, \ldots, T^{(i)}\}$. It is important to note that although estimating the M-vector thresholds $\boldsymbol{\gamma}$ when we already know the failure times is not useful, they need to be estimated since other unknown parameters depend on them. Also, determining the best values for hyperparameters γ_0 and σ_γ strongly depends on the threshold vector $\boldsymbol{\gamma}$. Assuming that hyperparameters are known, the full posterior based on the hierarchical model, which is proportional to the product of the likelihood and priors' probability, can be written as follows:

$$p(W, \sigma_N, N_1, \ldots, N_L, \gamma^{(1)}, \ldots, \gamma^{(M)}, \beta_1, \ldots, \beta_p | \mathcal{D}, \boldsymbol{\vartheta}) \propto \tag{9}$$
$$p(\mathcal{D}|\boldsymbol{\theta}, \boldsymbol{\vartheta}) \times \prod_{m=1}^{M} p(\gamma^{(m)}|\gamma_0, \sigma_\gamma) \times \prod_{p=1}^{P} p(\beta_p|\sigma_{\boldsymbol{\beta}})$$
$$\times \prod_{l=1}^{L} p(N_l|W, \sigma_N) \times p(W|\sigma_w) \times p(\sigma_N|a).$$

We should point out that since the damage index only changes at discrete time points and other variables are not time dependent, the posterior was written based on discrete time points. The first term is the likelihood probability that includes the probability of survival up to time $T^{(i)} - 1$ and failure at time $T^{(i)}$ for $i \in \{1, \ldots, M\}$, which can be written as follows:

$$p(\mathcal{D}|\boldsymbol{\theta}, \boldsymbol{\vartheta}) = \prod_{m=1}^{M} \prod_{t=1}^{T^{(m)}} p(o_t^{(m)}|N_1, \ldots N_L, \boldsymbol{\beta}, \gamma^{(m)}, \boldsymbol{x}^{(m)}) \tag{10}$$
$$= \left\{ \prod_{i} \prod_{t=1}^{T^{(i)}-1} p(c_t^{(i)} < \gamma^{(i)}) \right\} \times \left\{ \prod_{i} p(c_{T^{(i)}}^{(i)} \geq \gamma^{(i)}) \right\}.$$

Theoretically, one may employ Eq. (9) to find the maximum-a-posteriori (MAP) estimation in order to obtain point estimates of the model parameters. The posterior distribution given in Eq. (9) does not have a closed-form and optimizing it analytically (i.e., by taking a gradient over each unknown parameter and setting it equal to zero to solve it for each parameter) is not a feasible option given the possibility of having a large number of parameters. Below we provide a Markov chain Monte Carlo (MCMC) framework taking advantage of the hierarchical structure of the model for model training. This method will use the hierarchical structure to recursively estimate model parameters while accounting for uncertainty in the estimation.

3.1 A Metropolis-Within-Gibbs (MWG) Approach for Parameter Estimation

We use Markov chain Monte Carlo (MCMC) to approximate the elements of $\boldsymbol{\theta}$ recursively, specifically through the random walk Metropolis–Hastings algorithm. The algorithm employs a Gaussian proposal distribution $J(x, x')$, which proposes a new parameter set x' given the current parameter set x. It should be noted that given a node's parents in the directed acyclic graph, that node is conditionally independent of its grandparents and any other ancestors. Thus, we can generate simple forms for the conditionals that drive the MCMC sampler much faster and more efficiently. Each iteration of MCMC cycles over a subset of components of the parameter vector $\boldsymbol{\theta}$, keeping everything else fixed. The main steps of the algorithm are summarized in Algorithm 1 (Appendix). For large-scale applications, blocked sampling approaches may be necessary. We use $\langle k \rangle$ to refer to the estimates at the kth iteration of the MCMC algorithm. After running the MCMC algorithm, we can compute a Bayesian point estimate using the mean or mode of the posterior distribution. To better account for uncertainty, we can either use Bayesian interval estimates for parameters or calculate the posterior predictive distribution for any parameter/measure of interest.

3.2 Important Remarks for Parameter Estimation

Interval-Censored Failure Data In some deteriorating systems under inspection with hidden/silent failures, one of the key practical challenges for reliability analysis is the uncertainty about the actual time of failure. The inspection data in such cases are known to be interval-censored, that is, the time of failure is within a range between the two last inspection points $T^{(i)} - \Delta$ and $T^{(i)}$. Interval failure data will then have effects on the likelihood function given in Eq. (10). In order to be able to use our model in the presence of interval-censored data, we can rewrite the likelihood function as follows:

$$p(\mathcal{D}|\boldsymbol{\theta}, \boldsymbol{\vartheta}) = \left\{ \prod_{i=1}^{M} \prod_{t=1}^{T^{(i)}-\Delta} p(c_t^{(i)} < \gamma^{(i)}) \right\} \tag{11}$$
$$\times \left\{ \prod_{i=1}^{M} \sum_{t=T^{(i)}-\Delta+1}^{T^{(i)}} \prod_{t_1=T^{(i)}-\Delta+1}^{t-1} p(c_{t_1}^{(i)} < \gamma^{(i)}) \prod_{t_2=t}^{T^{(i)}} p(c_{t_2}^{(i)} \geq \gamma^{(i)}) \right\}.$$

The above likelihood function is computationally much more difficult to evaluate compared to Eq. (10) due to integrals. Approximation techniques may be necessary to efficiently evaluate Eq. (10).

Long Life Cycle and Missing Points Many deteriorating systems have a very long lifetime, and monitoring them at every cycle (for degradation monitoring) is totally unnecessary. For example, wind turbine gearboxes can age for many years, and as a result, inspections often occur every few months or whenever needed. Also, due to problems such as sensor errors the data collected are occasionally lost resulting in missing values. The model presented in this work is less volatile to missing points due to the fact that the required part of the missing data can be recovered by cycle ratios assuming that the load spectrum during the missing interval can be achieved. In other words, one may use the last observed (or the history of) cycle ratios to figure out what stress levels the system has been operating at up to the missing point and then use these cycle ratios to estimate the number of cycles spent at each stress level during a missing interval. This is also an important computational benefit for systems with long cycle times and/or missing points. For example, let us assume that the operating data from time t to time $t + \Delta$ are missing. One can use the updated cycle ratio as

$$\alpha_{l,t}^{(i)} = \frac{n_{l,t}^{(i)}}{\sum_{l'} n_{l',t}^{(i)}}, \quad l \in \{1, \ldots, L\}, i \in \{1, \ldots, M\}, t \in \{1, \ldots, T^{(i)}\},$$

to estimate the number of cycles spent at each stress level during the time interval $[t, t + \Delta]$. Using the history of cycle ratios to find the number of cycles spent at each stress level during an interval is also a reasonable solution for cases in which only a fraction of cycles at each loading value is known as a percentage rather than the actual cycle numbers. For example, if we have 10 years of data for a system with repeating load spectrums, one can use the cycle ratios of the last few years to estimate the number of cycles spent at each stress level for the first few years. In that sense, we do not necessarily have to know all the details about the working conditions from the beginning of the lifetime. Our numerical experiments using real wind turbine data verified that cycle ratios change significantly in the first stages of a turbine's lifetime. As a turbine ages, the cycle ratios remain almost constant and reach the steady state. Thus, it is reasonable to assume that past cycle ratios are good representatives of future operating conditions for a system.

4 Remaining Useful Life Prediction

Remaining useful life (RUL), also known as residual life (RL), is one of the most important measures used in online health monitoring of systems with gradual degradation. RUL of systems that operate under varying operating conditions can be better predicted if future operating conditions of the system are known. While this may be available for some systems, it is often not the case for systems under varying and rapidly changing operating conditions. We provide estimates for the RUL distribution under three different reasonable assumptions for future operating conditions based on the cycle ratio. We do not recommend using our model when none of the below cases is valid/verified for the system. We denote $\hat{\alpha}_{l,t}^{(i)}$ as the projected cycle ratio for future cycles and denote variable $r_t^{(i)}$ as the RUL of the system (computed) at time t, given that the system is still operating at time t. The following three cases are derived based on how much information is available on $\hat{\alpha}_{l,t}^{(i)}$.

Case I The future operating conditions are known (or approximately known) by cycle ratios. That is, $\hat{\alpha}_{l,t}^{(i)}$ is known for $l \in \{1, \ldots, L\}$.

Case II The future operating conditions are not known but the system is assumed to follow the same cycle ratios as the ones up to time t. That is, the system is going to spend $100\alpha_{l,t}^{(i)}\%$ under operating condition l. Thus, $\hat{\alpha}_{l,t}^{(i)} = \alpha_{l,t}^{(i)}$. Also, depending on the application, one may use the cycle ratio based on the last e (defined by users) cycles as the projection for future cycles $\hat{\alpha}_{l,t}^{(i)} = \frac{n_{l,t}^{(i)} - n_{l,t-e}^{(i)}}{\sum_{l'} n_{l',t}^{(i)} - n_{l',t-e}^{(i)}}$. This case is only valid when cycle ratios are almost constant.

Case III The future operating conditions are not known, but the loading cycles are assumed to be distributed in a uniform fashion during the future life cycles. Thus, we have $\hat{\alpha}_{l,t}^{(i)} = \frac{1}{L}$.

For all of the above cases, given the relationship between $c_t^{(i)}$ and $\gamma^{(i)}$, we have

$$r_t^{(i)} : \inf\left\{\mathcal{K};\, c_t^{(i)} + \mathcal{K} \cdot \exp(\boldsymbol{\beta x}^{(i)}) \left(\sum_{l=1}^{L} \frac{\hat{\alpha}_{l,t}^{(i)}}{N_l}\right) = \gamma^{(i)}\right\} \Rightarrow r_t^{(i)} = \frac{\gamma^{(i)} - c_t^{(i)}}{\exp(\boldsymbol{\beta x}^{(i)}) \sum_{l=1}^{L} \frac{\hat{\alpha}_{l,t}^{(i)}}{N_l}}.$$

Since $\gamma^{(i)}$ is not known a priori for a new system i, we can re-write the formula for $r_t^{(i)}$ in a stochastic form as

$$p\left(r_t^{(i)} = \mathcal{K}\right) = \frac{p\left(\gamma^{(i)} = c_t^{(i)} + \mathcal{K} \cdot \exp(\boldsymbol{\beta x}^{(i)}) \left(\sum_{l=1}^{L} \frac{\hat{\alpha}_{l,t}^{(i)}}{N_l}\right) | \gamma_0, a\right)}{p\left(c_t^{(i)} < \gamma^{(i)}\right)} \tag{12}$$

$$= \frac{\mathcal{N}\left(c_t^{(i)} + \mathcal{K} \cdot \exp(\boldsymbol{\beta x}^{(i)}) \left(\sum_{l=1}^{L} \frac{\hat{\alpha}_{l,t}^{(i)}}{N_l}\right); \gamma_0, a\right)}{1 - \mathcal{F}\left(c_t^{(i)}; \gamma_0, a\right)}, \quad \mathcal{K} > 0,$$

where $\mathcal{F}(c_t^{(i)}; \gamma_0, a)$ is the cumulative distribution function of the normal distribution with mean γ_0 and standard deviation a. Note that we can either use the point estimate of the parameters $\boldsymbol{\beta}$ and $\boldsymbol{N}$ from the posterior distribution to evaluate Eq. (12), or use the posterior predictive distribution of $r_t^{(i)}$ from the MCMC samples for $\boldsymbol{\beta}$ and $\boldsymbol{N}$. Equation (12) is a dynamic formula that gives the portability distribution of the RUL based on an estimate for the future cycle ratios. Many conventional models ignore such a consideration when predicting the remaining life. Now that the distribution of the remaining life is given, one can use it to study its stochastic behavior, such as its average, variance, and confidence interval. This formulation gives decision makers much more than a point estimate of the RUL as provided by deterministic Miner's rule models. Note that the RUL using our model is dynamic and changes over time, depending on the condition in which the system was operating over time. Therefore, the estimation is from the family of condition-based measures. We should point out that our model cannot directly handle a more complex case in which the projected future operating conditions are stochastic (such as in Liao and Tian (2013)). This can be done by imposing a probability distribution for the projected operating conditions represented by $\hat{\alpha}_{l,t}^{(i)}$ and incorporating it into Eq. (12).

5 A Dynamic Policy for Real-Time Alarm Generation

The ultimate value of statistical techniques for hazard monitoring and reliability analysis lies in their power for generating actionable insights that can help with maintenance decision making. Conventional maintenance decision-making models often optimize the time to terminate the operation to do maintenance (repair or replacement). Our model has a different focus, as it aims at generating real-time alarms/warnings to operators and decision makers who determine when to start preparing for expensive and time-consuming maintenance activities ahead of failure. This type of alarm is very common for critical and expensive mechanical systems, such as wind turbines. Our model also contributes to remote real-time monitoring and early fault warning strategies, from which a predictive maintenance service can be performed; thus, reliability of the units can be increased (Zhang et al., 2017). By triggering warnings using knowledge of the degradation state, decision makers may be able to issue better-timed maintenance decisions. For instance, our discussion with field experts and collaborators confirmed that wind turbine operators and decision makers would like to know approximately 30 days in advance when a turbine is going to have major failures. This 30-day notice will help them better prepare and plan for costly replacements, major repairs, and lengthy maintenance setups or parts ordering.

We define "warning generation" as a dynamic decision process that depends on the damage level of the system over time. Let us define d as the ideal time between the warning point and the failure point that is determined by decision makers. In other words, the warning generation system is considered efficient or perfect if it generates warnings when the actual time to failure (also called remaining useful life—RUL) is very close to d time units or cycles. To define the quality of the decision-making framework, we define a cost/risk function $g_d(\xi), \xi \geq 0$, to represent the cost of a warning at ξ units before the actual failure time, given d as the ideal value. It is clear that $g_d(d) = 0$, and there is a positive cost for warnings that are early or late (i.e., $g_d(\xi) > 0$, for $\xi \geq 0, \xi \neq d$), where $\xi < d$ means late warning and $\xi > d$ means early warning. The two extreme cases are (i) to generate a warning only at the failure with cost $g_d(0)$, that is, when the warning time equals the failure time, and (ii) to generate a warning at time zero with average cost of $g_d(E(T))$, where $E(T)$ is the mean time to failure. Our model is general in the sense that any mathematical form of cost function g can be considered depending on the application (e.g., hinge, quadratic, logistic, and exponential). The objective is to define a dynamic decision rule for the warning generation process that minimizes the expected total cost or total risk. It should be pointed out that depending on the application, decision makers may choose to have another form of objective function, such as average cost per unit of operation. The structure of the proposed policy is illustrated below.

Let us assume that at time t, we decide to generate a warning after a cycles. If the failure occurs at time $t+a+d$, then the policy is a perfect policy with cost 0. Given that the warning is generated at time $t+a$, three possible scenarios can occur with different cost forms: (i) Failure occurs between now (time t) and time $t+a$, that is, the remaining life is within a cycles; (ii) failure occurs between time $t+a$ and time $t+a+d$, that is, the remaining life is within $[a, a+d]$ cycles; and (iii) failure occurs after time $t+a+d$, that is, the remaining life is greater than $a+d$ cycles. Thus, the expected total cost for system i calculated at time t based on warning time $t+a$, denoted by $J_t^{(i)}(a)$, can be computed as follows:

$$J_t^{(i)}(a) = \int_0^a p\left(r_t^{(i)} = k\right) g_d(d)\mathrm{d}k + \int_a^{a+d} p\left(r_t^{(i)} = k\right) g_d(a+d-k)\mathrm{d}k + \int_{a+d}^{\infty} p\left(r_t^{(i)} = k\right) g_d(k-a-d)\mathrm{d}k. \tag{13}$$

Now the optimal warning time that is determined at time t (denoted by $R_t^{(i)}$) can be found by the following formula:

$$R_t^{(i)} = t + \underset{a}{\arg\min}\ J_t^{(i)}(a).$$

Assuming that the decision on whether to issue a warning is made every Δ cycles, we can define a dynamic decision policy as follows:

$$\text{Decision Policy at time } t: \begin{cases} \text{Issue a Warning Immediately}, & \text{if } 0 \leq R_t^{(i)} < \Delta \\ \text{Wait Until the Next Decision Epoch}, & \text{if } R_t^{(i)} \geq \Delta \end{cases}. \tag{14}$$

The corresponding problem is a 1-d optimization problem with one unknown (a) and can be solved with any single-variable optimization technique.

5.1 *Special Case—Linear Cost Function and Its Optimal Structure*

As mentioned before, any form of cost function g can be used in Eq. (13). Below, we discuss a special and reasonable form of cost function g and discuss the closed-form structure of the optimal policy. A very simple and typical form of function $g_d(\xi)$ is called the pinball loss or newsvendor cost (see for instance Rudin and Vahn, 2013), in which the cost of warning changes linearly with the number of late or early cycles as follows:

$$g_d(\xi) = \begin{cases} c_1(d - \xi), & \text{if } \xi < d \\ 0, & \text{if } \xi = d \\ c_2(\xi - d), & \text{if } \xi > d \end{cases}, \tag{15}$$

where c_1 is the unit cost of a late warning and c_2 is the unit cost of an early warning. We derive four useful remarks with regard to the optimal structure of the warning generation policy for this cost function below.

Remark 1 The optimal policy depends only on the ratios of $\frac{c_2}{c_1}$ and $\frac{c_1}{c_2}$ and not their individual numerical values.

Remark 2 If $c_2 \gg c_1$, then the optimal policy is to do nothing, that is, to issue the alarm at the failure point.

Remark 3 If $c_1 \gg c_2$, then the optimal policy is to issue the alarm immediately (at time zero).

Remark 4 The optimal warning time is 0, ∞, or the solution of the following equation:

$$p\left(r_t^{(i)} \leq a+d\right)(c_1+c_2) - c_1 p\left(r_t^{(i)} \leq a\right) - c_2 = 0.$$

Proofs of Remarks 1–4 The cost function in Eq. (13) can be simplified as

$$\begin{aligned} J_t^{(i)}(a) &= \int_0^a p\left(r_t^{(i)} = k\right) c_1 d\,\mathrm{d}k \\ &\quad + \int_a^{a+d} p\left(r_t^{(i)} = k\right) c_1(d-k+a)\mathrm{d}k \\ &\quad + \int_{a+d}^{\infty} p\left(r_t^{(i)} = k\right) c_2(k-a-d)\mathrm{d}k \\ &= c_1(a+d)\, p\left(r_t^{(i)} \leq a+d\right) - c_1 a\, p\left(r_t^{(i)} \leq a\right) \\ &\quad - c_2(a+d)\left(1 - p\left(r_t^{(i)} \leq a+d\right)\right) \\ &\quad - \int_a^{a+d} c_1 p\left(r_t^{(i)} = k\right) k\mathrm{d}k + \int_{a+d}^{\infty} c_2 p\left(r_t^{(i)} = k\right) k\mathrm{d}k, \end{aligned} \tag{16}$$

which gives

$$\frac{\mathrm{d}J_t^{(i)}(a)}{\mathrm{d}a} = p\left(r_t^{(i)} \leq a+d\right)(c_1+c_2) - c_1 p\left(r_t^{(i)} \leq a\right) - c_2. \tag{17}$$

It is clear from the optimization point of view that

$$\arg\min_a J_t^{(i)}(a) = \arg\min_a \frac{J_t^{(i)}(a)}{c_1} = \arg\min_a \frac{J_t^{(i)}(a)}{c_2},$$

which proves Remark 1. Now, if $c_2 \gg c_1$, then the first derivative given in Eq. (17) is always negative and thus $\arg\min_a J_t^{(i)}(a) = \infty$, which proves Remark 2. If $c_1 \gg c_2$, then the first derivative in Eq. (17) is always positive and thus $\arg\min_a J_t^{(i)}(a) = 0$, which proves Remark 3. Now, given that $a = 0$, $a = \infty$, and the solutions of $\frac{\mathrm{d}J_t^{(i)}(a)}{\mathrm{d}a} = 0$ are critical points of functions $J_t^{(i)}(a)$, we have the proof for Remark 4.

6 Numerical Experiments

To demonstrate the correctness of our model, its application in real-time damage monitoring, and its advantage over conventional models with regard to RUL estimation, several simulation-based numerical experiments were carried out in this section. We also used a real case to demonstrate the application of our model for the condition monitoring of wind turbines, which typically operate under varying operating conditions over their life cycles.

6.1 Simulation Experiments

We will show first that for the data generated from our model, the true data-generating parameters $\boldsymbol{\theta}$ can be recovered from the developed parameter estimation framework. We then show the effectiveness of the model in terms of predicting RUL and generating alarms. A fully stochastic framework was designed to simulate data for our numerical experiments. A system with five stress levels ($L = 5$) and four covariates ($P = 4$) are considered. Multiple trajectories of run-to-failure samples are generated for a single-unit degrading system according to the stochastic structure given in Fig. 1. For each sample, we recorded the time spent at each stress level up to the failure time, that is $[n_{l,t}^{(i)}]$, for $i \in \{1, \ldots, N\}, l \in \{1, \ldots, L\}$, and $t \in \{1, \ldots, T^{(i)}\}$. The hyperparameters are set throughout this section as given below:

$$\boldsymbol{\vartheta} = \left\{\sigma_W = 100,000, a = 10, \gamma_0 = 1, \sigma_\gamma - 0.1, \sigma_{\boldsymbol{\beta}} = 0.05\right\}.$$

Other model parameters are generated according to the model's hierarchy in Fig. 1. The values of other model parameters that are randomly generated based on the hierarchical model are given below:

$$\boldsymbol{\theta} = \Big\{W = 3,921, \sigma_N = 4, N_{1:5} = [259.3, 189.3, 160.2, 1284, 103.8],$$
$$\beta_{1:4} = [-0.01, -0.07, -0.07, 0.077]\Big\}.$$

Note that we ran all the numerical experiments with a few other sets of model hyperparameters and parameters, and the results were very similar. Figure 2 shows the relationship between the failure threshold and the true life/age of the simulated samples. As expected, samples with higher thresholds have higher lifetimes. This

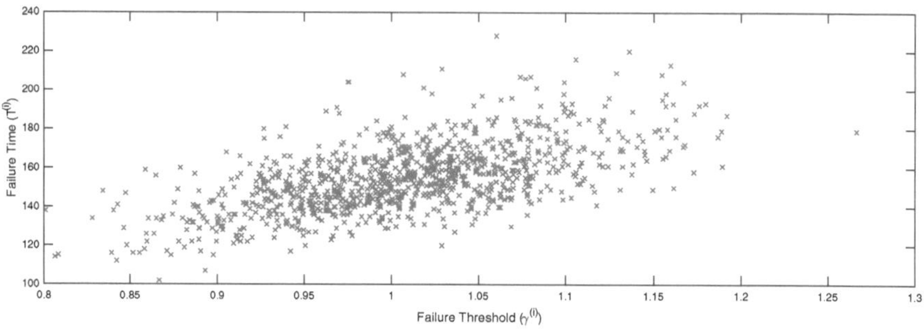

Fig. 2 Failure thresholds ($\gamma^{(i)}$) versus sample lifetimes $T^{(i)}$ for 1000 simulated samples

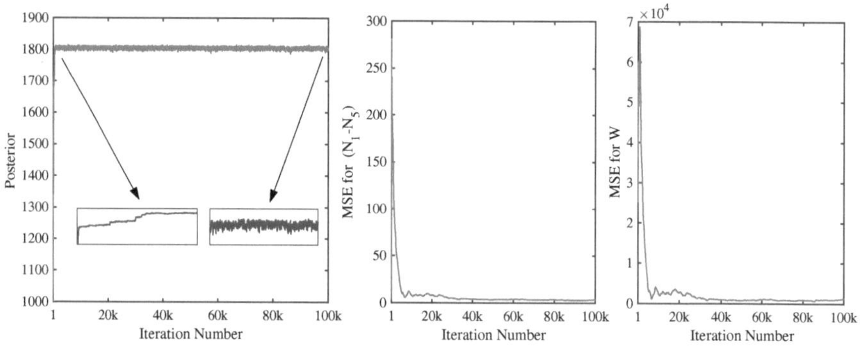

Fig. 3 MCMC iterations and changes in model posterior (left) and model estimation error for $N_1, \ldots, N_5$ (middle) and W (right)

relationship is not deterministic and is not entirely linear, as the age of the samples depends on the operating conditions and their covariates. Also, it can be seen that (as expected) the failure thresholds vary mainly between 0.8 and 1.2 and are centered around 1.

Parameter Estimation
To evaluate the efficiency of the parameter estimation procedure, we first simulated trajectories of run-to-failure data and then used our parameter estimation method to recover the true values of model parameters $\boldsymbol{\theta}$. For $M = 200$ samples, we showed in Fig. 3, the improvement in the posterior distribution (the left plot) and root mean squared errors of estimated parameters $N_1, \ldots, N_5$ and W over 100,000 iterations of MCMC (the right plots). It can be seen that the MCMC associated with model training converges relatively fast and has a reasonable mixing (the left plot). Also, the error of parameter estimation tends to zero as the number of MCMC iterations increases (the right two plots). We observed the same behavior for other parameters.

We then removed the first 5000 iterations (as burn-in samples) of MCMC sampling and kept only 95,000 samples to estimate the posterior estimates. Figure 4

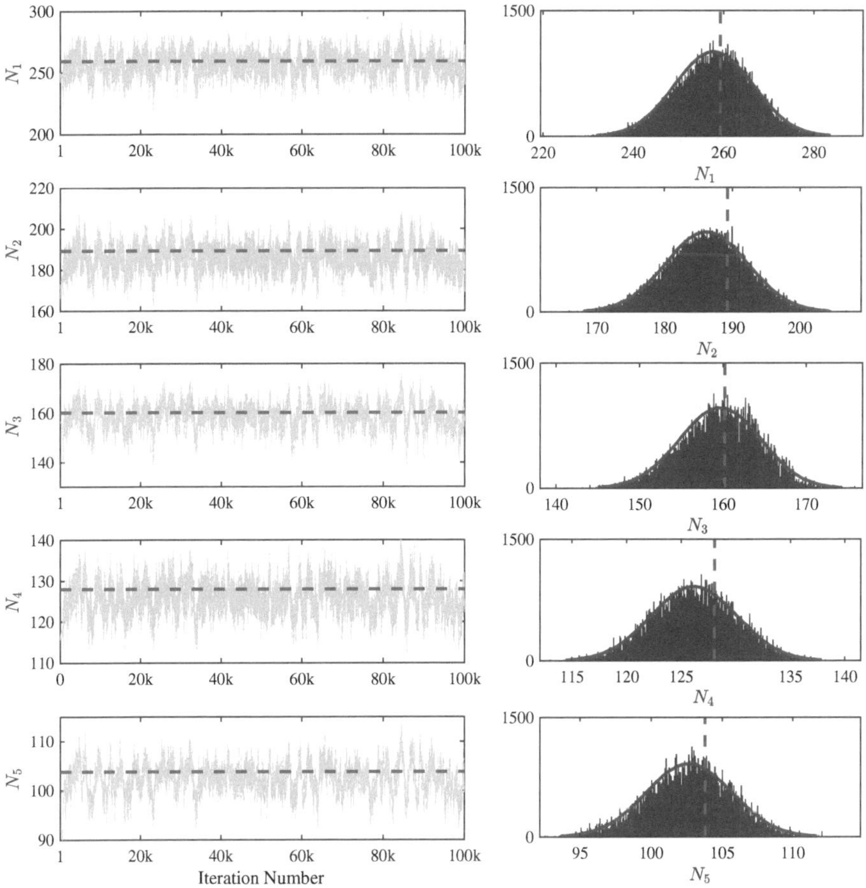

Fig. 4 Samples from the posterior of $N_1, \ldots, N_L$ (left), and their histograms (right). The first 5000 samples were discarded as burn-in time. The solid red lines are true values

shows samples from the posterior of variables $N_1, \ldots, N_L$ from 95,000 iterations of MCMC sampling. From this figure, we observe reasonable mixing and convergence. These figures show that the posterior samples were almost concentrated around the true values, and the posterior mean of each variable was generally close to its true value. To evaluate the effect of the number of samples (M) on the estimation results and to empirically assess the convergence rate for the underlying true values, we considered five cases for M, where M denotes the number of simulated samples ($M \in \{25, 50, 100, 200, 400\}$). We show the results here only for $N_1, \ldots, N_5$, σ_N, and $\beta_1, \ldots, \beta_4$, but the results were similar for other parameters. The true values of the parameters, as well as the means and the standard deviations (SD) of the estimated values from the conducted experiments, are presented in Table 1. Results shown in this table verify thatin almost all cases, estimates are very close

Table 1 Parameter estimation—summary of results for $M \in [25, 50, 100, 200, 400]$

Parameter		N_1	N_2	N_3	N_4	N_5	W	σ_N	β_1	β_2	β_3	β_4
TRUE		259.3	189.3	160.2	128.0	103.8	3921.6	4	−0.010	−0.069	−0.007	0.070
Mean	M=25	255.6	184.1	161.8	122.4	99.1	3853.0	8.26	−0.008	−0.057	0.000	0.067
	M=50	257.1	182.4	161.1	122.3	100.4	3858.8	5.7	−0.003	−0.056	0.002	0.069
	M=100	257.2	183.7	161.5	123.4	100.3	3865.8	5.35	−0.002	−0.058	−0.001	0.071
	M=200	257.2	184.6	160.3	124.1	101.5	3870.6	4.84	−0.005	−0.064	−0.001	0.071
	M=400	257.8	186.2	159.7	126.2	102.7	3888.0	4.69	−0.008	−0.057	−0.043	0.077
SD	M=25	12.7	9.0	7.8	5.5	4.1	196.7	8.55	0.031	0.031	0.031	0.030
	M=50	9.6	6.5	5.8	4.4	3.6	135.2	5.33	0.028	0.028	0.029	0.028
	M=100	9.2	6.4	5.7	4.2	3.3	139.8	5.19	0.028	0.028	0.028	0.028
	M=200	8.6	6.2	5.2	4.0	3.2	133.0	4.24	0.027	0.028	0.027	0.027
	M=400	8.6	6.0	4.9	3.9	3.0	111.0	4.14	0.024	0.025	0.025	0.024

to true values, particularly when the number of samples increases. It is clear that for parameter β_3 the estimation does not converge to the true values, which implies that more data are needed. Also, results verify that the standard error of estimation tends to zero as M increases. We can summarize that the parameter estimation method was computationally tractable and very effective in estimating unknown model parameters, particularly when we have more data.

Real-Time Remaining Useful Life Estimation
Here, we evaluate the power of the model to predict the conditional residual life, that is, the expected average number of cycles over which the system will continue to function computed at a certain time point, given the available information up to that time point. We first show how RUL is estimated for two random samples: one with a relatively long lifetime and one with a relatively short lifetime. In Figs. 5 and 6, the true RUL, the mean RUL estimated from our model, and the 95% prediction intervals are shown for the two selected samples. These values are calculated at 99 discrete points equivalent to 1%–99% of the lifetime of each sample. For instance, the first point on each line is calculated when the age of the system was at 1% of its total age. We have also shown the probability distribution of the remaining life at three points of 1%, 50%, and 99% of the lifetime for better visualization of the RUL prediction. It can be seen that the RUL estimates are mostly within the confidence intervals and are closer to true values as the system ages. Also, results show that as the system ages, the percentile (prediction) interval for the remaining life becomes a bit narrower, that is, the prediction uncertainty decreases. The plotted RUL distribution at three points of 1%, 50%, and 99% of the lifetime gives a better picture of life expectancy in the future based on the most updated set of operating signals. It can be seen that the distribution of remaining useful life centers more closely around true values (lower variance) as the system ages. In both Figs. 5 and 6, the RUL is calculated with the assumption that future operating conditions are unknown; thus, current cycle ratios are utilized to project future operating conditions (Case II in Sect. 4).

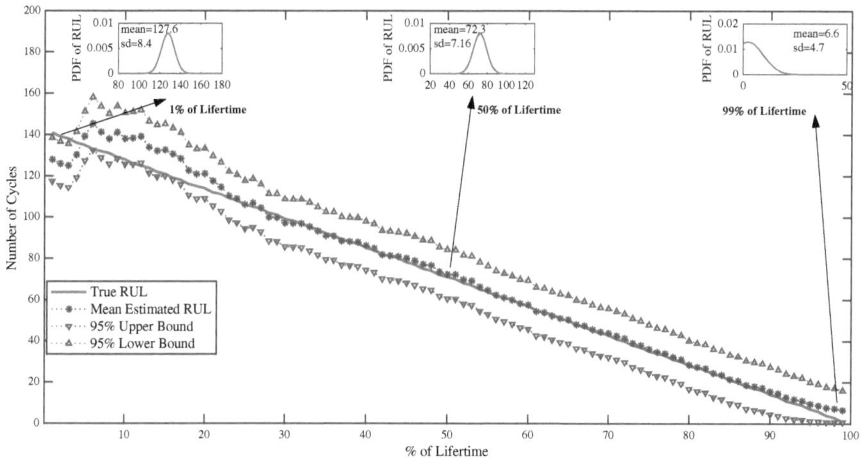

Fig. 5 RUL estimation and its uncertainty at different time points for a relatively short simulated sample

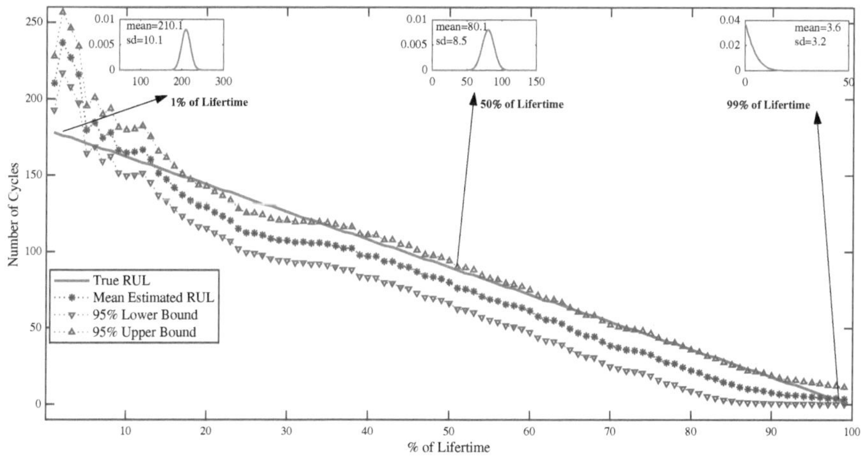

Fig. 6 RUL estimation and its uncertainty at different time points for a relatively long simulated sample; it can be seen that the mean RUL underestimates the actual RUL when the age of the system is relatively long, but the prediction interval covers true values reasonably well

We applied our model to 1000 samples and predicted RUL at certain proportions of each sample's life (1%–99%) using the mean of the predicted distribution (RUL). We calculated the relative error (%) and absolute error of RUL estimation as shown in Fig. 7. The plus signs in the middle are the mean of the errors over 1000 samples. Also, we plotted by circles the mean of the RUL estimation error under the assumption that the exact operating condition of future cycles is known. As shown in this figure, the data in the first few cycles could not accurately predict the

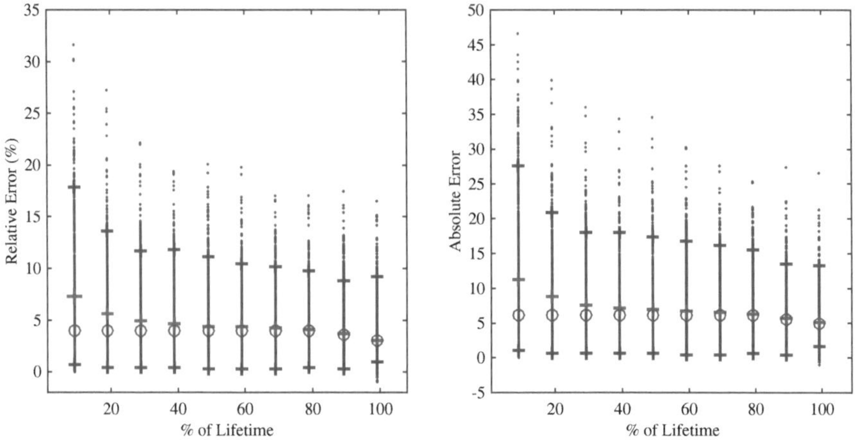

Fig. 7 The error of RUL estimation for 1000 samples calculated at 1%-100% of lifetime (each dot is for 1 sample). The plus signs in the middle are the mean of the errors when the future operating conditions are unknown, and the circles are the mean of the errors when the operating conditions are known. The 90% quantiles are shown as plus signs on the two sides of the vertical lines

remaining life. Results show that the estimation error decreases when the estimates are made closer to the end of the lifetime (that is mainly because more data are used for prediction). Although the estimates based on the true operating conditions are more accurate at the beginning of the lifetimes, they get closer to the case where the operating conditions are not known as the system ages, particularly due to the fact that the operating conditions become more stable. In practice, this phenomenon highly depends on the system, its operating conditions, and its variation over the system's life cycle.

Alarm/Warning Generation Policy

To show empirically the benefit of the developed cost-effective dynamic warning generation policy described in Sect. 5, we compared it with three other types of policies, namely a control limit policy and the two extreme cases of do-nothing and issue warning at the beginning of lifetimes (*Too Early Warning* policy). In the control-limit policy, the warning is generated when the damage index exceeds a predefined and optimized threshold. We applied each policy on 1000 simulated samples and reported the average cost for five different cases of c_2/c_1 (fixed $c_1 = 10$) and five different values of d (total of 25 combinations). Results shown in Fig. 8 verify that (i) the proposed policy performs as well as or better than other policies and (ii) as the cost of late warning increases, the proposed policy tends to recommend earlier replacement times and gets closer to the *Too Early Warning* policy. When the cost of late warning is zero, then our model and the control limit policy both act the same as the *Do Nothing Policy*. Among all policies, do nothing is the worst, especially when the cost of late warnings is high. Since the type of policy

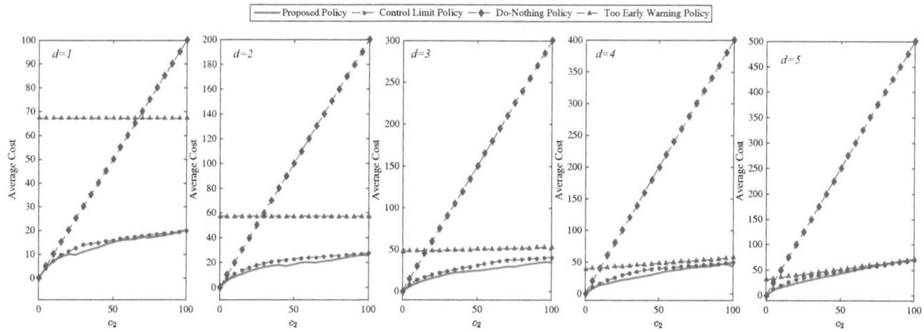

Fig. 8 Results for the developed cost policy and comparison with similar models for various combinations of d and c_2/c_1

developed in this chapter is naturally different from typical maintenance policies, there is no more apparent policy to compare it against.

6.2 A Case Study for the Condition Monitoring of Wind Turbines

Condition monitoring and fault diagnosis of wind turbines have received a high priority over the past years due to the continuous growth of wind energy generating sources and increasing demand for more careful planning and control of operation and maintenance costs (Herp et al., 2016). The wind power industry all over the globe is constantly seeking more cost-effective operations and maintenance (O&M) actions. As a result, various strategies have been proposed for wind turbines (Byon, 2013). Bearing failures are among the most costly types of failure for wind turbines, which can cause unplanned shutdowns, early bearing replacement, reduced availability, and increased cost of energy. A major part of the high price of clean energy produced by wind turbines is the result of the required preventive maintenance and costly breakdowns that result from unexpected gearbox failures. Wind turbines are usually equipped with supervisory control and data acquisition (SCADA) sensors that record various measurements of the dynamic environment every few minutes. Since SCADA measurements provide ample and cheap indirect information about the state of health of the turbine, leveraging these measurements for health monitoring is valuable and has gained more attention over the last decade (Long et al., 2011). Here, we show how wind turbine SCADA data can be used to monitor the degradation status of a wind turbine gearbox and predict its RUL. Through our collaboration with a large utility company in the USA, we have had access to SCADA data for a group of wind turbines operating at five different wind farm locations. It should be noted that due to the confidentiality of the data provided to us, we normalized the raw data and removed all cost and location information as well as detailed failure data from the chapter.

SCADA Data Available for the Study
Partial SCADA data for 15 turbines' life trajectories for a period of 2 years collected every 10 min were available for our research project. Each collection instance for each turbine included the date and time of collection, approximate location of the wind farm, average wind speed, pitch angles, average blade rotor revolutions per minute (rpm), and active power generated in kilowatts over the 10-min interval. Also, environmental features, such as temperature and humidity, could be extracted based on the locations of the turbines (not used in the chapter). Because missing values were rare and were largely resulting from lapses in turbine operation for maintenance or inspection, they were removed from the dataset. First, the total cycles until major gearbox failure ($T^{(i)}$) were calculated for each of the 15 available wind turbines by summing the average RPM in each collection instance before the turbine failed multiplied by 10 min. Throughout this work, one operation cycle is considered one revolution of a turbine's blade, and the cycle ratio is calculated by dividing the number of cycles in each stress bin (will be defined later) by the total number of cycles. It is assumed that the initial degradation level is minimal, but one can utilize any available information for the initial degradation using our model. We used location (with 5 possible outcomes) as the only covariate in our analysis. We conducted a 5-fold cross-validation for parameter estimation and performance evaluation. We randomly divided our dataset into five folds of equal size (3 turbines per fold). We trained the model with 4 folds (12 turbines) and used the last fold (3 turbines) for testing. This process was repeated 5 times so that all folds were used for testing. All results in the remainder of this chapter are based on the testing set.

Stress Levels and Data Preparation
Due to the stochastic nature of wind, a wind turbine gearbox experiences a wide loading variation during its life, which results in premature failure of various forms (Long et al., 2011). Similar to the work of McVittie and Errichello (1990), cyclic loading conditions caused by wind speed variation and torque are considered for defining stress levels. Various cycle counting methods (such as Rainflow and Racetrack methods) can be used to convert complicated load spectrum into a simplified histogram (McVittie and Errichello, 1990). In this chapter, the loads are simply grouped into 10 levels, and the individual loads are assumed to be the same values as the average of the maximum and minimum load for that group. The torque was calculated for each collection instance of each turbine using the formula below:

$$P(kW) = \frac{T(Nm) \times w(rpm)}{60 \times 1000/2\pi} \rightarrow T(Nm) = \frac{P(kw) \times 30{,}000}{\pi \times w(rpm)},$$

where P represents the power generated, T represents the torque, and w represents the rotational speed. Ten levels of torque bins were defined with upper and lower limits as percentages of the maximum torque observed among the turbines. The

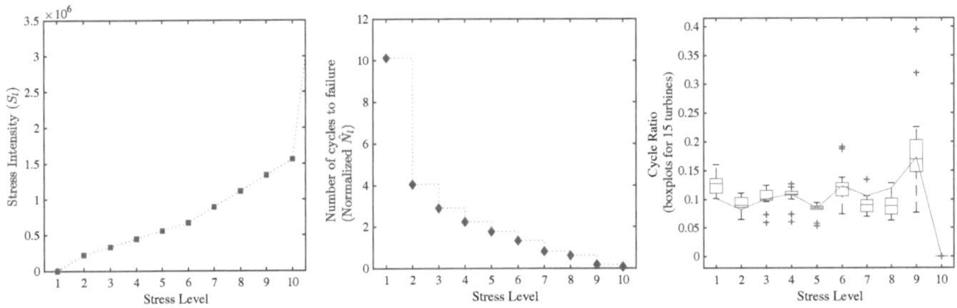

Fig. 9 Stress levels (left), expected normalized estimated lifetime at each stress level (middle plot), and the variation of operating conditions among 15 turbines (right). Note that although peak loads operate for a short period of time, they contribute the most damage to the system, that is, the life of the system at higher stress levels is lower than at lower stress levels

final stress intensities $S_1, \ldots, S_{10}$ are shown in Fig. 9(left plot). The upper and lower limits for each bin were designed to capture the varying operating conditions of the turbine, with smaller bins around the stress levels containing the bulk of operation cycles. Each collection instance (cycle) for each turbine was then categorized into one of the ten defined stress levels. For each turbine and each time point, the cumulative number of operation cycles at each stress level ($n_{l,t}^{(i)}$) was calculated by summing the RPM in each stress category and multiplying by 10 min. Finally, the percentage of operation cycles until failure at each stress level ($\alpha_{l,t}^{(i)}$) for each turbine was calculated. We then employed the MCMC algorithm developed earlier for parameter estimation. In Fig. 9(middle plot), the estimated values of $N_1, .., N_{10}$ are shown based on the mean of the MCMC samples. For confidentiality of the data, we multiplied all $N_1, .., N_{10}$ by a constant, so the numbers presented in this section do not reflect the real failure times. The relationship between the cycle ratio and the stress level for 15 turbines is shown in the right plot in Fig. 9. It can be seen that the turbines operate in some stress levels more than others.

6.3 Degradation Monitoring Using the Proposed Model

Here, we show the use of our model to monitor the degradation of the turbines over their life cycles. We first calculated the damage index $c_t^{(.)}$ for each turbine given the estimated parameters and then grouped based on the five locations. The damage index was calculated at 101 different time points representing 0%–100% of the actual lifetime. It can be seen from Fig. 10 that the damage index monotonically increases as the turbine ages, and the system fails when this index gets close to 1 (which is assumed to be the mean of the damage thresholds). Also, it can be seen that

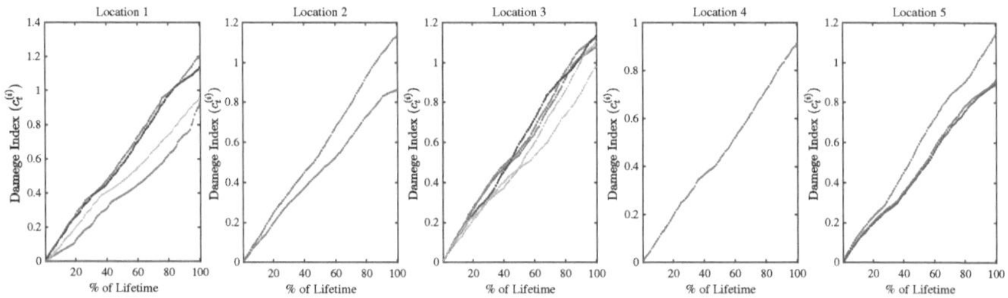

Fig. 10 Damage progression monitoring based on $c_t^{(i)}$ for $i \in \{1, \ldots, N\}$; turbines at the same locations are plotted together. One may find better results by including more covariates

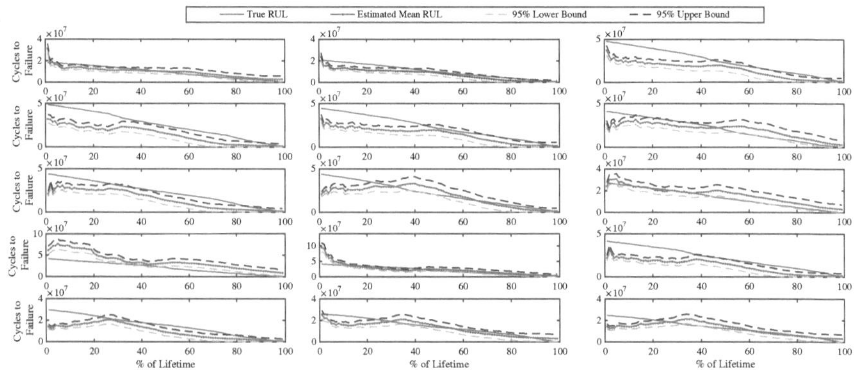

Fig. 11 RUL estimation and its prediction interval for 15 turbines

turbines at the same location behave similarly to each other, implying the usefulness of the location as a covariate in the model.

6.4 Remaining Useful Life Prediction

As discussed before, one of the main applications of our proposed framework is in RUL prediction. In Fig. 11, we plotted the actual cycles to failure (solid line), the estimated remaining life based on the mean (dotted line), and the 95% confidence intervals (dashed lines) for the 15 available turbines at 100 different time instances equivalent to 1%–100% of each turbine's true total age. It can be observed that the estimates are not good in the first half of the lifetime; however, they become closer to the true values as the system ages. We believe this is mainly due to the fact that we used the past cycle ratios as the estimate to project future cycle ratios. This type of estimation is more efficient when the system has already operated for a considerable amount of time so that cycle ratios are more stable and close to the steady state. To

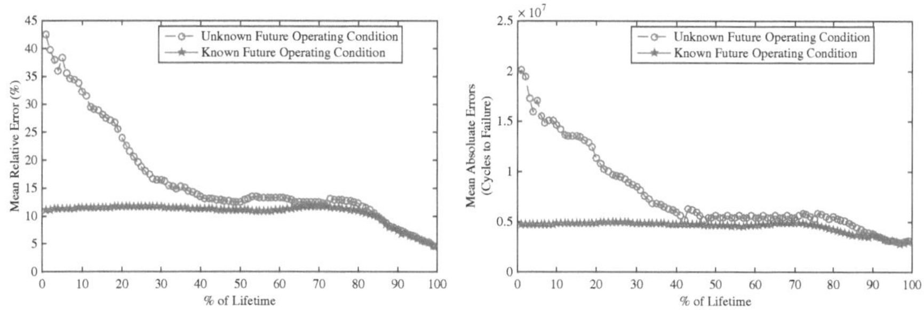

Fig. 12 RUL mean estimation error for 15 turbines vs. % of lifetime considering two cases for future operating conditions

better observe the effect of knowing future operating conditions and also the error of estimation as the system ages, we calculated the mean relative estimation error (%) and the mean absolute error for 15 turbines based on the two first scenarios described in Sect. 4: (i) The future operating conditions are known a priori and (ii) the future operating conditions are not known and are estimated from past history. It can be seen from Fig. 12 that the estimation error for RUL decreases as the system ages, and the estimates based on known future operating conditions are better, particularly during the first half of the life of the turbines. Since the projection of future operating conditions and true operating conditions approach each other as the system ages, the estimation errors based on these two cases become very close to each other.

6.5 *The Warning/Alarm Generation Process*

One of the main applications of our framework is to generate warnings based on an ideal lead time d. We conducted various experiments based on five different values for d and five different values for $\frac{c_1}{c_2}$. For each experiment, we applied the proposed warning generating policy to generate the optimal warning time. We then calculated the mean of the number of early and late cycles for the turbines with early warning and late warning separately as shown in Fig. 13. Each plot in Fig. 13 has two lines, one for early and one for late warnings. It is clear that the closer each point is to zero, the better is the recommended alarm time. It can be seen from this figure that as the cost of late warning with respect to early warning ($\frac{c_1}{c_2}$) increases, the proposed policy tends to suggest earlier warning time to avoid late warnings. Thus, when $\frac{c_1}{c_2}$ is large, the mean cycles for cases with late warning tend to zero (because a late warning is more expensive) and increase for cases with early warning. Similar to section "Alarm/Warning Generation Policy", we compared our policy with other similar policies. It is also clear that the model cannot always recommend the best alarm time; thus, we have early or late warnings. Results also show that when the ideal warning time with respect to the time to failure (denoted by d) is larger, the

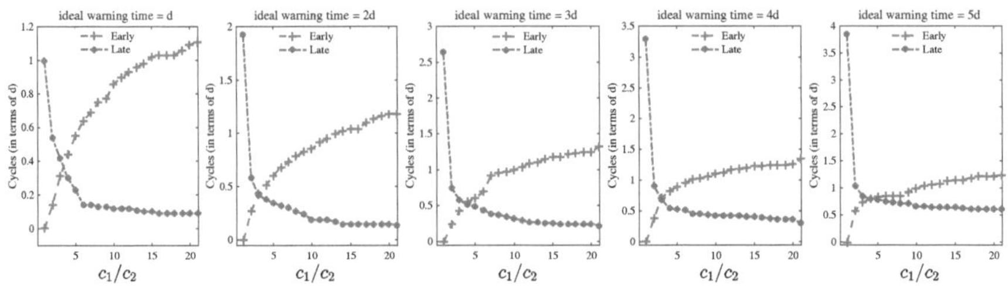

Fig. 13 The mean (for 15 turbines) of the early and late warnings versus $\frac{c_1}{c_2}$ for 5 cases of d. Note that each point shows the mean absolute value of the late or early warning cycles

magnitude of early and late cycles is smaller with respect to d. This means that the results are better if we have more time to generate the alarm (i.e., when d is large). Compared to the control limit policy and the two extreme cases of do nothing and warning immediately, results also verify that the proposed policy can provide better costs compared to similar policies and provide reasonable warning times as well. However, results are not reported here due to the similarity to Fig. 8.

7 Conclusion and Future Work

This chapter presents a new stochastic model for degradation modeling of mechanical systems operating under varying conditions, which is applicable to many types of systems, such as wind turbines. The model is based on the well-known and well-established Miner's rule approach; however, it exhibits much more complex behaviors and is able to accommodate uncertainty and parameter estimation reasonably well. A hierarchical Bayesian framework is first proposed to model the degradation process over time based on the number of cycles spent at each operating cycle. Then a parameter estimation model that can use historical data to train the structure of the model is developed. Finally, a cost-effective optimization-based warning generation policy is proposed to determine the optimal time to issue a warning given an ideal number of cycles to failure. From an implementation point of view, the framework is easy to apply and can incorporate prior knowledge on the damage process potentially through subjective priors in the hierarchical model. The stochastic nature of the model and the proposed alarm generation policy can extend the application of the widely used Miner's rule in real-time decision making under uncertainty. Future work will consist of considering the effects of loading history and loading sequence, time dependent covariates, and stochastic operation conditions.

Appendix: Algorithm 1

Algorithm 1: Steps for parameter estimation using MCMC

Step 0 Set $k = 0$ and generate randomly an initial state for model parameters as

$$\boldsymbol{\theta}^{\langle 0\rangle} = \left\{W^{\langle 0\rangle}, \sigma_N^{\langle 0\rangle}, N_1^{\langle 0\rangle}, \ldots, N_L^{\langle 0\rangle}, \beta_1^{\langle 0\rangle}, \ldots, \beta_P^{\langle 0\rangle}, \gamma^{(1)\langle 0\rangle}, \ldots, \gamma^{(m)\langle 0\rangle}\right\},$$

with a positive posterior probability value as in Eq. (9). Repeat the following steps until stationary distribution and the desired number of samples are reached considering optional burn-in and/or thinning.

Step 1 Sample $\boldsymbol{\theta}^*$ (for each parameter) from the symmetric Gaussian proposal distribution as $\boldsymbol{\theta}^* = \mathcal{N}(\boldsymbol{\theta}^{\langle k\rangle}, \delta)$, where δ shows the standard deviation of the random walk.

Step 2 Calculate the acceptance probability as

$$\alpha = \min\left(1, \frac{p(\boldsymbol{\theta}^*|\mathcal{D}, \boldsymbol{\vartheta})}{p(\boldsymbol{\theta}^{\langle k\rangle}|\mathcal{D}, \boldsymbol{\vartheta})}\right). \tag{18}$$

Step 3 Draw a random number u from Unif(0, 1). If $u \leq \alpha$, accept the proposal state $\{\boldsymbol{\theta}^*\}$ and set $\boldsymbol{\theta}^{\langle k+1\rangle} = \boldsymbol{\theta}^*$, else set $\boldsymbol{\theta}^{\langle k+1\rangle} = \boldsymbol{\theta}^{\langle k\rangle}$. Set $k : k + 1$.

It should be noted that during our implementation phase, we use a component-wise sampling approach where parameters are sampled sequentially based on their conditional distribution, that is, the posterior probability in Eq. (18) will be replaced by the corresponding conditional distribution as shown below:

- For W, we have

$$p(W|\mathcal{D}, \boldsymbol{\vartheta}) \propto \prod_{l=1}^{L} p\left(N_l|W, \sigma_N\right) \times p(W|\sigma_w).$$

- For $\sigma_N^{\langle k+1\rangle}$, we have

$$p(\sigma_N|\mathcal{D}, \boldsymbol{\vartheta}) \propto \prod_{l=1}^{L} p\left(N_l|\frac{W}{S_l}, \sigma_N\right) \times p(\sigma_N|a).$$

- For N_l ($l \in \{1, \ldots, L\}$), we have

$$p(N_l|\mathcal{D}, \boldsymbol{\vartheta}) \propto p\left(N_l|\frac{W}{S_l}, \sigma_N\right) \times \prod_{m=1}^{M}\prod_{t=1}^{T^{(m)}} p\left(o_t^{(m)}|N_1, \ldots N_L, \boldsymbol{\beta}, \gamma^{(m)}, \boldsymbol{x}^{(m)}\right).$$

  Although it may be very unlikely to observe non-decreasing trend in N_ls due to the increasing trend of S_l and relatively smart standard deviation σ_N compared to $\frac{W}{S_l}$, to fully control this decreasing trend, we reject all the samples that violate this requirement during the MCMC process.

- For $\gamma^{(m)}$, where $m \in 1, \ldots, M$, we have

$$p(\gamma^{(m)}|\mathcal{D}, \boldsymbol{\vartheta}) \propto p(\gamma^{(m)}|\gamma_0, \sigma_\gamma) \times \prod_{t=1}^{T^{(m)}} p(o_t^{(m)}|N_1, \ldots N_L, \boldsymbol{\beta}, \gamma^{(m)}, \boldsymbol{x}^{(m)}), \text{ and}$$

- For β_p, where $p \in \{1, \ldots, P\}$, we have

$$p(\beta_p|\mathcal{D}, \boldsymbol{\vartheta}) \propto p(\beta_p|\sigma_\beta) \times \prod_{m=1}^{M}\prod_{t=1}^{T^{(m)}} p(o_t^{(m)}|N_1, \ldots N_L, \boldsymbol{\beta}, \gamma^{(m)}, \boldsymbol{x}^{(m)}).$$

References

Akbarzadeh S, Khonsari MM (2016) On the applicability of miner's rule to adhesive wear. Tribol Lett 63(2):1–10

Basquin OH (1910) The exponential law of endurance tests. Am Soc Test Mater 10:625–630

Blason S, Correia JAFO, De Jesus AMP, Calcada RAB, Fernandez-Canteli A (2016) A probabilistic analysis of miner's law for different loading conditions. Struct Eng Mech 60(1):71–90

Byon E (2013) Wind turbine operations and maintenance: A tractable approximation of dynamic decision making. IIE Trans (Institute of Industrial Engineers) 45(11):1188–1201

Christensen RM (2008) A physically based cumulative damage formalism. Int J Fatigue 30(4):595–602

Ciavarella M, D'Antuono P, Demelio GP (2017) A simple finding on variable amplitude (Gassner) fatigue SN curves obtained using miner's rule for unnotched or notched specimen. Eng Fract Mech 176:178–185

Deb S, Claudio D (2015) Alarm fatigue and its influence on staff performance. IIE Trans Healthc Syst Eng 5(3):183–196

Fatemi A, Yang L (1998) Cumulative fatigue damage and life prediction theories: A survey of the state of the art for homogeneous materials. Int J Fatigue 20(1):9–34

Gu J, Barker D, Pecht M (2007) Prognostics implementation of electronics under vibration loading. Microelectron Reliab 47(12):1849–1856

Gu J, Barker D, Pecht M (2009) Health monitoring and prognostics of electronics subject to vibration load conditions. IEEE Sensors J 9(11):1479–1485

Herp J, Ramezani MH, Bach-Andersen M, Pedersen NL, Nadimi ES (2016) Bayesian state prediction of wind turbine bearing failure. Renew Energy 16(B):164–172

Liang Y, Chen W (2016) A regularized miner's rule for fatigue reliability analysis with Mittag-Leffler statistics. Int J Damage Mech 25(5):691–704

Liao H, Tian Z (2013) A framework for predicting the remaining useful life of a single unit under time-varying operating conditions. IIE Trans (Institute of Industrial Engineers) 45(9):964–980

Liu X-W, Lu D-G, Hoogenboom PCJ (2017) Hierarchical Bayesian fatigue data analysis. Int J Fatigue 100:418–428

Long H, Wu J, Matthew F, Tavner P (2011) Fatigue analysis of wind turbine gearbox bearings using SCADA data and Miner's rule. In: European wind energy conference and exhibition 2011 (EWEC 2011), pp 334–337

McVittie DR, Errichello RL (1990) Application of miner's rule to industrial gear drives. Gear Technol 7(1):1–12

Miner MA (1945) Cumulative damage in fatigue. J Appl Mech 12:A159–A16A

Novaes Menezes EJ, Arajo AM, Bouchonneau da Silva NS (2018) A review on wind turbine control and its associated methods. J Clean Prod 174:945–953

Ortegon K, Nies LF, Sutherland JW (2013) Preparing for end of service life of wind turbines. J Clean Prod 39:191–199

Palmgren A (1924) Die lebensdauer von kugellagern. Verfid-mwsrrchinik 68:339–341

Paolino DS, Cavatorta MP (2014) On the application of the stochastic approach in predicting fatigue reliability using Miner's damage rule. Fatigue Fract Eng Mater Struct 37(1):107–117

ReliaSoft (2007) Accelerated life testing. ReliaSoft Publishing, New York

Rudin C, Vahn G (2013) The big data newsvendor: Practical insights from machine learning analysis. In: SSRN

Stephens RI, Stephens RR, Fatemi A, Fuchs HO (2000) In: Metal fatigue in engineering. Wiley, New York

Stillinger CJ, Brekken TKA, Von Jouanne A, Paasch R, Naviaux D, Rhinefrank K, Prudell J, Schacher A, Hammagren E (2011) WEC prototype advancement with consideration of a real-time damage accumulation algorithm. In: 2011 IEEE PES trondheim powertech: the power of technology for a sustainable society (POWERTECH 2011)

Stillinger CJ, Brekken TKA, Von Jouanne A (2012) Furthering the study of real-time life extending control for ocean energy conversion. In: IEEE Power and Energy Society General Meeting, pp 1–9

Suhir E, Ghaffarian R, Yi S (2017) Probabilistic Palmgren-Miner rule, with application to solder materials experiencing elastic deformations. J Mater Sci Mater Electron 28(3):2680–2685

Sun YS (1994) Revised miner's rule and its application in calculating equivalent loads for components. Reliab Eng Syst Saf 43(3):319–324

Sun Q, Dui H-N, Fan X-L (2014) A statistically consistent fatigue damage model based on Miner's rule. Int J Fatigue 69:16–21

Vera-Tudela L, Khn M (2017) Analysing wind turbine fatigue load prediction: the impact of wind farm flow conditions. Renew Energy 107:352–360

Zhang Y, Ren S, Liu Y, Sakao T, Huisingh D (2017) A framework for big data driven product lifecycle management. J Clean Prod 159:229–240

Zuo FJ, Huang HZ, Zhu SP, Lv Z, Gao H (2015) Fatigue life prediction under variable amplitude loading using a non-linear damage accumulation model. Int J Damage Mech 24(5):767–784

Making Mission Abort Decisions for Systems Operating in Random Environment

Gregory Levitin and Maxim Finkelstein

Acronyms

cdf	Cumulative distribution function
MSP	Mission success probability
SSP	System survival probability
HPP	Homogeneous Poisson process

Notations

L	System lifetime
λ_M, λ_R	Shock rates during primary mission and rescue procedure, respectively
$F(t)$	*cdf* of time to internal failure during the mission
T_m	Random time of the m-th shock occurrence
τ	Mission time
ξ	Time from the start of a mission since which the mission is not aborted
$\varphi(t)$	Duration of the rescue procedure activated at time t

G. Levitin (✉)
The Israel Electric Corporation, Haifa, Israel

University of the Free State, Bloemfontein, South Africa
e-mail: levitin@iec.co.il

M. Finkelstein
University of the Free State, Bloemfontein, South Africa

ITMO University, St. Petersburg, Russia
e-mail: FinkelM@ufs.ac.za

A. T. de Almeida et al. (eds.), *Multicriteria and Optimization Models for Risk, Reliability, and Maintenance Decision Analysis*, International Series in Operations Research & Management Science 321, https://doi.org/10.1007/978-3-030-89647-8_13

S	SSP
R	MSP
C_F, C_L	Costs of mission failure and system loss, respectively
$c(u)$	Penalty cost function for uncompleted part of the mission u
C_p	Expected penalty
$P(t,i,\lambda)$	Probability of occurrence of i shocks in $[0, t)$ given the shock rate is λ
$q(i)$	Probability that the system survives the i-th shock
δ	Cumulated lifetime deceleration/acceleration factor (for internal failures)
Ω	Probability of the first shock survival
ω	Shock survival sensitivity factor

1 Introduction

Mission success probability (MSP), i.e., the probability of successfully completing a specific mission with or without a deadline (Rausand and Høyland 2003; Levitin et al. 2016a), is an important reliability characteristic for many engineering systems that perform specific tasks. However, at many instances primary mission goals can be sacrificed if continuation of the mission is associated with high risks. This often happens when survival of a system, due to safety- or cost-related reasons, may have a higher priority than accomplishing the defined mission (e.g., for aircrafts, submarines, or complex costly technological processes). In these cases, a system can implement a mission abort policy to improve its survivability and thus to decrease the risk of casualties and/or of substantial economic losses.

Mission abort policy can be an effective tool for enhancing system survival probability (SSP) of many real-world systems when a failure of a system during a mission results in a substantial economic loss. For instance, for aircrafts and spaceships, a failure can lead to a damage or loss of these objects, while in the case of complex technological or production processes, a failure can lead to substantial monetary losses. Then, as a preventive action, a rescue or recovery procedure can be initiated to enhance SSP and therefore, to decrease losses.

For making a mission abort decision, usually some degradation parameter should be observed that characterizes the current state of a system. For instance, this parameter can be the number of external impacts that decrease resilience ability of a system. Then upon reaching a certain predetermined value, a mission should be aborted and a safe rescue or recovery procedure should be initiated (Bell and Bearden 2014). A real-world example of the described scenario is an aircraft that can be required to abort the mission after certain number of external impacts associated with, for example, a malicious activity or nature conditions (e.g., lightning inducing electrical peaks in the electrical circuits). These impacts can cause deterioration of critical systems that make the risk associated with the mission completion unacceptable.

Reliability analysis of systems with mission abort policies is a rather new and practically important topic addressed only in a couple of papers so far. In the pioneering paper by Myers (2009), the author considered standby systems with an abort policy and a rescue procedure to be initiated upon the failure of a fixed number of components. The method was developed only for homogeneous hot standby systems with components having identical exponential time-to-failure distributions. In (Levitin et al. 2018a), the model was extended to heterogeneous systems and adaptive abort policy. In (Levitin et al. 2018b) the optimal mission abort policy is combined with the optimal loading of the system components. However, these papers do not take into consideration the influence of a stochastic environment on operational characteristics of systems and the corresponding abort policy. Neglecting the effect of a random environment and considering only static models can lead to serious discrepancies in assessing reliability and safety characteristics of various engineering systems.

As traditional reliability models are not applicable for addressing effects of mission aborts in evaluating and optimizing system reliability, we had to develop a new approach for modeling and evaluating the MSP and SSP of systems operating in a random environment and subject to mission aborts. In (Levitin and Finkelstein 2018a, c; Levitin et al. 2018c), we used external shock processes for modeling an impact of a random environment, which is an approach widely adopted in the literature.

There is an extensive literature on shocks modeling in reliability and risk analysis (see, e.g., the monographs (Nakagawa 2007; Finkelstein 2008; Finkelstein and Cha 2013) mostly devoted to shocks modeling). Traditionally, one distinguishes between two major types of shock models: cumulative shock models, when systems fail due to some cumulative effect and extreme shock models when systems can fail with certain probabilities upon any shock (see Klefsjo 1981; Mallor and Omey 2001; Gut and Husler 2005; Cha and Finkelstein 2011). The approach in (Levitin and Finkelstein 2018a, c; Levitin et al. 2018c) is based on the generalized extreme shock model (Cha and Finkelstein 2011; Cha and Mi 2007), in which the probability of a failure upon a shock increases with each experienced shock.

To the best of our knowledge, there are only a few papers in the literature that consider the number of shocks experienced by a system as a decision parameter for some optimization problems (see, e.g., (Finkelstein and Gertsbakh 2015)). In this chapter, we present the probabilistic models suggested in (Levitin and Finkelstein 2018a; Levitin and Finkelstein 2018c; Levitin et al. 2018c) that take into account deterioration in reliability characteristics with experienced shocks for systems with a possibility of a mission abort. We consider a policy when a mission is aborted, and the rescue procedure is activated immediately after the m-th shock.

Systems implementing the mission abort policy can be considered as the special cases of the phased-mission systems when a primary task and the rescue procedures correspond to different phases. The existence and duration of the rescue phase is not predetermined when a mission begins. Moreover, unlike traditional models, where the probability that all phases are performed without failures is considered as the single success criterion, in systems with the rescue option, the success probabilities for each phase are considered separately as they constitute different, though interdependent, metrics, namely, MSP and SSP (Fig. 1). A number of recent

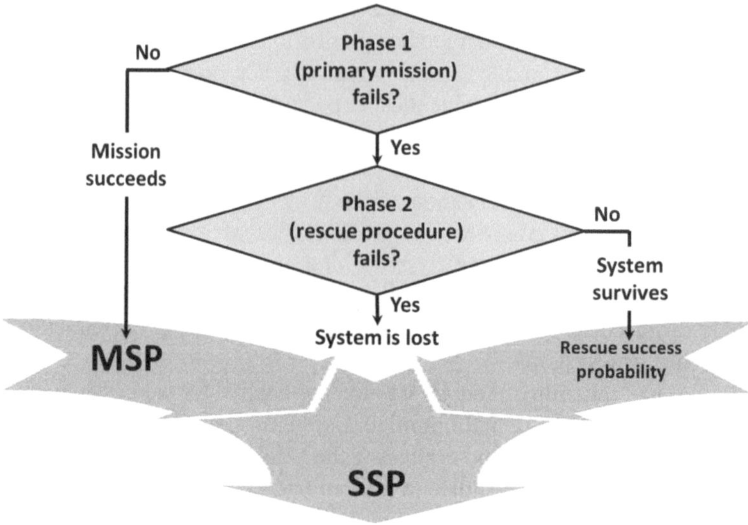

Fig. 1 Two success metrics for system with rescue option

publications are devoted to analysis of phased-mission systems (see, e.g., (Ma and Trivedi 1999; Levitin et al. 2013; Peng et al. 2014; Wang et al. 2015, 2017; Lu et al. 2015)). In these papers, the phases have the fixed durations, whereas in the case of a mission abort, the duration of phases depends on random factors. Note that the phased-mission systems with variable duration of phases have been studied in (Levitin et al. 2016b). However, all referenced papers did consider neither abort policies and MSP–SSP tradeoff nor the influence of random shocks.

Section 2 of this chapter describes the considered settings and gives general formulation of the mission abort policy optimization problems. Section 3 presents a derivation of the basic system success metrics (MSP and SSP) and considers tradeoff between them. It also presents examples of the optimal mission abort policies. Section 4 derives the expected penalty associated with uncompleted mission and gives illustrative examples of abort policies minimizing the expected total losses, associated with the uncompleted mission and the system loss. Section 5 considers more flexible mission abort policy in which the mission time is divided into several intervals, and a specific number of allowed shocks is determined for each interval. The advantage of such multi-interval abort policy compared to the single-interval policy is demonstrated. Section 6 discusses directions of further research.

2 Problem Formulation

Let a system perform a mission task that requires continuous operation during the fixed time τ. Let the system random lifetime in a static, deterministic environment (to be called baseline) be described by the *cdf* $F(t)$. In addition, a system can be exposed to random shocks that decrease its lifetime. Assume that shocks during the mission time occur in accordance with the homogeneous Poisson process (HPP) $\{N_M(t), t \geq 0\}$, with rate λ_M, where $N_M(t)$ is the number of shocks in $[0, t)$ and $T_1 < T_2 < \ldots$ are the arrival times of shocks. The results obtained in this chapter can be generalized to the case of the non-homogeneous Poisson process of shocks; however, for illustration and practical applications, the assumption of HPP is more relevant. In the model to be described, each shock can result in a failure of a system with probability that increases with the number of experienced shocks (and is survived with the complementary probability), which is an assumption that is often met in practice for degrading systems.

As it was stated in the "Introduction" section, at some instances, survival of a system, due to safety- or cost-related reasons, may have a higher priority than accomplishing the defined mission, as it is obviously the case for safety critical technological processes, some experiments, aircrafts, manned space missions, and submarines. In these cases, a mission abort policy can be implemented to improve SSP. Thus, when the successful mission completion becomes unlikely, it should be aborted, and a rescue procedure should be implemented. We assume that shocks are observable and, therefore, aborting a mission upon experiencing m shocks can increase the SSP. Moreover, the value of m should be obtained in an optimal way.

It is also reasonable from the practical point of view to consider a case when the environment for the rescue procedure differs from that for the mission. Thus, let the rate of the HPP during rescue, λ_R, differ from that for the mission, λ_M, i.e., $\lambda_R \neq \lambda_M$ (see example of changing environment in Sect. 4).

It is natural to assume that the duration of the rescue procedure is a function of the occurrence time of the m-th shock, i.e., $\varphi = \varphi(t_m)$, where t_m is the realization of the random T_m, $m = 1, 2, \ldots$, (see Sect. 4 for the example of this function). The larger m in our model corresponds to the larger level of deterioration of a system and, therefore, to the larger risks of failure. When t_m increases, the remaining mission time decreases. Thus, it may become unreasonable to start the rescue procedure if the mission is close to termination and the system has good chances to complete it. Therefore, we assume that the system continues executing the mission if $t_m \geq \xi$, where ξ is a time after which the mission should never be aborted, which, along with m, can be considered as a decision variable that can be chosen to achieve a proper balance between the MSP and the SSP.

Let L denote a lifetime of a system for the described scenario. A mission succeeds if the system does not fail in $[0, \tau)$ and less than m shocks occur in $[0, \xi)$ (no mission abort). In accordance with this description, the MSP can be defined as

$$R(\tau, \xi, m) = \Pr(L \geq \tau, T_m \geq \xi). \tag{1}$$

The system survives if it completes either the mission or the rescue procedure. The rescue procedure is activated only if $t_m < \xi$. To complete the rescue procedure activated at a random time T_m, the system lifetime must be not less than $T_m + \varphi(T_m)$. Thus, the SSP is

$$S(\tau, \xi, m) = R(\tau, \xi, m) + \Pr(L \geq T_m + \varphi(T_m), T_m < \xi), \tag{2}$$

where the second term corresponds to the probability that the rescue procedure saves the system.

When ξ is fixed and the decision parameter m is increasing, T_m is increasing in the sense of the usual stochastic order (Finkelstein 2008; Shaked and Shantikumar 2007) and, therefore, the MSP $R(\tau,\xi,m)$ is increasing (because of the decrease of the abort probability), whereas the SSP $S(\tau,\xi,m)$ is decreasing. Specifically, when $m = 0$ ($T_0 = 0$), the system does not perform the mission task and only executes the rescue procedure, which results in $R(\tau,\xi,0) = 0$ and $S(\tau,\xi,0) = \Pr(L \geq \varphi(0))$. On the other hand, for $m = \infty$, the system never performs the rescue procedure and survives only if the mission is successfully completed, which gives $R(\tau,\xi,\infty) = S(\tau,\xi,\infty) = \Pr(L \geq \tau)$.

When m is fixed and the decision parameter ξ is increasing, the time when the mission abort is allowed increases, which results in the smaller MSP and the larger SSP. For $\xi = 0$, the mission abort is totally prohibited and $R(\tau,\xi,m) = S(\tau,\xi,m)$.

In practice, it is desirable to achieve a balance (tradeoff) between $R(\tau,\xi,m)$ and $S(\tau,\xi,m)$. For example, the problem of obtaining the optimal m and ξ that achieve the maximum MSP subject to providing a desired level of the SSP S^* can be solved, i.e.

$$\max R(\tau,\xi,m) \text{ s.t.} S(\tau,\xi,m) > S^*. \tag{3}$$

When the mission failure and the loss of a system are associated with the corresponding costs C_F and C_L, the cost minimization problem with respect to the decision parameters m and ξ can be considered. The probability of the system loss is $1 - S(\tau,\xi,m)$. In the case of a system loss (due to its failure during the mission or the rescue procedure), the mission also fails and the total cost of losses is $C_F + C_L$. The probability that the system survives, but the mission fails, is $S(\tau,\xi,m) - R(\tau,\xi,m)$. In this case, the total cost of losses is C_F. Thus, the expected cost of the total losses that should be minimized is

$$\begin{aligned} C(\tau,\xi,m) &= (1 - S(\tau,\xi,m))(C_F + C_L) + (S(\tau,\xi,m) - R(\tau,\xi,m))C_F \\ &= C_F(1 - R(\tau,\xi,m)) + C_L(1 - S(\tau,\xi,m)). \end{aligned} \tag{4}$$

When a penalty is associated with the amount of the uncompleted work (uncompleted part of the mission), the expected penalty depends on the time of the mission abort. If the mission fails or is aborted at time $t \leq T_m$, a system completes the fraction of the mission t/τ and leaves the fraction of the mission $u = 1 - t/\tau$ uncompleted. Taking into account the penalty cost function associated with the uncompleted part of the mission $c(u)$, one can obtain the expected penalty for any mission abort policy ξ, m as

$$C_P(\tau,\xi,m) = \int_0^{\tau} c(1 - t/\tau)\,\Theta(t,\xi,m)\,dt, \tag{5}$$

where $\Theta(t, \xi, m)dt$ is the probability that the mission is terminated (due to aborting or a system failure) in $[t, t + dt)$. For instance, for the simplest case of the linear penalty cost function $c(u) = cu,\ c > 0,\ 0 \le u \le 1$, we see that $c(1 - t/\tau)$ is linearly decreasing in $t \in [0, \tau]$ from c to 0. In this case, the expected cost of losses that should be minimized is

$$\min C(\tau, \xi, m) = C_P(\tau, \xi, m) + C_L(1 - S(\tau, \xi, m)). \tag{6}$$

3 Mission Success Probability and System Survival Probability

Denote by $P(t,i,\lambda)$ for $i = 0,1,2,\ldots$ the probability of occurrence of i shocks affecting the system in $[0, t)$ given the shock rate is λ. Thus, for the homogeneous Poisson process

$$P(t, i, \lambda) = \exp\{-\lambda t\} \frac{(\lambda t)^i}{i!}. \tag{7}$$

Our approach is based on the generalized extreme shock model (Cha and Finkelstein 2011) when the probability of a failure upon a shock increases with each experienced shock. Let the shock survival probability of the system depend on the number of shocks it has survived in the past, which is a meaningful generalization of the simplest extreme shock model. Indeed, often the resistance of elements to shocks decreases with the number of experienced shocks. Thus, if the probability that the system survives the i-th shock is $q(i)$, then the probability of surviving all n shocks is $\prod_{l=0}^{n} q(l)$, where $q(0) \equiv 1$ by definition.

The probability that i shocks have occurred in $[0,\xi)$ and that additional k shocks have occurred in $[\xi,\tau)$ during the mission is in accordance with the property of independent increments for HPP

$$P(\xi, i, \lambda_M) P(\tau - \xi, k, \lambda_M). \tag{8}$$

The probability that less than m shocks have occurred in $[0, \xi)$ and the system survives all shocks and “internal failure” during its mission time τ in this case is

$$
\begin{aligned}
R(\tau,\xi,m) &= \Pr(L > \tau, T_m > \xi) \\
&= (1 - F(\tau)) \sum_{i=0}^{m-1} P(\xi, i, \lambda_M) \sum_{k=0}^{\infty} P(\tau-\xi, k, \lambda_M) \prod_{l=0}^{i+k} q(l) \\
&= (1 - F(\tau)) \sum_{i=0}^{m-1} \exp\{-\lambda_M \xi\} \tfrac{(\lambda_M \xi)^i}{i!} \sum_{k=0}^{\infty} \exp\{-\lambda_M(\tau-\xi)\} \tfrac{(\lambda_M(\tau-\xi))^k}{k!} \prod_{l=0}^{i+k} q(l) \\
&= (1 - F(\tau)) \exp\{-\lambda_M \tau\} \sum_{i=0}^{m-1} \tfrac{(\lambda_M \xi)^i}{i!} \sum_{k=0}^{\infty} \tfrac{(\lambda_M(\tau-\xi))^k}{k!} \prod_{l=0}^{i+k} q(l).
\end{aligned}
\tag{9}
$$

The computational aspects of obtaining the infinite sum in (9) are addressed in (Levitin and Finkelstein 2018a).

In accordance with (2), the system survival probability is the sum of $R(\tau,\xi,m)$ and the probability that the rescue procedure was activated and succeeded. The latter was denoted as $\Pr(L \geq T_m + \varphi(T_m), T_m < \xi)$. We will now obtain this probability.

If the m-th shock occurs at time $t < \xi$, the operating system immediately starts the rescue procedure. The probability that the m-th shock from the HPP with rate λ_M occurs in $[t, t+dt)$ is

$$
P(t, m-1, \lambda_M)\lambda_M dt = \lambda_M \exp\{-\lambda_M t\} \frac{(\lambda_M t)^{m-1}}{(m-1)!} dt, \tag{10}
$$

where $P(t,m-1,\lambda_M)$ is the probability that exactly $m-1$ shocks have happened in $[0,t)$ and $\lambda_M dt$ is the probability and that the additional shock has happened in $[t, t+dt)$. The probability that the system has survived the first m shocks is $\prod_{l=0}^{m} q(l)$. The probability that the system survives any number of shocks during the rescue procedure is

$$
\sum_{k=0}^{\infty} P(\varphi(t), k, \lambda_R) \prod_{l=0}^{k} q(m+l). \tag{11}
$$

Besides the differences in external shock rates during the primary mission and rescue procedure, the system operation conditions (loading) can also be different for these two phases, causing different "rate of aging." To account for this effect, we apply the cumulative exposure model (Nelson 1990; Sedjakin 1966). In accordance with this model, and given that the *cdf* of the time to internal failure is $F(t)$, the probability that the system does not fail because of internal failure until the termination of the rescue procedure activated at time t is $1 - F(t + \delta\varphi(t))$ where δ is the acceleration/deceleration factor used to reflect the change of the system loading from the regular mission phase to the rescue one.

Thus, as the rescue procedure is activated if m-th shock happens at any $t \in [0,\xi)$, we obtain

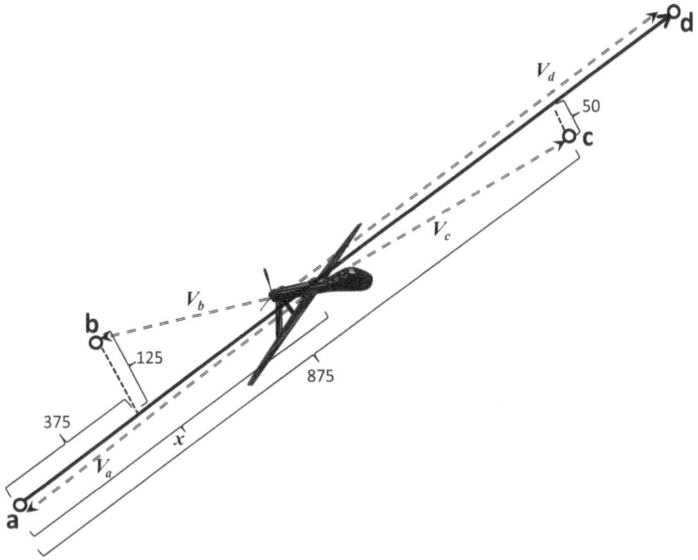

Fig. 2 Locations of the landing fields and emergency landing distances along the UAV route

$$
\begin{aligned}
&\Pr\left(L > T_m + \varphi\left(T_m\right), T_m < \xi\right) \\
&= \int_0^{\xi} \left(1 - F\left(t + \delta\varphi(t)\right)\right) \lambda_M P\left(t, m-1, \lambda_M\right) \prod_{l=0}^{m} q(l) \sum_{k=0}^{\infty} P\left(\varphi(t), k, \lambda_R\right) \prod_{l=0}^{k} q\left(l+m\right) dt \\
&= \frac{\lambda_M{}^m}{(m-1)!} \int_0^{\xi} \left(1 - F\left(t + \delta\varphi(t)\right)\right) \exp\left\{-\lambda_M t - \lambda_R \varphi(t)\right\} t^{m-1} \sum_{k-0}^{\infty} \frac{\left(\lambda_R \varphi(t)\right)^k}{k!} \prod_{l-0}^{m+k} q(l) dt.
\end{aligned} \tag{12}
$$

Illustrative Example 1 Consider an unmanned aerial vehicle (UAV) that should fly from location **a** to location **d** performing a surveillance mission (Fig. 2). The distance between the locations, which should be covered by the UAV to fulfill the mission, is 1250 km. The UAV speed during the mission is 212.5 km/h. Thus the mission time is $\tau = 1250/212.5 = 5.88$ h. There are two safe landing fields ***b*** and ***c*** that can be used for emergency landing along the route. The locations of these fields are depicted in Fig. 2. If the flight mission is aborted when the distance covered from the airport **a** is $x = 212.5 \cdot t$, the airplane has to cover distances $Va = x$, $Vb = \sqrt{(375 - x)^2 + 125^2}$, $Vc = \sqrt{(875 - x)^2 + 50^2}$ and $Vd = 1250\text{-}x$ to reach locations **a**, **b**, **c**, and **d**, respectively. The distance to the closest location for the emergency landing is min(*Va*,*Vb*,*Vc*,*Vd*). Figure 3 presents *Va*, *Vb*, *Vc*, *Vd*, and min(*Va*,*Vb*,*Vc*,*Vd*) as functions of the time elapsed from the mission beginning when the decision about the mission abort is made.

The UAV has its failure rate λ and is exposed to external shocks caused by lightning and electronic interference. The electronic equipment of the UAV is protected by the interference filters, which deteriorate with each impact because of

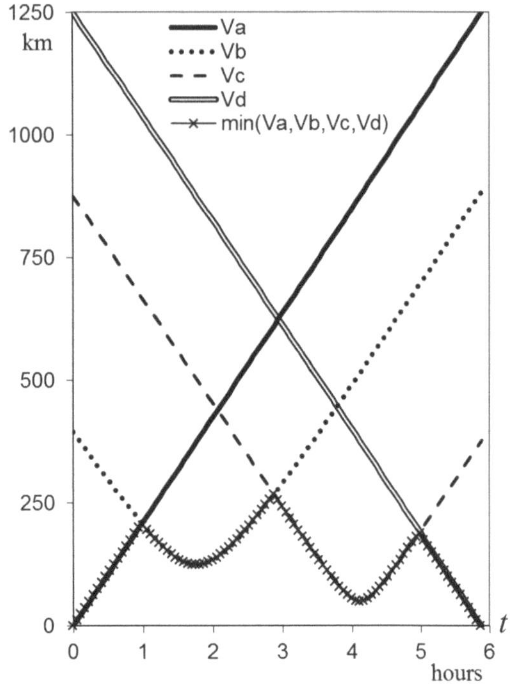

Fig. 3 *Va*, *Vb*, *Vc*, *Vd*, and min(*Va*,*Vb*,*Vc*,*Vd*) as functions of the time elapsed since the mission beginning

voltage surges causing their overheating. If filters reach some level of deterioration, they cannot protect the UAV equipment, and it does not survive the next shock. Besides external shocks, the failure of the electronic equipment can be caused by internal reasons (mainly, by memory overflow). The baseline time to internal failure abbeys Weibull distribution with *cdf* $F(t) = 1-\exp(-(0.01\ t)^{1.1})$. The internal failures and external shocks are totally independent. During the surveillance mission, the UAV should remain on the altitude where the shock rate is λ_M. To perform the rescue procedure, the UAV descends to the altitude where the shocks have lower rate λ_R and reduces its speed to 160 km/h. Having min(*Va*,*Vb*,*Vc*,*Vd*) and the UAV speed, one can obtain the function $\varphi(t) = \min(Va,Vb,Vc,Vd)/160$. Notice that $\varphi(t)$ can exceed the time needed to complete the mission $\tau - t$. During the flight with reduced speed, the UAV cumulative time deceleration factor is $\delta = 0.9$.

Assume, following (Cha and Finkelstein 2011), that UAV shock resistance function takes the form $q(0) = 1$, $q(l) = \Omega\omega(l)$, $l > 0$, where $\omega(l)$ is a decreasing function of its argument: $\omega(0) = 1$, $\omega(l) = \omega^{l-1}$, $0 < \omega < 1$, and Ω is the probability of survival under the first shock. Thus, the survival probability of the system at each shock decreases as the number of survived shocks in $[0, t)$ increases. In this case

$$\prod_{l=0}^{n} q(l) = \Omega^n \omega^{n(n-1)/2} \tag{13}$$

Fig. 4 *R* and *S* as functions of the decision parameters *m* and ξ for $\Omega = 0.98$, $\omega = 0.95$, $\lambda_M = 0.7$, $\lambda_R = 0.1$

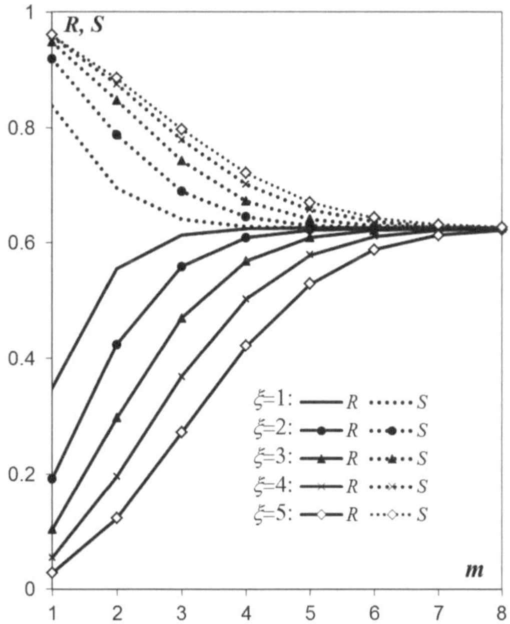

Figure 4 presents the MSP *R* and SSP *S* as functions of the decision parameters *m* and ξ for $\Omega = 0.98$, $\omega = 0.95$, $\lambda_M = 0.7$, $\lambda_R = 0.1$.

With the increase of *m*, the influence of parameter ξ on *R* and *S* decreases as the occurrence of *m* shocks, and, therefore, activation of the rescue procedure becomes unlikely for any interval $[0,\xi)$ for $\xi < \tau$. The decrease in ξ has the similar effect as the increase in *m*. In both cases, activation of the rescue procedure becomes less probable, which causes the increase in the MSP and the decrease in SSP.

To solve the optimization problems (3) and (4), one has to determine optimal value of discrete parameter *m* and continuous parameter ξ. As the value of *m* in practical applications cannot be very large, the optimization algorithm can enumerate *m* finding for each *m* optimal values of ξ by applying any algorithm for one-dimensional optimization. In this work, we enumerated *m* up to value of 20 and applied golden section search algorithm (Press et al. 1992) for finding ξ.

Figure 5 presents the optimal *m*, ξ solutions for the optimization problem (3) and the corresponding values of *R* and *S* as functions of *S** for $\Omega = 0.98$, $\omega = 0.95$, $\lambda_M = 0.7$, and $\lambda_R = 0.1$. The optimal *m* solutions when aborts are allowed during the entire mission ($\xi = \tau$) are given for comparison.

The greater SSP (corresponding to greater *S**) can be achieved by increasing the time interval ξ when the abort is allowed and/or decreasing the number of shocks *m* after which the mission should be aborted. As ξ can vary continuously, whereas *m* takes only discrete values, small increase of the SSP can be achieved by a small increase of ξ without changing *m*. However, when increase of ξ becomes insufficient for achieving the desired level of the SSP, *m* decreases, which causes sharp increase

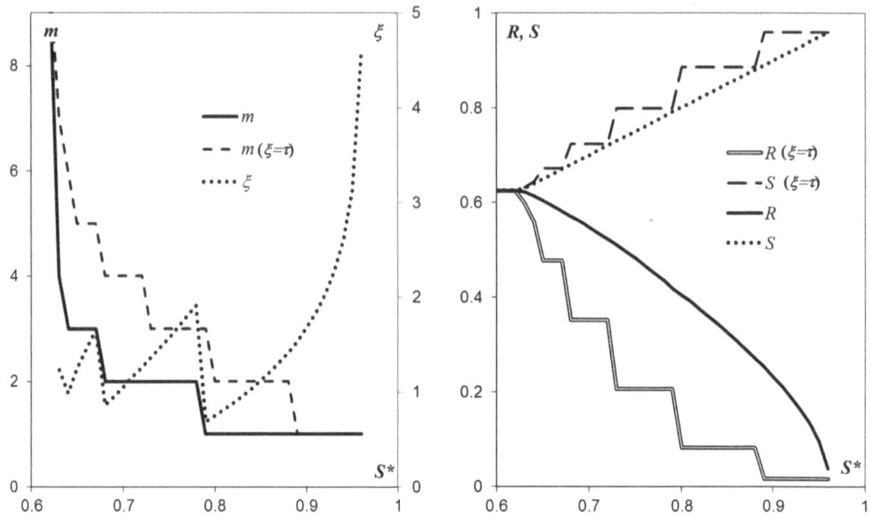

Fig. 5 Optimal solutions for (3) as functions of S^* for $\Omega = 0.98$, $\omega = 0.95$, $\lambda_M = 0.7$, $\lambda_R = 0.1$

of the SSP and decrease of the MSP. To prevent the excessive decrease of the MSP, ξ should decrease to the lowest level for which inequality $S(\tau,\xi,m) > S^*$ still holds. Further, small increase of the SSP is achieved by increasing ξ again while keeping m constant. This explains the monotonic behavior of m and the non-monotonic behavior of ξ.

For $S^* < 0.62$, the system can provide this SSP level without aborting the mission, which corresponds to $m = \infty$ and $R = S$. When the abort is allowed during the entire mission, the maximum MSP can be provided only by allowing more shocks before aborting the mission than it was allowed for the optimal ξ. The combination of the optimal m and ξ achieves greater MSP subject to providing the desired SSP level than the optimal m with $\xi = \tau$.

Figure 6 presents optimal solutions for optimization problem (4) as functions of the system loss cost C_L for $C_F = 1$, $\Omega = 0.98$, $\omega = 0.99$, $\lambda_M = 0.7$, and $\lambda_R = 0.1$. For small values of C_L, the SSP is less important than the MSP and the minimal cost solutions presume no aborts, which results in the maximal possible MSP and $R = S$. With the increase in C_L, the SSP should increase at the cost of MSP. Like in the previous example (Fig. 4), the increase of the SSP is achieved by monotonic decrease of m and increase of ξ when m remains constant with drops of ξ when m decreases.

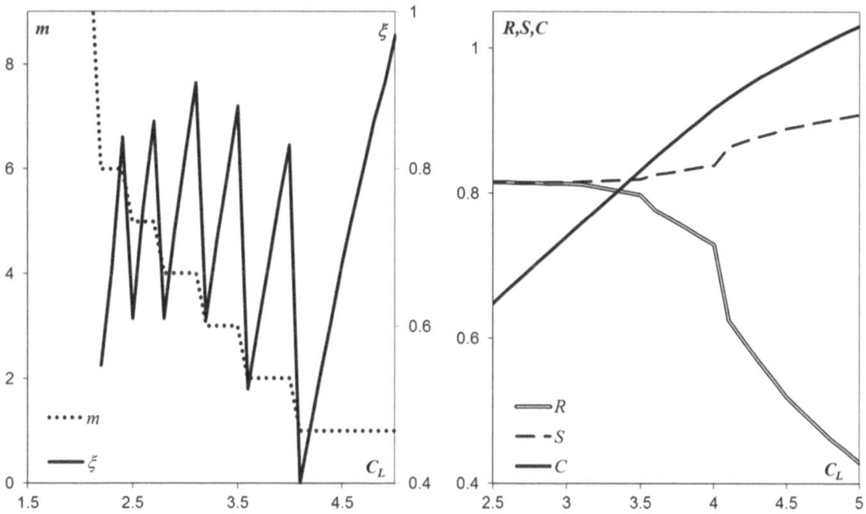

Fig. 6 Optimal solutions of problem (4) as functions of system loss cost C_L for $C_F = 1$, $\Omega = 0.98$, $\omega = 0.99$, $\lambda_M = 0.7$, $\lambda_R = 0.1$

4 Expected Penalty Associated with Uncompleted Mission

To solve the optimization problem (6), one has to obtain the expected penalty associated with uncompleted mission. In what follows, we consider the events resulting in a mission failure or abortion and derive the corresponding probabilities:

A mission can fail at time $t < \xi$ due to the internal failure if by the time of the internal failure not more than $m-1$ shocks have happened and all of these shocks have been survived. Thus, the probability that a mission fails in the time interval $[t,t + dt)$ $t < \xi$ as a result of an internal failure is

$$\Theta_1 (t, \xi, m) dt = F'(t) \sum_{i=0}^{m-1} P (t, i, \lambda_M) \prod_{l=0}^{i} q(l)dt. \tag{14}$$

A mission can fail at time $t < \xi$ due to the i-th shock if $i \le m$, no internal failures, and $i-1$ shocks happen (survived) by this time. Thus, the probability that a mission fails in the time interval $[t,t + dt)$ $t < \xi$ due to a shock is

$$\Theta_2 (t, \xi, m) dt = (1 - F(t)) \lambda_M \sum_{i=0}^{m-1} P (t, i, \lambda_M) (1 - q (i + 1)) \prod_{l=0}^{i} q(l)dt. \tag{15}$$

A mission is aborted at time $t < \xi$ if it does not fail before this time, the m-th shock happens at this time and all m shocks are survived. Thus, the probability that a mission is aborted in the time interval $[t,t + dt)$ $t < \xi$ as a result of the m-th shock is

$$\Theta_3(t,\xi,m)\,dt = (1 - F(t))\,\lambda_M P(t, m-1, \lambda_M)\prod_{l=0}^{m} q(l)dt. \tag{16}$$

A mission can fail at time $\xi \le t < \tau$ as a result of an internal failure if less than m shocks happen in $[0,\xi)$, any number of shocks happen in $[\xi,t)$, and all these shocks are survived. Thus, the probability that a mission fails in the time interval $[t,t + dt)$ $t < \xi$ as a result of an internal failure is

$$\Theta_4(t,\xi,m)\,dt = F'(t)\sum_{i=0}^{m-1} P(\xi, i, \lambda_M)\sum_{k=0}^{\infty} P(t-\xi, k, \lambda_M)\prod_{l=0}^{i+k} q(l)dt. \tag{17}$$

A mission can fail due to a shock at time $\xi \le t < \tau$ if no internal failures happen by this time, less than m shocks happen in $[0,\xi)$, any number of shocks happen in $[\xi,t)$, and all shocks that happened before time t are survived. The probability that a mission fails in the time interval $[t,t + dt)$ $\xi \le t < \tau$ as a result of a shock is

$$\begin{gathered}\Theta_5(t,\xi,m)\,dt = (1 - F(t))\,\lambda_M \\ \times\sum_{i=0}^{m-1} P(\xi, i, \lambda_M)\sum_{k=0}^{\infty} P(t-\xi, k, \lambda_M)\,(1 - q(i+k+1))\prod_{l=0}^{i+k} q(l)dt.\end{gathered} \tag{18}$$

Using the obtained probabilities and the penalty function $c(u)$, we can now obtain the expected penalty associated with uncompleted mission as

$$\begin{aligned}C_P(\tau,\xi,m) = &\int_0^{\xi} c(1 - t/\tau)\Big(\Theta_1(t,\xi,m) + \Theta_2\big(t,\xi,m\big) + \Theta_3(t,\xi,m)\Big)\,dt \\ &+ \int_{\xi}^{\tau} c(1 - t/\tau)\Big(\Theta_4(t,\xi,m) + \Theta_5\big(t,\xi,m\big)\Big)\,dt.\end{aligned} \tag{19}$$

Illustrative Example 2 Consider the UAV mission described in Example 1 and define the penalty function as follows. The surveyed objects are allocated unevenly along the UAV route. Therefore, the penalty cost function (which is a function of the fraction of the uncompleted part of a mission) takes the form.

$$c(x) = \begin{cases} 1.25x, & 0 \le x < 0.8 \\ 1, & x \ge 0.8 \end{cases}$$

In this case, the penalty, associated with totally unaccomplished work, is $c(1) = 1$.

Figure 7 presents the system loss probability $1-S$, the expected uncompleted mission cost penalty C_P, and the expected cost of losses C as functions of the decision parameters m and ξ for $C_L = 3.5$, $\Omega = 0.95$, $\omega = 0.999$, $\lambda_M = 0.7$, and $\lambda_R = 0.1$.

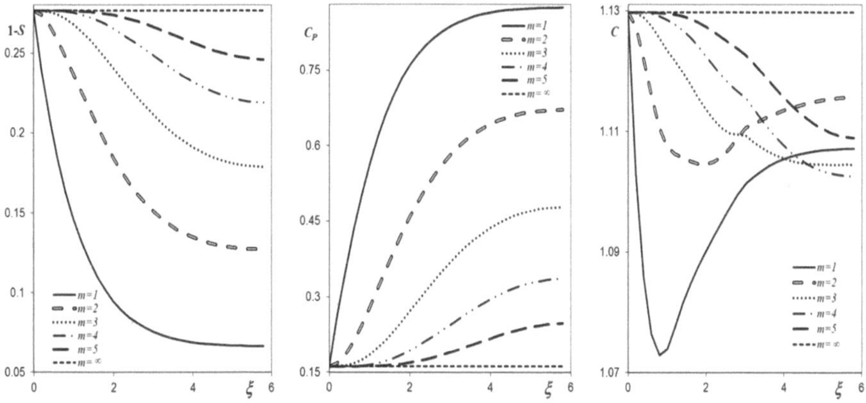

Fig. 7 The system loss probability 1-S, expected uncompleted mission penalty C_P, and expected cost of total losses C as functions of the decision parameters m and ξ for $C_L = 3.5$, $\Omega = 0.95$, $\omega = 0.99$, $\lambda_M = 0.7$, $\lambda_R = 0.1$

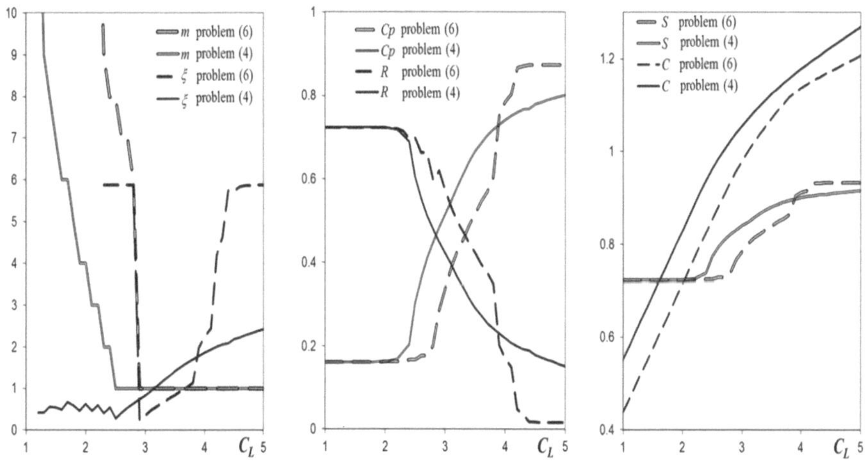

Fig. 8 Optimal solutions of problem (6) and problem (4) as functions of C_L for $C_F = 1$, $\Omega = 0.95$, $\omega = 0.99$, $\lambda_M = 0.7$, $\lambda_R = 0.1$

Note that, for $m = \infty$, no mission abort is allowed. Therefore parameter ξ has no meaning and does not influence the mission metrics. The best abort decision for $C_L = 3.5$ is to abort the mission only if the first shock occurs in the first 0.8 h.

Figure 8 presents the optimal m, ξ solutions for the optimization problem (6) and the corresponding values of C, C_P, and S as functions of C_L for $C_F = 1$, $\Omega = 0.95$, $\omega = 0.99$, $\lambda_M = 0.7$, and $\lambda_R = 0.1$. When C_L is small, the UAV survivability is not important and no mission aborts are allowed ($m = \infty$) to maximize the expected mission duration.

When C_L increases, the number of allowed survived shocks decreases, whereas the mission abort remains allowed during the entire mission ($\xi = \tau = 5.88$). Then, when C_L increases above the value of 2.9, the mission should be aborted after the first shock if it happens in the very beginning of the mission ($\xi = 0.24$). With the further increase of C_L, the time interval ξ when the abort is allowed increases, whereas the allowed number of shocks remains $m = 1$. Eventually, for $C_L > 4.9$, when the UAV survivability becomes much more important than the mission completion, the mission should be aborted if any shock happens any time ($m = 1$, $\xi = \tau$).

The solutions of the optimization problem (6) can be compared with solutions of the optimization problem (4) with $C_F = 1$ also presented in Fig. 8. In problem (4), the work performed before the mission abort has no value (e.g., when the surveillance information can be transferred to users only upon the mission completion because of absence of an onboard radio transmitter) and $c(0) = 1$, $c(x) = 0$ for $x > 0$.

It can be seen that when the system loss is not very costly ($C_L < 3.9$), the solutions of problem (6) provide lower expected uncompleted mission penalty cost than the corresponding solutions of problem (4). When the cost of the system loss increases, the solutions of problem (6) provide larger system survivability and larger expected uncompleted mission penalty cost than the solutions of problem (4). The overall expected loss C is larger for problem (4), because it does not take into account the "profit" earned by the system before the mission termination or failure, as it is done for problem (4).

5 Multiple Threshold Generalization

The approach when the optimal number m is obtained for the entire time period when a mission can be aborted is rather crude. A more flexible mission abort policy can be obtained by dividing the mission time in a number of adjacent intervals and using a rule when a mission is aborted (and a rescue procedure is activated) if the number of shocks exceeds the predetermined value m_h specific for each interval h. Such policy can be considered as a multivariate generalization of the policy considered above. As the algorithm for evaluating the MSP and the SSP suggested in Sect. 3 cannot be used for the case of multiple intervals and shock number thresholds, we further suggest another algorithm for obtaining these metrics.

Let us divide the total mission time τ into H intervals $[\xi_{h-1}, \xi_h)$ for $h = 1, \ldots, H$, where $\xi_0 = 0$ and $\xi_H = \tau$ by definition. If the m_h-th shock occurs before time ξ_h, the mission is aborted and the rescue procedure is initiated, otherwise the mission continues. The vectors $\boldsymbol{m} = [m_1, \ldots, m_H]$ and $\boldsymbol{\xi} = [\xi_1, \ldots, \xi_H]$ determine the mission abort policy. Notice that, for any h, obviously, $m_{h-1} \le m_h$, as the shocks are counted sequentially.

The special case of such abort policy when the mission is aborted if the time of the m-th shock is less than some $\xi < \tau$, i.e.,$T_m < \xi$, was considered in Sect. 2. Formally it means in this case that $H = 2$, $\boldsymbol{m} = [m,\infty]$, with $\boldsymbol{\xi} = [\xi,\tau]$.

A mission succeeds if the system does not fail in $[0, \tau)$ and less than m_h shocks occur until time ξ_h for $h = 1,\ldots,H$ (no mission abort). In accordance with this description, the MSP can be defined as

$$R(t, \boldsymbol{\xi}, \boldsymbol{m}) = \Pr\left(L \geq \tau, T_{m_h} \geq \xi_h \text{ for } h = 1, \ldots, H\right) \tag{20}$$

The system survives if it completes either the mission or the rescue procedure. The rescue procedure is activated in the time interval $[\xi_{h-1},\xi_h)$ only if $T_{m_h} < \xi_h$ and $T_{m_i} \geq \xi_{m_i}$ for $i = 1, \ldots, h-1$. To complete the rescue procedure activated at a random time T_{m_h}, the system lifetime must be not less than $T_{m_h} + \varphi\left(T_{m_h}\right)$, where $\varphi\left(T_{m_h}\right)$ is a duration of a rescue procedure activated at T_{m_h}. As a mission can be aborted in different time intervals $[\xi_{h-1},\xi_h)$ and the corresponding events are mutually exclusive, the system survivability can be defined as

$$\begin{aligned} S(t, \boldsymbol{\xi}, \boldsymbol{m}) = R(\tau, \xi, m) + \sum_{h=1}^{H} \Pr\Big(L \geq T_{m_h} + \varphi\left(T_{m_h}\right), \\ T_{m_h} < \xi_{m_h}, T_{m_i} \geq \xi_{m_i} \text{ for } i = 1, \ldots, h-1\Big), \end{aligned} \tag{21}$$

The probability that i shocks have occurred in $[\xi_{h-1},\xi_h)$ and that additional k shocks have occurred in $[\xi_h,\xi_{h+1})$ during the mission is in accordance with the property of independent increments for HPP,

$$P\left(\xi_h - \xi_{h-1}, i, \lambda_M\right) P\left(\xi_{h+1} - \xi_h, k, \lambda_M\right). \tag{22}$$

Let $W(h,n)$ be the joint probability of events that exactly n shocks have occurred before time ξ_h, and the number of shocks before time ξ_i,$i = 1,\ldots,h-1$ had never exceeded m_i. $W(0,0) = 1$, $W(0,n) = 0$ for $n > 0$ by definition. For $h > 0$, $W(h,n)$ can be obtained as

$$\begin{aligned} W(h, n) = \sum_{i_1=0}^{\max\{n,m_1-1\}} P\left(\xi_1, i_1, \lambda_M\right) \sum_{i_2=0}^{\max\{n,m_2-1\}-i_1} P\left(\xi_2 - \xi_1, i_2, \lambda_M\right) \times \ldots \\ \times \sum_{i_{h-1}=0}^{\max\{n,m_{h-1}-1\}-\sum_{k=1}^{h-2} i_k} P\left(\xi_{h-1} - \xi_{h-2}, i_{h-1}, \lambda_M\right) P\left(\xi_h - \xi_{h-1}, n - \sum_{k=1}^{h-1} i_k, \lambda_M\right). \end{aligned} \tag{23}$$

The mission success probability is the probability that the system did not fail during the mission time because of shocks or internal failures and it was not aborted in any time interval, i.e., for $i = 1,\ldots, H$ and the number of shocks before time ξ_i never exceeded m_i. Thus

$$R(\tau, \boldsymbol{\xi}, \boldsymbol{m}) = (1 - F(\tau)) \sum_{n=0}^{m_H-1} W(H, n) \prod_{l=0}^{n} q(l). \tag{24}$$

In what follows, we obtain the probability that the rescue procedure was activated and succeeded. The probability $Z_h(t)dt$ that the mission is aborted in $[t, t+dt) \subset [\xi_{h-1}, \xi_h)$ equals to the probability that the m_h-th shock happens in $[t, t+dt) \subset [\xi_{h-1}, \xi_h)$ and less than m_i shocks have happened in intervals $[\xi_{i-1}, \xi_i)$ for $i = 1, \ldots, h-1$ and the system survived all m_h experienced shocks:

$$Z_h(t)dt = \lambda_M Q(m_h) \sum_{n=0}^{m_{h-1}-1} W(h-1, n) P(t - \xi_{h-1}, m_h - n - 1, \lambda_M) dt. \tag{25}$$

The probability that the system survives all shocks during the rescue procedure when this procedure was activated at time instance $t \in [\xi_{h-1}, \xi_h)$ is

$$\sum_{k=0}^{\infty} P(\varphi(t), k, \lambda_R) \prod_{l=0}^{k} q(m_h + l). \tag{26}$$

As the rescue procedure is activated if the m_h-th shock happens at any time $t \in [\xi_{h-1}, \xi_h)$ and the mission abort events in different intervals $[\xi_{h-1}, \xi_h)$ are mutually exclusive, for the case under consideration, we obtain based on (10) the probability that the rescue procedure was activated and succeeded as:

$$\begin{aligned}&\sum_{h=1}^{H} \Pr\left(L > T_{m_h} + \varphi\left(T_{m_h}\right),\quad T_{m_h} < \xi_{m_h}, T_{m_i} \geq \xi_{m_i} \text{ for } i = 1, \ldots, h-1\right)\\ &= \sum_{h=1}^{H} \int_{\xi_{h-1}}^{\xi_h} (1 - F(t + \delta\varphi(t))) Z_h(t) \sum_{k=0}^{\infty} P(\varphi(t), k, \lambda_R) \prod_{l=0}^{k} q(m_h + l) dt\end{aligned} \tag{27}$$

and

$$\begin{aligned}S(t, \boldsymbol{\xi}, \boldsymbol{m}) &= R(\tau, \boldsymbol{\xi}, \boldsymbol{m}) + \sum_{h=1}^{H} \int_{\xi_{h-1}}^{\xi_h} (1 - F(t + \delta\varphi(t))) Z_h(t)\\ &\quad \times \sum_{k=0}^{\infty} P(\varphi(t), k, \lambda_R) \prod_{l=0}^{k} q(m_h + l) dt\\ &= R(\tau, \boldsymbol{\xi}, \boldsymbol{m}) + \lambda_M \sum_{h=1}^{H} \int_{\xi_{h-1}}^{\xi_h} (1 - F(t + \delta\varphi(t)))\\ &\quad \times \sum_{n=0}^{m_{h-1}-1} W(h-1, n) P(t - \xi_{h-1}, m_h - n - 1, \lambda_M)\\ &\quad \times \sum_{k=0}^{\infty} P(\varphi(t), k, \lambda_R) \prod_{l=0}^{m_h+k} q(l) dt,\end{aligned} \tag{28}$$

where $W(0,0) = 1$ and $W(0,n) = 0$ for $n > 0$ by definition.

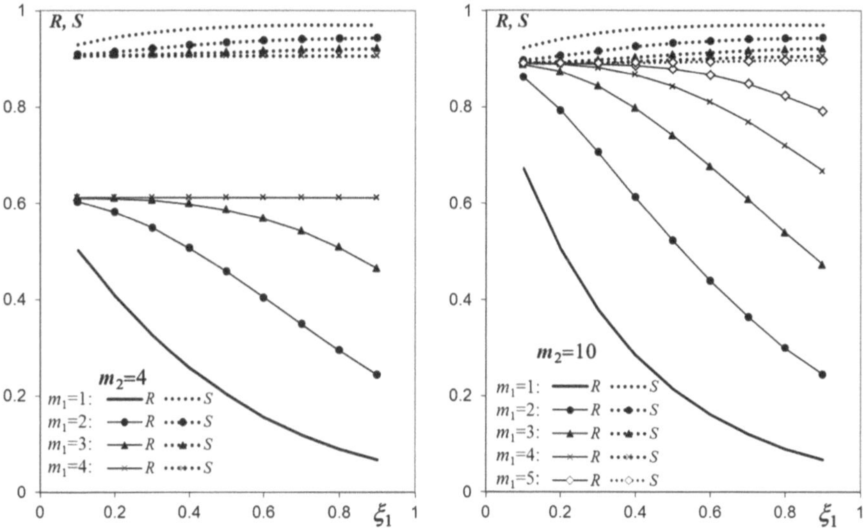

Fig. 9 MSP and SSP as functions of the decision parameters m_1 and ξ_1 for $m_2 = 4$ and $m_2 = 10$

Illustrative Example 3 Consider the previous example of the UAV mission and assume that the shock rates are $\lambda_M = 0.5$ and $\lambda_R = 0.1$ and the shock resistance function parameters are $\Omega = \omega = 0.99$.

Figure 9 presents the MSP and SSP for $H = 2$ mission abort policy (the mission time is split into two intervals) as functions of the decision parameters m_1 and ξ_1 for $m_2 = 4$ and $m_2 = 10$. It can be seen that with increase of the first interval ξ_1, the SSP increases whereas the MSP decreases. In the case of $m_1 = m_2 = 4$, the MSP and SSP do not depend on ξ_1. Indeed, when the same number of shocks is allowed in both intervals (i.e., during the entire mission), the duration of each interval does not matter. With increase of the number of allowed shocks m_1, MSP and SSP get closer because the mission abort becomes unlikely. The difference between MSP and SSP decreases also with the increase of m_2, which also makes the mission abort in the second interval less probable.

Figure 10 presents the MSP and SSP for $H = 2$ mission abort policy as functions of the decision parameters m_2 and ξ_1 for $m_1 = 1$. It can be seen that the influence of m_2 on R and S increases when ξ_1 decreases and the interval in which m_2 is used as mission abort criteria increases.

Table 1 presents the optimal solutions of problem (3) for different S^* and number of decision intervals H. It can be seen that transition from $H = 1$ to $H = 2$ considerably improves the solutions, whereas transition from $H = 2$ to $H = 3$ improves them slightly. No considerable improvement has been obtained for $H > 3$. Notice that for $S^* = 0.9$, the optimal solution for $H = 2$ limits the number of shocks in the final mission stage by 9, which gives higher MSP than the optimal abort policy

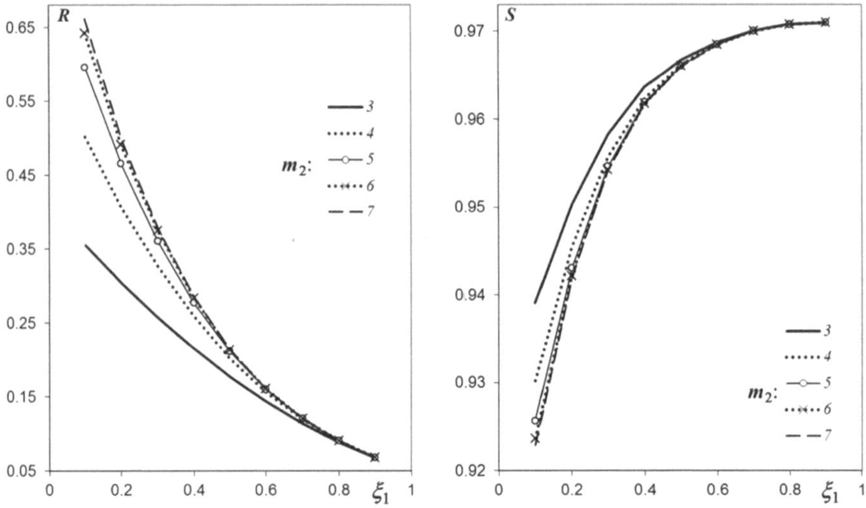

Fig. 10 R and S as functions of the decision parameters m_2 and ξ_1 for $m_1 = 1$

Table 1 Optimal abort policies for different S^* and H

S^*	H	m_1	m_2	m_3	ξ_1/τ	ξ_2/τ	ξ_3/τ	S	R
0.90	1	4	–	–	1	–	–	0.9065	0.6115
	2	2	9	–	0.14	1	–	0.9000	0.8366
	3	2	4	∞	0.13	0.33	1	0.9000	0.8384
0.92	1	3	–	–	1	–	–	0.9216	0.4099
	2	1	∞	–	0.09	1	–	0.9201	0.6909
	3	1	1	∞	0.0775	0.316	1	0.9200	0.6937
0.95	1	1	–	–	1	–	–	0.9710	0.0506
	2	1	11	–	0.27	1	–	0.9509	0.4141
	3	1	2	∞	0.24	0.374	1	0.9500	0.4256

$\boldsymbol{m} = [m,\infty]$ and $\boldsymbol{\xi} = [\xi,\tau]$ considered in Example 1, which achieves $R = 0.8311$ and $S = 0.9004$ for $m = 3$, $\boldsymbol{\xi} = 0.33$.

The improvement in the MSP due to the increase in the number of decision intervals can be intuitively illustrated by considering the corresponding changes when comparing the cases $H = 1$ and $H = 2$. For the first scenario, the decision to abort usually arrives relatively late and the level of degradation is already substantial. This also means that the mission failures could have occurred before with significant probabilities. On the other hand, aborting the mission at the earlier stage in $[0,\xi_1)$ results in an earlier detection of the unfortunate realization of the shock process, which eventually leads to the better value of the MSP, as these realizations are more likely to result in mission failures afterward. This effect seems to be maximal when comparing the first ($H = 1$) and the second ($H = 2$) scenarios, as on the subsequent steps most of unfortunate realizations are already detected and, therefore, the MSP is not significantly improved.

6 Further Research

Further research on the mission abort policy analysis and optimization can employ models of dependency between shocks and internal failures. The impact of shocks on a system time-to-failure (e.g., the direct influence of the shock process on the system baseline failure rate) can be modeled based on the approach suggested in (Levitin and Finkelstein 2018b), whereas the impact of the internal system state on shock consequences can be modeled based on the approach suggested in (Yang et al. 2017). Scenarios when probabilities of failures under shocks are different during the mission time and the rescue procedures can be also considered.

More complex abort policies also deserve further investigation. For example, the mission can be aborted when the number of shocks exceeds the thresholds in several consecutive time intervals.

The optimal protection problem can be solved when the cost of protection that improves the system shock resistance is compared with benefits of the increased MSP and SSP (given the optimal abort policy).

The optimal mission abort policy can be obtained for complex systems in which the sets of components performing the primary mission and rescue procedure are different, but overlapping.

Combination of the mission abort policy with optimal mission scheduling or routing can also constitute a topic for further research.

References

Bell JL, Bearden SD (2014) Reliability growth planning based on essential function failures. In: Proceedings of Annual Reliability and Maintainability Symposium (RAMS), Colorado Springs, CO, January

Cha JH, Finkelstein M (2011) On new classes of extreme shock models and some generalizations. J Appl Probab 48:258–270

Cha JH, Mi J (2007) Study of a stochastic failure model in a random environment. J Appl Probab 44:151–163

Finkelstein M (2008) Failure rate modelling for reliability and risk. Springer, London

Finkelstein M, Cha JH (2013) Stochastic modelling for reliability: shocks, burn-in, and heterogeneous populations. Springer, London

Finkelstein M, Gertsbakh I (2015) On 'time-free' preventive maintenance of systems with structures described by signatures. Appl Stoch Model Bus Ind 31:836–845

Gut A, Husler J (2005) Realistic variation of shock models. Stat Prob Lett 74:187–204

Klefsjo B (1981) Survival under the pure birth shock model. J Appl Probab 18:554–560

Levitin G, Finkelstein M (2018a) Optimal mission abort policy for systems in a random environment with variable shock rate. Reliab Eng Syst Saf 169:11–17

Levitin G, Finkelstein M (2018b) Optimal mission abort policy for systems operating in a random environment. Risk Anal 38(4):795–803
Levitin G, Finkelstein M (2018c) Optimal mission abort policy with multiple shock number thresholds, to appear in J Risk Reliab 232(6): 607–615
Levitin G, Xing L, Amari S, Dai Y (2013) Reliability of non-repairable phased-mission systems with propagated failures. Reliab Eng Syst Saf 119:218–228
Levitin G, Xing L, Zhai Q, Dai Y (2016a) Optimization of full vs. incremental periodic backup policy. IEEE Trans Dependable Secure Comput 13(6):644–656
Levitin G, Xing L, Zonouz AE, Dai Y (2016b) Heterogeneous warm standby multi-phase systems with variable mission time. IEEE Trans Reliab 65(1):381–393
Levitin G, Xing L, Dai Y (2018a) Mission abort policy in heterogeneous non-repairable 1-out-of-N warm standby systems. IEEE Trans Reliab 67(1):342–354
Levitin G, Xing L, Zhai Q, Dai Y (2018b) Co-optimization of state dependent loading and mission abort policy in heterogeneous warm standby systems. Reliab Eng Syst Saf 172:151–158
Levitin G, Finkelstein M, Dai Y (2018c) Mission abort policy balancing the uncompleted mission penalty and system loss risk. Reliab Eng Syst Saf 169:11–17
Lu J, Wu X, Liu Y, Lundteigen M (2015) Reliability analysis of large phased-mission systems with repairable components based on success-state sampling. Reliab Eng Syst Saf 142:123–133
Ma Y, Trivedi K (1999) An algorithm for reliability analysis of phased-mission systems. Reliab Eng Syst Saf 66:157–170
Mallor F, Omey E (2001) Shocks, runs and random sums. J Appl Probab 38:438–448
Myers A (2009) Probability of loss assessment of critical k-Out-of-n: G systems having a mission abort policy. IEEE Trans Reliab 58(4):694–701
Nakagawa T (2007) Shocks and damage models in reliability theory. Springer, London
Nelson W (1990) Accelerated testing, statistical models, test plans and data analysis. Wiley, New York
Peng R, Zhai Q, Xing L, Yang J (2014) Reliability of demand-based phased-mission systems subject to fault level coverage. Reliab Eng Syst Saf 121:18–25
Press W, Teukolsky S, Vetterling W, Flannery B (1992) Numerical recipes in C. Cambridge University Press, New York
Rausand M, Høyland A (2003) System reliability theory: models, statistical methods, and applications, 2nd edn. Wiley
Sedjakin N (1966) On one physical principle of reliability theory (in Russian). Techn Kibernetika 3:80–82
Shaked M, Shantikumar J (2007) Stochastic orders. Springer, London
Wang C, Xing L, Levitin G (2015) Probabilistic common cause failures in phased-mission systems. Reliab Eng Syst Saf 144:53–60
Wang C, Xing L, Peng R, Pan Z (2017) Competing failure analysis in phased-mission systems with multiple functional dependence groups. Reliab Eng Syst Saf 164:24–33
Yang L, Ma X, Peng R, Zhai Q, Zhao Y (2017) A preventive maintenance policy based on dependent two-stage deterioration and external shocks. Reliab Eng Syst Saf 160:201–211

Towards Prognostics and Health Management of Multi-Component Systems with Stochastic Dependence

Roy Assaf, Phuc Do, and Phil Scarf

1 Introduction

The degradation processes of system components are inevitable and could lead to faults and failures which jeopardise reliability, and causes unexpected downtime which results in lower efficiency and high maintenance costs. However, these degradation processes can be slowed or even in some cases stalled through the act of maintenance.

In order to conduct maintenance effectively, condition based maintenance (CBM) (Jardine et al., 2006; Peng et al., 2010) is usually considered. This is a major field in maintenance which in contrast to older maintenance strategies is proactive in nature and aims to carry out maintenance interventions only when needed. Another major field in maintenance is prognostics and health management (PHM), and it is getting substantial attention from the maintenance community recently, see Wang et al. (2017); Lei (2016); Kim et al. (2017). It is seen as a key enabler for CBM (Sun et al., 2012). According to Uckun et al. (2008) it can be described as an emerging engineering discipline which studies and associates the degradation processes to system lifecycle management.

PHM allows system health state assessment in real-time, as well as predicting its future health states. Thus in contrast to CBM, PHM is more concerned with

R. Assaf
School of Computing, Science and Engineering, Autonomous Systems and Robotics Centre, University of Salford, Salford, United Kingdom

P. Do (✉)
Research Centre for Automatic Control, University of Lorraine, Nancy, France
e-mail: phuc.do@univ-lorraine.fr

P. Scarf
Cardiff Business School, Cardiff University, Cardiff, United Kingdom

A. T. de Almeida et al. (eds.), *Multicriteria and Optimization Models for Risk, Reliability, and Maintenance Decision Analysis*, International Series in Operations Research & Management Science 321, https://doi.org/10.1007/978-3-030-89647-8_14

the actual health indicator extraction from the acquired signals, and puts a lot of emphasis on the prognostics step which is essential to performing optimal maintenance decision making.

PHM comprises three main elements. These are health indicator extraction; diagnostics and prognostics; and maintenance decision making. However the key idea of PHM is prognostics, whereby the end of life (EOL) of components is predicted, and consequently the remaining useful lifetime (RUL) can be evaluated as follows:

$$RUL_k = t_{eol} - t_k \tag{1}$$

where RUL_k represents the remaining useful life at a time t_k, and t_{eol} denotes the predicted end of life.

Recently an increasing number of manufacturing requirements is pushing for more complex systems to meet industrial needs. These systems have more components which might bear stochastic dependencies. Therefore maintaining such systems is becoming more of a challenge. Often however, components in a system are assumed to be independent, see Bouvard et al. (2011); Nguyen et al. (2014); Van Noortwijk (2009). But since real world systems are usually complex and include multiple interacting components, such interactions can potentially affect overall system availability, and jeopardise the effectiveness of PHM and CBM.

In Frei et al. (2013), the authors express their interest in investigating the claim that failures in a system are mostly mutually independent. They explain that it seems more likely that failures are correlated and that failures in some components might lead to failures in others. Also, recent CBM literature has been showing a growing interest in multi-component systems and their dependencies (Keizer et al., 2017). The modelling of stochastic dependence, whereby the health state of some components can be affected by the health states of other components remains the least explored (Keizer et al., 2017; Nicolai et al., 2009). This falls more under the PHM aspect, and literature on the topic is scarce.

In this chapter we present a methodology that leads towards PHM of multi-component systems. We cover how to extract health indicators from multi-component systems. And we present a methodology which makes use of these indicators within a prognostics approach. We show that this methodology effectively considers stochastic dependence between the different components by more accurately predicting their t_{eol}.

This chapter is organised as follows. In Sect. 2, we cover our methodology for implementing prognostics for multi-component systems with stochastic dependence. This consists of presenting a degradation model and a methodology for estimating the model parameters. In Sect. 3, we present an approach for effectively extracting health indicators from a multi-component gearbox platform. In Sect. 4, we apply our methodology and predict the t_{eol} of the components and discuss our results. Finally Sect. 5 concludes this chapter.

2 Prognostics for Multi-Component Systems with Stochastic Dependence

Similarly to forecasting, whereby past and present available data are analysed in order to predict future trends; prognostics follows the same process yet instead of just projecting trends into the future, it is more concerned with predicting the EOL time at which a specific failure threshold is reached, and consequently extracting RUL estimation as seen in Eq. (1). This is depicted in Fig. 1, where at time $t = k$ an attempt to predict the EOL is made. Since EOL is uncertain, it is usually represented by a probability density function (PDF), and consequently so is the RUL.

RUL estimation relates to a common question in industry, which is how long can a component operate correctly before reaching a certain failure threshold. Then, based on the RUL estimation, appropriate actions can be taken. Therefore it is the remaining time to maintenance from current time. Moreover, when consulting the literature RUL is usually more addressed than EOL; however, as Eq. (1) suggests, these terms are strongly related. Furthermore, the lower bound of a confidence interval of the RUL is usually considered for conservative purpose (Kim et al., 2017). This is crucial from both a cost-effective and a safety point of view, especially for critical equipment, such as aircraft engines, inertial navigation platforms used in aerospace and integral equipment on a production line.

Traditional methods for RUL estimation are heavily dependent on the time-to-failure data. However, these data are sometimes unavailable, as it is not always possible to have runs until system failure because of economic and safety issues. In such cases, data from degradation of components can be used as an alternative

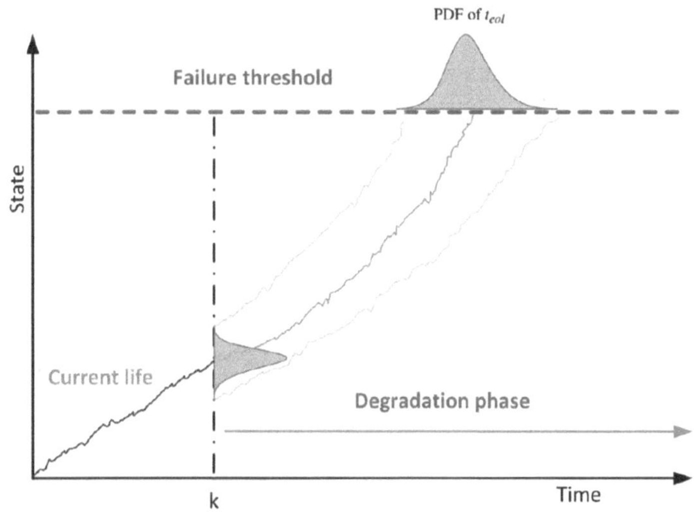

Fig. 1 An illustration of prognostics and PDF of t_{eol}

resource for RUL estimation. Several papers have reviewed and compared the main probabilistic prognostic methods for RUL estimation, see for example (Si et al., 2011), and (Le Son et al., 2013).

From the above, we conclude that prognostics is a vital step that helps industries manage their risks and prevents the occurrence of unforeseen components failures. And based on these predictions of future fault occurrences, maintenance and downtime costs can be minimised with CBM.

A holistic view of prognostics shows that it builds upon the following three aspects:

- State estimation: Based on the collected data, this step is used to give an estimation of the degradation state of the component.
- State prediction: The task of state prediction is to predict the degradation tendency according to the information of the historical data.
- EOL and RUL prediction: It serves for determining the time left under the degradation curve before final failure or before a predetermined failure threshold.

An extensive body of literature exists on prognostics approaches and applications, and therefore a considerable amount of review papers can be found where the classification of different prognosis approaches is presented (Jardine et al., 2006; Peng et al., 2010; Si et al., 2011; Sikorska et al., 2011). Mainly, prognostics approaches can be categorised into three types: physics based prognosis, data-driven prognosis, and hybrid prognostics.

The generic degradation model that is presented next can be used for performing predictions of t_{eol} for multi-component systems. The procedure for performing prognostics is achieved by first doing parameter identification which can be done using different approaches. We suggest the use of particle filter; this choice will be motivated later in this work.

Once the model parameters are identified, the model can be easily used to simulate and predict the health state of a component $X^i_{t_k}$ at a future time $r > k$. This is done until the degradation trajectory hits the failure threshold which indicates t_{eol}. At that point the RUL can easily be extracted, and a maintenance decision can be taken accordingly.

2.1 Degradation Model

Here we present a degradation model that will enable the modelling of the stochastic dependence between multiple components (Assaf et al., 2018). This model is later used jointly with a particle filter for performing prognostics.

Consider a multi-component system with n_c number of components. The degradation state of each component i is represented by an accumulation of wear over time which is assumed to be described by a scalar random variable X^i_t. Component i fails if its degradation state reaches a threshold value L^i. If any of the components fails we consider the system to have failed, and if a component is not operating for whatever reason, no change occurs to its degradation state unless a maintenance intervention is carried out.

We assume the evolution of the degradation state of component i is represented by

$$X_{t+1}^{i} = X_{t}^{i} + \Delta X_{t}^{i} \tag{2}$$

where ΔX_{t}^{i} represents the degradation increment of component i during one time step.

The degradation of a component i at time step t may depend on the operating conditions, the state of component i, and also the state of other components to a varying degree. Thus we suggest a general stationary model for the increment ΔX_{t}^{i}:

$$\Delta X_{t}^{i} = \Delta O_{t}^{i} + \Delta X_{t}^{ii} + \sum_{j \neq i} \Delta X_{t}^{ji} \tag{3}$$

where ΔO_{t}^{i} represents the degradation increment of component i that is caused by the operating conditions during one time step t. ΔO_{t}^{i} can be specified as a deterministic or as a random variable. ΔX_{t}^{ii} represents the degradation increment which is intrinsic to i at time step t. In other words ΔX_{t}^{ii} depends on the degradation state of component i at time step t. ΔX_{t}^{ii} can also be specified to be a deterministic or random variable. $\sum_{j \neq i} \Delta X_{t}^{ji}$ represents the sum of all degradation increments which are caused by the stochastic dependence of component i with the other components of the system. The stochastic dependence between a component i and another component j may be considered to be a deterministic or random variable.

In this work we will consider a case where $\Delta X^{ll} > 0$ and $\Delta X^{ji} > 0$, the components are stochastically dependent, and the increment in the degradation level of component i may depend not only on the state of component i but also on the state of the other component.

We will use the following model for the quantification of degradation influence between multiple components:

$$\Delta X_{t}^{ji} = \mu^{ji} \times (X_{t}^{i})^{\sigma^{ji}} \tag{4}$$

where X_{t}^{ji} represents the degradation impact of component j on component i at time t. And where μ^{ji} and σ^{ji} are non-negative real numbers which are used to quantify component j's influence on component i. According to this model, several special cases are specified as follows:

- $\mu^{ji} = 0$: Component j does not have any influence on the degradation behaviour of component i.
- $\mu^{ji} = 0$ and $\mu^{ij} = 0$: Component j and i are independently subject to gradual degradation.

- $\mu^{ji} > 0$ and $\sigma^{ji} = 0$: The impact of component j on the degradation of component i does not depend on the health state of component j.

Although the proposed degradation model can encompass as many components as we would like, the number of components to be considered when performing degradation modelling should be kept to the minimum. This is because adding more components would lead to an increase in model complexity and computation, and the composition of optimal CBM policies becomes more difficult and computationally complex. This is an already identified issue in multi-component maintenance literature (Alaswad and Xiang, 2017).

2.2 Parameter Estimation with Particle Filters

After deciding on a degradation model which is capable of modelling the stochastic dependence between multiple components, we now need an approach to estimate the parameters of this model.

There exists an extensive body of literature on the topic of parameter identification, see for example (An et al., 2015; Gebraeel and Pan, 2008; Lorton et al., 2013). In practice, if the degradation model is not too complex, we can fit the model parameters using maximum likelihood estimation (MLE). However, in the case of multi-component systems this is highly unlikely. Therefore if the model is too complex, or if we are collecting online observation on the health condition of the components and want to achieve real-time prognostics, we suggest to use sequential Monte Carlo methods, specifically the Particle Filter (PF) method which is a very popular approach for parameter estimation (Doucet and Johansen, 2009).

Particle filter draws upon stochastic filtering, Bayesian statistics, and Monte Carlo techniques. It is usually also referred to as sequential Monte Carlo; however, it should be distinguished from sequential Monte Carlo (SMC), since SMC methods encompass a broader range of algorithms (Doucet and Johansen, 2009), such as the well-known Gibbs sampling and Metropolis–Hastings algorithms. PF is also called a Bootstrap filter (Green, 1995); this is the case of the standard PF, see Algorithm 1.

PF allows for an online numerical estimation of the parameter values by means of a recursive Bayesian inference approach. The posterior distribution of the model parameters can be then obtained using a number of particles and their corresponding weights. This method is very flexible and can be used for non-linear models where the noise is not necessarily Gaussian. Such an approach has been successfully used in the field of prognostics for model parameter estimation (Jouin et al., 2016). Also, it is worth noticing that compared to typical works on PF, whereby filtering is mainly considered, prognostics concerns itself with future time horizons; this means that this field tries to go beyond the filtering step. In view of this, PF for prognostics should be used in accordance to the necessity of forecasting the state at future times, mostly without additional observations, adjusting the weights if necessary. Moreover, recent reviews about PF for PHM such as Jouin et al. (2016) suggest an

Algorithm 1: Particle filter algorithm

```
input : n_p number of particles
Initialisation
t = 0
for i ← 1 to n_p do
 | Sample x_0^i ~ p(x_0)
end
for t ← 1 to t_end do
 |  Importance Sampling
 |  for i ← 1 to n_p do
 |   | Sample x̃_t^n ~ p(x_t | x_{t-1}^i)
 |   | Set x̃_{0:t}^i = (x_{0:t-1}^i, x̃_t^i)
 |  end
 |  for n ← 1 to n_p do
 |   | Evaluate importance weights w̃_t^n = p(y_t | x̃_t^n)
 |  end
 |  Normalise importance weights w̃_t^n
 |
 |  Particle Selection
 |  for n ← 1 to n_p do
 |   | Considering w̃_t^n, re-sample with replacement n_p particles
 |  end
end
```

increasing amount of work on PF in PHM. Therefore this approach is considered as a state-of-the-art technique in PHM and will be used in this work.

In order to successfully implement prognostics in a multi-component system context, we need to effectively extract health indicators that reflect the state of the components. We show how this is done in the next section on data generated from gearbox accelerated life testing platform.

3 Case Study and Health Indicator Extraction

3.1 Case Study

In an industrial setting, gearboxes are present in virtually any mechanical system, playing the essential role of torque and speed conversion. We therefore consider the dataset generated from a gearbox accelerated life test platform from Assaf et al. (2018), Fig. 2.

To demonstrate the stochastic dependence between components, we will only consider a two gear system. Gear 1 and gear 2, referred to as C1 and C2, respectively.

The experimental runs of the gearbox were designed for accelerated life testing, thus achieving failure in a shorter amount of time than it would usually take under normal operating conditions. These runs are an alternating sequence of two types of

Fig. 2 Gearbox accelerated life testing platform

cycles: the first cycle is a low speed low load cycle, referred to as LSLL; and the second type is a high speed high load cycle, referred to as HSHL.

Due to the nature of the HSHL cycle, a high level of noise is present in the acceleration data. We therefore only use vibration data that are collected in the LSLL cycles in order to improve the signal to noise ratio. These LSLL cycles last for 3 min.

The dataset used here is obtained from an experimental scenario which consisted of three runs/tests to failure and was conducted in the following manner: The first run consisted of a new C1 and a new C2. The gearbox was run alternating between HSHL and LSLL until high levels of vibration were observed in the gearbox (meshing frequency magnitude exceeding 1800) at which point the experimental run was terminated. In run 2, C1 was replaced with a new gear, while C2 remained unchanged, so the second run consisted of a new C1 and a worn out C2. The gearbox was ran alternating between the HSHL and LSLL cycles until high vibration was observed; on this run high system vibration occurred in a shorter amount of time, and after terminating the run, C2 showed more severe damage on its teeth surface than that observed after the termination of run 1. In the third run, C1 was replaced with a new gear, while C2 remained unchanged, so we find ourselves with a similar condition scenario as in run 2, this time however with a more worn out C2. The gearbox ran alternating between the HSHL and LSLL cycles until high vibration was observed. This run lasted an even shorter amount of time than in run 2, and so the run was terminated earlier than in run 1 and run 2.

3.2 Health Indicator Extraction for Multi-Component Systems

Health indicator extraction sits at the heart of PHM; it is responsible for refining the raw condition monitoring data so that diagnostics and prognostics can make efficient use of it. This aspect of PHM is done in three main phases: data acquisition, pre-processing, and processing of the data. The following details our methodology for obtaining component health state data from multi-component systems.

In a multi-component system setting, it is wise to use multiple sensors. The sensors chosen for this gearbox platform are accelerometers, and their placement is scattered evenly around the system. This grants different vantage points for data collection and is done so that the different components of the system can be easily differentiated, especially when dealing with similar components that can emit signals around similar frequencies.

Since vibration data has been acquired from this gearbox platform, we would like to apply some signal pre-processing which aims to eliminate the noise in the signal and increase the signal-to-noise ratio (SNR). We start by removing outliers from the vibration data. First a window of data points based on the operating profile of the system should be specified. Then the median value or geometric mean of the data and the median absolute deviation (MAD) are computed for that window. The values that exceed the median plus or minus the MAD value are then filtered by replacing them with a random variable sampled as $X \sim \mathcal{N}(med, mad)$, thus preserving as much as possible the true nature of the signal. This is important for diagnostics and prognostics.

Data detrending and centring should then follow. Scaling should be done if necessary, depending on the presence of dissimilar sensors or if different data ranges are used. Filtering the data can follow after this step; this depends on the frequency band of interest and whether other unnecessary frequencies can be rejected without loss of information.

Finally the physical meaning of the signal should be obtained, so that an engineering perspective can be added. This depends on the specifics of the sensors used. For example if accelerometers are used, this step should be applied and would result with a signal that has its acceleration denoted in acceleration of gravity (G), instead of the generic digital signal value.

The final step of health indicator extraction is signal processing which is used to extract health indicators or fault-related information from machinery (Lei et al., 2013, 2014). These indicators would help us accurately diagnose and predict the future states of the system.

A major challenge for modelling existing stochastic dependency in a multi-component system is the complex nature of the signals acquired. Each signal may represent a mixture of the signals of all components at once, but to varying degrees. Therefore, we use time-frequency analysis. This is usually used in blind source separation (Yilmaz and Rickard, 2004; Abrard et al., 2001), in which mixed signals are separated without the aid of other information. This is performed by exploiting the difference in the time-frequency signatures of the sources to be separated. Much of the literature in this field focuses on audio applications (Puigt and Deville, 2005) and machine sound signals (Zhong et al., 2006). Applications of time-frequency analysis for identifying various sources of vibrations data can be found in Vulli et al. (2009); Dekys et al. (2017).

Consequently, an STFT can be applied on the cleaned signal and allow for the analysis to be performed in both time and frequency domains, isolating the frequency components of interest all while representing the evolution of their energy through time.

The STFT can be applied over the time-waveform data of a component i in the following manner:

$$s_i^{'} = STFT\{s_i[n]\}(\tau, \omega) = \sum_{n=-\infty}^{+\infty} h[n-\tau]s[n]\exp^{-j\omega n} \tag{5}$$

where $s^{'}$ represents the short-time Fourier transform of the input signal $s(t)$, and $h(t)$ the window function. Optimum window length depends on the application. A high resolution in time and in frequency cannot be accomplished simultaneously. If high resolution in time domain is needed, the size of the window should be reduced. If the application demands frequency domain information to be more specific, then the size of the window should be increased (Kadambe, 1992; Satish, 1998). Therefore, if we want to resolve the fundamental and harmonics of a signal, a long window should be used. If it is needed to detect the onset or presence of some events, a short window should be used. Some examples of window functions are Gaussian and a Hamming windows (Harris, 1978; Jones and Baraniuk, 1994).

After the STFT is applied on the signal, the frequency root mean square (FRMS) can be computed over the frequency band of interest. This is done in order to estimate how the magnitude of the frequency band of interest evolves in time. This is applied as such:

$$X_{FRMS} = \sqrt{\frac{1}{N}\sum_{i=1}^{N} s_i^{'2}} \tag{6}$$

where N is the number of data points, and n is the nth value.

In this way, we can study a time series signal that describes the evolution of the health condition of the components over time. This makes the prognostics aspect of PHM easier and more effective.

And so after performing the health indicator extraction step described here, we acquire the RMS degradation trajectories for C1 and C2. These are presented in Fig. 3. The RMS values inform us of the vibration energy in the machine that originates from the gears. Therefore they represent the degradation level of the components since the higher the vibration energy, the more the gears are degraded and the more prone the gearbox is to damage. This proness of a gear to damage is the manifestation in reality of the terms ΔX_t^{ii} and ΔX^{ji} in the model Eq. (3), the

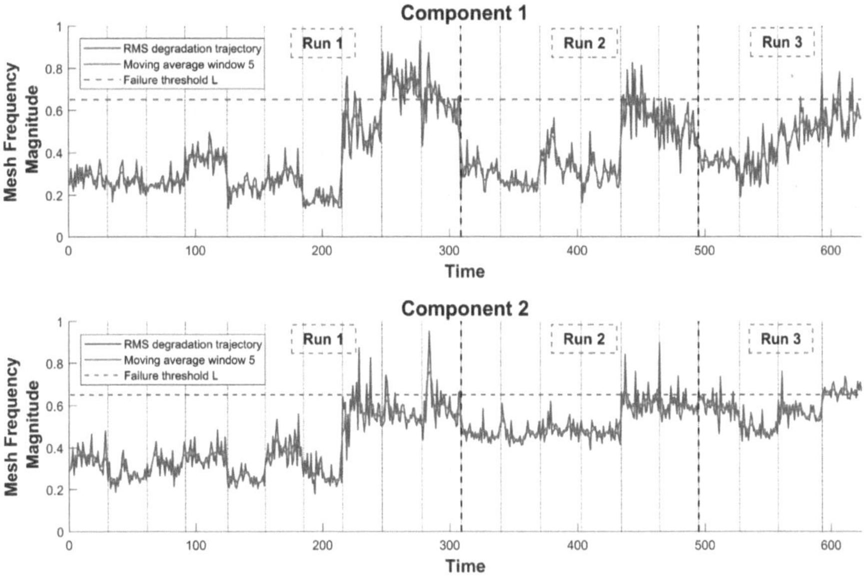

Fig. 3 Evolution of degradation of the gears in all three runs, represented by the mesh frequency magnitude

former because the gear itself is worn, and the latter because the other gear is worn. Based on these results, we consider a component to be severely worn out or to have failed once it reaches the threshold vibration magnitude of $L^i = 0.65$ for $i = 1, 2$.

4 Predicting End of Life of Components

Here we fit the generic multi-component degradation model to the data generated from the gearbox accelerated life testing platform presented earlier. We refer to gear 1 and 2 as component 1 (C1), and component 2 (C2), respectively.

After performing the health indicator extraction step, we acquire the RMS degradation trajectories for C1 and C2. These are presented in Fig. 3. The RMS values inform us of the vibration energy in the machine that originates from the gears. Therefore they represent the degradation level of the components since the higher the vibration energy, the more the gears are degraded and the more prone the gearbox is to damage. This proness of a gear to damage is the manifestation in reality of the terms ΔX_t^{ii} and ΔX^{ji} in the model (3), the former because the gear itself is worn, and the latter because the other gear is worn. Based on the results of Sect. 3, we consider a component to be severely worn out or to have failed once it reaches the threshold vibration magnitude of $L^i = 0.65$ for $i = 1, 2$.

Due to the physical characteristics of the gears, we know that the degradation level of components C1 and C2 increases with time, and that this degradation level cannot decrease without maintenance intervention. Therefore, both components are considered to have inherent degradation that increases with time. Consequently we assume that these degradation increments are gamma-distributed:

$$X \sim \Gamma(\alpha^i, \beta^i)$$

The corresponding probability density function (PDF) is

$$f_{\alpha^i,\beta^i}(x) = \frac{1}{\Gamma(\alpha^i)}(\beta^i)^{\alpha^i} x^{\alpha^i - 1} e^{-\beta^i x} \mathscr{I}_{\{x \geq 0\}}$$

where $\Gamma(\alpha^i) = \int_0^{+\infty} u^{\alpha^i - 1} e^{-u} du$ denotes a complete gamma function; and $\mathscr{I}_{\{x \geq 0\}}$ is an indicator function. $\mathscr{I}_{\{x \geq 0\}} = 1$ if $x \geq 0$, $\mathscr{I}_{\{x \geq 0\}} = 0$ and otherwise.

These increments are denoted by ΔX^{11} and ΔX^{22} for C1 and C2, respectively. Thus, $\Delta X^{11} \sim \Gamma(\alpha^1, \beta^1)$ and $\Delta X^{22} \sim \Gamma(\alpha^2, \beta^2)$.

Next, we model the degradation interactions between the two components. From Fig. 3 it appears that the state of C2 affects the rate of degradation of C1. This can be seen when we observe the time to failure of C1 when coupled with a worn out C2 in both runs 2 and 3, and that in run 3, where C2 was more worn out, the time to failure of C1 was shorter than run 2. Thus the degradation rate of C1 appears to be dependant on the degradation level of C2 and vice versa. This has been further analysed in Sect. 5.

ΔX^{21} is used to denote the increment in the degradation level of C1 due to C2, and ΔX^{12} the increment in the degradation level of C2 due to C1.

We denote the degradation states for C1 and C2 at time t by X_t^1 and X_t^2, respectively. Thus, the evolution of degradation for C1 is described as

$$\Delta X_t^1 = \Gamma(\alpha^1, \beta^1) + \mu^{21} \times (X_{t-1}^2)^{\sigma^{21}} \tag{7}$$

and for C2 as

$$\Delta X_t^2 = \Gamma(\alpha^2, \beta^2) + \mu^{12} \times (X_{t-1}^1)^{\sigma^{12}} \tag{8}$$

There are four parameters to be estimated for each component from the data; these sets of parameters are denoted by Θ^1 and Θ^2. Where $\Theta^1 = (\alpha^1, \beta^1, \mu^1, \sigma^1)$ and $\Theta^2 = (\alpha^2, \beta^2, \mu^2, \sigma^2)$. We use a PF to estimate these parameters using $n_p = 1000$.

We obtain the mean estimated value of each parameter in Table 1. Note that since the degradation level is normalised between 0 and 1, the greater the value

Table 1 Estimated parameter values

Component	α^i	β^i	μ^i	σ^i
C1	0.0233	0.0425	0.0995	7.6659
C2	0.0125	0.0914	0.0493	9.7375

Table 2 Actual time of end of life, and average predicted time of end of life ($\widehat{t_{eol}}$) for components 1 and 2

	Actual t_{eol}		$\widehat{t_{eol}}$ with interaction		$\widehat{t_{eol}}$ no interaction	
	C1	C2	C1	C2	C1	C2
Run 1	248	227	239	259	301	429
Run 2	133		157		301	
Run 3	111		118		301	

of the parameter σ^i the smaller the impact that is to be considered from the other component on component i.

To further validate the parameter values of the degradation model considering the interactions between the 2 components, we compute the R^2 values for the fit of the average estimated degradation trajectory resulting from the PF to the real degradation trajectories. For component 1 this is $R_1^2 = 0.792$ and for component 2 it is $R_2^2 = 0.753$. If we were to consider a reduced model whereby no stochastic dependence is considered between the two components and we were left with a gamma process describing the evolution of the degradation level, the average fit of such models would result in a $R_1^2 = 0.671$ and $R_2^2 = 0.575$.

Now we use the degradation model with the estimates obtained in Table 1 and generate 1000 simulation using the model in order to predict the degradation trajectories of the components. In the following figures these are referred to as "with interaction". These simulations are also performed using the reduced model, whereby no stochastic dependence is considered; these are referred to as "no interaction". This is done so that we can compare the prognostic performance difference between the case where we consider degradation dependence in degradation modelling, and in the case where we do not.

The simulations are performed for C1 in run 1, and for C2 in run 1. Then, since C2 remains unchanged for run 2, we only simulate the degradation trajectory for C1 in runs 2 and 3, all while considering the state of C2 in those runs.

Table 2 summarises the different t_{eol} estimates that are extracted from these simulations. It is clear that considering degradation dependencies provides an advantage when attempting to predict the real degradation trajectories of the components. This is clearly seen when considering the time instance where the degradation of a component is supposed to reach the failure threshold.

From Table 2 we see that the difference between the actual observed t_{eol} and the average predicted $\widehat{t_{eol}}$ for C1 when not considering stochastic dependence shows a strict growth trend. It starts from 53 in run 1, to 168 in run 2, and then 190 in run 3. This is because the parameters of the models are estimated in run 1 using PF. Therefore the reduced model cannot account for the accelerated degradation that is due to a new C1 being coupled with a worn out C2. On the other hand this difference does not show this trend when we consider the stochastic dependence. The difference is 9 in run 1, then 24 in run 2, and then just 7 in run 3. This clearly indicates the criticality of modelling stochastic dependencies between components when attempting to do prognostics.

Finally, a note regarding the prognostics using the generic model provided. These prediction of life $\widehat{t_{eol}}$ are simulated at $t = 0$ in runs 2 and 3. Therefore, if PF is used for an online update of the parameters after receiving new observations of the component health, we assume that the predictions would then be even more accurate. This would also allow for considering break points in the component's health state, in the likes of shocks that might occur due to environmental effects or sudden excess loading.

5 Conclusion

In this chapter we presented work that leads to PHM of multi-component systems. We have highlighted the importance of considering the stochastic dependence between components when performing prognostics for these systems. We started by presenting an approach for effectively extracting health indicators from multi-component systems. We then made use of a generic degradation model that is capable of capturing the stochastic dependence between different components, and used particle filters to estimate the parameters of this model on data generated using a gearbox life testing platform. The results confirmed the importance of modelling the stochastic dependence for achieving accurate predictions of the end of life of components in multi-component systems.

References

Abrard F, Deville Y, White P (2001) A new source separation approach based on time-frequency analysis for instantaneous mixtures. In: Proceedings of the ECM2S, pp 259–267

Alaswad S, Xiang Y (2017) A review on condition-based maintenance optimization models for stochastically deteriorating system. Reliab Eng Syst Saf 157:54–63

An D, Kim NH, Choi J-H (2015) Practical options for selecting data-driven or physics-based prognostics algorithms with reviews. Reliab Eng Syst Saf 133:223–236

Assaf R, Do P, Nefti-Meziani S, Scarf P (2018) Wear rate–state interactions within a multi-component system: a study of a gearbox-accelerated life testing platform. Proc. Inst. Mech. Eng. O J Risk Reliab 232(4):425–434

Bouvard K, Artus S, Bérenguer C, Cocquempot V (2011) Condition-based dynamic maintenance operations planning and grouping. application to commercial heavy vehicles. Reliab Eng Syst Saf 96(6):601–610

Dekys V, Kalman P, Hanak P, Novak P, Stankovicova Z (2017) Determination of vibration sources by using STFT. Procedia Eng 177:496–501

Doucet A, Johansen AM (2009) A tutorial on particle filtering and smoothing: Fifteen years later. Handbook of Nonlinear Filtering 12(656–704):3

Frei R, McWilliam R, Derrick B, Purvis A, Tiwari A, Serugendo GDM (2013) Self-healing and self-repairing technologies. Int J Adv Manuf Technol 69(5–8):1033–1061

Gebraeel N, Pan J (2008) Prognostic degradation models for computing and updating residual life distributions in a time-varying environment. IEEE Trans Reliab 57(4):539–550

Green PJ (1995) Reversible jump Markov chain Monte Carlo computation and Bayesian model determination. Biometrika 82(4):711–732

Harris FJ (1978) On the use of windows for harmonic analysis with the discrete Fourier transform. Proc IEEE 66(1):51–83

Jardine AK, Lin D, Banjevic D (2006) A review on machinery diagnostics and prognostics implementing condition-based maintenance. Mech Syst Signal Process 20(7):1483–1510

Jones DL, Baraniuk RG (1994) A simple scheme for adapting time-frequency representations. IEEE Trans Signal Process 42(12):3530–3535

Jouin M, Gouriveau R, Hissel D, Péra M-C, Zerhouni N (2016) Particle filter-based prognostics: Review, discussion and perspectives. Mech Syst Signal Process 72:2–31

Kadambe S (1992) On the window selection and the cross terms that exist in the magnitude squared distribution of the short time Fourier transform. In: Conference Proceedings of the IEEE sixth SP workshop on statistical signal and array processing, 1992, pp 22–25. New York, IEEE

Keizer MCO, Flapper SDP, Teunter RH (2017) Condition-based maintenance policies for systems with multiple dependent components: A review. Eur J Oper Res 261(2):405–420

Kim N-H, An D, Choi J-H (2017) Prognostics and health management of engineering systems. Springer, Berlin

Le Son K, Fouladirad M, Barros A, Levrat E, Iung B (2013) Remaining useful life estimation based on stochastic deterioration models: a comparative study. Reliab Eng Syst Saf 112:165–175

Lei Y (2016) Intelligent fault diagnosis and remaining useful life prediction of rotating machinery. Butterworth-Heinemann, Oxford

Lei Y, Li N, Lin J, Wang S (2013) Fault diagnosis of rotating machinery based on an adaptive ensemble empirical mode decomposition. Sensors, 13(12):16950–16964

Lei Y, Lin J, Han D, He Z (2014) An enhanced stochastic resonance method for weak feature extraction from vibration signals in bearing fault detection. Proc Inst Mech Eng C J Mech Eng Sci 228(5):815–827

Lorton A, Fouladirad M, Grall A (2013) A methodology for probabilistic model-based prognosis. Eur J Oper Res 225(3):443–454

Nguyen K-A, Do P, Grall A (2014) Condition-based maintenance for multi-component systems using importance measure and predictive information. Int. J. Syst. Sci. Oper Logist 1(4):228–245

Nicolai RP, Frenk JBG, Dekker R (2009) Modelling and optimizing imperfect maintenance of coatings on steel structures. Struct Saf 31(3):234–244

Peng Y, Dong M, Zuo MJ (2010) Current status of machine prognostics in condition-based maintenance: a review. Int J Adv Manuf Technol 50(1–4):297–313

Puigt M, Deville Y (2005) Time–frequency ratio-based blind separation methods for attenuated and time-delayed sources. Mech Syst Signal Process 19(6):1348–1379

Satish L (1998) Short-time Fourier and wavelet transforms for fault detection in power transformers during impulse tests. IEE Proc Sci Meas Technol 145(2):77–84

Si X-S, Wang W, Hu C-H, Zhou D-H (2011) Remaining useful life estimation–a review on the statistical data driven approaches. Eur J Oper Res 213(1):1–14

Sikorska J, Hodkiewicz M, Ma L (2011) Prognostic modelling options for remaining useful life estimation by industry. Mech Syst Signal Process 25(5):1803–1836

Sun B, Zeng S, Kang R, Pecht MG (2012) Benefits and challenges of system prognostics. IEEE Trans Reliab 61(2):323–335

Uckun S, Goebel K, Lucas PJ (2008) Standardizing research methods for prognostics. In: International conference on prognostics and health management 2008 (PHM 2008), pp 1–10. New York, IEEE

Van Noortwijk, J (2009) A survey of the application of gamma processes in maintenance. Reliab Eng Syst Saf 94(1):2–21

Vulli S, Dunne J, Potenza R, Richardson D, King P (2009) Time-frequency analysis of single-point engine-block vibration measurements for multiple excitation-event identification. J Sound Vib 321(3–5):1129–1143

Wang D, Tsui K-L, Miao Q (2017) Prognostics and health management: A review of vibration based bearing and gear health indicators. IEEE Access 6:665–676

Yilmaz O, Rickard S (2004) Blind separation of speech mixtures via time-frequency masking. IEEE Trans Signal Process 52(7):1830–1847

Zhong Z, Chen J, Zhong P, Wu J (2006) Application of the blind source separation method to feature extraction of machine sound signals. Int J Adv Manuf Technol 28(9–10):855–862

Multi-objective Bayesian Optimal Design for Accelerated Degradation Testing

Xiao-Yang Li

1 Introduction

Products with high reliability are designed and manufactured to operate for a long lifetime before their failures. Because of the short research-development period, reliability testing is required to be conducted with rigorous time constraints. Since there are few or even no failure during tests, traditional reliability testing that can only record failure data is obviously not suitable for the reliability and lifetime predictions of such products. An accelerated degradation testing (ADT) utilizes more severe stress conditions to accelerate the performance degradation of a product; and then, the collected degradation data are used to recognize the performance degradation law and predict reliability and lifetime under normal conditions. Thus, ADT has drawn much attention and been widely used. Theoretically, the entropy production in the second law of thermodynamics shows the following: on the one hand, it seems that irreversible degradation is the basic law of nature and this process will be promoted by external conditions (McPherson 2010); on the other hand, only when the behavior of a system has sufficient randomness, the degradation can exist. From this point of view, ADT can also be regarded as a scientific experimental method, which aims to recognize a performance degradation law under the influence of uncertainties by using an experimental way. How to quantify and control the uncertainties embedded in an ADT to better realize a degradation law is the core task of an ADT. This chapter will focus on the control of uncertainties.

Generally, there are two different kinds of uncertainties in an ADT, i.e., the inherent uncertainty and the model selection uncertainty (Liu et al. 2017). The inherent

X.-Y. Li (✉)
Science and Technology on Reliability and Environmental Engineering Laboratory, School of Reliability and Systems Engineering, Beihang University, Beijing, China
e-mail: leexy@buaa.edu.cn

A. T. de Almeida et al. (eds.), *Multicriteria and Optimization Models for Risk, Reliability, and Maintenance Decision Analysis*, International Series in Operations Research & Management Science 321, https://doi.org/10.1007/978-3-030-89647-8_15

uncertainties come from the unit-to-unit variation and the errors of measurement instruments and testing equipment. The model selection uncertainties are due to the accelerated degradation model assumptions and the testing objective selections. If these uncertainties can be better controlled, the more precise recognition of degradation law and the more accurate reliability and lifetime evaluations could be guaranteed. The mathematical optimization is always used as a design approach to plan an ADT, in which the balanced plan will be made to decrease the negative influences of uncertainties and improve the accuracy of evaluations by the optimized periodic and repetitive measurement arrangement. Unfortunately, to the best of our knowledge, there are few publications about how to simultaneously control the inherent uncertainty and the model selection uncertainty.

The classical optimal design for ADT is based on an accelerated degradation model with specified parameter values in the accelerated degradation testing (ADT), in which only the inherent uncertainties from the unit-to-unit variation and the errors of measurement instruments and testing equipment are considered (Ye et al. 2014; Wang et al. 2016; Tseng et al. 2009; Liao and Tseng 2006; Huang et al. 2016; Hu et al. 2015). The optimal test plan designed is, then, generally referred to as the local optimal solution. The uncertainties in specified parameter values cause the optimal test plan to be suboptimal. Therefore, the Bayesian optimal design of ADT is developed. Based on available historical data and expert information, prior distributions can be assigned to account for the model parameter uncertainties in a Bayesian optimal design; and then, the optimal test plan can be obtained by averaging over the parameter space and the sample space. Compared to a classical ADT optimal design, in which the crisp values are taken for the model parameters, a Bayesian optimal design is a global optimal method. The existing Bayesian optimal designs of ADT lie on the two aspects: (i) the concerns are mainly on different expected utility functions (e.g., relative entropy (Hamada et al. 2001), quadratic loss function (Li et al. 2015; Xu and Tang 2015; Liu and Tang 2010; Peng et al. 2014), D-optimality (Shi and Meeker 2012; Zhang and Meeker 2006)); (ii) degradation models are the focuses (e.g., degradation models based on the Wiener process, gamma process, and inverse Gaussian (IG) process (Wang et al. 2016; Escobar 1995; Roy and Mukhopadhyay 2015; Meeker and Escobar 1998; Wang and Xu 2010)).

With decades of development, the fruitful achievements on optimal design of ADT have been made, and there are many optimal design methods to choose. From the perspective of engineering, different methods, however, bring confusion to engineers, i.e., which expected utility functions should be chosen? Actually, the model selection uncertainties appear in this situation. Multi-objective optimization methodology is utilized to solve this problem by generating a Pareto optimal frontier of solutions with the consideration of the dominant optimality among several optimization objectives. But there are few publications focusing on the multi-objective ADT design. For example, (Marseguerra et al. 2003) proposed a formulation of two-objective ADT optimal design problem which optimized both the estimation accuracy of the failure time distribution percentiles and the testing cost. The uncertainty of model assumption was ignored in (Marseguerra et al. 2003).

Moreover, even though we use priors to incorporate parameter uncertainties, is that enough to handle the uncertainty caused by model assumptions?

In order to bridge the existing gap and simultaneously control the inherent uncertainty and the model selection uncertainty, this chapter proposes a Bayesian multi-objective design method for ADT in which three objectives are considered and data envelopment analysis (DEA) is further used to prune the Pareto solutions so as to find out the plan with the highest relative efficiency. Since the inverse Gaussian process can well describe the monotonic degradation process and has been widely used, the proposed methodology in this chapter will be illustrated based on the inverse Gaussian process. In Sect. 2, the framework of Bayesian optimal design for ADT based on the IG process is presented. In Sect. 3, multi-objective Bayesian optimal model for ADT is presented and multi-objective optimal algorithm called NSGA-II (Wang et al. 2016) is employed to solve the optimal model. And DEA method is introduced to crop the Pareto solutions solved by NSGA-II in order to find out the plan with highest relative efficiency. In Sect. 4, a numerical case is utilized to illustrate the applicability and validity of the proposed method. Finally, Sect. 5 concludes the chapter.

2 IG Process in an ADT and Bayesian Inference

Assume that the degradation path of a product follows the inverse Gaussian (IG) process. The product fails when its degradation path $Y(t)$, $t \geq 0$ reaches a predefined threshold level Y_D and then the associated first-passage time is given by T_D.

2.1 The Preliminary of IG Process in an ADT

$Y(t)$ satisfies an IG process, if:

(1) $Y(0) = 0$ with probability one.
(2) $Y(t)$ has independent increments, i.e., $Y(t_2) - Y(t_1)$ and $Y(t_4) - Y(t_3)$ are independent, for $0 \leq t_1 < t_2 \leq t_3 < t_4$.
(3) Each increment follows an IG distribution, i.e., $\Delta Y(t) \sim \mathrm{IG}(\mu \Delta\Lambda(t), \lambda \Delta\Lambda^2(t))$, where $\mu > 0$, $\lambda > 0$, $\Delta\Lambda = \Lambda(t) - \Lambda(s)$, and $\Lambda(t)$ is a given monotone increasing function of time t with $\Lambda(0) = 0$.

And then for any $y > 0$, the probability density function (PDF) of IG (u, v), $u > 0$, $v > 0$, with mean u and variance u^3/v is defined by

$$f_{IG}(y, u, v) = \sqrt{\frac{v}{2\pi y^3}} \cdot \exp\left[-\frac{v(y-u)^2}{2u^2 y}\right] \tag{1}$$

According to the above assumptions and settings, the degradation process can be described by $Y(t) \sim \mathrm{IG}(\mu\ \Lambda(t), \lambda\Lambda^2(t))$, where $\mu\Lambda(t)$, $\mu^3\Lambda(t)/\lambda$ of $Y(t)$ are the mean and variance respectively. By substituting $u = \mu\Lambda(t)$ and $v = \lambda\Lambda^2(t)$ into (1), we can obtain

$$f_{IG}(y, \mu, \lambda) = \sqrt{\frac{\lambda(\Lambda\ (t))^2}{2\pi y^3}} \cdot \exp\left[-\frac{\lambda(\Lambda\ (t))^2(y - \mu\Lambda\ (t))^2}{2(\mu\Lambda\ (t))^2 y}\right] \tag{2}$$

The parameter μ, a function of the accelerated stress s, could be appropriately assumed as the degradation rate of a product and could be written as follows:

$$\mu(s) = \exp\left[a + b\varphi(s)\right] \tag{3}$$

where a and b are the parameters to be estimated in ADT. For simplification, the stress level can be standardized using the linear normalization method in this chapter. Specifically, assume that s_0 and s_H are the usage stress level and the highest stress level in the test, respectively. Then, let $\varphi(s_j)$ be a standardized function of s as

$$\varphi(s) = \left(\xi\left(s_j\right) - \xi\left(s_0\right)\right) / \left(\xi\left(s_H\right) - \xi\left(s_0\right)\right) \tag{4}$$

where $\xi(s)$ is the known function of s such as $\xi(s) = 1/s$ for temperature stress and $\xi(s) = \ln(s)$ for electric stress (Lim and Yum 2011).

As for the parameter λ, it is just a constant and will not change with the time, i.e., $\lambda_1 = \lambda_2 = \ldots = \lambda_K$ for K accelerated stress levels in an ADT. Because of the independences on the time for μ and λ, the degradation process is a homogeneous IG process, or called simple IG process. Since there are three different shapes of performance degradation trend, including linear, convex, and concave, it is appropriate to assume $\Lambda(t) = t^\beta$ and $\beta > 0$ (Zio 2016). Since an IG process strictly increases with time, the cumulative distribution function (CDF) of T_D can be expressed as

$$\begin{aligned} F_{Y_D}(t) &= P\left(Y(t) > Y_D\right) \\ &= \Phi\left[\sqrt{\tfrac{\lambda}{Y_D}}\left(t^\beta - \tfrac{Y_D}{\exp[a+b\varphi(s)]}\right)\right] - \\ &\exp\left(2\lambda t^\beta / \exp\left[a + b\varphi(s)\right]\right) \cdot \Phi\left[-\sqrt{\tfrac{\lambda}{Y_D}}\left(\tfrac{Y_D}{\exp[a+b\varphi(s)]} + t^\beta\right)\right] \end{aligned} \tag{5}$$

According to (Deb et al. 2002), $Y(t)$ approximately follows a normal distribution with mean $\mu(s)\Lambda(t)$ and variance $\mu^3(s)\Lambda(t)/\lambda$ as $\mu(s)\Lambda(t)$ and t are increased. Therefore, the formula (5) can be approximately rewritten as

$$F_{Y_D}(t) = 1 - \Phi\left[\frac{Y_D - \exp\left[a + b\varphi(s)\right] t^\beta}{\sqrt{\exp\left[a + b\varphi(s)\right]^3 t^\beta / \lambda}}\right] \tag{6}$$

Then, the q-quantile lifetime of Y_D is

$$t_q = \Lambda^{-1}\left(\frac{\exp[a + b\varphi(s)]}{4\lambda}\left(z_q + \sqrt{(z_q)^2 + 4Y_D\lambda/\exp[2a + 2b\varphi(s)]}\right)^2\right) \tag{7}$$

where z_p is the q-quantile of standard normal distribution and $\Lambda^{-1}()$ is the inverse function of $\Lambda()$.

Moreover, the parameters in formula (6), a, b, λ, and β, are assumed to be independent from each other, and they are supposed to consist of a parameter vector $\boldsymbol{\theta} = (a, b, \lambda, \beta)$. In engineering application, it is hard to give the crisp values for $\boldsymbol{\theta}$ when we design an ADT and the degradation model (2) is used to describe the degradation process, because there are some epistemic uncertainties embedded in these parameters. Naturally, it is appropriate to assume that these parameters follow some prior distributions; and then, the Bayesian inference could be used to handle this situation. Generally, the historical information and experts' knowledge are available before the implementation of an ADT; hence, they could be used to determine the prior distributions of $\boldsymbol{\theta}$.

For convenience, these parameters are supposed to be the parameter vector $\boldsymbol{\theta} = (a, b, \lambda, \beta)$ of random variables, and the parameters are independent from each other. Meanwhile, in practice, these parameters follows different distributions when the degradation increment x follows an IG distribution.

2.2 ADT Settings and Bayesian Inference

Let n test items be put into an ADT. And there are K accelerated stress levels arranged in the ADT, including $s_0 < s_{\min} \le s_1 < s_2 < \ldots < s_K \le s_{\max} \le s_H$, where $s_{\min}$ and $s_{\max}$ are the lowest and highest stress levels which will be used in an ADT respectively.

There are two popular stress loading patterns for an ADT, called constant stress accelerated degradation testing (CSADT) and step stress accelerated degradation testing (SSADT). For the former, the test items are divided into K groups and the lth group is tested under s_l. While in the latter SSADT, all test items are tested from the lowest stress level to the highest stress level step by step, i.e., $s_1 \to s_2 \to \ldots \to s_K$. In a SSADT, we always set up such an assumption that the residual degradation only depends on the current stress level and the current accumulated degradation damage with no memory of how it cumulates, and this assumption will make few differences of statistical analysis to CSADT and SSADT. Here, we only focus on the SSADT.

Let m_l be degradation measurements on s_l, $l = 1, 2, \ldots, K$, and the total measurement times of the whole test is M and $M=\sum_{l=1}^{k} m_l$.

We assume that τ is the non-overlapped interval of degradation measurement during a SSADT and keeps fixed. Then, the test duration t_l on s_l is $t_l = \tau m_l$, and the total test duration is $T_0 = \tau \times M$.

If $Y(t_{ilj})$ denotes the measurement result for the jth measurement of the ith item on the lth stress level at time t_{ilj} ($i = 1, 2, \ldots, n$, $l = 1,2,\ldots,K$, $j = 1, 2, \ldots, m_l$), the degradation increment is $x_{ilj} = Y(t_{il(j+1)}) - Y(t_{ilj})$ and follows the IG process shown in Eq. (2). Therefore, the likelihood function can be obtained by

$$p(x|\boldsymbol{\theta}) = \prod_{l=1}^{K}\prod_{i=1}^{n}\prod_{j=1}^{m_l}\left(\frac{\lambda\left((m_{il(j+1)}\tau)^{\beta} - (m_{ilj}\tau)^{\beta}\right)^2}{2\pi x_{ilj}^3}\right)^{1/2} \exp\left[-\frac{\lambda\left[x_{ilj} - \exp[a + b\varphi(s_l)]\left((m_{il(j+1)}\tau)^{\beta} - (m_{ilj}\tau)^{\beta}\right)\right]^2}{2(\exp[a + b\varphi(s_l)])^2 x_{ilj}}\right] \tag{8}$$

and the posterior distribution $p(\boldsymbol{\theta}|x)$ of $\boldsymbol{\theta}$ updated from its prior distribution $\pi(\boldsymbol{\theta})$ is

$$p(\boldsymbol{\theta}|x) = \frac{p(x|\boldsymbol{\theta})\,\pi(\boldsymbol{\theta})}{\int_{\Theta} p(x|\boldsymbol{\theta})\,\pi(\boldsymbol{\theta})\,d\boldsymbol{\theta}} \tag{9}$$

where the denominator $\int_{\Theta} p(x|\boldsymbol{\theta})\pi(\boldsymbol{\theta})d\boldsymbol{\theta}$ is defined as the marginal likelihood function and denoted as $p(x) = \int_{\Theta} p(x|\boldsymbol{\theta})\pi(\boldsymbol{\theta})d\boldsymbol{\theta}$.

According to Eqs. (7) and (9), the Bayesian posterior q-quantile lifetime of Y_D is denoted as follows:

$$t(q, \boldsymbol{\theta}|x) = t(q|\boldsymbol{\theta}) \cdot p(\boldsymbol{\theta}|x)$$

$$= \Lambda^{-1}\left(\frac{\exp[a + b\varphi(s)]}{4\lambda}\left(z_p + \sqrt{(z_p)^2 + 4Y_D\lambda/\exp[a + b\varphi(s)]^2}\right)^2\right) \cdot p(\boldsymbol{\theta}|x)\Big) \tag{10}$$

3 Multi-objective Bayesian Optimal Model for ADT

From Sect. 1, there are two kinds of uncertainties embedded in an ADT which are the inherent uncertainty and the model selection uncertainty. It is reasonable to regard that the stochastic IG process can quantify the uncertainties due to the errors of measurement instruments and testing equipment. When the Bayesian theory is further utilized, the uncertainties due to the unit-to-unit variation and model assumptions, e.g., Eqs. (2) and (3), are supposed to be appropriately quantified. Hence, these uncertainties are well controlled after we make an optimal design based on these Bayesian IG process (Li et al. 2017).

In addition, there are different expected utility functions (also known as optimal objectives) (e.g., relative entropy (Hamada et al. 2001), quadratic loss function (Li et al. 2015; Xu and Tang 2015; Liu and Tang 2010; Peng et al. 2014), D-optimality (Shi and Meeker 2012; Zhang and Meeker 2006)), and the different objective means the different focus. For instance, when the relative entropy is selected as the Bayesian optimal objective, it depicts the expected information gained during the test (Ye and Chen 2014; Lindley 1971). While when the quadratic loss (also known as the posterior variance) is selected, it can depict the accuracy of evaluation. No matter which objective is selected, the corresponding uncertainty will be caused.

In order to simultaneously control all the aforementioned uncertainties, a Bayesian multi-objective design method for ADT is proposed in which three objectives are considered, i.e., maximizing KL divergence, minimizing quadratic loss function, and minimizing testing cost, so that the objective selection uncertainties can be considered. By solving the multi-objective Bayesian optimal model, the Pareto solutions can be obtained. In other words, there are still some alternative plans, and it still brings the confusion to engineers on how to choose the right one. Actually, under this circumstance, the Pareto solutions are the mathematically optimal results with the given model and parameter assumptions. Although the Bayesian theory is adopted and the relative assumptions could be loosen, the prior assumptions still existed. Consequently, the data envelopment analysis (DEA) is further used to prune the Pareto solutions obtained by taking the consideration of engineering experiences from experts, so that the equilibrium optimal plan can be achieved.

3.1 Optimal Objectives

The optimal design can be obtained by maximizing the expected utility of an experiment (Peng et al. 2014), also known as the optimal objective. The utility function can be expressed as $U(d, \eta, x, \boldsymbol{\theta})$, where η is a design plan chosen from the possible plan set $\boldsymbol{D}$, x is the collected sample data, and a decision rule d from the decision rule set H is selected with the given η and observed x. Therefore, for any design plan η, the expected utility of the best decision can be expressed as

$$E(\eta) = \int_{\Omega} \max_{d \in \mathrm{H}} \int_{\Theta} U(d, \eta, x, \boldsymbol{\theta})\, p(\boldsymbol{\theta} \,|x\,, \eta)\, p(x\,|\eta)\, \mathrm{d}\boldsymbol{\theta}\mathrm{d}x \tag{11}$$

where $p(x|\eta)$ is the marginal likelihood function for the given η and $p(\boldsymbol{\theta}|\, x, \eta)$ is the posterior distribution of $\boldsymbol{\theta}$ under the given η and observed x. The pre-posterior expected utility $E(\eta)$ of the best decision rule is used to account for the uncertainty of the unknown $\boldsymbol{\theta}$ in the parameter space Θ and in the sample space Ω. Therefore, the Bayesian best plan η^* is the plan which can make $E(\eta)$ maximize

$$E\left(\eta^*\right) = \max_{\eta \in D} \int_{\Omega} \max_{d \in \mathrm{H}} \int_{\Theta} U(d, \eta, x, \boldsymbol{\theta})\, p(\boldsymbol{\theta} \,|x\,, \eta)\, p(x\,|\eta)\, \mathrm{d}\boldsymbol{\theta}\mathrm{d}x \tag{12}$$

Table 1 Optimal objectives in an ADT

Perspectives	Measures	Manipulations
Information	KL Divergence	Maximization
Accuracy	Quadratic loss	Minimization
Economy	Cost	Minimization

In the case of SSADT, a design plan η is written as η (n, M, $\mathbf{s}$, $\mathbf{m}$), where n is the number of test items, M denotes the total number of degradation measurements, $\mathbf{s} = (s_1, s_2, \ldots s_K)$ is the specified accelerated stress levels, and the number of degradation measurements on each accelerated stress level is demonstrated by $\mathbf{m} = (m_1, m_2, \ldots m_K)$.

The KL divergence and the quadratic loss represent the two different aspects of the test's utility; hence, both of them are chosen as the objectives for our multi-objective problem. In addition, test costs always exert great impact on the conduction of a test. From the economic point of view, it is supposed that the less the cost is, the better the test is. We, therefore, take it as the third objective. To sum up all these three objectives, the associated information is listed in Table 1.

The details of these three objectives are given as follows.

(1) KL Divergence.

In the Bayesian theory, KL divergence is always used as a measure of distance between a prior distribution and a posterior distribution. From the perspective of Shannon information, it also represents the information gain provided by a test, also known as relative entropy and is defined as (Li et al. 2015)

$$KL(\eta) = \iint \log\left(p\left(x\,|\boldsymbol{\theta},\eta\right)\right) p\left(x\,|\boldsymbol{\theta},\eta\right) \mathrm{d}\boldsymbol{\theta}\mathrm{d}x - \int \log\left(p\left(x|\eta\right)\right) p\left(x|\eta\right) \mathrm{d}x \tag{13}$$

where $p(x|\eta)$ is the marginal likelihood function as mentioned above and the likelihood function with known parameter vector $\boldsymbol{\theta}$ is $p(x|\boldsymbol{\theta}, \eta)$. Furthermore, Eq. (13) can be written as

$$KL(\eta) = E_x E_{\boldsymbol{\theta}} \log\left[p\left(x\,|\boldsymbol{\theta},\eta\right)\right] - E_x \log\left(p\left(x|\eta\right)\right) \tag{14}$$

Since $\boldsymbol{\theta}$ is the known parameter, it is easy to calculate $E_x E_{\boldsymbol{\theta}}(p(x|\boldsymbol{\theta}, \eta))$ using Monte Carlo simulation. However, because $p(x|\eta)$ is the marginal likelihood function, the Markov chain Monte Carlo (MCMC) sampling method is implemented to solve the problem, for example, using a software like WinBUGS (Liu and Tang 2010). The harmonic mean estimator illustrated by Newton and Raftery (Newton and Raftery 1994) will be used to estimate $p(x)$, denoted as

$$p\left(x|\eta\right) \approx \left\{\frac{1}{N}\sum_{r=1}^{N}\left[p\left(x|\boldsymbol{\theta},\eta\right)\right]^{-1}\right\}^{-1} \tag{15}$$

According to the above, the KL divergence also explained as the information gain from a test should be obtained by maximizing (14) and written as max $KL(\eta)$.

(2) Quadratic Loss Function.

The typical loss functions could be the quadratic loss, the absolute error loss, the 0–1 loss, and others (Robert 2007), and they all depict the evaluation accuracies of estimators. Among these choices of loss functions, the quadratic loss is the most popular one, because of its straightforward computation and relationship to the classical least square method. It is convex, so the corresponding decision will be unique.

Based on Eq. (10), the posterior variance of $t\ (q, \boldsymbol{\theta}|\ x, \eta)$ is the quadratic loss of q-quantile lifetime of Y_D on the usage condition; hence, it depicts the evaluation accuracy of the q-quantile lifetime. We make expectation for Var($t\ (q, \boldsymbol{\theta}|\ x, \eta)$) with respect to $\boldsymbol{\theta}$ and x; then, we can get the pre-posterior variance as follows:

$$Q\,(\eta) = E_x E_\theta\,[Var\,(t\,(q, \boldsymbol{\theta}|x))] \tag{16}$$

Therefore, the optimal plan should be obtained by minimizing (16) or maximizing $-Q(\eta)$, written as max $-\ Q(\eta)$.

(3) Test Cost.

The ADT cost mainly includes the costs of test items and test operation. The cost of test items is expressed as the product of the sample unit price of test items and its quantity, and the operation cost mainly includes the cost of the consumed test resources, i.e., test labor, power resource, and others. For the sake of simplicity, the operation cost is captured by the product of operation unit price and total test duration. Therefore, the total test cost $C(\eta)$ of plan η is computed by

$$C\,(\eta) = C_1 n + C_2 \tau \sum_{l=1}^{K} m_l \tag{17}$$

where C_1 is the unit price of a test item and C_2 is the unit price of the operation in an ADT. The optimal plan should be obtained by minimizing $C(\eta)$, written as min $C(\eta)$.

3.2 Optimal Model

Generally, the decision variables are constrained in advance according to the actual conditions of an ADT, and the main practical constraints involved in the proposed model include four parts:

(1) Sample size n. With the considerations of statistical requirement and engineering reality, there is a range for n, say $n_{\min} \le n \le n_{\max}$ (e.g., $n \in$ (Ye et al. 2014; Tseng et al. 2009)).
(2) Total measurements M. With the considerations of statistical requirement and engineering reality, there is a range for M, say $M_{\min} \le M \le M_{\max}$.
(3) The number of accelerated stress levels K. Generally, in order to ensure the accuracy of the extrapolation in the stress dimension, $K = 3 \sim 5$ is a reasonable range with the consideration of engineering application (Xiaoyang and Tongmin 2009).
(4) The number of degradation measurements on each accelerated stress level. During the same test duration, more information will be obtained under the higher stress levels. Correspondingly, more measurements should be arrange on the lower stress levels, and we have $m_1 > m_2 > \ldots > m_K$ $(l = 1,2,\ldots,K)$ (Xiaoyang and Tongmin 2009).

Thus, the multi-objective Bayesian optimal model for Bayesian SSADT design can be expressed as

$$\begin{cases} \max & KL(\eta) = E_x E_{\boldsymbol{\theta}} \log\left(p\left(x|\boldsymbol{\theta}, \eta\right)\right) - E_x \log\left(p(x)\right) \\ \min & Q(\eta) = E_x E_{\boldsymbol{\theta}} \left[\mathrm{Var}\left(t\left(q, \boldsymbol{\theta}|x, \eta\right)\right)\right] \\ \min & C(\eta) = C_1 n + C_2 \tau \sum_{l=1}^{K} m_l \\ \text{s.t.} & n_{\min} \le n \le n_{\max} \\ & M_{\min} \le M \le M_{\max} \\ & s_{\min} < s_1 < s_2 < \cdots < s_K \le s_{\max} \\ & m_1 \ge m_2 \ge \cdots \ge m_K > 0. \end{cases} \tag{18}$$

The Pareto optimal test plans $\boldsymbol{\eta}^*(n^*, M^*, \boldsymbol{s}^*, \boldsymbol{m}^*)$ can be obtained by solving the above proposed multi-objective programming.

3.3 Optimization Procedure for Multi-objective Bayesian Optimal Design of ADT

The proposed Bayesian SSADT multi-objective optimization model in (18) can be recognized as a typical multi-objective optimization problem consisting of three objectives aiming at finding the Pareto optimal set of solutions. As NSGA-II (Ntzoufras 2009) has good global search ability and well-distributed non-dominated solutions in the Pareto optimal front, we introduce NSGA-II to report a Pareto optimal set of solutions to the SSADT design problem. Detailed descriptions on the processes of NSGA-II are summarized as follows. Firstly, fast non-dominated sorting method is adopted to rank the population fronts and a parameter called crowding distance is calculated in the same front. Then tournament selection is made between two individuals randomly selected from parent population. The feasible solution with lower front number is selected if the two solutions come from different fronts. The feasible solution with higher crowding distance is selected if the two solutions are from the same front. Genetic operators including crossover and mutation are used to generate a new offspring population. Finally, the parent and offspring populations are combined together after being ranked by the procedures of fast non-dominated sorting and crowding distance assignment. The goal of greedy NSGA-II is to iteratively find a set of solutions ordered by fronts under the concept of Pareto dominance (see Definition 1). A greedy NSGA-II method is used to find the Pareto optimal set of solutions to (18).

The general flowchart for solving (18) is shown in Fig. 1.

Definition 1 Deb et al. (2002), Newton and Raftery (1994). Given two solutions $\boldsymbol{\eta}_1$ and $\boldsymbol{\eta}_2$, solution $\boldsymbol{\eta}_1$ dominates solution $\boldsymbol{\eta}_2$ if the following conditions are satisfied: (i) solution $\boldsymbol{\eta}_1$ is not worse than $\boldsymbol{\eta}_2$ for all objectives; (ii) solution $\boldsymbol{\eta}_1$ is strictly better than $\boldsymbol{\eta}_2$ for at least one objective.

In NSGA-II, the child population $Q(g)$ is first created from the parent population $P(g)$, and then a combination set $R(g) = Q(g) \cup P(g)$ is sorted according to the above fast non-dominated sorting method and crowding distance.

Now, the remaining of this section will present the specific optimization process.

(1) Initialization.

Initially, parameters of NSGA-II including $G_{\max}$ (the maximum generation number), *nPop* (the population number), p_c (the crossover proportion), p_m (the mutation proportion), and μ (the mutation probability) are given. Randomly create a feasible parent population P_g based on the range of the testing design space and corresponding constraints. And set generation number $g = 1$. Each chromosome $\boldsymbol{\eta} \in P_g$ is represented by a test plan $\boldsymbol{\eta} = (n, M, \mathbf{s}, \mathbf{m})$.

(2) Fast Non-dominated Sorting.

The current feasible parent population P_g is sorted into different non-dominated fronts based on fast non-dominated sorting algorithm according to the concept of Pareto dominance given by *Definition 1*. The first front denotes a completely non-dominant set in the current population, and the second front consists of chromosomes being dominated by these in the first front only and the front goes so on. Each chromosome in the each front are assigned rank (fitness) values or based on front in which they belong to. Chromosomes in the first front are given a fitness value of 1 and chromosomes in the second are assigned fitness value as 2 and so on. For each feasible chromosome, we compute: domination count η, the number of feasible chromosomes which dominate the feasible chromosome $\boldsymbol{\eta}$, and H_η, a set of chromosomes that the chromosome $\boldsymbol{\eta}$ dominates. The fast non-dominated sorting algorithm can be explained with the pseudocodes in (Deb et al. 2002).

(3) Crowding Distance Assignment.

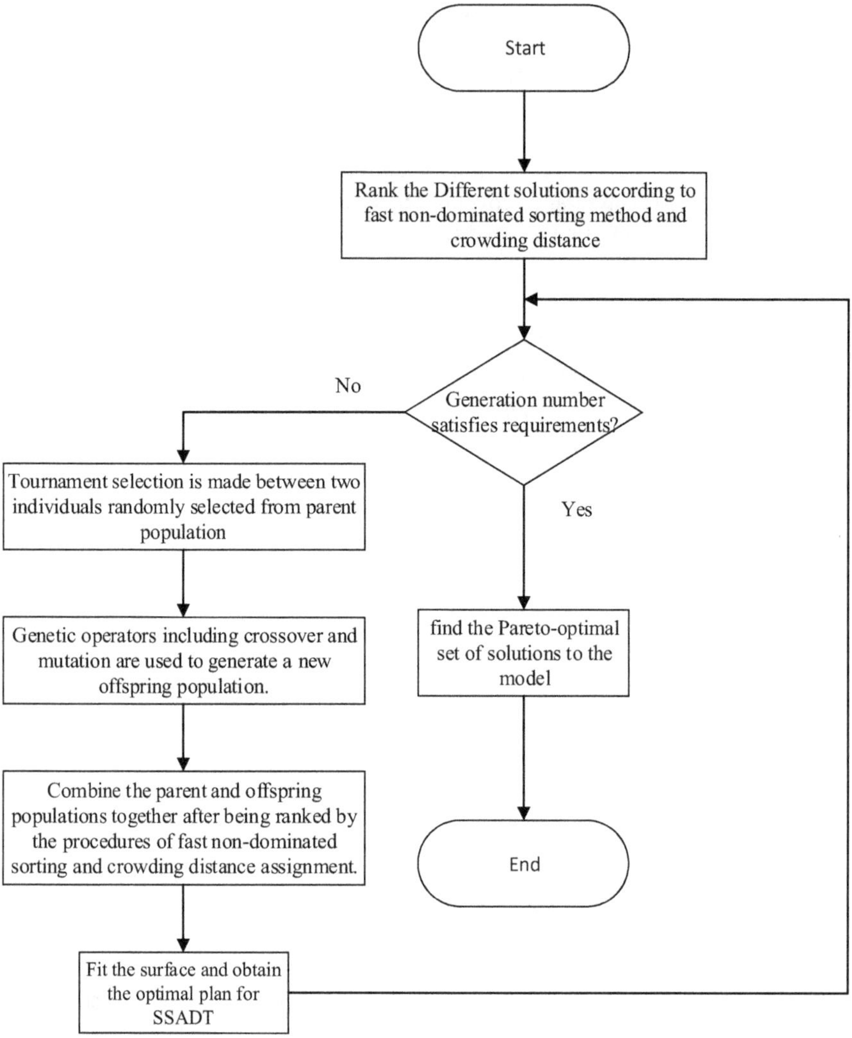

Fig. 1 Flowchart of dynamic Bayesian SSADT optimal model

The optimal chromosomes are selected based on the rank and crowding distance. The crowding distance assignment after the fast non-dominated sort aims to find the Euclidian distance between each chromosome in a front based on their d objective functions in the d dimensional hyper space. The chromosomes in the boundary are always selected as they are assigned infinite distances. In addition, comparing the crowding distance between two chromosomes in different fronts is meaningless. The crowding distance assignment algorithm is just as implemented in (Deb et al. 2002).

(4) Selection Process.

Once the individuals are sorted based on the fast sorting non-domination method, the selection process is carried out with crowding distance assignment method. The crowded comparison operator ($\prec_n$) is defined to guide the selection process at the associated procedures of NSGA-II toward a uniformly spread-out Pareto optimal front. Given that every feasible chromosome $\boldsymbol{\eta}$ in the population P_g has two attributes: non-dominated rank ($\eta_{\rm rank}$) and crowding distance ($\eta_{\rm dis}$). The partial order operator $\prec_n$ is defined as $\boldsymbol{\eta}\prec_n \iota$ if $\eta_{\rm rank} < \iota_{\rm rank}$ or ($\eta_{rank} = \iota_{rank}$ and $\eta_{dis} < \iota_{dis}$). That is, the solution with the lower (better) rank between two solutions with differing non-domination ranks is preferred. Otherwise, if both solutions belong to the same front, then the solution located in a lesser crowded region is preferred. The chromosomes are selected by using a binary tournament selection with crowded comparison operator.

(5) Greedy Genetic Operators.

Since there are lots of decision variables as well as constraints in the proposed Bayesian multi-objective ADT design model (18), the NSGA-II are not suitable anymore. In order to improve the convergence performance of NSGA-II, we implement the idea of binary tournament selection to design greedy genetic operators (including crossover and mutation), such that the winner of each tournament (the one with the best fitness) can be selected with priority in the greedy genetic operations. The process of binary tournament selection involves running two "tournaments" (or chromosomes) randomly selected from the current population. For example, the winner of the population P_g can be obtained by binary tournament selection *BT Select* (P_g) given as follows in Pseudo 1.$\eta^{1.}_{\rm rank}$

Pseudo 1: Binary Tournament Selection $\boldsymbol{\eta} = BTSelect(Pg)$			
Given two feasible solutions $\boldsymbol{\eta}^1, \boldsymbol{\eta}^2 \in Pg$			
If $.\boldsymbol{\eta}^1_{rank} < \boldsymbol{\eta}^2_{rank}$	$\boldsymbol{\eta} = \boldsymbol{\eta}^1$		
elseif $\boldsymbol{\eta}^2_{rank} < \boldsymbol{\eta}^1_{rank}$	$\boldsymbol{\eta} = \boldsymbol{\eta}^2$		
if $.\boldsymbol{\eta}^1_{dis} < \boldsymbol{\eta}^2_{dis}$	$\boldsymbol{\eta} = \boldsymbol{\eta}^1$	else	$\boldsymbol{\eta} = \boldsymbol{\eta}^2$
endif			
endif			

Crossover Operator Crossover is a genetic operator used to vary the programming of a chromosome or chromosomes from one generation to the next. With the consideration of the proposed formulation programming and its constraints structure, we give the following crossover operator in Algorithm 1 embedded with the concept of greedy choice.

Algorithm 1: Greedy Crossover

```
C=p_C * |Pg|                                              //number of solutions
for any i=1 : C
  η=BTSelect(Pg)=(s_1,···,s_K, m_1,···,m_K, n)            // father chromosome
  ι=BTSelect(Pg)=(s'_1,···, s'_K, m'_1,···,m'_K, n')      // mother chromosome
  c_1=(s_1,···,s_K, m_1,···,m_K, n')  c_2=(s'_1,···, s'_K, m'_1,···,m'_K, n)
  c_3=(s_1,···,s_K, m'_1,···,m'_K, n)  c_4=(s'_1,···, s'_K, m'_1,···,m'_K, n')
  F_C=fast-non-dominated-sort(c_1, c_2, c_3, c_4)         // non-dominated fronts
  crowding-distance-assignment(F_C)
  η'=F_1, ι'=F_2                                          // the best two chromosomes
End for
```

Algorithm 2: Greedy Mutation

```
B=p_C * |Pg|                                   //number of solutions
for any i=1 : B
r=rand()                                       //random number r ∈ (0,1)
if r <=μ
  η_1=BTSelect(Pg)                             // parent chromosome
    =(s_1,···,s_K,m_1,···,m_K,n)
  while isfeasible(η_2)                        // check feasibility of η_2
    η_2=random(η_1)                            // two-point mutation endwhile
  F_m=fast-non-dominated-sort(η_1, η_2)        // non-dominated fronts of η_1, η_2
  crowding-distance-assignment(F_m)            // crowding-distances in F_m
  η=F_1                                        // the best chromosome
end if
end for
```

Mutation Operator Mutation is an important genetic operator used to maintain the genetic diversity from one generation of a population of chromosomes to the next. The mutation operator involves a readjustment of the fitness values of solutions to sustain a steady selective pressure in the population and to prevent the premature convergence of the population to suboptimal solutions. Here we propose the following greedy mutation operator as shown in Algorithm 2.

(6) Recombination and Selection.

In order to ensure the elitism of the optimal front population, the offspring population Q_g by the greedy genetic operators is combined with the current generation population P_g, that is, all the previous and the current best chromosomes are added in the population. Then the combined population $R_g = P_g \cup Q_g$ is sorted based on non-domination given in Definition 1. The new generation is filled by each front subsequently until the population size exceeds the current population size (see Fig. 2 as an illustration). If the population exceeds the current population size by adding all the chromosomes in front F_i, then chromosomes in front F_i are selected

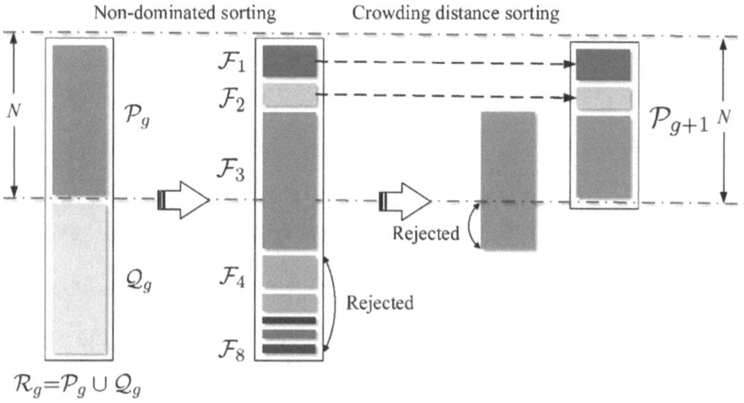

Fig. 2 Illustration on recombination and selection

based on their crowding distance values in the descending order until the population size reaches the current population size. And, hence, the process repeats to generate the subsequent generations. The whole recombination and selection process is just as summarized in Fig. 2.

3.4 Pruning of Pareto Solution by Data Envelopment Analysis

In the Pareto optimal solutions obtained by the above methodology, there are too many alternatives for decision-makers to select. Essentially, the Pareto solutions are the mathematically optimal results with the given assumptions, e.g., Eqs. (2), (3), and (9). The prior distributions substitute the crispy parameter values, but the assumptions for prior distributions still existed. In engineering application, experts and engineers always have lots of experiences and judgments about the actual conduction of a test, such as how good are the stress levels, measurements, and so on. These experiences can help with the control of the uncertainties from the assumptions of prior distributions. Since we can assess the optimal plans in the Pareto solutions from different aspects and the different plan has its results of KL divergence, quadratic loss, and cost, it is a typical multiple objective selection optimization (MOSO) problem. Consequently, the data envelopment analysis (DEA) is further used to prune the Pareto solutions, which is a linear programming-based technique for measuring and comparing the relative performance of decision-making units (DWUs) with multiple inputs and outputs.

Assume that a problem involves N DMUs containing A inputs and B outputs for each one and then the relative efficiency of the wth DMU can be denoted as

$$\mathrm{RE}_w = \frac{\sum_{j=1}^{B} u_j y_{wj}}{\sum_{i=1}^{A} v_j x_{wi}}, \ w = 1, 2, \ldots, N \tag{19}$$

where x_{wi} and y_{wj} denote the output and input of the wth DMU and u_j and $v_i \geq \varepsilon$ are the weights of x_{wi} and y_{wj} with ε as a small positive quantity to guarantee the non-negativity of the weights.

According to engineering practice, for a SSADT, model include two parts:

(1) The wider the range of accelerated stress levels, the more the sample size n and the total number of measurements M; then, the more comprehensive information collected and the higher the prediction precision.
(2) The closer the lowest accelerated stress level to the normal stress level, the less extrapolation in the stress dimension and the more credible the prediction.

Based on the above analysis, we can further select three variables for DEA model, and they are n, M, and $s_1 - s_0$. And considering the variables $KL(\eta)$, $Q(\eta)$, and $C(\eta)$, there are six variables in total. In the DEA analysis, all minimization type variables should be regarded as the inputs, while all maximization type ones should be regarded as the outputs. Consequently, we say $Q(\eta)$, $C(\eta)$, and $s_1 - s_0$ consist of the inputs and n, M, and KL consist of the outputs in our DEA model. Then the DEA model has $A = 3$, $B = 3$, $x_{w1}{=}Q(\eta)$, $x_{w2}{=}C(\eta)$, $x_{w3}{=}s_1{-}s_0$, $y_{w1} = n$, $y_{w2}{=}M$, $y_{w3}{=}KL(\eta)$. A specific weight set for each solution w_0 can be found by maximizing the relative efficiency of the solution with the constraint that the relative efficiency of other solutions is less than 1, which is given as (20)

$$\begin{cases} \max \ RE_{w_0} = \dfrac{\sum_{j=1}^{3} u_{w_0 j} y_{w_0 j}}{\sum_{i=1}^{3} v_{w_0 i} x_{w_0 i}}, \ w = 1, 2, \ldots, N \\ \text{s.t.} \quad \dfrac{\sum_{j=1}^{3} u_{w_0 j} y_{w_0 j}}{\sum_{i=1}^{3} v_{w_0 i} x_{w_0 i}} \leq 1, \ w = 1, 2, \ldots, N, \ w \neq w_0 \end{cases} \tag{20}$$

The decision variables in model (20) are the weights of the inputs and outputs $v_{w0}(i = 1, 2, 3)$ and u_{w0j} $(j = 1, 2, 3)$ for the specific solution w_0, and the solution contains a weight set which is the most favorable to w_0. In order to simplify the solution, the above fractional linear programming problem can be equivalently transformed into a general linear programming problem (as shown in (21)), while we can obtain the same optimal results

$$\begin{cases} \max \ RE_{w_0} = \sum\limits_{j=1}^{3} u_{w_0 j} y_{w_0 j} \\ \text{s.t.} \quad \sum\limits_{i=1}^{3} v_{w_0 i} x_{w_0 i} = 1 \\ \quad \sum\limits_{j=1}^{3} u_{w_0 j} y_{w_0 j} - \sum\limits_{i=1}^{3} v_{w_0 i} x_{w_0 i} \leq 0, \quad w = 1, 2, \ldots, N, w \neq w_0 \end{cases} \tag{21}$$

Finally, decision-makers can take the highest relative efficiency as the final choice according to the ranking of relative efficiencies in DEA, which is favorable both to the mathematical and the practical extent.

4 Numerical Case of Multi-objective SSADT Bayesian Optimal Model

For an electrical connector, it is required to maintain good contact force throughout its functional lifetime. Contact force is generated by the deflection of the electrical contacts within the connector, and any plastic (unrecoverable) strain occurred to the metal contacts will cause the performance degradation of the connector. One of the typical failure mechanisms due to the plastic strain is the stress relaxation, which is defined as a time-dependent loss in stress under constant strain.

In this section, we will employ the stress relaxation data of electrical connectors in reference (Chhikara and Folks 1989) to illustrate the proposed multi-objective SSADT Bayesian optimal methodology.

In general, an electrical connector is defined as failure when the stress relaxation is over 30%, i.e., $Y_D = 30$. The accelerated degradation data were obtained under the conditions of $s_1 = 65$ °C, $s_2 = 85$ °C, and $s_3 = 100$ °C, respectively. These collected ADT data and the corresponding measurement times are shown in Tables 2 and 3.

In keeping with (Ye et al. 2014), the collected stress relaxation data are assumed to follow the IG process, and the stress function $\xi(s)$ is written as $1/s$. Mean and the variance of the model parameters can be obtained by employing the maximum likelihood estimation method and square roots of the diagonal of the Fisher matrix, as reported in Table 4.

Based on the parameter estimations in Table 3, the prior distributions of parameters can be determined; then, the uncertainties of model parameters are quantified. In this section, parameters a and b are assumed to follow normal distributions, and parameters λ and β follow gamma distributions (Li et al. 2017). The details are reported in Table 5.

Table 2 Stress relaxation degradation data of electrical connectors under different accelerated stress levels

Temperature	Group number	Stress loss										
65 °C	1	2.12	2.7	3.52	4.25	5.55	6.12	6.75	7.22	7.68	8.46	9.46
	2	2.29	3.24	4.16	4.86	5.74	6.85	*	7.40	8.14	9.25	10.55
	3	2.4	3.61	4.35	5.09	5.5	7.03	8.24	8.81	9.629	10.27	11.11
	4	2.31	3.48	5.51	6.2	7.31	7.96	8.57	9.07	10.46	11.48	12.31
	5	3.14	4.33	5.92	7.22	8.14	9.07	9.44	10.09	11.2	12.77	13.51
	6	3.59	5.55	5.92	7.68	8.61	10.37	11.11	12.22	13.51	14.16	15
85 °C	7	2.77	4.62	5.83	6.66	8.05	10.61	11.2	11.98	13.33	15.64	
	8	3.88	4.37	6.29	7.77	9.16	9.9	10.37	12.77	14.72	16.8	
	9	3.18	4.53	6.94	8.14	8.79	10.09	11.11	14.72	16.47	18.66	
	10	3.61	4.37	6.29	7.87	9.35	11.48	12.4	13.7	15.37	18.51	
	11	3.42	4.25	7.31	8.61	10.18	12.03	13.7	15.27	17.22	19.25	
	12	5.27	5.92	8.05	9.81	12.4	13.24	15.83	17.59	20.09	23.51	
100 °C	13	4.25	5.18	8.33	9.53	11.48	13.14	15.55	16.94	18.05	19.44	
	14	4.81	6.16	7.68	9.25	10.37	12.4	15	16.2	18.24	20.09	
	15	5.09	7.03	8.33	10.37	12.22	14.35	16.11	18.7	19.72	21.66	
	16	4.81	7.5	9.16	10.55	13.51	15.55	16.57	19.07	20.27	22.4	
	17	5.64	6.57	8.61	12.5	14.44	16.57	18.7	21.2	22.59	24.07	
	18	4.72	8.14	10.18	12.4	15.09	17.22	19.16	21.57	24.35	26.2	

Table 3 Measurement time under different stress levels

Temperature	Performance measurement time										
65 °C	108	241	534	839	1074	1350	1637	1890	2178	2513	2810
85 °C	46	108	212	408	632	764	1011	1333	1517	2586	
100 °C	46	108	212	344	446	626	729	972	1005	1218	

Table 4 Estimated results of the model parameters

Estimated parameters	$\hat{a}$	$\hat{b}$	$\hat{\lambda}$	$\hat{\beta}$
Mean	−1.8966	1.7379	0.6337	0.4493
Variance	0.1903	0.1738	0.1968	0.0178

Table 5 Prior distributions of parameters

Parameter	a	b	λ	β
Distribution I	*Normal* (−1.89, 0.19)	*Normal* (1.74, 0.17)	*Gamma* (2.04, 0.13)	*Gamma* (11.34, 0.04)

Referring to reference (Yang and Guangbin 2008), we set the normal stress level s_0 with 40 °C, and $s_{\min}$ and $s_{\max}$ are set to 50 °C and 100 °C, respectively. Without loss of generality, assume K to be 3. The cost of test (C_1, C_2) is assumed to be (2, 0.02) × 10^2 dollars, and the measurement interval τ is 10 h. Meanwhile, in order to get the optimal plan, the constraints for the decision variables (i.e., n, M, $\mathbf{s}$ and $\mathbf{m}$) are set as follows: (i) Without loss of generality, $n_{\max}$ is set to be 5; (ii) the total degradation measurements M directly reflect the actual test duration and cost constraints for a SSADT. The larger the M is, the longer duration the test takes, and the more cost the test needs. In practice, limits on both the test duration and cost are realistically necessary. Therefore, here M is assumed to vary between 100 and 150; (iii) the lowest stress level s_1 is assumed to vary between 50 °C and 80 °C, and the highest stress level s_3 is set as 100 °C, and then the middle stress level s_2 can be set by using the interval between $\xi(s_l)$ and $\xi(s_{l+1})$ constant; (iv) as mentioned in Sect. 3.3, performance degrades slower on the lower stress levels than that on the higher stress levels. In order to guarantee that there is enough useful degradation information under all stress levels, more degradation measurements should be assigned under the lower stress level than those on higher stress levels, which leads to the constraints of $m_1 - m_2 \geq 10$ and m_2-$m_3 \geq 5$.

According to the basic knowledge of Sect. 3 and the degradation data of this section, we can obtain the following multi-objective SSADT Bayesian optimal model:

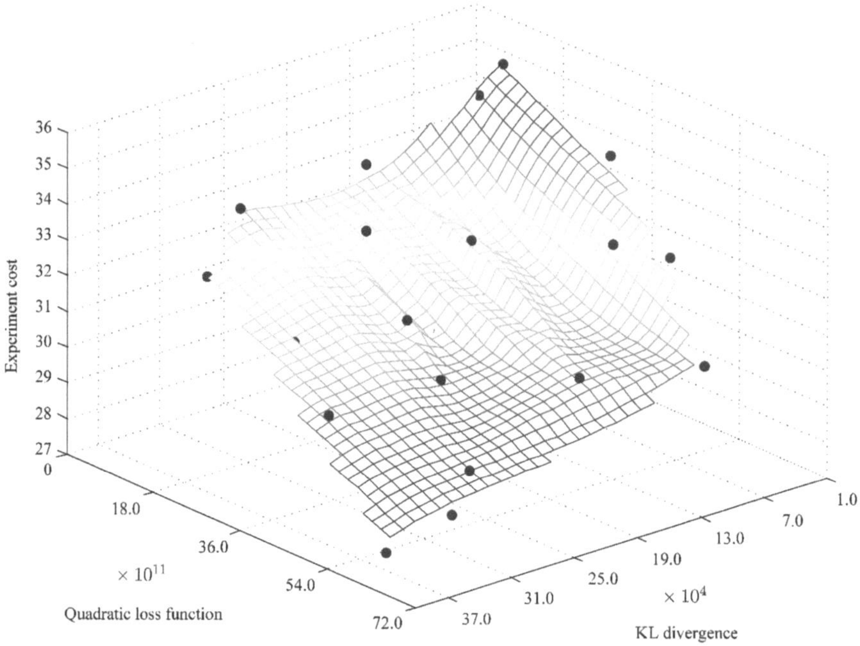

Fig. 3 Pareto frontier of Bayesian SSADT multi-objective design model

$$\begin{cases} \max & KL(\eta) = E_x E_{\theta,} \log\left(p\left(x|\theta, \eta\right)\right) - E_x \log\left(p(x)\right) \\ \min & Q(\eta) = E_x E_{\theta,} \left[Var\left(t\left(q, \theta, |x, \eta\right)\right)\right] \\ \min & C(\eta) = C_1 n + C_2 \tau \sum_{l=1}^{3} m_l \\ s.t. & 3 \le n \le 5, \quad 50 \le S_1 \le 80 \\ & \sum_{l=1}^{3} m_l = M, \quad 100 \le M \le 150 \\ & s_3 = s_H = 100, \quad 1/s_2 = 1/2\left(1/s_1 + 1/s_3\right) \\ & m_1 - m_2 \ge 10, \;\; m_2 - m_3 \ge 5. \end{cases} \tag{22}$$

Then, based on Sect. 4, we can obtain the Pareto frontier of multi-objective Bayesian SSADT design model, as shown in Fig. 3.

The plans on the Pareto frontier are the optimal ones in the mathematical sense. In other words, they are optimal only when the assumptions in Table 4 are true. As mentioned above, the model selection uncertainties are then caused. In order to further reduce and control the negative effect from these uncertainties, the proposed DEA mentioned in Sect. 3.4 is applied.

By using the proposed DEA method, the relative efficiencies can be calculated for the plans on the Pareto frontier, as reported in Table 6.

Table 6 The relative efficiencies of plans on the Pareto frontier

No.	$\eta S1(°C)$ $S2(°C)$ $S3(°C)$		$m1$	$m2$ $m3$ n	$KL(\eta)$	$Q(\eta)$	$C(\eta)$	$RE(\eta)$
Plan 1	70.00	82.00100.00	97	16 7 3	3.70×10^5	2.41×10^{12}	30.00	0.8415
Plan 2	60.00	75.00100.00	80	55 11 4	6.86×10^4	6.83×10^{12}	37.20	0.6352
Plan 3	75.00	85.71100.00	117	21 8 4	2.69×10^5	1.77×10^{11}	37.20	1.058
Plan 4	**74.00**	**85.00100.00**	**67**	**31 22 3**	8.8877×1100^4	2.2299×1100^{11}	**30.00**	**2.0304**
Plan 5	**56.00**	**71.79100.00**	**53**	**43 26 5**	5.3311×1100^4	2.6611×1100^{11}	**34.40**	**2.2576**
Plan 6	68.00	80.95100.00	76	45 22 3	1.08×10^5	4.18×10^{11}	34.60	0.9079
Plan 7	57.00	72.61100.00	71	50 13 4	8.37×10^4	4.26×10^{11}	34.80	1.0154
Plan 8	77.00	87.01100.00	63	38 7 3	1.45×10^5	6.62×10^{11}	27.60	1.2140
Plan 9	79.00	88.27100.00	67	23 11 3	1.53×10^5	1.20×10^{12}	26.20	0.3177
Plan 10	79.00	88.27100.00	88	52 5 4	2.50×10^5	1.71×10^{12}	37.00	0.6187
Plan 11	62.00	76.54100.00	86	48 15 4	6.29×10^4	1.95×10^{12}	37.80	0.7192
Plan 12	75.00	85.00100.00	53	39 28 3	7.43×10^4	2.14×10^{12}	30.00	0.8002
Plan 13	77.00	87.01100.00	94	25 12 4	1.79×10^5	2.15×10^{12}	34.20	0.6456
Plan 14	**51.00**	**67.55100.00**	**61**	**40 30 3**	4.2200×1100^4	2.2244×1100^{12}	**32.20**	**1.4767**
Plan 15	**58.00**	**73.42100.00**	**68**	**20 13 5**	2.4400×1100^4	2.3366×1100^{12}	**30.20**	**1.4228**
Plan 16	76.00	86.36100.00	69	27 4 5	1.98×10^4	2.82×10^{12}	30.00	0.8071
Plan 17	75.00	85.71100.00	117	21 8 4	2.19×10^4	3.55×10^{12}	37.20	0.5631
Plan 18	**50.00**	**66.67100.00**	**90**	**18 6 5**	1.6622×1100^5	3.8888×1100^{12}	**32.80**	**1.6**
Plan 19	59.00	74.00100.00	43	40 37 5	2.77×10^4	4.48×10^{12}	34.00	0.8028
Plan 20	72.00	83.72100.00	64	28 12 5	2.22×10^5	6.59×10^{12}	30.80	0.7553

From Table 6, five test plans (i.e., Plan 4, Plan 5, Plan 14, Plan 15, Plan 18) with the higher relative efficiencies remain in the pruned solution set for selection in engineering practice.

5 Conclusions

In this chapter, a Bayesian multi-objective design method for an ADT is proposed in which three objectives are considered and the data envelopment analysis (DEA) is further used to crop the Pareto solutions so as to find out the plan with the highest relative efficiency. By the illustrative case study on the electrical connector's SSADT, our conclusions are summarized as follows:

(i) In the proposed Bayesian SSADT multi-objective optimal model, the objectives of the proposed optimization model consists of maximizing the KL divergence, minimizing the quadratic loss function of the q-quantile lifetime at usage condition, and minimizing the test cost. With the practical constraints, the greedy NSGA-II is applied to get the optimal SSADT plans. By simultaneously considering these three objectives, the uncertainty due to the objective model selection could be better controlled.

(ii) After the Pareto optimal solutions are generated, the DEA method is applied to effectively clear the choice confusion induced by the algorithm of greedy NSGA-II. Then, the experiences from experts and engineers can be incorporated into the optimal decision. Consequently, the uncertainties from the assumptions of prior distributions can be well controlled by using the experiences and judgments from experts and engineers.

In summary, an ADT can be regarded as a scientific experimental method, which aims to recognize a performance degradation law under the influence of uncertainties by using an experimental way. How to control the uncertainties embedded in an ADT to better realize a degradation law is one of the core tasks of an ADT. The methodology proposed in this chapter can effectively contribute the control of the uncertainties in ADT by using the Bayesian theory, multi-objective optimization, and data envelopment analysis. Although the proposed methodology is illustrated based on the inverse Gaussian process, the other degradation processes (e.g., the Wiener process or gamma process) could be also the applications.

Actually, there should be more considerations and constraints in reality, and only testing cost and some engineering experiences are considered. Different Bayesian alphabet optimality, the consumed power energy during tests, etc. should be considered so that the more realistic balanced plan can be optimized. Moreover, multiple-stress cases are quite common in reality, and the corresponding researches should be extended from the proposed single stress condition to such general cases.

References

Chhikara RS, Folks JL (1989) The inverse Gaussian distribution: theory, methodology, and applications. Marcel Dekker, New York

Deb PA, Agarwal S, Meyariyan TAMT (2002) A fast and elitist multiobjective genetic algorithm: NSGA-II. IEEE Trans Evol Comput 6:182–197

Escobar LA (1995) Planning accelerated life tests with two or more experimental factors. Am Soc Qual Control Am Stat Assoc 37:411–427

Hamada M, Martz HF, Reese CS, Wilson AG (2001) Finding near-optimal Bayesian experimental designs via genetic algorithms. Am Stat 55:175–181

Hu CH, Lee MY, Tang J (2015) Optimum step-stress accelerated degradation test for Wiener degradation process under constraints. Eur J Oper Res 241:412–421

Huang J, Golubović DS, Koh S, Yang D, Li X, Fan X et al (2016) Lumen degradation modeling of white-light LEDs in step stress accelerated degradation test. Reliab Eng Syst Saf 154:152–159

Li X, Rezvanizaniani M, Ge Z, Abuali M, Lee J (2015) Bayesian optimal design of step stress accelerated degradation testing. J Syst Eng Electron 26:502–513

Li X, Hu Y, Zio E, Kang R (2017) A Bayesian optimal design for accelerated degradation testing based on the inverse gaussian process. IEEE Access 5:5690–5701

Liao CM, Tseng ST (2006) Optimal design for step-stress accelerated degradation tests. IEEE Trans Reliab 55:59–66

Lim H, Yum B-J (2011) Optimal design of accelerated degradation tests based on Wiener process models. J Appl Stat 38:309–325

Lindley DV (1971) Bayesian statistics, a review. J Account Res 2:108–116

Liu X, Tang LC (2010) A Bayesian optimal design for accelerated degradation tests

Liu L, Li X, Zio E, Kang R, Jiang T (2017) Model uncertainty in accelerated degradation testing analysis. IEEE Trans Reliab 66:603–615

Marseguerra M, Zio E, Cipollone M (2003) Designing optimal degradation tests via multi-objective genetic algorithms. Reliab Eng SystSaf 79:87–94

McPherson JW (2010) Reliability physics and engineering: time-to-failure modeling

Meeker WQ, Escobar L (1998) Statistical methods for reliability data. Wiley, New York

Newton MA, Raftery AE (1994) Approximate Bayesian inference with the weighted likelihood bootstrap. J R Stat Soc 56:3–48

Ntzoufras I (2009) Bayesian model and variable evaluation. In: Bayesian modeling using WinBUGS, pp 389–433

Peng W, Liu Y, Li YF, Zhu SP, Huang HZ (2014) A Bayesian optimal design for degradation tests based on the inverse Gaussian process. J Mech Sci Technol 28:3937–3946

Robert C (2007) The Bayesian choice: from decision-theoretic foundations to computational implementation. Springer, New York

Roy S, Mukhopadhyay C (2015) Bayesian D-optimal accelerated life test plans for series systems with competing exponential causes of failure. J Appl Stat 43:1477–1493

Shi Y, Meeker WQ (2012) Bayesian methods for accelerated destructive degradation test planning. IEEE Trans Reliab 61:245–253

Tseng ST, Balakrishnan N, Tsai CC (2009) Optimal step-stress accelerated degradation test plan for gamma degradation processes. IEEE Trans Reliab 58:611–618

Wang X, Xu D (2010) An inverse Gaussian process model for degradation data. Technometrics 52:188–197

Wang H, Wang GJ, Duan FJ (2016) Planning of step-stress accelerated degradation test based on the inverse Gaussian process. Reliab Eng Syst Saf 154:97–105

Xiaoyang L, Tongmin J (2009) Optimal design for step-stress accelerated degradation testing with competing failure modes. In: Annual reliability and maintainability symposium, pp 64–68

Xu A, Tang Y (2015) A Bayesian method for planning accelerated life testing. IEEE Trans Reliab 64:1383–1392

Yang, Guangbin (2008) Life cycle reliability engineering. Wiley, Hoboken

Ye ZS, Chen N (2014) The inverse Gaussian process as a degradation model. Technometrics 56:302–311

Ye ZS, Chen LP, Tang LC, Xie M (2014) Accelerated degradation test planning using the inverse Gaussian process. IEEE Trans Reliab 63:750–763

Zhang Y, Meeker WQ (2006) Bayesian methods for planning accelerated life tests. Technometrics 48:49–60

Zio E (2016) Some challenges and opportunities in reliability engineering. IEEE Trans Reliab PP:1–14

Part V
Optimization and Multiobjective Models for Maintenance Modelling

Maintenance Requirements Analysis and Whole-Life Costing

Richard Dwight and Peter Gordon

1 Introduction: Fundamental Concepts

We seek to set out some basic ideas to provide the foundation for ongoing research and application of ideas to the practical problem of determining effective and efficient maintenance programs as faced by industry and, inferring from these, whole-life costs. This is achieved through setting out of the activities that are applied in practice to manage the impact of physical asset deterioration in-service against the imperative of delivering products or services. Perhaps in contrast with the main thrust of this publication, the emphasis here is on examining the overall problem to which optimisation and multi-objective models may be applied. Of particular interest is the nature of failure causes and resulting relevant data to the decision of maintenance action selection.

There remains some confusion: particularly in the research-based literature but also flowing to practice, on the purpose and limitations of maintenance requirements analysis, MRA. The narrow focus of any such process is here considered to be on facilitating the design of a 'preventive maintenance' or PM program for a physical asset. (It is acknowledged that such ideas have been somehow massaged into the area of so-called software maintenance. Such application is not addressed here.) Another scourge of this relatively unestablished academic discipline of maintenance management, and indeed asset management, is the imprecise, confused or even misrepresentation of word, phrase and acronym definitions. Reviews of recent

R. Dwight (✉)
School of Mechanical Materials Mechatronic and Biomedical Engineering, University of Wollongong, Wollongong, NSW, Australia
e-mail: radwight@uow.edu.au

P. Gordon
Sydney, NSW, Australia

A. T. de Almeida et al. (eds.), *Multicriteria and Optimization Models for Risk, Reliability, and Maintenance Decision Analysis*, International Series in Operations Research & Management Science 321, https://doi.org/10.1007/978-3-030-89647-8_16

papers confirm complex jumbling of concepts and words. Here we use what we maintain to be the generally accepted understanding of 'preventive maintenance' as the name of a general type of maintenance policy. The term 'maintenance policy' is used here to define a prescribed set of actions to be taken with the aim of maintaining: retaining a system in or restoring it to the condition required to fulfil its intended purpose. A PM policy prescribes a set of maintenance actions that have the intent of 'preventing' the in-service failure of an item: component or system. Failure may be defined as a condition of an item where it cannot perform the function required of it. Sometimes, and wrongly in our view, PM is differentiated from predictive, proactive or condition-based maintenance, which we maintain are overlapping subsets of PM. Further, PM is often contrasted with 'corrective maintenance' which we maintain is not the name of a maintenance policy but rather a label for one form of maintenance action: one which is triggered by a perception that an item has 'failed', i.e. restoration to an 'available' condition of an item after it has failed.

For complex systems, a significant task is to determine the maintenance policies to be applied: when components making up the system should be refurbished or renewed. The task of determining PM policies, the analysis process itself, we call here 'maintenance requirements analysis, MRA'. We propose that whole-life costing is founded on assumptions concerning the maintenance policies, including PM policies, that will be followed and projections of the effects of these policies on asset performance and whole of life costs.

There are four circumstances under which maintenance actions are demanded by a system: i.e. action-determinants. Only some of these are amenable to the prescription of a PM policy. These are essentially defined by either the scope of available maintenance actions or the nature of the degradation and failure process that constrains what might be effective and efficient actions:

1. The requirement for lubrication and servicing activities. These are purely a function of, and so part of, the design. The satisfactory operation of many systems is contingent on the regular re-lubrication, servicing and adjustment of components. Without this attention, elements of the system are liable to faulty operation and premature failure.
2. The need for refurbishment or replacement following mechanical wearing-out of an item: so-called wear-out failures. These are failures occurring through the gradual loss of strength: resistance to applied 'forces', associated with the degradation, wearing or relative displacement between components with time or usage. Carter (1986) draws attention to the large range of possible mechanisms leading to mechanical failure and by way of illustration sets out some of them: erosion, corrosion, fatigue, surface degradation, de-fastening, creep, ageing, fouling, contamination, leaking and thermal effects. He builds on this further by drawing attention to the likely occurrence of interaction between any number of these phenomena resulting in failure. There are similarly wear mechanisms that can lead to the failure of electronic components. Maricau and Gielen (2013) describes the following mechanisms for CMOS transistors: hot carrier injection,

time-dependent dielectric breakdown, bias temperature instability and electro-migration. Perhaps the point here is that the range of phenomena we are seeking to model and control, through failure time alone is incredible.

3. The need for refurbishment or replacement following overloading of a component: overload failures. These are failures that occur under load conditions for which the component was not designed. As such differentiating them from wear-out failures is a matter of definition. They may arise from mal-operation, foreign object damage, vandalism, extreme environmental loading and secondary damage from the failure of another component or system. They typically result from a single load event leading to failure immediately or shortly after the overload. Generally this is synonymous with 'random failure' relative to the item's 'age'. Overloads may alternatively not give rise to an immediate failure. In such cases, 'preventive maintenance' may include activities to identify unexpected damage before it progresses to failure.
4. The need for refurbishment or replacement arising from induced defects. These are failures associated with errors during any of the manufacturing and assembly, installation or maintenance of the item. They include the installation of a defective or incorrect part, improper installation or repair of a part, omission of parts on reassembly of a component and failure to remove tools or materials from the equipment being maintained. These failures typically result from a single error that increases the susceptibility of the component to failure: reduces its strength. Subsequent failure is likely to occur early in the life of the item. Failure later in the item's life is also possible as observed by Hobbs (2008).

The first of these four action-determinants, lubrication and servicing activities, seeks to assure that items function as designed and achieve the maximum life possible given the way the system is used. The requirement for these tasks and their frequency and timing does need to be identified as part of any process for establishing the overall maintenance program. It could be that the lubricant is treated as an item to which maintenance is applied; however, consideration of the required actions need not be informed by failure analysis even though the lack of effective actions may be highlighted in an engineering-based analysis of failure events. Logically, a priori identification of the required actions comes from analysis and understanding of the design of the components and how the system will be used. Coincidentally the typically effective actions are of relatively insignificant cost negating the need for detailed cost analysis.

The remaining three action-determinants are causes of system failure. An understanding of how they manifest and the effect they have on the system informs the basic nature of the failure process and its cause. They define whether there will be indicators or warnings that the failure is imminent and what maintenance actions are possible. Importantly they prescribe the relevant set of feasible actions that would control the occurrence of failure arising from such phenomena. Overload and defect generated failures are better managed in other ways. Recommendations on defect, or error, management are widely available (e.g. Reason and Hobbs, 2003). The strategies here are more appropriate for addressing induced defects and overloads caused by operator error. Other overload failures are better addressed by system and component design.

2 A General Process

We have previously highlighted (Dwight et al. 2012) that the general nature of the process of developing a PM program via maintenance requirements analysis (MRA) generally relies on adaptation of an existing program and should be multidimensional. The multidimensional nature of the process derives from three considerations in constructing a program: the state of knowledge of the objectives of the program; cognisance of the relevant attributes of the available maintenance system resources; and the information available that would inform any decision process and any supporting models.

It is held that there are only three possible maintenance policies that may be selected and applied to an item in response to the inevitability of its deterioration and failure. These are distinguished, as per the previous discussion of action-determinants, from actions to address lubrication, overload and defect initiated failure. These maintenance policies are identified here as condition-based PM, usage-based PM and failure-based maintenance. This classification system is founded on the trigger for the refurbishment or replacement action for the item. Perhaps unnecessarily, these three possible policies are defined here by refurbishment or renewal of an as yet un-failed item being triggered by distinctly different information: condition-based PM by condition information indicating imminent failure; usage-based PM by the cumulated load volume indicating imminent failure; and failure-based maintenance, quite obviously, by the realisation that the item has failed. Feasible options derive from the possible behaviour of the item itself. Expediency may mean a fourth trigger, particularly opportunity, may be applied; however, thoughtful acceptance of any opportunity demands purposeful rejection of the three preventive maintenance policies.

Various MRA processes have been devised by a range of authors over many years now (Nowlan and Heap 1978; Kelly 1983, 1989; Gits 1992; MILSTD 2173 1986; Smith 1993; Moubray 1997; NAVAIR 00-25-403 2005; UK Ministry of Defence Standard JAP(D) 100C-22 2009). All such processes have been termed RCM processes with the exception of those proposed by Kelly, and by Gits which is termed maintenance concept design (aside: many authors wrongly suggest that RCM is a maintenance action when in fact it is simply an analysis process by which maintenance actions may be selected). While the novelty and efficacy of these proposed approaches is claimed, it is suggested here that there is a commonality between them. This gives rise to a general MRA process which may be expressed as:

1. System selection and system boundary definition
2. Determination of system functions and manifestations of failure of these functions
3. Identification of the mechanisms by which the functional failures may be caused by the system: sometimes called the failure modes or 'root' causes
4. Description of the effects of the identified failure modes
5. Classification of the consequences of the occurrence of each of the failure modes into categories that instruct both a method of analysis and action prescription
6. Determination of parameters and their behaviour that provides some capability to predict the occurrence of failure
7. Selection of a particular maintenance policy and detailing of the associated actions accounting for both consequences and predictors available for failure
8. Grouping or clustering of actions prescribed for the system as a whole given analysis of constituent items and iteration in the light of revised action costs given actions may share resources
9. Implementation and monitoring of resulting actions

3 Practical Application of the General Process

The laborious nature of applying such a process has been highlighted and is a distinct barrier to logical determination of maintenance programs. Practical application of this basic logic depends on a clear focus on the objectives of its application and the current status of information informing it: the situation. The general steps outlined may usefully be aggregated into broader stages within which methods, models and approaches can be devised that suit the situation at hand. Such a set of three fundamental stages is arranged in Fig. 1 and proposed as:

1. System definition: establishing the context for maintenance action selection and design – involving Steps 1 and 2. This encapsulates what the asset is and what it's meant to be doing. Perhaps this uncovers valuable information relevant to the operation as well as the maintenance of the system.
2. Failure analysis incorporating failure mode definition: identifying possible failure modes and their effects and categorising them according to consequences – involving Steps 3–6. This involves determining and characterising how the system fails, where failure is defined by the functions established at Stage 1.
3. Action determination: selecting effective and efficient task clusters, i.e. developing the maintenance program – involving Steps 7–9. This is concerned with identifying the maintenance actions to be applied at failure mode level and then aggregated to item and system level. It is conducted given knowledge of the function and modes of failure of the system. The need for complex analysis will depend on how much is already known about the possible tasks and their effects. We may have a good understanding about all the possible tasks that we would want to consider for that sort of asset. These may be a limited set of

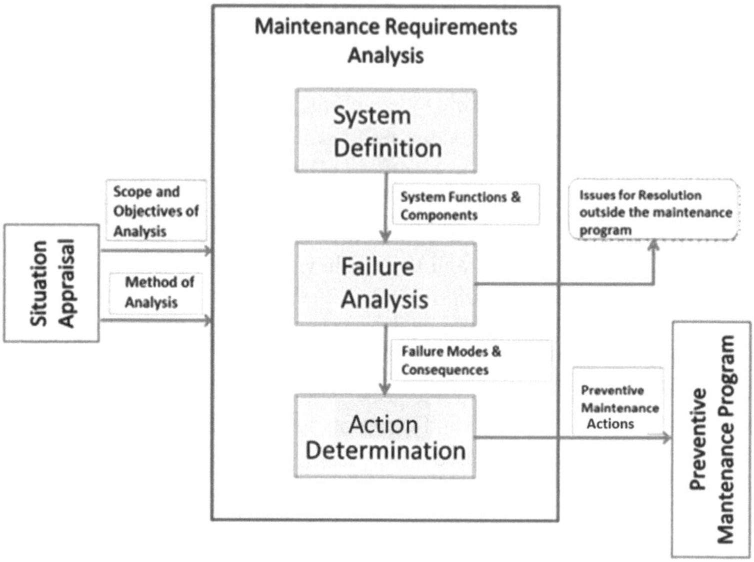

Fig. 1 Basic elements of maintenance requirements analysis highlighting the role of context for the analysis as determined through 'Situation Appraisal'

possible tasks. Perhaps only cleaning activities or unit replacement are feasible options. It may be that there is a considerable range of possible tasks or that the possibilities are unknown resulting in the need for more complex analysis of all the failure modes that could possibly occur and reviewing what sort of tasks might be appropriate to them.

The situation may dictate that not all of the three stages require detailed consideration. Attention is drawn to the general nature of application of the analysis which would normally be adaptive. Certainly the 'situation' determines what approach is taken in conducting each of the required stages. Given the adaptive situations commonly confronted in practice, some prescription of suitable processes can be made. They are considered as alternatives to the formulaic process initially set out.

Some alternate approaches have been suggested distinguished by the starting point for any analysis:

3.1 Investigating an Existing Set of Policies

Although identification of tasks is a latter step in the logical process, there is no reason why logic cannot be traced backwards through the process to determine whether the important element of the logic outlined earlier in this section has been met. Where an established set of tasks are suspected of being ineffective or of low efficiency analysis may be focused on those tasks. The relevant objective is to 'optimise' the existing set of tasks as opposed to improving the in-service reliability of the system.

The consistency of the task with the existing knowledge of the failure behaviour is often lacking. By understanding the reasoning behind the policy, the failure mechanism that is implied can be established. This provides the focus for consideration of competing preventive maintenance policies or other action. Gains in the performance of the equipment are incidental when using this approach although still likely due to the removal of counterproductive tasks. Experience with this approach indicates that an initial analysis can be completed in a few hours.

3.2 Investigating a Known List of Failure Modes

Where an overall review is required on systems that have a reasonable history of use, the relevant failure modes experienced may be known or be discoverable. Rather than attempting to generate: imagine, possible failure modes from first principles, the effort required is to ensure that the history is known. This is suggested by some of the published approaches to maintenance requirements analysis (e.g. Moubray 1997) through the engagement of a broad group of system stakeholders.

3.3 Application of a Safety Net

The previous suggested approaches in this section rely to a large extent on knowledge of events that have already occurred. This could be seen as dangerous and not taking advantage of the proactive aspects of requirements analysis. To ensure that particularly unwanted events are covered adequately, be they operational, safety or environment related, a similar directed search is advocated to that outlined with the proposal for the acquisition phase where the process needs to commence at Stage 1.

It is useful to consider techniques such as fault tree analysis to guide the search for failure modes relevant to the consequences of concern: particular system behaviour resulting in safety or other outcomes of interest. Many systems have single critical functions. A reasonable example is a sewage pumping station whose function could be stated simply as being 'to prevent overflow of sewage'. This being the case, a directed search for basic causes of such an event appears logical.

4 Challenges in Developing a Maintenance Policy: Failure Analysis

Failure of items is generally associated with uncertainty. The prediction of events is the central issue in prescribing activities directed at preventing them. A list of challenges in the application of proposed processes and indeed reliability-based decision models proposed in the literature includes the following:

1. Available data: system data easily available – availability and reliability; subsystem and component data is more difficult as information is required on what has been done at that level (functional location) of the system. Reasons for actions need to be coded to allow basic reliability data to be achieved.
2. Potential volume of data, which is often a preoccupation of industry despite its disconnection from relevant issues faced in achieving business outcomes.
3. Complex interactions between life cycle decisions; operations and maintenance policies; design, manufacturing, construction, operation and maintenance actions; and resulting errors.
4. Conflict between managing work through the support systems and achieving component lives.
5. Changing patterns of use or short-term requirements to operate the system: short-notice variations in the cost of undertaking maintenance. Also changing resource capabilities and opportunities to undertake maintenance. Integration of maintenance actions into asset management as a system.
6. Complex relationships between specific components and specific subsystems and systems that they are temporarily a part of: component tracking and reliability data collection resulting for both maintenance management and maintenance analysis.
7. Dealing with potentially useful sensor technologies and communication systems providing real-time condition information: false positives; maintenance programs for these sensors; actions triggered by condition information.

The desire for quantitative decision-making heightens the importance of data availability and transposes the prediction problem into one of obtaining relevant data as the basis for prediction. Wishing for the data is perhaps unavoidable but futile.

Correlating measurable or identifiable condition variables with remaining component life is a central concern in practice. Maintenance engineers are generally satisfied to get condition information that enables them to decide whether a component should be immediately replaced or whether it will survive beyond the next convenient replacement time. However with expensive components subject to slow deterioration, better estimates of remaining life are valuable in planning for spares availability and the timing of major system shutdowns.

5 Focus on Failure Analysis

There are two aspects to failure analysis:

1. Understanding the cause of the failure
2. Understanding the consequences of the failure

Consequences are normally self-evident where failures are analysed. They are normally also multifactorial having safety and environmental, operational, maintenance, statutory and community dimensions, for example. The cause of failure however may require careful investigation. For ongoing improvement of a PM program there is a challenge in recording in sufficient detail the cause of component and system failures and relating this back to the prospective failure analysis conducted for the original program.

Poor data records are often called out as an impediment to failure analysis in practice. Increasingly the literature contains attempts to make use of whatever information is around (e.g. Arif-Uz-Zaman et al. 2017). Advances in engineering and engineering approaches: driven by necessity as system complexity and performance expectations and imperatives increase, result in fewer events being generated from which failure data may be recorded, poorly or otherwise. This forces considerations of precursors of failure and involvement with the degradation process.

The mechanisms for failure may not be amenable to failure data analysis. Given the determining factors underpinning the selection of a PM policy, failure analysis is directed at determining how the reliability of an item may change over time, what drives such a change and how any change may be predicted or detected.

Failure analysis is central to MRA. However PM is not the best way to address all failure modes. Wear-out failure modes should be the predominant focus of PM program development.

When performing MRA, failure analysis is prospective. The analysis is to uncover likely future failure modes. Failure cause for prospective studies can come from manufacturers, from experience within the industry with similar systems or from engineering analysis of load and environmental stresses on components.

Understanding the way component reliability changes with age or usage informs PM selection. It is prime evidence of whether PM is effective and efficient in addressing that failure mode. Conversely, overstress occurs with the component otherwise capable of resisting the applied loads, and defective items are unable to resist normal loads from the outset. Recognition of these fundamentally different determinants of required actions: PM or otherwise, using failure data is problematic. Overstress and defect-induced failure may give rise to recognisable failure distributions. Such distributions do not however provide conclusive evidence. Identification of the existence of these determinants essentially involves ruling out of wear-out behaviour. Whether a wear-out mechanism is evident in any failure data analysis depends on the basis for that data: whether or not the dependent variable is relevant to the deterioration process.

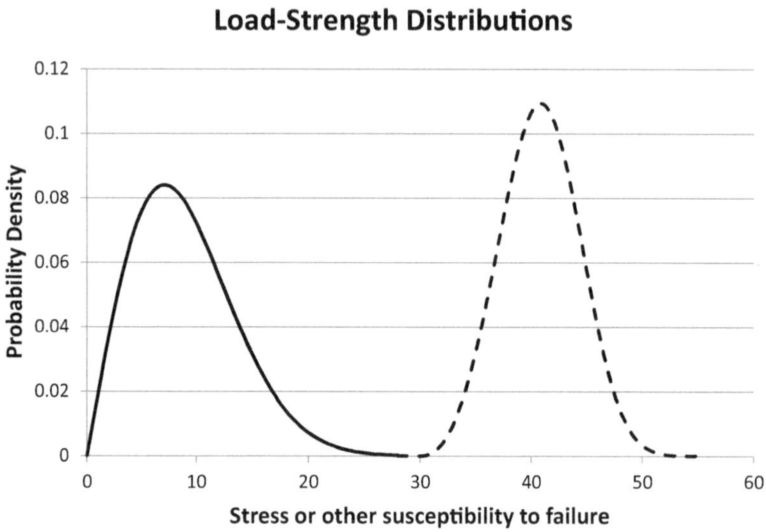

Fig. 2 Load and strength distributions at typically representing a safe situation. The solid curve represents a load distribution, $L(s)$, and the dashed curve a strength (resistance to load) curve, $S(s)$

The typical representation of failure is set out in Fig. 2 which depicts a load, $L(s)$, and a strength, $S(s)$, distribution as a function of a stress or other susceptibility to failure, s, where the strength is defined as the sustainable load. The progression of a mechanical wearing-out of an item has these distributions increasingly overlapping as a function of use. Defect-generated failure implies some form of overlapping of the distributions from the time of initial commissioning of the item. Reliability may be conceptualised as determined by the relationship:

$$R(t) = \int_0^{\infty} \left[S(s) \left(\int_0^{s} L(s) ds \right)^{t} \right] ds \qquad \text{(Carter 1986)}$$

where the probability that the strength is greater than a nominated value of strength, s, is given by the area under the function between that value and infinity. The probability that the load has a value of s is $L(s).ds$

Figure 3 sets out a depiction of an overload situation. In that case loads exceeding the strength are applied from a separate mechanism to the normally applied load.

Significant and ongoing work to describe a reliability function that matches observed data from such situations ignores the reality that the failure data is generated from a number of underlying mechanisms and so is not amenable to modelling with a single distribution. Traditional reliability literature advocates modelling of early-life failures using a Weibull distribution with a shape parameter value of less than 1. This models a situation where all items will eventually fail by the failure mode but the probability of failure reduces the longer the component has been in use. This suggests a situation where the resistance to failure increases with

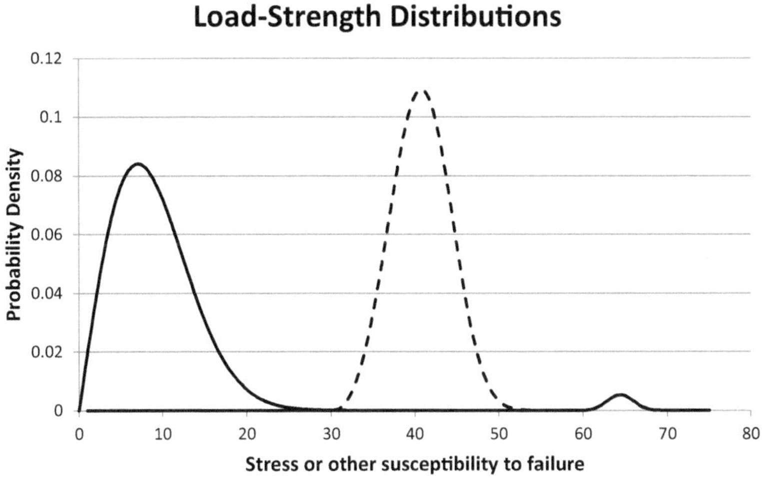

Fig. 3 Load and strength distributions illustrating an overload failure situation. The solid curve represents a load distribution, $L(s)$, and the dashed curve a strength (resistance to load) curve, $S(s)$

usage. This is an improbable failure mechanism and is not consistent with an early-life failure mechanism. It may be a reasonable approximation if competing failure modes are being modelled and other failure modes dominate after an initial period of usage.

In practical terms, the different action-determinants are not amenable to the same treatments. It is questionable whether there is any value in trying to model overload or defect-related failure. If there is an emerging problem then it will be difficult to ascertain the proportion of the population affected by the defect that is causing the failures and in any case the priority is to determine and eliminate the cause of failure.

6 Views on the State of the Art and New Advances

6.1 Advances in MRA Thinking

It is unclear that any significant advances have been made to the processes first devised with the advent of complex aircraft. The so-called reliability-centred maintenance (RCM) approach attributed to Nowlan and Heap (1978) has similar objectives and has been popularly adopted as a standard approach to maintenance requirements analysis. A process labelled maintenance concept design by Gits (1992) approached the problem in arguably similar ways and has been developed further by others (particularly Waeyenbergh and Pintelon 2002, 2009). That work incorporated modelling and optimisation more directly into the process which had only been implicit previously. There have been many applications of such processes (e.g. Carretero 2003; Rosqvist et al. 2009). The nature of the process has remained relatively static.

6.2 A Word on 'Multi-criteria Decision Analysis'

Perhaps the normal decisions in developing a maintenance policy require consideration of a diverse set of factors, some of which are dynamic. Satisfying the need to minimise impacts on available resources, operations, safety and environment traditionally has been converted to an overall cost impact leading to cost rate-based models. Where outcomes were not amenable to conversion to a cost: perhaps safety and environmental and longer-term business survival are amongst these, then the criterion may be to attain a particular reliability. It seems though that these are inextricably linking and are not independent or even different criteria. A system with a particular reliability will give rise to events that can be described in terms of their cost. The availability of multi-criteria techniques does not alter this. Perhaps they provide a different way of dealing with the range of inputs to decisions as suggested by, for example, Shyjith et al. (2008) and Emovon et al. (2018). Treatments generally do not come along with identification of specific criteria or practical examples. The publish-or-perish imperative is perhaps driving the literature in this area and MRA commentary in general. Perishing is perhaps an option that should be given more consideration.

6.3 Has the Time for MRA Passed?

Considering how complex system failure occurs, it is more normal that the unexpected occurs, driven by unknown system behaviour or unrecognised damage events.

The 'predictable' modes of failure will normally have been designed out or fully accounted for, of course through MRA analysis. One might expect that there is limited novelty remaining. Methods for constructing resilient systems may be more where advances are required. Valiant efforts to inject more sophisticated qualitative (e.g. Chemweno et al. 2016) as well as quantitative approaches may not be as purposefully directed to problems experienced in industry as is needed to make real progress. Such calls are not new. Geraerds (1972) drew attention to this problem.

6.4 Identification and Characterisation of Feasible Modes of Failure and Their Behaviour

Further consideration needs to be given to failure analysis particularly given that:

1. Increasing prominence of overload and induced defects negates common reliability models found on the expectation that failures are wear-out in nature.
2. Approaches to characterisation include projections using engineering judgement and identification of similar situations through statistical analysis.

7 Overall Thoughts

1. The main thrust of publications on MRA has been to provide a basis for the logical development of preventive maintenance programs for complex systems. This essentially involves ensuring that proposed actions are applicable to the failure mode they are designed to address. This does mean understanding how the reliability of the components is expected to change with usage but does not require statistical analysis of failure data. As explained in this chapter, it simply requires an understanding of the failure mechanisms involved.
2. The other focus for analysis is on selecting maintenance actions that optimise the cost effectiveness of the overall program. While it may seem that there are numerous options open to the maintenance engineer, in practice the options are often constrained and the decisions to be made are not difficult and do not require complex decision modelling.
3. For researchers seeking to add value to the field of maintenance requirements analysis, the message is to ensure that the methods being investigating are not more complex than the problems being solved and are grounded in the issues arising in practice.

References

Arif-Uz-Zaman et al (2017) Extracting failure time data from industrial maintenance records using text mining. Adv Eng Inform 33(August):388–396

Carretero C (2003) Applying RCM in large scale systems: a case study with railway networks. Reliab Eng Syst Saf 82:257–273

Carter ADS (1986) Mechanical reliability. Macmillan Education, Basingstoke

Chemweno et al (2016) Development of a novel methodology for root cause analysis and selection of maintenance strategy for a thermal power plant: a data exploration approach. Eng Fail Anal 66(August):19–34

Dwight R, Gordon P, Scarf PA (2012) Dynamic maintenance requirements analysis in asset management. In: European safety and reliability conference: advances in safety, reliability and risk management, ESREL 2011, pp 847–852

Emovon et al (2018) Hybrid MCDM based methodology for selecting the optimum maintenance strategy for ship machinery systems. J Intell Manuf 29(3):519–531

Geraerds WMJ (1972) Towards a THEORY OF MAINTENANCE. In: Bureau R (ed) The organisation of logistic support systems. The English University Press, London, pp 297–329

Gits C (1992) Design of maintenance concepts. Int J Prod Econ:217–226

Hobbs A (2008) An overview of human factors in aviation maintenance: Japan Airlines Boeing 747, 1985, p 3. Australian Transport Safety Bureau Safety Report (AR-2008-055). https://protect-au.mimecast.com/s/ePFZC91WrrTN4PJ1tol5ZC?domain=atsb.gov.au; https://www.atsb.gov.au/media/27818/hf_ar-2008-055.pdf

Kelly A (1983) Notes from specialist short course – management of industrial, maintenance

Kelly A (1989) Maintenance procedures and their selection. In: Maintenance and its management, conference communication. Farnham, Surry, pp 76–83

Maricau E, Gielen G (2013) Analog IC reliability in nanometer CMOS

MIL-STD-2173(AS) (1986) Reliability-centered maintenance requirements for Naval Aircraft, weapons systems and support equipment

Moubray J (1997) Reliability-centered maintenance, 2nd edn. Industrial Press

NAVAIR 00-25-403 (2005) Guidelines for the naval aviation reliability-centered maintenance process

Nowlan SF, Heap H (1978) Reliability centred maintenance. National Technical Information Service, Springfield

Reason J, Hobbs A (2003) Managing maintenance error: a practical guide, Kindle edn. CRC Press

Rosqvist T, Laakso K, Reunanen M (2009) Value-driven maintenance planning for a production plant. Reliab Eng Syst Saf 94:97–110

Shyjith et al (2008) Multi-criteria decision-making approach to evaluate optimum maintenance strategy in textile industry. J Qual Maint Eng 14(4):375–386

Smith AM (1993) Reliability-centered maintenance. McGraw-Hill, New York

UK Ministry of Defence Standard JAP(D) 100C-22 (2009) Guide to developing and sustaining preventive maintenance programmes, August. https://www.gov.uk/government/uploads/system/.../japd100c-22.pdf

Waeyenbergh G, Pintelon L (2002) A framework for maintenance concept development. Int J Prod Econ 77:299–313

Waeyenbergh G, Pintelon L (2009) CIBOCOF: a framework for industrial maintenance. Int J Prod Econ 121:633–640

Maintenance Policies for Non-repairable Components

Bram de Jonge

1 Introduction

We consider maintenance policies for non-repairable components. We consider a component as a part of a system that is subject to maintenance interventions and for which no further subdivisions are made into sub-components that are individually subject to any maintenance interventions. The condition of a non-repairable component cannot be partially improved by carrying out a repair; maintenance of a non-repairable component is therefore always a replacement. In most cases, such a replacement will result in a component that is as-good-as-new. Only if a heterogeneous set of spare components is considered, the quality of a new components differs per replacement.

A component can be replaced either after its failure or before its failure. In the first case we talk about corrective, reactive, or failure-based maintenance; the second case is referred to as preventive maintenance. It is generally preferred to perform maintenance interventions preventively, for instance because failure of a component can result in damage to other components, and because it can lead to unplanned downtime. However, performing preventive maintenance too often is also undesirable and costly. Therefore, a balance has to be found between the preventive maintenance frequency and the risk of failures.

A maintenance policy describes when to carry out preventive maintenance. A distinction can be made between time-based maintenance policies and condition-based maintenance (CBM) policies. The former is based on the time that a component is in service, the latter allows for maintenance activities that are performed based on degradation information.

B. de Jonge (✉)
Department of Operations, University of Groningen, Groningen, The Netherlands
e-mail: b.de.jonge@rug.nl

A. T. de Almeida et al. (eds.), *Multicriteria and Optimization Models for Risk, Reliability, and Maintenance Decision Analysis*, International Series in Operations Research & Management Science 321, https://doi.org/10.1007/978-3-030-89647-8_17

Time-based maintenance is easy to implement as only the time that a component is in service has to be recorded. However, substantial remaining useful life is wasted if the machine is still in reasonable condition when preventive maintenance is performed, and a breakdown might occur if it happens to deteriorate faster than expected. Condition-based maintenance, on the other hand, generally results in more effectively scheduled preventive maintenance, and, in the ideal case, preventive maintenance that is performed just before failure. However, applying condition-based maintenance is only possible if there are conditions that are related to the moment of failure, and if it is technically possible to monitor these conditions. Furthermore, condition-based maintenance should only be applied if its benefits outweigh the efforts and costs required to apply it. These requisites include condition monitoring equipment and software to store, analyze, and initiate maintenance actions.

2 Time-Based Maintenance

Traditionally, two time-based preventive maintenance policies can be distinguished, viz. age-based maintenance and block-based maintenance (Barlow and Proschan, 1965; Gertsbakh, 2000). Under the age-based maintenance policy, corrective maintenance is performed when the component fails, and preventive maintenance is performed when the age of the component reaches T, whichever occurs first (see Fig. 1). The maintenance age T is the decision variable of this policy. Under the block-based maintenance policy (sometimes also called periodic maintenance), preventive maintenance is performed at fixed times kT, $k = 1, 2, \ldots$. Corrective maintenance is performed when the component fails, but this does not affect the preventive maintenance schedule (see Fig. 2). The maintenance interval T is the decision variable of this policy. The disadvantage of block-based maintenance is that preventive maintenance is sometimes performed shortly after a failure. The main advantages, on the other hand, are the easier planning as it is known in advance when preventive maintenance will be performed, and the clustered maintenance actions if the same block-based policy is used for multiple components.

We let F denote the (cumulative) distribution function of the time until failure of the component. We will consider time-based maintenance from a cost perspective. The cost of performing a preventive maintenance action is denoted by c_{pm}, the cost

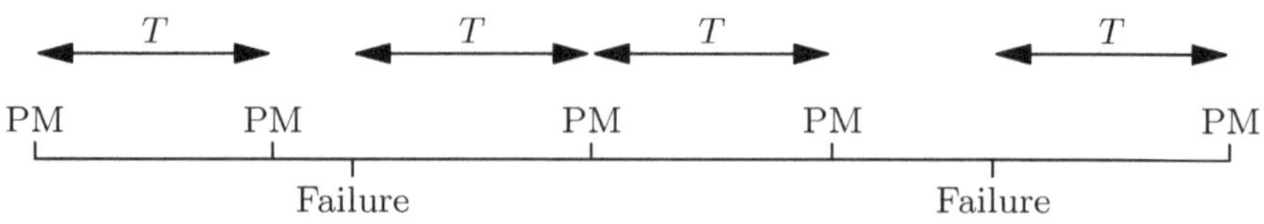

Fig. 1 Scheme of the age-based maintenance policy

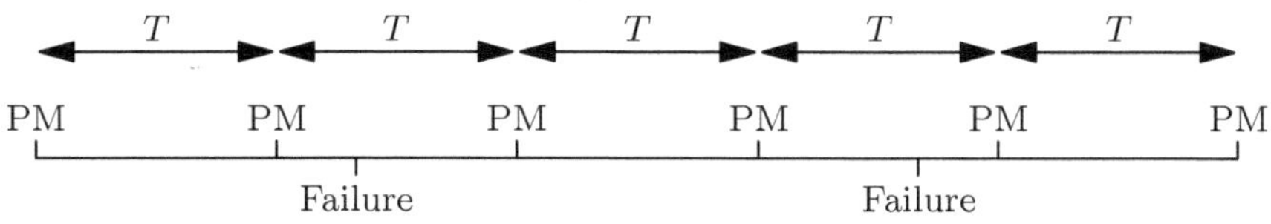

Fig. 2 Scheme of the block-based maintenance policy

of a corrective maintenance action by $c_{\rm cm}$. The cost of preventive maintenance is assumed to be lower than the cost of corrective maintenance, i.e., $c_{\rm pm} < c_{\rm cm}$, implying that preventive maintenance can be beneficial when scheduled effectively. In the basic models both preventive and corrective maintenance actions are assumed to require a negligible amount of time and to make the component as-good-as-new. The cost of performing corrective maintenance is often normalized to 1, so that only one cost parameter c for the relative cost of performing preventive maintenance is required.

2.1 *Age-Based Maintenance*

The (long-run) cost rate (i.e., the long-run mean cost per unit of time) of the age-based maintenance policy depends on the maintenance age T and is denoted by $\eta_{\rm age}(T)$. Because both types of maintenance make the component as-good-as-new, standard renewal theory can be used to evaluate this cost rate. By referring to the time between consecutive maintenance actions as a cycle, the cost rate can be written as

$$\eta_{\rm age}(T) = \frac{\text{Mean cost per cycle}}{\text{Mean cycle length}}$$
$$= \frac{c_{\rm cm}F(T) + c_{\rm pm}(1 - F(T))}{\int_0^T (1 - F(x))\,{\rm d}x}.$$

Studies that consider the age-based maintenance policy typically assume that the lifetime distribution is known with certainty. De Jonge et al. (2015) acknowledge that this is often not realistic, and they consider the optimal age-based maintenance policy under uncertainty in the lifetime distribution. They assume a certain parametric lifetime distribution and include uncertainty in its parameters.

In general, they represent the vector of parameters of the lifetime distribution by s and denote the joint density function that models the uncertainty in s by $g(s)$, which is defined on $\mathbb{R}^n$. Instead of the cost rate we can now talk about the expected cost rate $\eta_{\text{age}}^{E}(T)$ as a function of the maintenance age T:

$$\eta_{\text{age}}^{E}(T) = \int_{s \in \mathbb{R}^n} g(s) \frac{c_{\text{cm}} F(T; s) + c_{\text{pm}}(1 - F(T; s))}{\int_0^T (1 - F(x; s))\, \mathrm{d}x}\, \mathrm{d}s_1 \cdots \mathrm{d}s_n.$$

The preventive maintenance age T_{opt}^{E} that minimizes this expected cost rate is considered as the optimal maintenance age.

De Jonge et al. (2015) start to consider a uniform lifetime distribution with uncertainty in its right end point; this uncertainty is modeled by a second uniform distribution. Although the uniform distribution is not the most realistic lifetime distribution, this setting has the advantage that it can be analyzed algebraically.

The authors continue to consider a Weibull lifetime distribution, which is the most commonly used distribution to model lifetimes. The Weibull distribution has a shape parameter k and a scale parameter λ. Because the failure mode of a component often provides an accurate estimation for the shape parameter k, there is in practice generally most uncertainty in the scale parameter λ. The authors model the uncertainty in λ by using a uniform distribution on the interval $[1 - \alpha, 1 + \alpha]$. The value of $\alpha \in [0, 1]$ can be interpreted as a measure for the level of uncertainty in λ. This setting needs to be analyzed numerically.

Figure 3 shows the optimal maintenance age as a function of the level of uncertainty α in the scale parameter λ of a Weibull lifetime distribution with

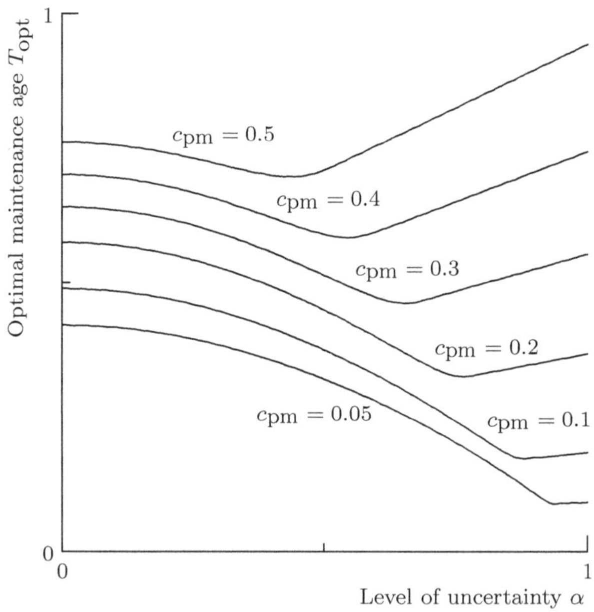

Fig. 3 Optimal preventive maintenance age under uncertainty in the scale parameter λ of a Weibull lifetime distribution with shape parameter $k = 5$, corrective maintenance cost $c_{\text{cm}} = 1$, and for various preventive maintenance costs c_{pm}

$k = 5$. It turns out that the optimal maintenance age first decreases in the level of uncertainty. If the level of uncertainty exceeds a certain threshold the optimal maintenance age starts to increase. The initial decrease is expected; more uncertainty in the lifetime distribution results in earlier preventive maintenance. However, if the uncertainty increases further, it becomes too expensive to prevent very early failures. Longer lifetimes also become more likely when the uncertainty increases; this results in an increasing maintenance age.

A similar pattern is observed when a uniform lifetime distribution with uncertainty in its right end point is considered. This parameter basically also is the scale parameter of this distribution. We also expect a similar pattern when uncertainty in the scale parameter of other parametric lifetime distributions is considered, and when the uncertainty in the scale parameter itself is modeled by a different distribution. We would also like to mention that parametric bootstrapping has also been used to obtain the probability distribution of an estimator for the optimal maintenance age (Tokumoto et al., 2014).

In the setting above a static decision is considered that is not updated when more information becomes available. However, the distribution that models the uncertainty can be updated when more data becomes available. When a failure occurs an event duration is obtained, whereas a preventive maintenance action results in censored durations. Both types of durations can be used to update the uncertainty in a Bayesian manner.

Event durations are more informative than censored durations, and long censored durations are more informative than short censored durations. In other words, the choice of a maintenance age influences the information that becomes available. This is acknowledged by De Jonge et al. (2015); they suggest to postpone preventive maintenance actions at the start of the lifespan of a component. This will result in an increase in costs during the first phase of the lifespan of the components, but it also results in reduced uncertainty and thereby in more effectively scheduled maintenance actions during the remainder of the lifespan. The aim is to find a balance so that the total costs during the entire lifespan is minimized. In the literature, this tradeoff is also referred to as the exploration–exploitation dilemma.

Because (De Jonge et al., 2015) are the first to recognize that the choice of the maintenance policy influences the information that becomes available, they have considered a simple setting with only two component types, viz., weak and strong components. Both component types have a Weibull lifetime distribution with a common value of the shape parameter k; the values of the respective scale parameters λ are different. The knowledge is modeled by the estimated probability that the component is strong, and a threshold policy is used that postpones preventive maintenance as long as this probability exceeds a certain threshold, i.e., as long as it is not sufficiently sure that the component is weak. This threshold policy is compared to a policy that minimizes the expected cost rate based on the current knowledge as described above. It turns out that the threshold policy can offer substantial cost reductions as opposed to the policy that minimizes the expected cost rate.

The previous analysis is based on a Weibull lifetime distribution with uncertainty only in its scale parameter. Although most uncertainty is often in the scale parameter, there also exist situations in which uncertainty in the shape parameter is expected. This can be the case if the failure mode of equipment is not known, or if there are multiple competing failure modes. This may lead to interesting results because a shape parameter $k < 1$ corresponds to a decreasing failure rate, implying that preventive maintenance is never beneficial. The optimal policy in settings where it is not known whether there is an increasing or a decreasing failure rate is of interest.

Another avenue for future research is to assume that the parametric distribution itself is not known, i.e., to assume model uncertainty instead of parameter uncertainty. A difficulty of such settings is that a selection of candidates for the true parametric distribution has to be made, and that prior probabilities need to be specified. Moreover, other optimality criteria instead of the expected cost rate could be considered. Minimization of the expected cost rate leads to the best decisions on average, but these decisions may be unacceptable for certain values of the unknown parameters.

2.2 *Block-Based Maintenance*

For the block-based maintenance policy the renewal points are the times at which preventive maintenance is performed. Renewal cycles thus always have length T, and the preventive maintenance cost is incurred once per cycle (at the end of each cycle). We let $m(t)$ denote the expected number of failures during a period with length t that starts with a component that is as-good-as-new, and during which no preventive maintenance is performed. The cost rate $\eta_{\text{block}}(T)$ of the block-based maintenance policy as a function of the preventive maintenance interval T equals

$$\eta_{\text{block}}(T) = \frac{c_{\text{pm}} + c_{\text{cm}} m(T)}{T}. \tag{1}$$

The main difficulty in evaluating $\eta_{\text{block}}(T)$ is that it requires the evaluation of the mean number of failures $m(T)$ during a time period with length T. The function $m(t)$ is called a renewal function and can be calculated as

$$m(t) = \sum_{n=1}^{\infty} F_n(t),$$

in which F_n represents the nth convolution of the lifetime distribution function F. The first convolution F_1 equals the distribution function F itself; the other convolutions can be determined recursively:

$$F_n(t) = \int_0^t f(x)F_{n-1}(t-x)\mathrm{d}x, \quad n = 2, 3, \ldots$$

In practice, $m(t)$ is often approximated numerically by using the first few convolutions. This generally results in good approximations because the number of failures to expect in between consecutive preventive maintenance actions is typically low.

Studies that consider a block-based maintenance policy generally assume that machines or components are either used continuously, or that the deterioration does not depend on the actual usage. In practice, however, this is often not realistic. De Jonge and Jakobsons (2018) consider the block-based maintenance policy for a component that is not used continuously and for which the actual usage is random. Furthermore, the component is assumed to only deteriorate when it is active. Although the future usage is stochastic, it is assumed that all maintenance actions have to be scheduled in advance, and therefore a block-based maintenance policy is considered.

The authors model the random component usage by a Markov switching. The component is alternately active and idle, and the lengths of these periods are modeled by exponential durations. Active periods are exponentially distributed with rate parameter α_1, whereas idle periods are exponentially distributed with rate parameter α_0. It follows that active periods have mean length $1/\alpha_1$ and that idle periods have mean length $1/\alpha_0$, from which it follows that the usage rate ρ of the component is given by

$$\rho = \frac{\frac{1}{\alpha_1}}{\frac{1}{\alpha_1} + \frac{1}{\alpha_0}} = \frac{\alpha_0}{\alpha_0 + \alpha_1}.$$

As mentioned before, the main difficulty in evaluating the cost rate (1) of the block-based maintenance policy is the evaluation of the renewal function $m(t)$. In the current setting with random component usage there is not even a closed-form expression for the distribution function F of the time until failure. There are, however, two limiting cases that can be analyzed using the renewal function of the lifetime distribution. We will denote this renewal function by $m_W(t)$. If, for instance, the component has a Weibull lifetime distribution, then m_W is the renewal function of the Weibull distribution.

The two limiting cases are those with a very high and with a very low switching frequency. If the switching frequency is very high, the usage in between two preventive maintenance actions is very stable. Approximately, the component will be active during time period ρT in between two consecutive preventive maintenance actions, and failures can only occur during this time period. Thus, the expected number of failures during the maintenance interval can be approximated by $m_W(\rho T)$, and the cost rate (1) by

$$\eta_{\mathrm{freq}}(T) = \frac{c_{\mathrm{pm}} + c_{\mathrm{cm}}m(T)}{T} \approx \frac{c_{\mathrm{pm}} + c_{\mathrm{cm}}m_W(\rho T)}{T}.$$

In the other limiting case the switching frequency is very low. This implies that, in between two consecutive preventive maintenance actions, it is very likely that the component is either entirely active, or entirely idle. The corresponding probabilities are ρ and $1-\rho$, respectively, with ρ equal to the usage rate (2). Failures can only occur if the component is active, implying that the expected failure cost during a maintenance interval is $c_{\text{cm}}\rho m_W(T)$. In this case the cost rate can be approximated by

$$\eta_{\text{rare}}(T) = \frac{c_{\text{pm}} + c_{\text{cm}} m(T)}{T} \approx \frac{c_{\text{pm}} + c_{\text{cm}} \rho m_W(T)}{T}.$$

Because the usage is quite stable for high switching frequencies, this limiting case results in a relatively long preventive maintenance interval. In order to avoid failures during long active periods, a much more conservative preventive maintenance interval is optimal for low switching frequencies. De Jonge and Jakobsons (2018) analyze the general case of the problem by formulating it as a set of integral equations. They show that the optimal maintenance interval and the corresponding cost rate for more moderate switching frequencies are in between the two bounds obtained from the two limiting cases. Furthermore, they also show that, for moderate switching frequencies, it is important to choose the maintenance interval based on the actual usage pattern, instead of only based on the usage rate of the component.

Future research in this area could consider active and idle periods that are not exponentially distributed. In such a setting one has to keep track of the time that the component is already active or idle, which complicates the analysis. Instead of analyzing this setting algebraically, it would also be possible to use simulations. Another possibility for future developments could be to consider multiple component speeds, instead of only on and off. This means that more sophisticated stochastic models are needed to model the random usage of the component. Random component usage can also be relevant in settings with condition-based maintenance. In such settings there is often a planning time between initiating and performing preventive maintenance. A component that is not used continuously during the planning time is expected to result in a higher optimal deterioration level at which preventive maintenance is scheduled. Finally, in the above, it is assumed that the component usage is dictated externally. However, if there is some flexibility in the usage, the performance of the system would benefit from the possibility to simultaneously optimize maintenance and usage decisions.

3 Condition-Based Maintenance

Because of the increasing possibilities to monitor, store, and analyze condition information of equipment, condition-based maintenance (CBM) policies are gaining popularity. A prerequisite for analyzing and optimizing condition-based maintenance policies is the modeling of deterioration processes of components.

Distinctions between deterioration processes can be made based on the state space (either discrete or continuous), and on the time scale (also either discrete or continuous).

Another important distinction that can be made is that between continuous condition monitoring and condition monitoring based on inspections. The first case is applicable if a sensor is used for condition monitoring; in this case we continuously know what the actual deterioration level of the component is. When inspections are needed to obtain condition information, we do not only need to determine when to carry out maintenance, but we also need to determine an inspection schedule or policy.

Inspection schedules are either periodic or aperiodic. The advantage of periodic inspections is that the entire inspection schedule is fully specified by a single decision variable, namely the time between consecutive inspections. This eases both the optimization and the implementation in a practical industrial context. However, when acceptable in practice, aperiodic inspections are often preferred because failure becomes more likely as the deterioration level increases. A final note is that an entire aperiodic inspection schedule can be fixed in advance, but that the next inspection can also be scheduled dynamically based on the currently observed deterioration level.

3.1 Delay-Time Model

The most simple deterioration model is the so-called delay-time model. It is a continuous-time model that adds a "deteriorated" state in between the operating state and the failed state. Thus, the model has three states. It is called the delay-time model because a delayed failure occurs after reaching the deteriorated state. When considering the delay-time model, probability distributions have to be specified for the time until reaching the deteriorated state, and for the time in between reaching the deteriorated state and failure. Most studies that adopt the delay-time model assume that an inspection is required to observe the deteriorated state and that failures are self-announcing. Analysis is easiest if the exponential distribution is used to model the time until reaching the deteriorated state. In that case, if immediate preventive maintenance is carried out when an inspection reveals the deteriorated state, all inspections are renewal points.

Although the delay-time model is proposed by Christer (1976) in 1976, there are new developments in delay-time modeling to date. For instance, (Van Oosterom et al., 2014) consider a periodic inspection schedule, but they relax the common assumption that preventive maintenance should be carried out immediately when an inspection reveals the deteriorated state. Instead, they allow the maintenance action to be delayed. The advantage is twofold. First, the utilization of the useful life of the component is improved, and second, the maintenance cost is reduced as a result of a longer time window to prepare maintenance resources. Wang et al. (2017) allow for a delayed first inspection, and a periodic inspection schedule thereafter. Furthermore, they initially schedule a replacement at a certain age. If an inspection reveals the deteriorated state and the time until the age-based replacement is less than a certain threshold level, then the preventive replacement action will be delayed. Otherwise, the component will be replaced immediately.

3.2 Gamma Deterioration Process with Continuous Monitoring

A commonly used continuous-time continuous-state stochastic deterioration process is the stationary gamma process. The gamma process was introduced in the area of reliability by Abdel-Hameed (1975). It has the property that the deterioration increments, within any time interval of any length, are gamma distributed with identical scale parameter.

The density function f of the gamma distribution with shape parameter $\alpha > 0$ and scale parameter $\beta > 0$ equals

$$f_{\alpha,\beta}(x) = \frac{1}{\Gamma(\alpha)\beta^{\alpha}} x^{\alpha-1} \mathrm{e}^{-\frac{x}{\beta}}, \quad x > 0,$$

in which $\Gamma(\alpha) = \int_0^{\infty} z^{\alpha-1}\mathrm{e}^{-z}\,\mathrm{d}z$ denotes the gamma function. The stationary gamma process has a shape function at with shape parameter $a > 0$ and a scale parameter $b > 0$. It is a continuous-time process $\{X(t) : t \geq 0\}$ with the following properties:

1. $X(0) = 0$ with probability 1.
2. $X(\tau) - X(t) \sim f_{a(\tau-t),b}$ for $\tau > t \geq 0$.
3. $X(t)$ has independent increments.
4. $X(t)$ is a jump process with infinitely many jumps in any time interval.

The process is stationary because the increments $X(\tau) - X(t)$ depend only on $\tau - t$ for all τ and t. The expectation and the variance of the process $X(t)$ are given by

$$\mathrm{E}(X(t)) = abt \quad \text{and} \quad \mathrm{Var}(X(t)) = ab^2t,$$

respectively. Thus, the variance of the deterioration process, relative to its mean, is small if a is large and b is small, and is large if a is small and b is large. We will use the standard deviation $\sigma = \sqrt{a} \cdot b$ as a measure for the amount of volatility in the stationary gamma deterioration process. Figure 4 shows sample paths of stationary gamma processes with $\sigma = 0.05$, $\sigma = 0.5$, and $\sigma = 5$.

De Jonge et al. (2017) consider a single maintainable component that is monitored continuously and for which the deterioration is modeled by a stationary gamma process. Failure occurs when the amount of deterioration exceeds a given level L. After such a failure an immediate corrective maintenance action will

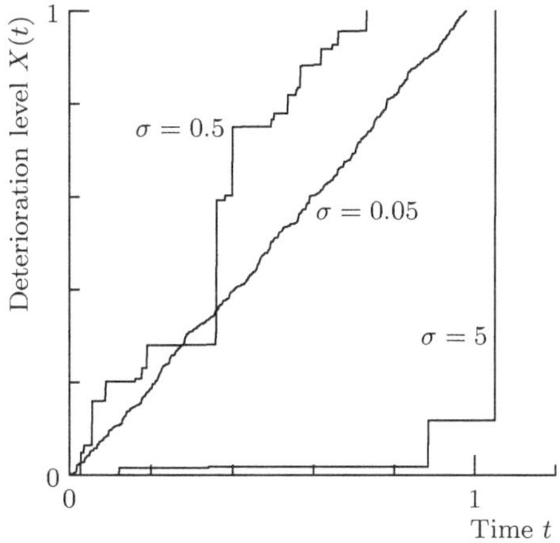

Fig. 4 Sample paths of stationary gamma processes for various standard deviations σ

be carried out. Furthermore, as long as the component is functioning, preventive maintenance can be carried out. The costs of preventive and corrective maintenance are again denoted by c_{pm} and c_{cm}, respectively. Both types of maintenance are assumed to require a negligible amount of time and to make the component as-good-as-new, i.e., they will bring the deterioration level back to 0.

The aim of the study is to compare the performance of condition-based maintenance to the performance of time-based maintenance. The condition-based maintenance policy is prescribed by a single deterioration threshold level M. Preventive maintenance is performed when the deterioration level exceeds this level M. This commonly used policy is called the control-limit policy. The threshold M should not be chosen too close to the failure level L because the deterioration process is a jump process. In other words, when M is close to L and when the deterioration level exceeds M, it may also immediately jump over L, resulting in failure. The time-based maintenance policy that is considered is the age-based maintenance policy. Thus, preventive maintenance is carried out if a certain maintenance age T is reached, see also Sect. 2.1.

Figure 5 shows the cost rate of the condition-based maintenance (CBM) policy as a function of the preventive maintenance threshold M, and the cost of the time-based maintenance (TBM) policy as a function of the maintenance age T. The gamma process is specified by $a = 5$ and $b = 0.22$ (this results in a mean time to failure of 1), the failure threshold equals $L = 1$, and the cost parameters are $c_{\text{pm}} = 0.2$ and $c_{\text{cm}} = 1$. Simulation has been used to make the figure. It turns out that the cost rate under the optimal CBM policy is substantially lower than the cost rate under the optimal TBM policy. In other words, the availability of condition information results in substantial cost savings. It can also be observed that the optimal preventive maintenance threshold M is much smaller than the failure threshold $L = 1$. As explained before, this is caused by the fact that the deterioration process is a jump process.

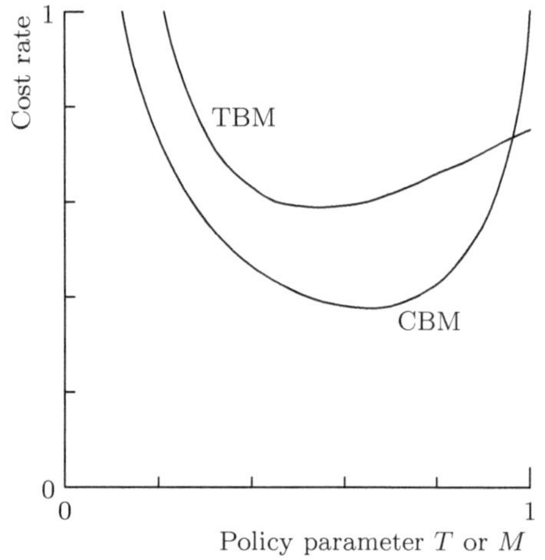

Fig. 5 The cost rate under the CBM policy (as a function of the M) and under the TBM policy (as a function of T)

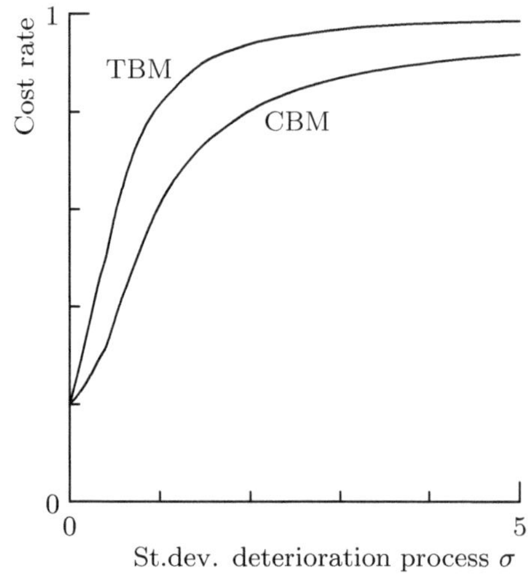

Fig. 6 The cost rate under the optimal CBM and the optimal TBM policy for a varying standard deviation σ of the gamma deterioration process

Figure 6 shows the effect of the level of volatility σ of the gamma deterioration process on the cost rates of the optimal policies. For very low levels of volatility there is almost no randomness in the moment of failure, and both CBM and TBM are very effective, i.e., both are able to carry out preventive maintenance just before failure. For very high levels of volatility, on the other hand, failure is almost always caused by a sudden very large deterioration increment. Both the CBM and the TBM policy cannot prevent this from happening. Note that, in this case, the lifetime distribution is close to an exponential distribution, and that the optimal age-based maintenance policy is to never carry out preventive maintenance (because of the constant failure rate). The benefit of CBM compared to TBM is largest for moderate levels of volatility.

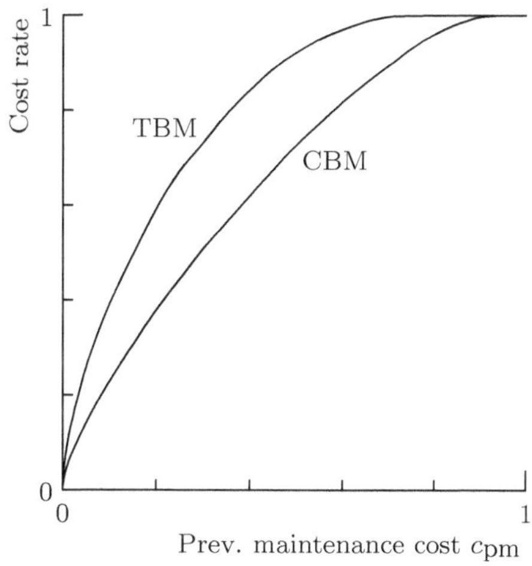

Fig. 7 The cost rate under the optimal CBM and the optimal TBM policy for a varying preventive maintenance cost c_{pm}

Figure 7 shows the effect of the preventive maintenance cost c_{pm} on the cost rates of the optimal policies. For extremely small preventive maintenance costs, both policies will use a very high maintenance frequency (at very low cost), and almost no failures will happen. This results in a very low cost rate for both policies. For extremely high preventive maintenance costs, carrying out preventive maintenance is not beneficial anymore, and the cost rates of both policies are very high. Again, the cost saving of CBM as opposed to TBM is largest for moderate preventive maintenance costs.

De Jonge et al. (2017) continue to consider the effect of various practical factors that influence the benefit of condition-based maintenance compared to time-based maintenance. The factors that they consider are a required planning time that is needed to carry out preventive maintenance, noise in the observed deterioration information, and uncertainty in the deterioration level at which failure occurs.

In practice there is often a planning time needed between initiating and performing maintenance. Here we assume that a fixed planning time s is required for carrying out preventive maintenance. Furthermore, if failure occurs during the planning time we assume that corrective maintenance will be carried out immediately and that only the high corrective maintenance cost is incurred. The preventive maintenance cost of the maintenance action that was already planned does not need to be paid anymore. We note that a planning time does not influence the time-based maintenance policy. However, for the condition-based maintenance policy, the decision is no longer to determine the deterioration level at which preventive maintenance should be carried out, but it is now the deterioration level at which preventive maintenance should be planned. During the planning time the condition information cannot be used anymore, and, as a consequence, the performance of the condition-based maintenance policy decreases. Figure 8 shows the cost rates of both policies as a function of the planning time s. When the planning

Fig. 8 The cost rate under the optimal CBM and the optimal TBM policy for a varying planning time s

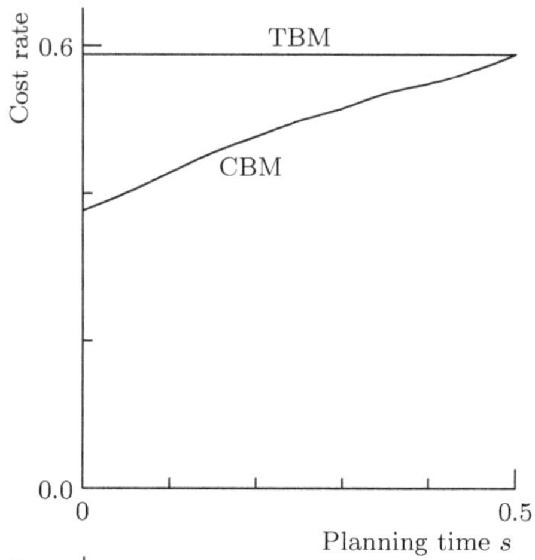

Fig. 9 The cost rate under the optimal CBM and the optimal TBM policy for a varying level of noise σ_p in the condition monitoring

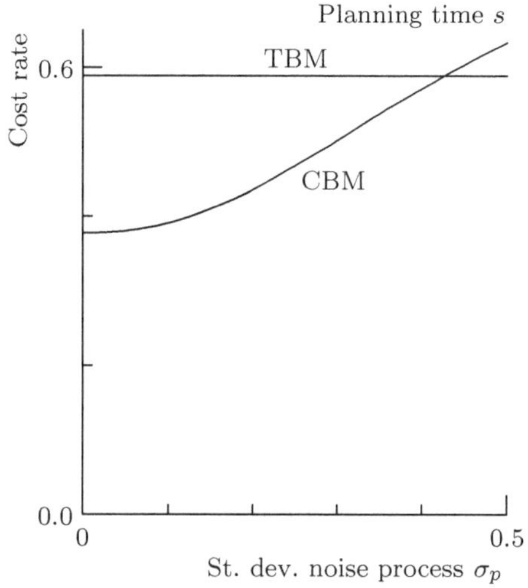

time equals the optimal maintenance age of the time-based maintenance policy, all benefits of condition-based maintenance have vanished.

Another factor that is likely to exist in practice is imperfect condition information due to noise. The difference between the actual deterioration level and the observed deterioration level has been modeled by a Brownian motion, multiplied by a parameter σ_p. The value of σ_p can be interpreted as a measure for the amount of noise. Because the time-based maintenance policy does not use any condition information, noise does not influence the performance of this policy. Condition-based maintenance, on the other, is negatively influenced by imperfect condition monitoring because the obtained information has a lower value. Figure 9 shows the optimal cost rates of both policies as a function of the amount of uncertainty σ_p. We observe that small amounts of noise only have a minor influence. However, if the amount of noise is substantial, it can even be the case the obtained condition information should not be used at all anymore.

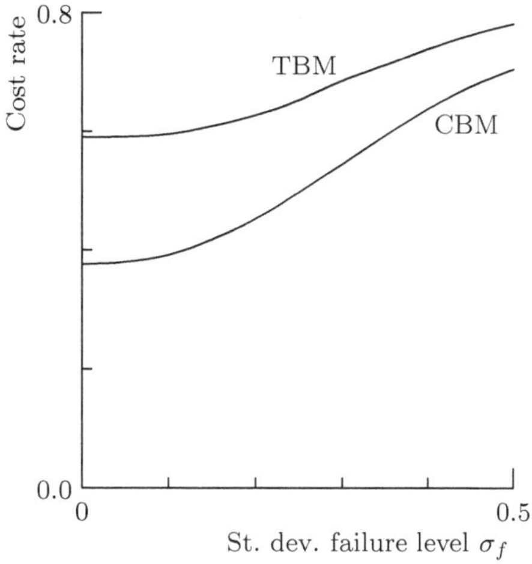

Fig. 10 The cost rate under the optimal CBM and the optimal TBM policy for a varying level of uncertainty σ_f in the deterioration failure level

Studies on condition-based maintenance typically assume that failure occurs when a certain fixed level of deterioration is exceeded. In practice, however, there are also many situations where this assumption is not realistic. The randomness in the failure deterioration level has been modeled by a normal distribution with mean 1. The standard deviation σ_f of this normal distribution can be seen as a measure for the amount of uncertainty in the failure level. In contrast to the imperfect condition information does the random failure level also affect the time-based maintenance policy. Randomness in the failure level leads to a higher variance in the time until failure, which has a negative impact on the performance of time-based maintenance. The condition-based maintenance policy also suffers from an uncertain failure level as it lowers the value of the condition information. Figure 10 shows that the effect on condition-based maintenance is larger than on time-based maintenance, implying that the benefit of condition-based maintenance is reduced if there is uncertainty in the failure level.

When deciding to switch from time-based maintenance to condition-based maintenance it is important to assess whether the benefits outweigh the additional costs for monitoring equipment and for collecting, storing, and analyzing condition data. It is important that both the volatility of the deterioration process and the cost of preventive maintenance compared to that of corrective maintenance are not extremely low or extremely high. Furthermore, it is important to realize that a required planning time, imperfect condition monitoring, and an uncertain failure level negatively impact the cost saving of condition-based maintenance as opposed to time-based maintenance.

3.3 Gamma Deterioration Process with Periodic Inspections

In this section we reconsider the setting of Park (1988), in which a periodic inspection policy is considered for a component that deteriorates according to a stationary increasing continuous-time continuous-state deterioration process. The stationary gamma process is an example of such a process. If an inspection reveals a deterioration level that exceeds a certain threshold level, an immediate preventive maintenance action is carried out. Failure is assumed to occur if a certain fixed failure threshold L is exceeded. Failures are assumed to be self-announcing and are followed by an immediate corrective maintenance action. Furthermore, the inspection schedule is reset after a failure. Both types of maintenance are assumed to make the component as-good-as-new, and to require a negligible amount of time. The cost of preventive maintenance is denoted by c_{pm}, the cost of corrective maintenance by c_{cm}, and the cost of an inspection by c_{i}. We make the reasonable assumptions that $c_{\mathrm{i}} < c_{\mathrm{pm}} < c_{\mathrm{cm}}$ and that $c_{\mathrm{i}} + c_{\mathrm{pm}} < c_{\mathrm{cm}}$.

The maintenance policy in the above setting is described by two decision variables, the time between two consecutive inspections, denoted by T, and the preventive maintenance deterioration threshold, denoted by M. Initially, we consider the time between inspections as fixed in our analysis, and, for ease of notation, we scale time such that the time between two consecutive inspections is 1. In other words, the ith inspection is performed at time i. Later on, the time between inspections can be varied to investigate how this influences the optimal cost rate, and to search for the optimal inspection interval.

Given the fixed inspection interval 1, the aim is to obtain an expression for the cost rate $\eta(M)$ as a function of the preventive maintenance threshold M. Because the component is as-good-as-new after each maintenance action, standard renewal theory can be applied. We call the time between two consecutive maintenance actions a cycle, and we calculate the cost rate $\eta(M)$ as the mean cost per cycle, denoted by $C(M)$, divided by the mean cycle length, denoted by $D(M)$. That is,

$$\eta(M) = \frac{C(M)}{D(M)}.$$

We will continue to derive expressions for $C(M)$ and $D(M)$, both as a function of the preventive maintenance threshold M, which can be evaluated numerically.

We will denote the deterioration process by $X(t)$ with $X(0) = 0$. We let $G_t(x)$ denote the distribution function of the deterioration level at time t, i.e., $G_t(x)$ equals the probability that the deterioration level has not exceeded x at time t:

$$G_t(x) = P(X(t) < x).$$

We have that $G_0(x) = 1$ for all $x \geq 0$, and $G_t(0) = 0$ for all $t > 0$. The derivative of $G_t(x)$ with respect to x is the density function of the deterioration level at time t and will be denoted by $g_t(x)$:

$$g_t(x) = \frac{\mathrm{d}}{\mathrm{d}x} G_t(x).$$

We use the following expression for the mean cost per cycle $C(M)$:

$$C(M) = c_{\mathrm{pm}} + \sum_{i=1}^{\infty} i \cdot c_{\mathrm{i}} \cdot \mathrm{P}\,(\text{PM at inspection } i)$$
$$+ \sum_{i=1}^{\infty} ((i-1)c_{\mathrm{i}} + c_{\mathrm{cm}} - c_{\mathrm{pm}}) \cdot \mathrm{P}\,(\text{Failure between inspections } i-1 \text{ and } i)\,.$$

Thus, we first incur the preventive maintenance cost c_{pm} and subtract it if a cycle ends with failure. The mean cost per cycle can be written as

$$C(M) = c_{\mathrm{pm}} + \sum_{i=1}^{\infty} i \cdot c_{\mathrm{i}} \cdot \mathrm{P}\,(X(i-1) \le M \text{ and } M < X(i) \le L)$$
$$+ \sum_{i=1}^{\infty} ((i-1)c_{\mathrm{i}} + c_{\mathrm{cm}} - c_{\mathrm{pm}}) \cdot \mathrm{P}\,(X(i-1) \le M \text{ and } X(i) > L)\,.$$

Because the deterioration level at time 0 is degenerate ($X(0) = 0$) we take the first term out of the two summations. Furthermore, by letting $\Delta X_i = X(i) - X(i-1)$ denote the additional amount of deterioration between inspection $i-1$ and inspection i, it follows that $C(M)$ can be written as

$$C(M)$$
$$= c_{\mathrm{pm}} + c_{\mathrm{i}} \cdot \mathrm{P}\,(M < X(1) \le L) + (c_{\mathrm{cm}} - c_{\mathrm{pm}}) \cdot \mathrm{P}\,(X(1) > L)$$
$$+ \sum_{i=2}^{\infty} i \cdot c_{\mathrm{i}} \cdot \mathrm{P}\,(X(i-1) \le M \text{ and } M < X(i-1) + \Delta X_i \le L)$$
$$+ \sum_{i=2}^{\infty} ((i-1)c_{\mathrm{i}} + c_{\mathrm{cm}} - c_{\mathrm{pm}}) \cdot \mathrm{P}\,(X(i-1) \le M \text{ and } X(i-1) + \Delta X_i > L)$$
$$= c_{\mathrm{pm}} + c_{\mathrm{i}} \cdot \mathrm{P}\,(M < X(1) \le L) + (c_{\mathrm{cm}} - c_{\mathrm{pm}}) \cdot \mathrm{P}\,(X(1) > L)$$
$$+ \sum_{i=2}^{\infty} i \cdot c_{\mathrm{i}} \cdot \mathrm{P}\,(X(i-1) \le M \text{ and } M - X(i-1) < \Delta X_i \le L - X(i-1))$$
$$+ \sum_{i=2}^{\infty} ((i-1)c_{\mathrm{i}} + c_{\mathrm{cm}} - c_{\mathrm{pm}}) \cdot \mathrm{P}\,(X(i-1) \le M \text{ and } \Delta X_i > L - X(i-1))\,.$$

Because the deterioration level $X(i-1)$ at time $i-1$ is independent of the additional amount of deterioration ΔX_i between time $i-1$ and time i, and because the density function of ΔX_i equals g_1, we have that $C(M)$ can be written as

$$
\begin{aligned}
C(M) = {} & c_{\text{pm}} + c_{\text{i}} \cdot \mathrm{P}(M < X(1) \leq L) + (c_{\text{cm}} - c_{\text{pm}}) \cdot \mathrm{P}(X(1) > L) \\
& + \sum_{i=2}^{\infty} i \cdot c_{\text{i}} \cdot \int_0^M g_{i-1}(x) \int_{M-x}^{L-x} g_1(y)\, \mathrm{d}y\, \mathrm{d}x \\
& + \sum_{i=2}^{\infty} ((i-1)c_{\text{i}} + c_{\text{cm}} - c_{\text{pm}}) \cdot \int_0^M g_{i-1}(x) \int_{L-x}^{\infty} g_1(y)\, \mathrm{d}y\, \mathrm{d}x \\
= {} & c_{\text{pm}} + c_{\text{i}}(G_1(L) - G_1(M)) + (c_{\text{cm}} - c_{\text{pm}})(1 - G_1(L)) \\
& + \sum_{i=2}^{\infty} i \cdot c_{\text{i}} \cdot \int_0^M g_{i-1}(x)(G_1(L-x) - G_1(M-x))\, \mathrm{d}x \\
& + \sum_{i=2}^{\infty} ((i-1)c_{\text{i}} + c_{\text{cm}} - c_{\text{pm}}) \cdot \int_0^M g_{i-1}(x)(1 - G_1(L-x))\, \mathrm{d}x.
\end{aligned}
$$

By rearranging the two sums and combining terms with the variable i in one summation and without it in another summation we obtain

$$
\begin{aligned}
C(M) = {} & c_{\text{pm}} + c_{\text{i}}(G_1(L) - G_1(M)) + (c_{\text{cm}} - c_{\text{pm}})(1 - G_1(L)) \\
& + \sum_{i=2}^{\infty} i \cdot c_{\text{i}} \int_0^M g_{i-1}(x)(1 - G_1(M-x))\, \mathrm{d}x \\
& + \sum_{i=2}^{\infty} (c_{\text{cm}} - c_{\text{pm}} - c_{\text{i}}) \cdot \int_0^M g_{i-1}(x)(1 - G_1(L-x))\, \mathrm{d}x \\
= {} & c_{\text{pm}} + c_{\text{i}}(G_1(L) - G_1(M)) + (c_{\text{cm}} - c_{\text{pm}})(1 - G_1(L)) \\
& + c_{\text{i}} \sum_{i=2}^{\infty} i \int_0^M g_{i-1}(x)\, \mathrm{d}x - c_{\text{i}} \sum_{i=2}^{\infty} i \int_0^M g_{i-1}(x) G_1(M-x)\, \mathrm{d}x \\
& + (c_{\text{cm}} - c_{\text{pm}} - c_{\text{i}}) \sum_{i=2}^{\infty} \int_0^M g_{i-1}(x)\, \mathrm{d}x \\
& - (c_{\text{cm}} - c_{\text{pm}} - c_{\text{i}}) \sum_{i=2}^{\infty} \int_0^M g_{i-1}(x) G_1(L-x)\, \mathrm{d}x.
\end{aligned}
$$

By realizing that $\int_0^M g_{i-1}(x)G_1(M-x)\ \mathrm{d}x$ equals the probability that the deterioration level at time $i-1$ is below M, and that it is still below M one time period later, we have that

$$\int_0^M g_{i-1}(x) G_1(M-x)\, \mathrm{d}x = \int_0^M g_i(x)\, \mathrm{d}x = G_i(M),$$

which allows us to write $C(M)$ as

$$\begin{aligned}
C(M) &= c_{\mathrm{pm}} + c_{\mathrm{i}}(G_1(L) - G_1(M)) + (c_{\mathrm{cm}} - c_{\mathrm{pm}})(1 - G_1(L)) \\
&\quad + c_{\mathrm{i}} \sum_{i=2}^{\infty} i G_{i-1}(M) - c_{\mathrm{i}} \sum_{i=2}^{\infty} i G_i(M) \\
&\quad + (c_{\mathrm{cm}} - c_{\mathrm{pm}} - c_{\mathrm{i}}) \sum_{i=2}^{\infty} G_{i-1}(M) - (c_{\mathrm{cm}} - c_{\mathrm{pm}} - c_{\mathrm{i}}) \\
&\quad \times \int_0^M \sum_{i=2}^{\infty} g_{i-1}(x) G_1(L-x)\, \mathrm{d}x \\
&= c_{\mathrm{pm}} + c_{\mathrm{i}}(G_1(L) - G_1(M)) + (c_{\mathrm{cm}} - c_{\mathrm{pm}})(1 - G_1(L)) \\
&\quad + c_{\mathrm{i}} G_1(M) + c_{\mathrm{i}} \sum_{i=1}^{\infty} G_i(M) \\
&\quad + (c_{\mathrm{cm}} - c_{\mathrm{pm}} - c_{\mathrm{i}}) \sum_{i=1}^{\infty} G_i(M) - (c_{\mathrm{cm}} - c_{\mathrm{pm}} - c_{\mathrm{i}}) \\
&\quad \times \int_0^M \sum_{i=1}^{\infty} g_i(x) G_1(L-x)\, \mathrm{d}x.
\end{aligned}$$

For ease of notation we let the function $H(x)$ be defined as

$$H(x) = \sum_{i=1}^{\infty} G_i(x),$$

and the function $h(x)$ as the derivative of $H(x)$, i.e.,

$$h(x) = \frac{\mathrm{d}}{\mathrm{d}x} H(x) = \sum_{i=1}^{\infty} g_i(x).$$

We then have that $C(M)$ can be expressed as

$$\begin{aligned}
C(M) &= c_{\mathrm{pm}} + c_{\mathrm{i}}(G_1(L) - G_1(M)) + (c_{\mathrm{cm}} - c_{\mathrm{pm}})(1 - G_1(L)) \\
&\quad + c_{\mathrm{i}} G_1(M) + c_{\mathrm{i}} H(M) \\
&\quad + (c_{\mathrm{cm}} - c_{\mathrm{pm}} - c_{\mathrm{i}}) H(M) - (c_{\mathrm{cm}} - c_{\mathrm{pm}} - c_{\mathrm{i}}) \int_0^M h(x) G_1(L-x)\, \mathrm{d}x,
\end{aligned}$$

which can be rewritten to our following final expression for the mean cost per cycle $C(M)$:

$$C(M) = c_{\text{cm}} + (c_{\text{cm}} - c_{\text{pm}})H(M) - (c_{\text{cm}} - c_{\text{pm}} - c_{\text{i}})(G_1(L) + \int_0^M h(x)G_1(L-x)\,\mathrm{d}x).$$

We will now continue with the mean cycle length $D(M)$, which can be expressed as

$$D(M) = \sum_{i=1}^{\infty} i \cdot \mathrm{P}\,(\text{PM at inspection } i) + \sum_{i=0}^{\infty} \mathrm{E}\,(\text{Time until failure between inspections } i \text{ and } i+1)\,.$$

Similar to the determination of $C(M)$, we can show that

$$\begin{aligned}
&\sum_{i=1}^{\infty} i \cdot \mathrm{P}\,(\text{PM at inspection } i)\\
&= G_1(L) - G_1(M) + \sum_{i=2}^{\infty} i \cdot \int_0^M g_{i-1}(x)(G_1(L-x) - G_1(M-x))\,\mathrm{d}x\\
&= G_1(L) - G_1(M) + \sum_{i=1}^{\infty} (i+1) \int_0^M g_i(x)(G_1(L-x) - G_1(M-x))\,\mathrm{d}x. \qquad (2)
\end{aligned}$$

We let $F_x(t)$ and $f_x(t)$ respectively denote the distribution and density function of the time t at which deterioration level x is reached. We have

$$F_x(t) = P(X(t) \geq x) = 1 - P(X(t) \leq x) = 1 - G_t(x).$$

We can now write

$$\begin{aligned}
&\sum_{i=0}^{\infty} \mathrm{E}\,(\text{Time until failure between inspections } i \text{ and } i+1)\\
&= \int_0^1 t f_L(t)\,\mathrm{d}t + \sum_{i=1}^{\infty} \int_0^M \int_0^1 (i+t) g_i(x) f_{L-x}(t)\,\mathrm{d}t\,\mathrm{d}x\\
&= F_L(1) - \int_0^1 F_L(t)\,\mathrm{d}t + \sum_{i=1}^{\infty} \int_0^M \int_0^1 i g_i(x) f_{L-x}(t)\,\mathrm{d}t\,\mathrm{d}x
\end{aligned}$$

$$
\begin{aligned}
&+\sum_{i=1}^{\infty}\int_0^M\int_0^1 t g_i(x) f_{L-x}(t)\,\mathrm{d}t\,\mathrm{d}x\\
&=\int_0^1 G_t(L)\,\mathrm{d}t - G_1(L) + \sum_{i=1}^{\infty}\int_0^M i g_i(x) F_{L-x}(1)\,\mathrm{d}x\\
&+\int_0^M\int_0^1 t h(x) f_{L-x}(t)\,\mathrm{d}t\,\mathrm{d}x\\
&=\int_0^1 G_t(L)\,\mathrm{d}t - G_1(L) + \sum_{i=1}^{\infty} i\int_0^M g_i(x)(1-G_1(L-x))\,\mathrm{d}x\\
&+\int_0^M h(x)\left(\int_0^1 G_t(L-x)\,\mathrm{d}t - G_1(L-x)\right)\mathrm{d}x. \qquad (3)
\end{aligned}
$$

By combining (2) and (3) it can be shown that the mean cycle length $D(M)$ equals

$$
D(M)=\int_0^1 G_t(L)\,\mathrm{d}t+\int_0^M h(x)\int_0^1 G_t(L-x)\,\mathrm{d}t\,\mathrm{d}x.
$$

The cost rate $\eta(M)$ as a function of the preventive maintenance threshold is thus equal to

$$
\begin{aligned}
\eta(M) &= \frac{C(M)}{D(M)}\\
&= \frac{c_{\mathrm{cm}} + (c_{\mathrm{cm}} - c_{\mathrm{pm}})H(M) - (c_{\mathrm{cm}} - c_{\mathrm{pm}} - c_{\mathrm{i}})(G_1(L) + \int_0^M h(x)G_1(L-x)\,\mathrm{d}x)}{\int_0^1 G_t(L)\,\mathrm{d}t + \int_0^M h(x)\int_0^1 G_t(L-x)\,\mathrm{d}t\,\mathrm{d}x}. \qquad (4)
\end{aligned}
$$

We will now consider the specific stationary gamma deterioration process with parameter values $a = 2.5$ and $b = 0.5$. Furthermore, we assume a breakdown deterioration level $L = 4$, a corrective maintenance cost $c_{\mathrm{cm}} = 10$, a preventive maintenance cost $c_{\mathrm{pm}} = 1$, and an inspection cost $c_{\mathrm{i}} = 0.1$. Note that the inspection interval is still fixed at 1. Figure 11 shows the cost rate $\eta(M)$ as function of the preventive maintenance threshold M. It turns out to be optimal to carry out preventive maintenance if an inspection reveals a deterioration level of at least $M_{\mathrm{opt}} = 1.52$. The corresponding cost rate equals $\eta(M_{\mathrm{opt}}) = 0.81$.

In the case of a stationary gamma deterioration process, (4) basically provides us with a formula $\eta(M, L, a, b, c_{\mathrm{cm}}, c_{\mathrm{pm}}, c_{\mathrm{i}})$ for the cost rate, in which L, a, b, c_{cm}, c_{pm}, and c_{i} are model parameters. For a stationary gamma deterioration process with parameters a and b, the deterioration increment during a time period of length T is gamma distributed with parameters aT and b. Therefore, for an inspection interval with an arbitrary length T, the cost rate $\bar{\eta}$ can easily be expressed in terms

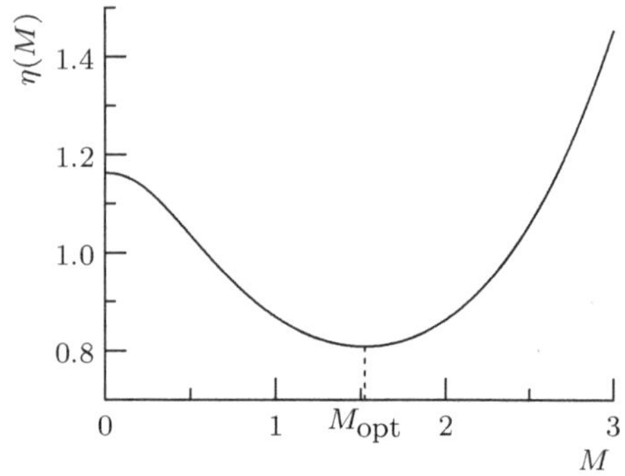

Fig. 11 The cost rate $\eta(M)$ as a function of the preventive maintenance threshold M

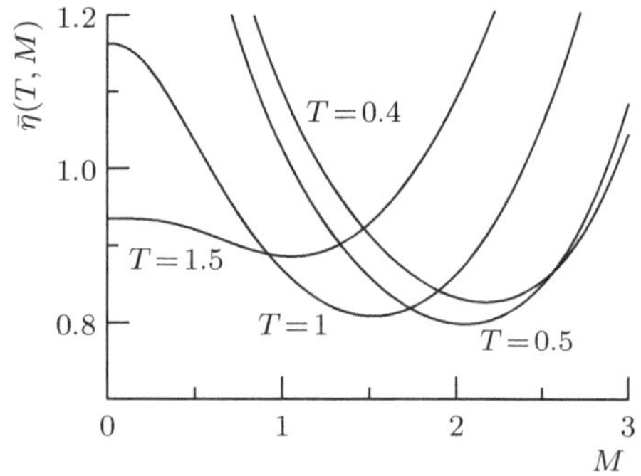

Fig. 12 The cost rate $\bar{\eta}(T, M)$ for various inspection intervals T and as a function of the preventive maintenance threshold M

of the cost rate η for an inspection interval with length 1:

$$\bar{\eta}(T, M, L, a, b, c_{\text{cm}}, c_{\text{pm}}, c_{\text{i}}) = \frac{\eta(M, L, aT_i, b, c_{\text{cm}}, c_{\text{pm}}, c_{\text{i}})}{T}.$$

Figure 12 shows the cost rate $\eta(M, T)$ for various inspection intervals T, again as a function of the preventive maintenance threshold M. Based on this figure it can be concluded that the optimal inspection interval should be somewhere between 0.4 and 1. If we optimize $\eta(M, T)$ numerically over both M and T, we find that the optimal inspection interval equals $T_{\text{opt}} = 0.68$, and that preventive maintenance should be carried out if an inspection reveals a deterioration level of at least $M_{\text{opt}} = 1.85$. Thus, by allowing an inspection interval with length different from 1, it is optimal to inspect the component more frequently, and, as a consequence, the preventive maintenance threshold will increase. The corresponding optimal cost rate decreases to $\bar{\eta}(T_{\text{opt}}, M_{\text{opt}}) = 0.78$.

3.4 Discretizing Continuous-Time Continuous-State Deterioration Processes

The drawback of modeling deterioration by a continuous-time continuous-state stochastic process is its complicated analytical tractability. The maintenance policies that we have considered in Sect. 3.2 are for instance difficult to evaluate numerically. The main reason for this is the overshoot behavior of the gamma process that is caused by the fact that it makes jumps. The analysis in Sect. 3.2 is therefore based on simulation.

An alternative method that can be used to analyze maintenance policies for a component that deteriorates according to a continuous-time continuous-state process is by discretizing this process. De Jonge (2019) presents an approach for discretizing stationary non-decreasing continuous-time continuous-state deterioration processes into discrete-time Markov chains with stationary increments. The first step of this approach is to discretize the continuous time into discrete time steps with a certain length Δt. Furthermore, the deterioration levels between 0 and the failure level L are subdivided into m deterioration intervals x_k, $k = 1, \dots, m$. These intervals correspond to states $1, \dots, m$ in the Markov chain. All deterioration levels above L are combined into the failed state $m + 1$. The transition probabilities of the Markov chain are calculated based on the assumption that the deterioration level is uniformly distributed on a certain interval x_k when it is within this interval at an arbitrary moment in time.

As an example, if we consider a stationary gamma deterioration process with parameters $a = 2$ and $b = 0.2$, a failure threshold level $L = 1$, a number of deterioration states before failure of $m = 4$, and time steps with length $\Delta t = 0.1$, we obtain the following transition probability matrix for the discrete-time Markov chain:

$$P = \begin{pmatrix} 0.861847 & 0.120864 & 0.013780 & 0.002713 & 0.000795 \\ 0.000000 & 0.861847 & 0.120864 & 0.013780 & 0.003509 \\ 0.000000 & 0.000000 & 0.861847 & 0.120864 & 0.017288 \\ 0.000000 & 0.000000 & 0.000000 & 0.861847 & 0.138153 \\ 0.000000 & 0.000000 & 0.000000 & 0.000000 & 1.000000 \end{pmatrix}.$$

De Jonge (2019) also points out how the initial maintenance policy considered in Sect. 3.2 can be evaluated based on the discretized deterioration process and on matrix algebra. Because failed components will remain failed as long as no maintenance is carried out, the Markov chain with transition probability matrix P is an absorbing Markov chain with state $m + 1$ the absorbing state. The matrix P can be written as

$$P = \begin{pmatrix} Q & \mathbf{r} \\ \mathbf{0} & 1 \end{pmatrix},$$

in which Q is an $m \times m$ matrix. The probability of going from a deterioration state $i \leq m$ to a deterioration state $j \leq m$ in exactly k time steps is equal to entry (i, j) of the matrix Q^k. The fundamental matrix R is given by

$$R = \sum_{k=0}^{\infty} Q^k = (I_m - Q)^{-1},$$

in which entry (i, j) equals the expected number of time periods that the process is in state j before it is being absorbed, given that it started in state i. After carrying out maintenance the component is as-good-as-new and the expected time until failure equals $\sum_j R_{1j}$.

We let $M \in \{1, \ldots, m\}$ denote the preventive maintenance threshold, and $\eta(M)$ the corresponding cost rate. Standard renewal theory can again be used to calculate this cost rate. We let $C(M)$ denote the mean cost per maintenance action and $D(M)$ the mean time until maintenance. Thus,

$$\eta(M) = \frac{C(M)}{D(M)}.$$

Because the deterioration process is non-decreasing, we have that R_{1j}, $j < M$, is also the expected number of time periods that the deterioration level is j before reaching a deterioration level of at least M, i.e., before maintenance is carried out. This results in the following expression for the mean time until maintenance:

$$D(M) = \sum_{j<M} R_{1j}.$$

Furthermore, because the probability of failure is $P_{j,m+1}$ if the deterioration level is j, it follows that the probability that a cycle ends with failure equals $\sum_{j<M} R_{1j} P_{j,m+1}$, implying that the mean cost per maintenance action equals

$$C(M) = c_{\text{pm}} + (c_{\text{cm}} - c_{\text{pm}}) \sum_{j<M} R_{1j} P_{j,m+1}.$$

Based on the above we now have the following expression for cost rate:

$$\eta(M) = \frac{C(M)}{D(M)} = \frac{c_{\text{pm}} + (c_{\text{cm}} - c_{\text{pm}}) \sum_{j<M} R_{1j} P_{j,m+1}}{\sum_{j<M} R_{1j}}.$$

By choosing a sufficiently high number of deterioration states m in the discretization, this formula provides us with a smooth graph of the cost rate as a function of the preventive maintenance threshold M. Figure 13 shows this cost rate for the case that we have considered in Sect. 3.2, i.e., a stationary gamma deterioration process with parameters $a = 5$ and $b = 0.22$, a failure threshold $L = 1$, and cost parameters

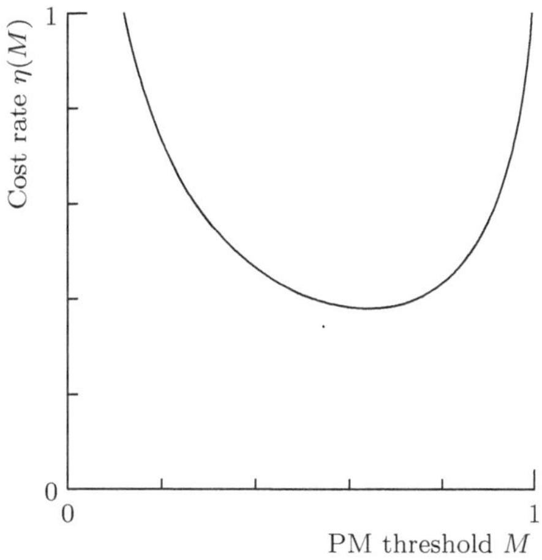

Fig. 13 The cost rate $\eta(M)$ as a function of the preventive maintenance threshold M

$c_{\text{pm}} = 0.2$ and $c_{\text{cm}} = 1$. Furthermore, $m = 100$ deterioration states before failure have been considered. We observe that this graph is virtually identical to the graph of CBM in Fig. 5. The main advantage of this approach is that we avoid the long calculation times that are required for simulation.

3.5 *Aperiodic Inspections*

Modeling deterioration by a discrete-time Markov chain is also useful when aiming to determine optimal policies by using the framework of Markov decision processes. This methodology is for instance applicable for determining maintenance policies with aperiodic inspections. We again consider a single component that deteriorates according to a discrete-time Markov chain with transition probability matrix P. There are m deterioration states before failure and a state $m + 1$ that represents failure. Failures are assumed to be self-announcing; all other deterioration states can only be observed by an inspection. Inspections can be performed at the start of each time period, the cost of an inspection is denoted by c_{i}, and an inspection is assumed to take a negligible amount of time. Furthermore, also at the start of each time period, preventive maintenance can be performed. This can done immediately after an inspection, based on the observed deterioration level, or without performing an inspection first. When failure occurs, corrective maintenance should be carried out immediately. Both preventive and corrective maintenance are assumed to take a negligible amount of time and to bring the component back to the as-good-as-new state. The costs of a preventive and of a corrective maintenance action are denoted by c_{pm} and c_{cm}, respectively. This setting is also considered by Maillart (2006), in particular for a small number of deterioration states.

The optimal inspection and maintenance decisions can be determined by formulating the above as a Markov decision process. Because the exact deterioration state of the components is uncertain as long as no inspection or maintenance is carried out, and because this uncertainty cannot be ignored, it is appropriate to formulate the problem as a partially observable Markov decision process (Monahan, 1982). A partially observable Markov decision process is a generalization of the standard Markov decision process, and can be formulated as a Markov decision process with an enlarged state space, namely the space of probability distributions over the underlying states. The states of a partially observable Markov decision process are typically called either knowledge states or belief states.

In the setting that we consider it is convenient to denote the knowledge states by, for instance, $\theta^{i,j}$, in which i denotes the last observed deterioration level, and j denotes the number of time periods ago that this deterioration level has been observed, $i = 1, \ldots, m,\ j = 0, 1, \ldots$ Thus, $\theta^{i,0}$ denotes the knowledge state if it is known with certainty that the current deterioration level is i. For $j > 0$ the actual deterioration level is uncertain, and in general, the probability of a sudden failure increases both in i and in j. The exact probabilities can be calculated based on the transition probability matrix P of the Markov chain. Another remark is that the number of knowledge state is infinite. However, we can fairly choose a sufficiently large N for which we can be reasonably sure that, under the optimal policy, the time between two consecutive actions (either inspection or maintenance) will never exceed N periods. This results in a finite number of states $\theta^{i,j}$, $i = 1, \ldots, m$, $j = 0, 1, \ldots, N$.

In any state the optimal action will always be either to do nothing, to carry out an inspection, or to perform preventive maintenance. Corrective maintenance is performed immediately when failure occurs and is therefore not really considered as an action. In other words, if failure occurs, we incur cost c_{cm} and we immediately move to state $\theta^{1,0}$. If we are in state $\theta^{1,0}$ the component is as-good-as-new with certainty, the optimal action will thus be to do nothing. For states $\theta^{i,0}$, $i = 2, \ldots, m$, the deterioration level is also known with certainty, implying that the optimal action will be either to do nothing or to carry out preventive maintenance. In all other states, any of the three actions can be chosen. Based on this reasoning, the value iteration algorithm (Puterman, 1994) can be applied, and the optimal inspection and maintenance policy can be determined.

We will continue to consider an example. We consider a component that deteriorates according to a stationary gamma deterioration process with parameters $a = 0.5$ and $b = 0.25$, and with failure deterioration level $L = 1$. We will discretize this gamma process by using the approach in Sect. 3.4, and we will use $m = 50$ deterioration states before the failed state, and time steps with length $\Delta t = 0.1$. The cost of corrective maintenance is $c_{\text{cm}} = 5$, that of preventive maintenance $c_{\text{pm}} = 1$, and that of an inspection $c_{\text{i}} = 0.1$.

Figure 14 shows the optimal inspection and maintenance policy. The horizontal axis shows the last revealed deterioration state (for the discrete-state deterioration process), and the vertical axis the number of periods between observing this state and the next preventive maintenance action or inspection. If an inspection reveals a

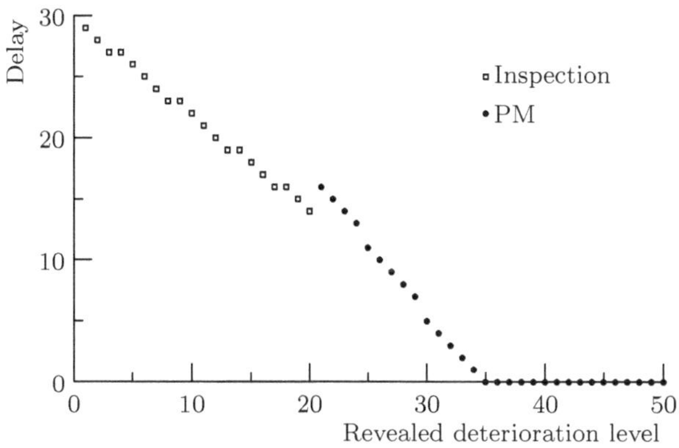

Fig. 14 Optimal action and delay time as a function of the currently revealed deterioration level

deterioration state of at most 20, we observe that a new inspection will be scheduled. The time until this next inspection is decreasing in the observed deterioration state, resulting in a dynamic aperiodic maintenance policy. If a deterioration state of at least 21 is revealed by an inspection, preventive maintenance will be carried out, either immediately or after a certain number of time periods. For deterioration states 21–34 an immediate failure is not that likely, but scheduling another inspection is not cost effective. In these cases, a delayed preventive maintenance action will be scheduled, with a delay time that is decreasing in the observed deterioration state. For an observed deterioration state of at least 35, the risk of a failure is deemed too high, and an immediate preventive maintenance action will be carried out.

4 Concluding Remarks

We have considered maintenance policies for non-repairable components, i.e., maintenance interventions can be seen as a replacement of the component. We started to consider two time-based maintenance policies, viz., age-based maintenance and block-based maintenance. For the age-based maintenance policy we have considered the effect of uncertainty in the scale parameter of the lifetime distribution on the optimal preventive maintenance age. This setting could be extended to uncertainty in other parameters of the lifetime distribution as well, or uncertainty in the parametric distribution itself (model uncertainty). For the block-based maintenance policy we have mainly focused on the optimal maintenance interval under random usage of the component. Suitable extensions of this setting would be to consider multiple component speeds, instead of only on and off, and some flexibility in the usage of the component. Furthermore, the effect of uncertainty in the lifetime distribution is also of interest in settings with block-based maintenance.

We continued to consider condition-based maintenance polices. First, we mentioned some recent developments in delay-time modeling. After that, we have adopted a continuously monitored stationary gamma deterioration process and we have considered the performance of condition-based maintenance as opposed to time-based maintenance. This analysis was based on simulation and studied the effect of the volatility of the deterioration process and of the relative cost of preventive maintenance. Furthermore, the presence of a planning time, of noise in the obtained deterioration information, and of uncertainty in the lifetime distribution have been considered. After this, we have considered a stationary gamma deterioration process combined with periodic inspections. We have obtained mathematical expressions to simultaneously optimize the inspection interval and the preventive maintenance deterioration threshold. Finally, we have provided an approach that can be used to discretize continuous-time continuous-state deterioration processes. We have first used the obtained Markov chain to reconsider the condition-based maintenance policy for a continuously monitored stationary gamma process. This analysis is based on matrix algebra. Thereafter, we have pointed out how the Markov chain and the concept of Markov decision processes can be used to determine optimal aperiodic inspection and maintenance policies.

The models with condition-based maintenance could be extended by considering various types of uncertainty. The parameters of the gamma deterioration process, or even the functional form of the deterioration process could be unknown. Furthermore, the degree of imprecision of the deterioration increments could be uncertain, or the distribution of a random failure level could be unknown. As a final suggestion, random usage of a component or production decisions could also be considered in settings with condition-based maintenance.

References

Abdel-Hameed M (1975) A gamma wear process. IEEE Trans Reliab R-24(2):152–153

Barlow RE, Proschan F (1965) Mathematical theory of reliability. Wiley, New York

Christer AH (1976) Innovative decision making. In: Bowen KC, White DJ (eds) Proceedings of the NATO Conference on the role and Effectiveness of Theories of Decision in Practice. Hodder and Stoughton, London, pp 368–377

De Jonge B (2019) Discretizing continuous-time continuous-state deterioration processes, with an application to condition-based maintenance optimization. Reliab Eng Syst Saf 188:1–5

De Jonge B, Jakobsons E (2018) Optimizing block-based maintenance under random machine usage. Eur J Oper Res 265(2):703–709

De Jonge B, Dijkstra AS, Romeijnders W (2015) Cost benefits of postponing time-based maintenance under lifetime distribution uncertainty. Reliab Eng Syst Saf 140:15–21

De Jonge B, Klingenberg W, Teunter RH, Tinga T (2015) Optimum maintenance strategy under uncertainty in the lifetime distribution. Reliab Eng Syst Saf 133:59–67

De Jonge B, Teunter RH, Tinga T (2017) The influence of practical factors on the benefits of condition-based maintenance over time-based maintenance. Reliab Eng Syst Saf 158:21–30

Gertsbakh I (2000) Reliability theory, with applications to preventive maintenance. Springer, Berlin

Maillart LM (2006) Maintenance policies for systems with condition monitoring and obvious failures. IIE Trans 38(6):463–475

Monahan GE (1982) A survey of partially observable markov decision processes: theory, models, and algorithms. Manag Sci 28(1):1–16
Park KS (1988) Optimal continuous-wear limit replacement under periodic inspections. IEEE Trans Reliab 37(1):97–102
Puterman ML (1994) Markov decision processes. Wiley, New York
Tokumoto S, Dohi T, Yun WY (2014) Bootstrap confidence interval of optimal age replacement policy. Int J Reliab Qual Saf Eng 21(4):1450018
Van Oosterom CD, Elwany AH, Çelebi D, Van Houtum G-J (2014) Optimal policies for a delay time model with postponed replacement. Eur J Oper Res 232(1):186–197
Wang H, Wang W, Peng R (2017) A two-phase inspection model for a single component system with three-stage degradation. Reliab Eng Syst Saf 158(1):31–40

Models of Imperfect Repair

Ming Luo, Shaomin Wu, and Phil Scarf

1 Introduction

It is accepted that no technical systems can last forever without any failures. As such, repair is needed in order to restore a failed item to a working state. For an asset management firm, it is vital to estimate the number of failures of a typical technical system and then to estimate the capital expenditure spending on repair and maintenance. For example, a water company may wish to estimate how many failures of each asset such as a water pumper or a mixer will have in the next 5 years, so it can plan their budget accordingly.

To ensure a system to operate and to reduce the probability of failures, three types of maintenance may be adopted: corrective maintenance, preventive maintenance and predictive maintenance. Corrective maintenance is a synonym of the term *repair*; preventive maintenance is carried out at pre-specified time points in order to reduce the probability of failure; and predictive maintenance is condition-based maintenance, with which maintenance is performed once the condition of the maintained system indicates the need for maintenance.

Once a failure occurs, repair upon the failure may end up with the following five situations:

M. Luo
Huddersfield Business School, University of Huddersfield, Huddersfield, United Kingdom
e-mail: m.luo@hud.ac.uk

S. Wu (✉)
Kent Business School, University of Kent, Canterbury, United Kingdom
e-mail: s.m.wu@kent.ac.uk

P. Scarf
Cardiff Business School, Cardiff University, Cardiff, United Kingdom
e-mail: p.a.scarf@salford.ac.uk

A. T. de Almeida et al. (eds.), *Multicriteria and Optimization Models for Risk, Reliability, and Maintenance Decision Analysis*, International Series in Operations Research & Management Science 321, https://doi.org/10.1007/978-3-030-89647-8_18

Better-than-perfect repair. In the case that the failed item is replaced with a new item, which is not identical to the failed one and is more reliable than the failed one, we say the repair is *better than perfect.* Due to technological advance, such a situation may happen when a more advanced item is used to replace the failed item.

Perfect repair. If the failed item is replaced with a new identical item, we say the repair is perfect, or a perfect repair. That is, the item used to replace the failed item has the same reliability as the failed one. In the reliability literature, perfect repair is also called as *good-as-new repair.*

Minimal repair. The minimal repair can restore the failed item to the status just as before it failed. In this case, the effectiveness of the repair is minimal as it simply brings the item back to an operating status but it does not improve the reliability of the repaired item. Hence, if the effectiveness of a preventive maintenance is minimal, then the maintenance is not needed as the purpose of a preventive maintenance is to improve the reliability of the maintained item.

Worse-than-minimal repair. If a repair unfortunately brings the maintained item to a worse status than the status just before its failure, then the repair is a worse-than-minimal repair. Such a repair may largely be caused by unskilled repairmen.

Imperfect repair. If the effectiveness of a repair is between that of the perfect repair and that of the minimal repair, the repair is said *imperfect repair.* Imperfect repair may occur more often than the above four scenarios. This is especially true for a complex system that is composed of many components. If a component fails and is then replaced, the reliability of the system is improved. That is, the repair effectiveness is better than that of the minimal repair. However, since the entire system is not replaced, the repair effectiveness is worse than the perfect repair.

Modelling the effectiveness of imperfect repair is an essential requirement in various scenarios, for example, when people plan maintenance strategies, or estimate the residual lifetime for some important systems, like nuclear power plants, aeroplanes and trains. Sometimes, these systems seem to be still in normal working conditions, when they come to the end of their planned life. To extend their functioning life, one must justify some reliability requirements. One way to do so is to take into account the effectiveness of repair actions or corrective maintenance. Repair is carried out after a failure and intends to put the system into a state in which it can perform its function again. Modelling the effect of these repair actions is of great practical interest and is the first step in order to be able to assess maintenance efficiency (Doyen and Gaudoin, 2004).

In the reliability literature, widely used methods of estimating the number of failures are stochastic processes. There are many models that have been proposed to model the effectiveness of imperfect repair, for example, the Brown–Proschan models (Brown and Proschan, 1983), the virtual age models (Kijima, 1989) and the geometric process models (Yeh, 1988). It is noted that models for preventive maintenance and corrective maintenance are essentially different in the sense that preventive maintenance is pre-scheduled and hence the methods to model the effectiveness of a series of preventive maintenance on a maintained item are deterministic models; corrective maintenance cannot be pre-scheduled and hence the methods to model the effectiveness of a series of corrective maintenance on a maintained item are stochastic processes (Doyen and Gaudoin, 2004). Nevertheless, the ways to depict the effectiveness of a maintenance action, no matter whether it is preventive or corrective maintenance, are similar. For example, age-reduction models are used in both preventive maintenance modelling (Wu and Zuo, 2010) and corrective maintenance modelling (Doyen and Gaudoin, 2004).

In this article, the term *item* and the term *system* are exchangeable.

There has been a lot of research on modelling the failure process of a repairable system, which mainly concentrates on modelling the repair effect of a repairable system through considering: (1) the working time probability functions after repairs (for example, the geometric process (Lam, 1988)); (2) the effective age of the maintained item (for example, the virtual age models (Kijima, 1989)); (3) the failure intensity of the maintained item (for example, the intensity modification model (Doyen and Gaudoin, 2004)) and (4) the virtual component methods (for example, (Wu and Scarf, 2017)). Those models can be categorised as the following.

Basic models. This category includes the renewal process (RP) and the nonhomogeneous Poisson process (NHPP). The RP is used in modelling perfect repair and the NHPP is used in modelling minimal repair. They are the bases of many further developments. That is, to a certain degree, many failure process models can be regarded as the extensions of those two models. The extensions of the RP include: the geometric process introduced by Lam (1988) and its many versions of extensions (Braun et al., 2005; Wu, 2018; Chan et al., 2006; Bordes and Mercier, 2013; Wu, 2018). The extensions of the NHPP include, for example, (Guida and Pulcini, 2009) introduce an intensity function that can depict a failure process exhibiting the bathtub curve pattern; (Syamsundar and Naikan, 2009; Guo et al., 2010) introduce segmented failure intensity functions; (Lindqvist et al., 2003) introduce the time-transformed renewal process (or the trend renewal process) that have both the ordinary renewal process and the NHPP as special cases. Lawless and Thiagarajah (1996) introduce a new model that incorporates both time trends and renewal-type behaviour.

Age reduction models. This class may have an intensity function (precisely, hazard function) $\lambda_0(a_1 t + a_2)$, where a_1, a_2 are estimable parameters, respectively. Examples include the virtual age models (Kijima, 1989; Wu and Scarf, 2015) and the ARA models (Doyen and Gaudoin, 2004). Work that extends this subclass also includes the model discussed in Dorado et al. (1997).

Intensity modification models. In this class, its intensity function $b_1\lambda_0(t)+b_2$. This class mainly modifies the intensity function after repair. It includes the arithmetic reduction of intensity (ARI) models.

Hybrid intensity models. In this class, an intensity function is obtained by combining different intensity functions (Brown and Proschan, 1983) or the same intensity with different arguments (Zhang and Jardine, 1998; Percy et al., 2010). There is a widely studied type of models, i.e., the $(p, 1-p)$ type, is originated from Brown and Proschan (1983) who assume that at the time of each failure a perfect maintenance/repair occurs with probability p and a minimal repair occurs with probability $1 - p$, independently of the previous history of repair and maintenance. Block et al. (1985) generalise the Brown–Proschan model by allowing the probability of a perfect repair to depend on the age of the failed item: assuming that at the time of each failure a perfect maintenance/repair occurs with probability $p(t)$ and a minimal repair occurs with probability $1 - p(t)$. Other extensions of the Brown–Proschan model have been made, see Zhang and Jardine (1998); Percy et al. (2010) for examples.

Virtual component models. Wu and Scarf (2017) proposed two models to model the failure process of a repairable series system composed of multiple components. Both models assume a real-world system can be analogised to virtual systems composed of multiple virtual components. Correspondingly, the failure intensity of each model is a mixture of two different failure intensities, which does not follow the $(p, 1 - p)$ rule. Wu (2019) proposed another model that integrates the failure intensity functions based on the exponential smoothing method, compared the proposed model with nine other models, and found the proposed model outperforms those existing models on 11 out of 15 real world datasets.

Other types. There are other types of failure process models that may not be categorised into the above classes, for example, the superimposed renewal process (Hoyland and Rausand, 2004), the branching Poisson process (Ascher and Feingold, 1984), the Markovian models (Bean et al., 2010), etc.

2 Existing Models of Imperfect Repair Models

In this section, we borrow the definitions of the symbols from Wu and Scarf (2017) (Table 1).

Denote the successive failure times of a repairable system by $\{T_k\}_{k\geq 1}$, from $T_0 = 0$. Denote the times between failures by $\{X_k\}_{k\geq 1}$ and $\{X_k = T_k - T_{k-1}\}$.

Table 1 Notations

Symbol	Description
T_k	The time of kth failure of a system.
$N(t)$	The number of failures of the system up to time t.
X_k	The time between $(k-1)$th and kth failures.
$\lambda(t)$	The failure intensity function.
$\lambda_I(t)$	The initial failure intensity function before the first failure.
$F(.)$	Cumulative Distribution Function of a random variable.
$f(.)$	Probability Distribution Function of a random variable.
ρ	The effectiveness of repair on failure intensity of a system in ARI/ARA models.
S_k	The effectiveness of the kth repair on failure intensity of a system in GRI/GRA models.

Assume a repair task is performed after each failure and the repair times are negligible. Let $N(t)$ denote the number of failures of the system up to time t. The failure process of the system can be defined equivalently by the random processes $\{X_k\}_{k\geq 1}$ or $\{N(t)\}_{t\geq 0}$ and is characterised by the intensity function,

$$\lambda(t) = \lim_{\Delta t\to 0} \frac{P\{N(t+\Delta t) - N(t) \geq 1|\mathscr{H}(t)\}}{\Delta t}, \tag{1}$$

where $P\{N(t+\Delta t)-N(t) \geq 1|\mathscr{H}(t)\}$ is the probability that the system fails within the interval $(t, t+\Delta t)$, given the history of failures up to time t, $\mathscr{H}(t)$ (Cox and Lewis, 1966).

Another basic assumption is that the initial intensity, i.e. the failure intensity before the first failure, is a deterministic and continuous function of time, $\lambda_I(t)$, and the system is wear-out continuously, i.e. the initial intensity is strictly increasing.

2.1 Geometric Process and Its Extensions

This section discusses the geometric process, its limitations, and its extension. We begin with an important definition on stochastic order.

Assume that X and Y are two random variables. If for every real number r, the inequality $P(X \geq r) \geq P(Y \geq r)$ holds, then X is stochastically greater than or equal to Y, or $X \geq_{st} Y$. Equivalently, Y is stochastically less than or equal to X, or $Y \leq_{st} X$ (p. 404 in Ross (1996)).

Given a sequence of non-negative random variables $\{X_k, k = 1, 2, \ldots\}$, if they are independent and the cdf of X_k is given by $F(a^{k-1}t)$ for $k = 1, 2, \ldots$, where a is a positive constant, then $\{X_k, k = 1, 2, \cdots\}$ is called a geometric process (GP) (Lam, 1988).

The above definition is given by Lam (1988), although it is likely that this definition was around earlier. For example, in Smith and Leadbetter (1963), it reads "*we consider the situation in which failing components are replaced by new ones with better statistical properties. Specifically, it is assumed that the nth replacement has a lifetime distribution* $F(a^k t)$" and also gives the GP-version renewal function. Nevertheless, most publications typically credit the geometric process to Lam (1988).

$\{X_k, k = 1, 2, \cdots\}$ in the GP may be stochastically increasing (decreasing) if $a < 1$ $(a > 1)$. If $a = 1$, then $\{X_k, k = 1, 2, \cdots\}$ reduces to a renewal process. That is, when $a \neq 1$, the GP offers an alternative that can model the effectiveness of imperfect maintenance.

Some authors either proposed similar definitions to that of the GP (Finkelstein, 1993; Wang and Pham, 1996) or made an attempt to extend the GP (Braun et al., 2005; Wu and Croome, 2006; Lam, 2007). Those different versions can be unified: They replace a^{k-1} with $g(k)$, where $g(k)$ is a function of k and is defined differently by different authors, as discussed below.

For a sequence of non-negative random variables $\{X_k, k = 1, 2, \ldots\}$, different consideration has been laid on the distribution of X_k, as illustrated in the following (in chronological order).

(a) Finkelstein (1993) proposes a process, named the *general deteriorating renewal process*, in which the distribution of X_k is $F_k(t)$, where $F_{k+1}(t) \leq F_k(t)$. A more specific model is defined such that $F_k(t) = F(a_k t)$ where $1 = a_1 \leq a_2 \leq a_3 \leq \ldots$ and a_k are parameters. In this model, $g(k) = a_k$.
(b) Wang and Pham (1996) defines a quasi-renewal process, which assumes $X_1 = W_1$, $X_2 = aW_2$, $X_3 = a^2 W_3, \ldots$, and the W_k are independently and identically distributed and $a > 0$ is constant. Here, $g(k) = a^{1-k}$.
(c) Braun et al. (2005) proposes a variant, which assumes that the distribution of X_k is $F_k(t) = F(k^{-a}t)$, or $g(k) = k^{-a}$. The authors argued that the expected number of event counts before a given time, or analogously, the Mean Cumulative Function (MCF) (or, the renewal function), does not exist for the decreasing GP. As such, they propose the process as a complement.
(d) Wu and Croome (2006) set $g(k) = \alpha a^{k-1} + \beta b^{k-1}$, where α, β, a and b are parameters. Their intention is to extend the GP to model more complicated failure patterns such as the bathtub shaped failure patterns.
(e) Chan et al. (2006) extends the GP to the threshold GP: A stochastic process $\{Z_n, n = 1, 2, \ldots\}$ is said to be a threshold geometric process (threshold GP), if there exists real numbers $a_i > 0, i = 1, 2, \ldots, k$ and integers $\{1 = M_1 < M_2 < \ldots < M_k < M_{k+1} = \infty\}$ such that for each $i = 1, \ldots, k$, $\{a_i^{n-M_i} Z_n, M_i \leq n < M_{i+1}\}$ forms a renewal process.
(f) Bordes and Mercier (2013) set $g(k) = a^{b_k}$ (where a and b_k are parameters) and discuss statistical properties of the process. The purpose of their extension is to overcome the limitation that the GP only allows for logarithmic or explosive growth.
(g) Wu and Wang (2017) extend the GP by relaxing the assumption that $\{X_k, k = 1, 2, \ldots\}$ are independent. They introduce a definition in which a sequence of non-negative random variables $\{X_k, k = 1, 2, \ldots\}$ in which $\{X_k, k = 1, 2, \ldots\}$ are dependent and the cdf of X_k is given by $F(a^{k-1}t)$ for $k = 1, 2, \ldots$.
(h) Wu (2018) proposes a definition, called *doubly geometric process*, in which a sequence of non-negative random variables $\{X_k, k = 1, 2, \ldots\}$ in which $\{X_k, k = 1, 2, \ldots\}$ are independent and the cdf of X_k is given by $F(a^{k-1}x^{h(k)})$ for $k = 1, 2, \ldots$, where $h(k)$ is a function of k and the likelihood of the parameters in $h(k)$ has a known closed form.

2.2 *Reduction of Intensity Models*

The reduction of intensity models is used when the effect of repair is considered to reduce the failure intensity. The reduction methods can be categorised into different groups such as arithmetic reduction of intensity (ARI) (Doyen and Gaudoin, 2004), geometric reduction (Doyen et al., 2017), etc.

The basic idea of ARI considers that each repair activity can reduce the failure intensity of an amount depending on the past of the failure process. In literature, the ARI models are constructed with two assumptions (Doyen and Gaudoin, 2004):

1. Each maintenance action reduces the failure intensity by sub-tracking an amount possibly depending on the past of the failure process.
2. After failure, the wear-out speed is the same as before failure.

By considering different effects of the past failure process on current failure intensity, the ARI models can be classified into ARI_∞, ARI_1 and ARI_m models. The ARI_∞ means the arithmetic reduction of intensity with infinite memory, which is built with the assumption that repair reduces the failure rate of an amount proportional to the current failure rate. With consideration of Assumption 1, the ARI_∞ failure intensity is

$$\lambda(t) = \lambda_I(t) - \rho \sum_{j=0}^{N_t-1} (1-\rho)^j \lambda_I(T_{N_t-j}). \tag{2}$$

The ARI_1 means the repair activity can only reduce the relative wear since the last repair. This model is called the arithmetic reduction of intensity with memory one. With consideration of Assumption 1, the ARI_1 failure intensity is

$$\lambda(t) = \lambda_I(t) - \rho\lambda_I(T_{N_t}). \tag{3}$$

The ARI_m is called the arithmetic reduction of intensity model with memory m, it means there are m previous failures involved in the current failure rate. With consideration of A1, the ARI_m failure intensity is

$$\lambda(t) = \lambda_I(t) - \rho \sum_{j=0}^{\mathbf{Min}(m-1, N_t-1)} (1-\rho)^j \lambda_I(T_{N_t-j}). \tag{4}$$

In the above models, the intensity is reduced arithmetically, they may not cop with some scenarios very well such as strong slowdown of the wear. Then, the geometric reduction of intensity is introduced by Doyen et al. (2017), recently. To build a geometric reduction of intensity (GRI) model, the subtractions and sums in ARI can be replaced by divisions and products, respectively. With consideration of Assumption 1, the GRI_∞ failure intensity is

$$\lambda(t) = \lambda_I(t) - \sum_{j=1}^{N_{t-1}} \frac{1 - \frac{1}{S_j}}{\prod_{k-j+1}^{N_{t-1}} S_k} \lambda_I(T_j). \tag{5}$$

2.3 Reduction of Age Models

This class of models principally consider that repair can restore the system's age as repair can reduce the failure intensity of the system at time t equal to the initial intensity at time A_t, where $A_t < t$. This class of models is also called virtual age models. In this class, the real age of a system is its functioning time t; and the virtual age of a system is defined as a positive function of its real age, possibly depending on past failures: $A_t = A(t; N_t, T_1, \ldots, T_{N_t})$. The failure intensity is a function of its virtual age: $\lambda_t = \lambda(A)$. This idea that repair activities can reduce the virtual age of the system is mainly based on Kijima's virtual age models (Tanwar et al., 2014; Kijima, 1989), which are on the basis of Generalised Renewal Process (GRP). In Kijima's first model, the nth repair is assumed can remove the wear incurred only during the time between $(n - 1)$th and nth repairs, then the virtual age is $A_n = A_{n-1} + \rho X_n$, where A_n is the virtual age after the nth repair, X_n is the time between the nth and the $(n - 1)$th repairs, and ρ is the effectiveness of repair. In Kijima's second model, the nth repair is assumed can reduce all wear accumulated up to the nth repair, then the virtual age is $A_n = \rho(A_{n-1} + X_n)$. In Kijima's models, when $\rho = 0$, the repair is perfect, when $\rho = 1$, the repair is minimal.

According to Doyen and Gaudoin (2004), the reduction of age can also be arithmetic or geometric. The arithmetic reduction of age (ARA) models can be classified into, by analogy with the ARI models, arithmetic reduction of age model with infinite memory (ARA_∞), arithmetic reduction of age model with memory one (ARA_1) and arithmetic reduction of age model with memory m (ARA_m). The ARA_1 model is similar to Kijima's first model, and the ARA_∞ model is similar to Kijima's second model.

The ARA_∞ model assumes the nth repair can reduce the virtual age of the system by a proportional amount of its age before the nth repair. Then, the failure intensity of ARA_∞ model is

$$\lambda(t) = \lambda_I\left(t - \rho\sum_{j=0}^{N_t-1}(1-\rho)^j T_{N_t-j}\right). \tag{6}$$

The ARA_1 model assumes the nth repair can reduce the virtual age of the system by a proportional amount of its age between the nth and the $(n - 1)$th repair. Then, the failure intensity of ARA_1 model is

$$\lambda(t) = \lambda_I(t - \rho T_{N_t}). \tag{7}$$

The ARA_m model assumes the nth repair can reduce the virtual age of the system by a proportional amount of its age between nth and $(n - m)$ repairs. Then, the failure intensity of ARA_m model is

$$\lambda(t) = \lambda_I \left(t - \rho \sum_{j=0}^{\mathbf{Min}(m-1, N_t-1)} (1-\rho)^j T_{N_t-j} \right). \tag{8}$$

Similar to the relationship between the ARI and GRI models, the ARA models can also be extended to geometric reduction of age (GRA) models (Doyen et al., 2017). The GRA_∞ failure intensity is Doyen et al. (2017)

$$\lambda(t) = \lambda_I \left(t - T_{n-1} + \frac{\sum_{j=1}^{n-1} \left[\prod_{k=1}^{j-1} S_k \right] X_j}{\prod_{j=1}^{n-1} S_j} \right). \tag{9}$$

After introduced by Doyen and Gaudoin (2004), the ARI and ARA models have been widely used and provided a good fit for many real maintenance datasets. Syamsundar and Naikan (2011) combine imperfect repair models and proportional intensity models to build imperfect repair proportional intensity models to cope with the field data consisting of times to failure and covariate data.

Nguyen et al. (2015) apply the ARA_∞ model on the failure dataset from a fleet of six load-haul-dump machines in a Swedish mine, as the model can help the researchers to quantify the effect of repair on each machine and to take into account the effect of the early missing data. The parameters are estimated through maximum likelihood method in this research.

Tanwar and Bolia (2015) model the imperfect repair by ARA models with incorporating the effect of imperfect corrective and preventive maintenance. In this research, four virtual age processes are introduced to describe the different repair patterns and restoration degrees for corrective and preventive maintenance. The parameters are estimated through maximum likelihood estimation.

Dauxois and Maalouf (2018) introduce a new imperfect maintenance model based on the ARI model. The arithmetic reduction of intensity is assumed on the interarrival times of failures on a system subject to recurrent failures instead of on the failure intensity.

The parameters of ARI and ARA models can be estimated through maximum likelihood estimates (Toledo et al., 2015). Corset et al. (2012) propose a Bayesian analysis of the ARA models and discuss the choice of prior distributions and the computation of posterior distributions. In this research, a single reliable repairable system which only has very few failures is considered. For this system, the quality of the maximum likelihood estimates is very poor because the number of observations is not enough. Then, the Bayesian analysis is employed to improve the accuracy of parameter estimations, as it can add the expert knowledge to operation feedback data. The expert knowledge on the system aging and repair efficiency can be reflected by the prior distributions.

2.4 Virtual Component Models

In the literature, widely used failure process models such as the generalised renewal process (GRP), geometric process (GP) and non-homogeneous Poisson process (NHPP) cannot distinguish the effect of repair upon failure of difference components in a complex system, as they consider the system as a one-component system (Wu and Scarf, 2017). To model the failure process of a multi-component system as a whole when the lifetime distribution of each component is unknown, (Wu and Scarf, 2017) introduce the concept of a virtual component. The idea of Wu and Scarf (2017) is: For a series system composed of multiple components, if the system fails, the failed component is replaced with an identical component and the replacement time is negligible. Assuming the times to failures of the system are known but upon each failure, which component causes the system to fail is unknown. With such data, it is not possible to estimate the failure process model for each individual component. Wu and Scarf (2017) assumes that the failure process of the real system is equivalent to that of a virtual system composed of virtual components. Whenever the real system fails, the virtual system is assumed to fail simultaneously and the failures are caused by the virtual components in turn. For example, assume a real series system is composed of three components A, B and C. We assume that the system is equivalent to a virtual series system composed of virtual components a, b and c. If we know the times of the first n failures, $T_1, \ldots, T_{10}$, say. Then the failure process of the virtual system is assumed to be caused by virtual components $a, b, c, a, b, c, a, b, c$ and a, respectively. Based on such assumptions, (Wu and Scarf, 2017) introduce two models and compares the performance of the models with several existing models on artistically simulated data. The results show that the proposed models have smaller AIC (Akaike Information Criterion).

3 Conclusions and Future Development

This paper reviewed some existing methods of modelling imperfect repair. The Geometric process and its extensions can be adapted to model the effectiveness of imperfect maintenance in various scenarios, as the $g(k)$ can be defined differently. However, the complexity of calculation (parameter estimation) should be considered in practice. The Reduction of Intensity and Reduction of Age models being constructed in a more intuitional way makes them more handy than the GP, but the strict assumptions should be minded and the interpretability of them in some complex scenarios also should be considered. Regarding the Virtual Component models, they provide a new gateway to model and interpret the reliability of multi-component systems; they can be developed with considering various interplays among the components.

There is much work needing further development in the future. The focus may be on the development of models for systems with different repair modes including: (1)

develop a method to model the failure process of a given complex system composed of many repairable components, while the repair effectiveness of each component is assumed unknown; (2) use modern machine learning techniques to model the failure process.

References

Ascher H, Feingold H (1984) Repairable systems reliability modeling, inference, misconceptions and their causes. Marcel-Dekker, New York

Bean MM, O'Reilly NG, Sargison JE (2010) A stochastic fluid flow model of the operation and maintenance of power generation systems. IEEE Trans Power Syst 25(3):1361–1374

Block HW, Borges WS, Savits TH (1985) Age-dependent minimal repair. J Appl Probab 22(2):370–385

Bordes L, Mercier S (2013) Extended geometric processes: Semiparametric estimation and application to reliability. J Iran Stat Soc 12(1):1–34

Braun JW, Li W, Zhao YQ (2005) Properties of the geometric and related processes. Nav Res Logist 52(7):607–616

Brown M, Proschan F (1983) Imperfect repair. J Appl Probab 20(4):851–859

Chan JSK, Yu PLH, Lam Y, Ho APK (2006) Modelling SARS data using threshold geometric process. Stat Med 25(11):1826–1839

Corset F, Doyen L, Gaudoin O (2012) Bayesian analysis of ARA imperfect repair models. Commun Stat Theory Methods 41(21):3915–3941

Cox DR, Lewis PAWL (1966) The statistical analysis of series of events. Wiley, London

Dauxois J, Maalouf E (2018) Statistical inference in a model of imperfect maintenance with arithmetic reduction of intensity. IEEE Trans Reliab 67(3):987–997

Dorado C, Hollander M, Sethuraman J (1997) Nonparametric estimation for a general repair model. Ann Stat 25(3):1140–1160

Doyen L, Gaudoin O (2004) Classes of imperfect repair models based on reduction of failure intensity or virtual age. Reliab Eng Syst Saf 84(1):45–56

Doyen L, Gaudoin O, Syamsundar A (2017) On geometric reduction of age or intensity models for imperfect maintenance. Reliab Eng Syst Saf 168:40–52

Finkelstein MS (1993) A scale model of general repair. Microelectron Reliab 33(1):41–44

Guida M, Pulcini G (2009) Reliability analysis of mechanical systems with bounded and bathtub shaped intensity function. IEEE Trans Reliab 58(3):432–443

Guo H, Mettas A, Sarakakis G, Niu P (2010) Piecewise NHPP models with maximum likelihood estimation for repairable systems. In: Reliability and Maintainability Symposium (RAMS), 2010 Proceedings-Annual. IEEE, New York, pp 1–7

Hoyland A, Rausand M (2004) System reliability theory: models, statistical methods, and applications. Wiley, New York

Kijima M (1989) Some results for repairable systems with general repair. J Appl Probab 26(1):89–102

Lam Y (1988) Geometric processes and replacement problem. Acta Math. Appl. Sinica 4:366–377

Lam Y (2007) The geometric process and its applications. World Scientific, Singapore

Lawless JF, Thiagarajah K (1996) A point-process model incorporating renewals and time trends, with application to repairable systems. Technometrics 38(2):131–138

Lindqvist BH, Elvebakk G, Heggland K (2003) The trend-renewal process for statistical analysis of repairable systems. Technometrics 45(1):31–44

Nguyen T, Dijoux Y, Fouladirad M (2015) On application of an imperfect repair model in maintenance scheduling. In: Proceedings—annual reliability and maintainability symposium, vol 2015

Percy DF, Kearney JR, Kobbacy KAH (2010) Hybrid intensity models for repairable systems. IMA J Manag Math 21(4):395–406
Ross SM (1996) Stochastic processes, 2nd edn. Wiley, New York
Smith WL, Leadbetter MR (1963) On the renewal function for the Weibull distribution. Technometrics 5(3):393–396
Syamsundar A, Achutha Naikan VN (2009) Sequential detection of change points for maintained systems using segmented point process models. Qual Reliab Eng Int 25(6):739–757
Syamsundar A, Naikan VNA (2011) Imperfect repair proportional intensity models for maintained systems. IEEE Trans Reliab 60(4):782–787
Tanwar M, Bolia N (2015) Maintenance modelling using generalized renewal process for sequential imperfect corrective and preventive maintenance. Int J Performability Eng 11(5):427–442
Tanwar M, Rai RN, Bolia N (2014) Imperfect repair modeling using kijima type generalized renewal process. Reliab Eng Syst Saf 124:24–31
Toledo MLGD, Freitas MA, Colosimo EA, Gilardoni GL (2015) ARA and ARI imperfect repair models: Estimation, goodness-of-fit and reliability prediction. Reliab Eng Syst Saf 140:107–115
Wang H, Pham H (1996) A quasi renewal process and its applications in imperfect maintenance. Int J Syst Sci 27(10):1055–1062
Wu S (2018) Doubly geometric processes and applications. J Oper Res Soc 69(1):66–77
Wu S (2019) A failure process model with the exponential smoothing of intensity functions. Eur J Oper Res 275(2):502–513
Wu S, Clements-Croome D (2006) A novel repair model for imperfect maintenance. IMA J Manag Math 17(3):235–243
Wu S, Scarf P (2015) Decline and repair, and covariate effects. Eur J Oper Res 244(1):219–226
Wu S, Scarf P (2017) Two new stochastic models of the failure process of a series system. Eur J Oper Res 257(3):763–772
Wu S, Wang G (2017) The semi-geometric process and some properties. IMA J Manag Math 29(2):229–245
Wu S, Zuo MJ (2010) Linear and nonlinear preventive maintenance models. IEEE Trans Reliab 59(1):242–249
Yeh L (1988) A note on the optimal replacement problem. Adv Appl Probab 20(2):479–482
Zhang F, Jardine AKS (1998) Optimal maintenance models with minimal repair, periodic overhaul and complete renewal. IIE Trans (Institute of Industrial Engineers) 30(12):1109–1119

Opportunistic Maintenance Policies for Multi-Components Systems

Phuc Do, Roy Assaf, and Phil Scarf

1 Introduction

In the framework of reliability theory and stochastic modelling, the system to be maintained is modelled from a functional point of view, that is to say we look at the way a main function is fulfilled. Then the sub-systems and components are described in light of this main function, to express how they interact and can contribute altogether to its achievement. Interactions between components and subsystems are usually classified into three main types, namely stochastic dependence, structural dependence and economic dependence (Do Van et al., 2013; Dekker et al., 1997; Nicolai and Dekker, 2008). Stochastic dependence implies that the state of a component may affect the lifetime distribution of other components. Structural dependence refers to systems where it is impossible to maintain a component without having an impact on others. This is principally the case when a component is not directly accessible and it is required to disassemble or stop other ones in order to execute its maintenance action. Economic dependence between components exists when the cost of joint maintenance of a group of components is different from the total costs of individual maintenance of these components.

Although the problem of stochastic, economic and structural dependencies have been widely studied for maintenance issues (Cho and Parlar, 1991; Dekker et al.,

P. Do (✉)
Research Centre for Automatic Control, University of Lorraine, Nancy, France
e-mail: phuc.do@univ-lorraine.fr

R. Assaf
School of Computing, Science and Engineering, Autonomous Systems and Robotics Centre, University of Salford, Salford, United Kingdom

P. Scarf
Cardiff Business School, Cardiff University, Cardiff, United Kingdom

A. T. de Almeida et al. (eds.), *Multicriteria and Optimization Models for Risk, Reliability, and Maintenance Decision Analysis*, International Series in Operations Research & Management Science 321, https://doi.org/10.1007/978-3-030-89647-8_19

1997; Ozekici, 1988; Thomas, 1986; Wildeman, 1996), the challenges for modelling are still very important, because of the diversity of situations that are arising from the industry. We must actually consider these dependencies either to opportunistically optimize maintenance intervention of several components or because we must take into account certain constraints (e.g. in the case of structural dependence). The dependencies may condition the optimal specific actions for each component and under which conditions it is advantageous to group or ungroup interventions. Taking into account dependencies between components in maintenance modelling and optimization of multi-component systems has recently received a lot of attention from researchers (Bian and Gebraeel, 2014a; Do Van et al., 2013; Golmakani and Moakedi, 2012; Iung et al., 2016; Nicolai and Dekker, 2008; Scarf and Deara, 2003). A review on recent advances on condition-based maintenance for multi-dependent systems is given in Keizer et al. (2017). Indeed, economic dependence has been studied and integrated in various multi-component maintenance models (Do Van et al., 2013; Liu et al., 2013; Nicolai and Dekker, 2008; van der Duyn Schouten and Vanneste, 1990). Note however that stochastic and structural dependence are not considered in these works. Failure dependence between components has been also investigated in inspection (Golmakani and Moakedi, 2012), maintenance and warranty optimization for two-component systems (Scarf and Deara, 2003; Zhang et al., 2017). In the latter, both economic and failure interaction are considered in several block replacement models.

Recently, called condition-based maintenance (CBM), in which the observed component/system condition is used for the preventive maintenance decision, has been introduced and has become nowadays efficient models in maintenance optimization frameworks. It should be noted that in CBM maintenance models, degradation level and predictive reliability are two main decision indicators for maintenance decision-making (Castanier et al., 2005; Huynh et al., 2014; Nguyen et al., 2014; Tian and Liao, 2011). Thus, the knowledge about the degradation evolution of a component/system is critical for CBM decision-making, especially in the context of multi-dependent component systems with interactions between components. In this way, a new type of stochastic dependence, called degradation interaction, whereby the degradation evolution of a component may depend not only on its degradation level but also on that of other components, has been introduced in Bian and Gebraeel (2014,a) for prognostics of system lifetime, and in Rasmekomen and Parlikad (2016) for CBM maintenance optimization. However, in Rasmekomen and Parlikad (2016) the work does not consider neither economic dependence nor intrinsic degradation dependence (whereby degradation evolution of a component depends on its own degradation level). To face with this issue, a more "complete CBM model" considering both stochastic dependence (intrinsic and extrinsic), through a model of degradation interactions, and economic dependence has been introduced in Phuc et al. (2019). In this work, the degradation level is used for both individual and opportunistic maintenance decision-making. In that way, two decision variables (one individual preventive threshold and one opportunistic preventive threshold) are required for each component. As a consequence the number of decision variables increases quickly with respect to the number of components. From a practical point of view, this may lead to some difficulties in optimization process, especially when these decision variables cannot be optimally obtained by an analytical method. To overcome this issue, the predictive reliability-based opportunistic maintenance models have been introduced in literature (Tian and Liao, 2011; Huynh et al., 2014; Nguyen et al., 2014, 2015). In these works, the predicted reliability is used for preventive maintenance decision-making. The advantages of these maintenance models are that only limited number of decision variables (mainly 3 decision variables) are needed for whatever the number of components. The remainder challenges of these maintenance policies are the prediction of the components/system reliability. This may become much more difficult in presence considering dependencies (stochastic and/or structural dependence) between components (Keizer et al., 2017; Bian and Gebraeel, 2014a). This is also an important research issue in different scientific communities such as PHM (prognostics and health management) society, PHM IEEE or PHM China.

This chapter focuses on both kinds of condition-based maintenance policies (degradation-based opportunistic maintenance policy and predictive reliability-based opportunistic maintenance policy) allowing to overcome the mentioned issues. In that way, we will, on one hand, highlight a modelling framework which can taking into account different types of dependencies between components. Two kinds of opportunistic maintenance polices are presented. On the other hand, we will highlight the uses and the performance of one opportunistic maintenance policy in the context of multi-dependence between components.

The remainder of this chapter is structured as follows. Section 2 gives the description of general assumptions, decision indicators (degradation and reliability metrics), maintenance operations and costs. Different dependencies between components are also discussed and formulated. Section 3 focuses on the presentation of two kinds of opportunistic condition-based maintenance policies. The application of the degradation-based opportunistic maintenance policy through an example is described in Sect. 4. Different numerical results and sensitivity analyses are also studied and discussed. Finally, the last section concludes the chapter with a discussion of topics for future research.

2 System Modelling and Maintenance Costs Structures

Consider a system consisting of N components in which a preventive or corrective maintenance action on one or more components needs a shutdown of the entire system. The system components are all subject to degradation. In that way, the degradation of component i (with $i = 1, \dots, N$) at time t can be summarized by an observable random scalar variable X_t^i. Component i ($i = 1, \dots, N$) is considered as failed if the component degradation level reaches its failure threshold L^i. When a component is not functioning for whatever reason, its degradation level is assumed to be unchanged during the stoppage period if no maintenance is performed. The

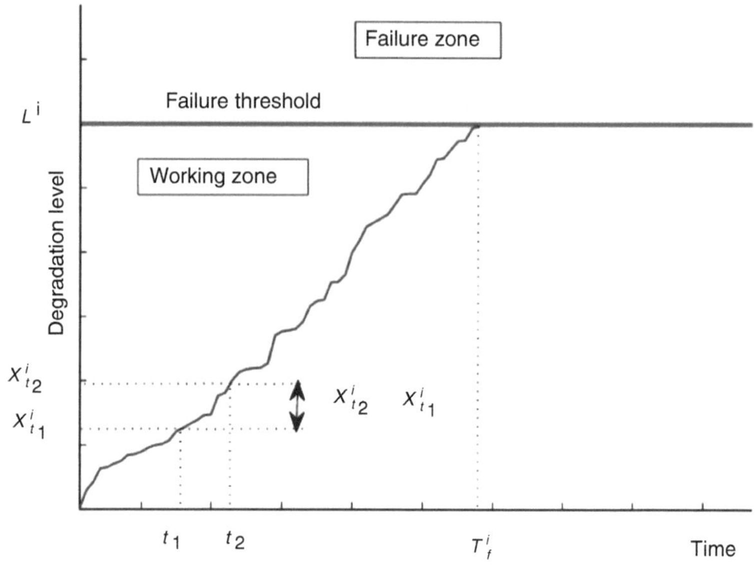

Fig. 1 Illustration of the degradation evolution of component i

degradation evolution and the corresponding state of component i are illustrated in Fig. 1. T_f^i is the random variable of time to failure for component i.

Note that the degradation level has been used as a decision indicator for condition-based maintenance (CBM) decision-making. Recently, it has been also used for opportunistic maintenance optimization (Castanier et al., 2005; Phuc et al., 2019). In this chapter, a degradation-based opportunistic CBM policy is presented in Sect. 3.

2.1 Reliability Metrics

The reliability $R^i(t)$ of component i at time t is defined as the probability that component i is in functioning state between times 0 and t:

$$R^i(t) = \mathbb{P}\left[T_f^i > t\right] = 1 - \mathbb{P}\left[T_f^i \leq t\right]. \tag{1}$$

For a deteriorating component i ($i = 1, \ldots, N$), its reliability at time t, $R^i(t)$, can be defined as probability that the component degradation level at time t ($t \geq 0$) is still lower than its failure threshold L^i. In that way, the reliability of component i at time t can be written as follows:

$$R^i(t) = \mathbb{P}\left[X_t^i < L^i\right] = 1 - \mathbb{P}\left[X_t^i \geq L^i\right]. \tag{2}$$

We assume now that component i is still functioning at time t and that its degradation level measured at time is t $X_t^i = x_t^i$ $(x_t^i < L^i)$. The conditional probability that component i can survive until time $t+u$ $(u > 0)$ can be determined by

$$R^i(t+u|t) = R^i\left(t+u|X_t^i = x_t^i\right) = \mathbb{P}\left[X_{t+u}^i < L^i|X_t^i = x_t^i < L^i\right]. \tag{3}$$

$R^i(t+u|t)$ is also called the predictive reliability of component i and estimated at time t. It is shown in the literature that the predictive reliability could be used as an efficient indicator for maintenance decision-making (Huynh et al., 2014; Nguyen et al., 2014; Tian and Liao, 2011). Note however that the estimation of the predictive reliability $R^i(t+u|t)$ may become difficult when the degradation processes of components are dependent (Bian and Gebraeel, 2014,a).

2.2 Maintenance Actions and Costs

From a practical point of view, to perform a maintenance action, the related maintenance resources (i.e., maintenance tools, repairmen, spare parts, etc.) need to be prepared in advance. In this chapter, it is assumed that all necessary maintenance resources for executing maintenance actions are always available at a planned inspection time. Maintenance actions (replacements and inspections) can be carried out only at discrete times. Note that replacements may be corrective (which is on failure of components/system) or preventive (prior to components/system failure).

A preventive cost is incurred if a preventive replacement is individually carried out. In a general manner, we denote C_p^i to be the preventive cost of component i. This can be divided into two parts:

$$C_p^i = c_p^i + c_d \cdot d_i, \tag{4}$$

where,

- $c_d \cdot d_i$ represents the downtime cost due to production loss during replacement which takes d_i time units.
- c_p^i represents all other costs (spares, labour, set-up).

In the same way, the corrective replacement cost of component i is $C_c^i = c_c^i + c_d \cdot d_i$, $(c_c^i \geq c_p^i)$.

It should be note that preventive replacement of a component indicates the replacement of the component when it is still functioning, and replacement of a component means the replacement of a component when after it fails.

To measure the degradation level of a surviving component, an inspection action is performed and each inspection action incurs an inspection cost, denoted c^I.

2.3 Dependence Modeling and Cost Saving

In this section, three kinds of dependencies between a group of components (stochastic, structural and economic dependence) are modelled and discussed.

Stochastic Dependence
Assume that the degradation processes of different components in the system may be dependent, i.e., the degradation evolution of a component may depend not only on its degradation level but also on that of other components. To model this kind of stochastic dependence, assume that evolution of the degradation level of component i is denoted by

$$X_{t+1}^i = X_t^i + \Delta X_t^i, \tag{5}$$

where ΔX_t^i is the increment in the degradation level of component i during one time unit (from t to $t+1$). For two components i and j that are deteriorating in a dependent manner, we suppose that the increment ΔX_t^i can be divided into two parts: one that arises intrinsically in the component and another that is due to (caused by) the degradation level of the other component. In that way, a general stationary model can be derived as follows:

$$\Delta X_t^i = \Delta X_t^{ii} + \Delta X_t^{ji} \text{ with } i, j = 1, 2 \text{ and } (i \neq j), \tag{6}$$

where ΔX_t^{ii} and ΔX_t^{ji} are such that:

- ΔX_t^{ii} is the increment in the degradation level of component i caused by itself during one time unit, i.e., ΔX_t^{ii} depends only on the degradation level of component i at time t. Note that ΔX_t^{ii} may be specified as deterministic or as a random variable.
- ΔX_t^{ji} is the increment in the degradation level of component i induced by component j during one time unit. ΔX_t^{ji} indicates the degradation interaction between the two components j, i and may be also specified as deterministic or as a random variable.

The reader can refer to Phuc et al. (2019) for a complete description of the degradation interaction model. A degradation interaction model for two-component system is presented in Sect. 4.

Structural Dependence
From a practical point of view, in a multi-component system, performing replacement action on a component implies a partial or fully dismantling of other components. In that way, the total maintenance duration of group of several components may be reduced when the group components are jointly replaced.

Assume now a group of several components, denoted $G = \{i, j, ..l\}$, are replaced together, and the reduction maintenance duration can be expressed as follows:

$$\Delta D(G) = b^G \cdot \sum_{i \in G} d_i, \tag{7}$$

where b^G $(0 \leq b^G \leq 1 - \max_{i \in G} d_i / \sum_{i \in G} d_i)$ is the duration-saving factor for joint replacement. The larger are b^G, the stronger is the structural dependence between components of the group. $b^G = 0$ means that all components of the group G are structurally independent. On the contrary, $b^G = 1 - \max_{i \in G} d_i / \sum_{i \in G} d_i$ means that all components of the group G are strongly dependent structurally. It should be noted that the structural dependence has also an important impact on the availability of a system (Do et al., 2015).

Economic Dependence

As when economic dependence exists between a group of components, the total maintenance cost can be reduced when the group components are jointly replaced. The cost reduction replies principally on the sharing of the set-up cost or preparation cost (Wildeman et al., 1997; Nicolai and Dekker, 2008; Do Van et al., 2013). In that way, it is assumed that all components of group G are replaced together, and the maintenance cost reduction of group G can be expressed as follows:

$$\Delta C(G) = a^G \cdot \sum_{i \in G} c_-^i, \tag{8}$$

where:

- c_-^i $(i \in G)$ could be either c_p^i or c_c^i, i.e. preventive or corrective.
- a^G $(0 \leq a^G < \min_{i \in G} c_-^i / \sum_{i \in G} c_-^i)$ is the cost-saving factor for joint replacement of group components. As is shown in Wildeman et al. (1997), the cost saving is typically equal to 5% of the total replacement cost of the components ($a^G = 0.05$). The larger are a^G, the stronger is the economic dependence between components of the group.

It is important to note that, the economic dependence is herein positive ($\Delta C(G) \geq 0$). However, a failure of a component or group components in parallel or complex structure systems does not necessarily lead to a failure of the system, and so the economic dependence may be positive or negative, see Nguyen et al. (2014); Vu et al. (2014).

Joint Maintenance and Total Cost Saving
From Eqs. (7) and (8), the total cost saving when all components of group G are jointly replaced can be evaluated as follows:

$$CS(G) = \Delta C(G) + \Delta D(G) \cdot c_d = a^G \cdot \sum_{i \in G} c_-^i + b^G \cdot \sum_{i \in G} d_i \cdot c_d. \quad (9)$$

Note that a^G, b^G indicate the economic and structural dependence degree between components of the group G. When $a^G = 0$ and $b^G = 0$, these components are economically and structurally independent.

3 Opportunistic Condition-Based Maintenance Policies

In this section, two opportunistic condition-based maintenance policies are presented. In the first policy, the degradation level is used as a decision indicator for both individual and opportunistic maintenance decision-making of each component. In that way, the degradation level of surviving each component needs to be measured at an inspection. In the second maintenance policy, the maintenance decision indicator is the predictive reliability which is estimated at each inspection time.

In both maintenance policies, it is assumed that inspection actions are instantaneous, perfect, and non-destructive. A failure of a component is also assumed to be immediately revealed by a self-announcing mechanism, but that the failed component can be replaced only at the next inspection. The preventive or corrective maintenance of a component can be an opportunity for maintaining other components.

3.1 Degradation-Based Opportunistic Maintenance Policy

At each regular time interval $T_k = k \cdot \Delta T$ ($k = 1, 2, \ldots$ and $\Delta T > 0$, the degradation levels of all the surviving components of the system are measured. It important to note that the inter-inspection interval ΔT is the first decision variable which needs to be optimized according to an optimization criterion such as cost, reliability, etc. Based on the current degradation level of component, a decision about whether or not the component should be replaced at time T_k will be taken. Two kinds of preventive maintenance rules are herein specified: individual preventive replacement and opportunistic preventive replacement. If a component fails between (T_{k-1}, T_k), then it is replaced at time T_k.

Individual Preventive Replacement

If the degradation level of component i ($i = 1, 2, ..., N$) at time T_k, $x^i_{T_k}$, is not lower than a fixed threshold m^i_p ($x^i_{T_k} \geq m^i_p$), component i is immediately replaced. m^i_p,

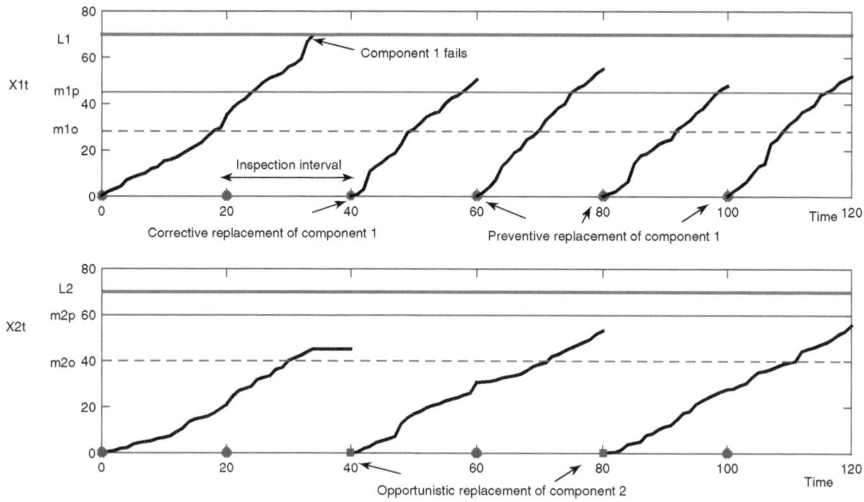

Fig. 2 Illustration of components' degradation evolution and the degradation-based opportunistic maintenance policy

called the individual preventive threshold of component i, is a decision variable to be optimized.

Opportunistic Replacement

For each component i, an opportunistic threshold, denoted m_o^i $(0 < m_o^i \leq m_p^i)$, is introduced. The opportunistic maintenance decision rules are the following. If one or more other components are preventively or correctively replaced at time T_k, component i is opportunistly replaced at time T_k if the degradation level of component i is greater or equal to the opportunistic threshold m_0^i, i.e., $x_{T_k}^i \geq m_0^i$. Note that m_o^i $(i = 1, 2)$ is also a decision variable which must be optimized.

An illustration of the degradation-based opportunistic maintenance policy for two components i and j $(i \neq j)$ is shown in Fig. 2.

Several variants of the degradation-based opportunistic maintenance policy are specified as follows:

- When $m_p^i = m_o^i$ $(\forall i = 1, 2, \ldots, N)$, there is no opportunistic replacement, the policy becomes a classical condition-based maintenance policy (Nguyen et al., 2014) with discrete inspections.
- When $m_o^i = m_o^j = 0$ $(\forall i \neq j)$, all components are jointly replaced together.

It should be note that in this maintenance policy, we need to optimize $2 * N + 1$ decision variables. From practical point of view, this may lead to some difficulties in the optimization process, especially when the optimal values of these decision variables cannot be analytically obtained. A numerical example of a two-component system is illustrated in Sect. 4.

3.2 Predictive Reliability-Based Opportunistic Maintenance Policy

In the same manner, all the surviving components of the system are inspected at each regular time interval $T_k = k \cdot \Delta T$ $(k = 1, 2, \ldots)$ with the inter-inspection interval ΔT being the first decision variable of the predictive reliability-based opportunistic maintenance policy. Based on the measured degradation levels, the predictive reliability of survival components, $R^i(T_{k+1}|T_k)$, is then estimated using Eq. (3). The latter is then used for both individual and opportunistic maintenance decision-making.

Individual Preventive Replacement

If the predictive reliability degradation level of component i $(i = 1, 2, .., N)$ at time T_k, $R^i(T_{k+1}|T_k)$, is lower or equal to a fixed threshold R_p $(0 \leq R_p \leq 1)$, $R^i(T_{k+1}|T_k \leq R_p$, component i is immediately replaced. R_p, called the individual preventive threshold, is the second decision variable which needs to be optimized.

Opportunistic Replacement

An opportunistic preventive threshold, denoted R_o $(0 < R_p \leq R_0)$, is introduced for opportunistic maintenance decision-making of all components of the system. More precisely, if one or several maintenance actions (corrective or preventive replacement) of one or more other components are performed at time T_k, component i is opportunistly replaced at time T_k if its prediction reliability is higher than the preventive threshold R_p but lower or equal to the opportunistic threshold R_0, i.e., $R_0 < R^i(T_{k+1}|T_k) \leq R_0$. Note that R_o is the third decision variable which needs be optimized.

An illustration of the predictive reliability-based opportunistic maintenance policy for two components i and j $(i \neq j)$ is shown in Fig. 3. At time T_1, only component i is replaced since its predictive reliability of component i is lower than the individual preventive threshold R_p; however, the preventive reliability of component j is higher than both individual and opportunistic preventive thresholds. At time T_2, the predictive reliability of component j is lower than the individual preventive threshold (R_p); it is firstly selected to be preventively replaced. The predictive reliability of component i is still higher than the individual preventive threshold but lower than the opportunistic preventive threshold (R_o); component j is also selected to be opportunistically replaced together with component i. No maintenance action is carried out at time T_3 since the predictive reliability of both components is higher than R_p.

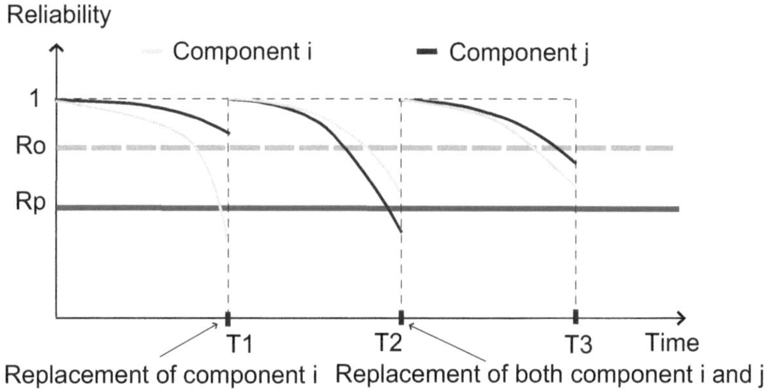

Fig. 3 Illustration of the predictive reliability-based opportunistic maintenance policy

According to the policy, only three decision variables are needed for a multi-component system. From a practical point of view, this can directly reduce the complexity of optimization process and makes the maintenance policy more applicable and efficient with large scale systems.

In the same manner with the degradation-based opportunistic maintenance policy, several variants of the predictive reliability-based maintenance policy are specified as follows:

- When $R_p = R_o$, there is no opportunistic replacement.
- When $R_o = 0$, replacement of a component leads to the replacement of the other components, i.e., all components of the system are jointly replaced together.

3.3 Cost Model for Optimization of Maintenance Policies

For each maintenance policy (degradation-based opportunistic policy or predictive reliability-based opportunistic one), its decision variables need to be optimized given some suitable criterion. For this purpose, a cost model is herein presented. In that way, the long-run expected cost per unit of time (or cost-rate) including replacement and inspection costs is usually used (Van Noortwijk, 2009; Vu et al., 2015; Castenier et al., 2005a; Phuc et al., 2019).

The cost-rate is generally defined as follows:

$$C^{\infty} = \lim_{t \to \infty} \frac{C(t)}{t}, \tag{10}$$

where $C(t)$ is the cumulative total maintenance (replacement and inspection) cost in period $(0\ t]$. Note that, according to the renewal theory (Ross, 1996), Eq. (10) can be expressed as

$$C^{\infty} = \frac{\mathbb{E}[C(T_r)]}{\mathbb{E}[T_r]}, \tag{11}$$

where $\mathbb{E}[.]$ is mathematical expectation and T_r is the length of the first renewal cycle of the system, i.e. all system components are replaced at time T_r. Without losses of generality, it is assumed that $T_r = \Delta T \cdot l$ (l is a positive integer), and so we get

$$C(T_r) = \frac{\sum_{k=1}^{l} \left(C_{ins}^k + C_{main}^k\right) + T_{down} \cdot c_d}{l \cdot \Delta T},$$

with:

- T_{down} is the total downtime of the system due to the failure of components within $[0, T_r]$.
- $C_{ins}^k = v \cdot c_I$ with v ($v = 0, 1, 2$) being the number of surviving components that are inspected at time T_k.
- C_{main}^k is the total maintenance cost at time T_k. Assume that a group of several components, denoted G^k, are replaced at time T_k. C_{main}^k can be evaluated as follows:

$$C_{main}^k = \sum_{i \in G^k} C_-^i - CS(G^k),$$

where $CS(G^k)$ is the cost saving when all components of group G^k are replaced together. $CS(G^k)$ can be obtained by Eq. (8).

It is shown in Grall et al. (2002); Castenier et al. (2005a); Phuc et al. (2019) that a closed-form expression for the cost-rate in Eq. (11) cannot be obtained. Monte Carlo simulation can be however used to evaluate the cost-rate with given the decision variables values. By changing the values of the decision variables and performing an exhaustive, the minimum cost-rate can then be identified.

4 Numerical Example

Consider a two-component system, a shutdown of a component, namely C1 or C2, leads to a failure of the system. Let us denote the degradation levels for C1 and C2 at time t by X_t^1 and X_t^2, respectively. As discussed in Sect. 2.3, the evolution of degradation for a component i ($i = 1, 2$) can be described by

$$X_t^i = X_{t-1}^i + \Delta X_t^i + \Delta X_t^{ij} \text{ with } j = 1, 2 \text{ and } j \neq i, \tag{12}$$

where,

Table 1 Degradation parameter values of a two-dependent component system

Component	α^i	β^i	μ^i	σ^i
C1	0.0233	0.0425	0.0995	7.6659
C2	0.0125	0.0914	0.0493	9.7375

- The intrinsic effect ΔX_t^{ii} is random and assumed to follow a Gamma distribution with shape parameter α^i and scale parameter β^i (see Appendix A for more details), i.e., $\Delta X_t^{ii} \sim \Gamma(\alpha^i, \beta^i)$.
- The interaction effect, $\Delta X_t^{ji} = \mu^j \cdot (X_t^j)^{\sigma^j}$ where μ^j, σ^j are non-negative real numbers that quantify the impact of component j on the degradation rate of component i. If $\mu^i = \mu^j = 0$, two components i and j are stochastically independent.

The degradation parameter values are reported in Table 1.

About the maintenance costs, when each component is individually replaced, its replacement cost and maintenance duration are $c_p^1 = c_c^1 = 500$ acu (arbitrary cost unit), $c_p^2 = c_c^2 = 600$ acu and $d_1 = d_2 = 1$ atu (arbitrary time unit). When both components are replaced together, the total maintenance cost is reduced by 5% ($a^G = 0.05$) and 50% of the total maintenance duration is saved ($b^G = 0.5$). In addition, when the system fails, a downtime cost rate $c_d = 100$ for each time unit is incurred. Each inspection action costs also 10 acu ($c_I = 10$).

To study the performance of the degradation-based opportunistic maintenance policy and the opportunistic maintenance rules, let us distinguish the general policy, namely policy P, with its two variant ones:

- Policy P1 (no opportunistic replacement) is obtained by setting $m_p^i = m_o^i$ ($\forall i = 1, 2, \ldots, N$).
- Policy P2 (both components are always replaced together) is given by setting $m_o^i = m_o^j = 0$ ($\forall i \neq j$).

4.1 Optimum Maintenance Policy

To evaluate the maintenance cost-rate, a large number of system life cycles were simulated with above data. To find the optimal decision variables $(\Delta T, m_p^1, m_o^1, m_p^2, m_o^2)$, the cost-rate C^∞ is evaluated for different values of ΔT ($\Delta T > 0$), m_p^1 ($0 < m_p^1 \leq L^1$), m_o^1 ($0 < m_o^1 \leq m_p^1$), m_p^2 ($0 < m_p^2 \leq L^2$) and m_o^2 ($0 < m_o^2 \leq m_p^2$) using Eq. (11). The step size is 0.05 for each preventive or opportunistic threshold and 5 for the inter-inspection time. With a precision of 0.010 specified for the cost-rate, from 10,000 renewal cycles, the convergence of the cost-rate is reached. The optimum values of the decision variables are obtained as the following: $\Delta T^* = 60$, $m_p^{1*} = 0.55$, $m_o^{1*} = 0.50$, $m_p^{2*} = 0.50$ and $m_o^{2*} = 0.40$ with the minimum cost-rate $C^\infty = 2.90$ acu.

Table 2 Proportion of maintenance actions

	Individual replacement of C1	Individual replacement of C2	Joint replacement of C1 and C2
Policy P	0.31	0.38	0.31
Policy P1	0.34	0.53	0.12
Policy P2	0	0	1

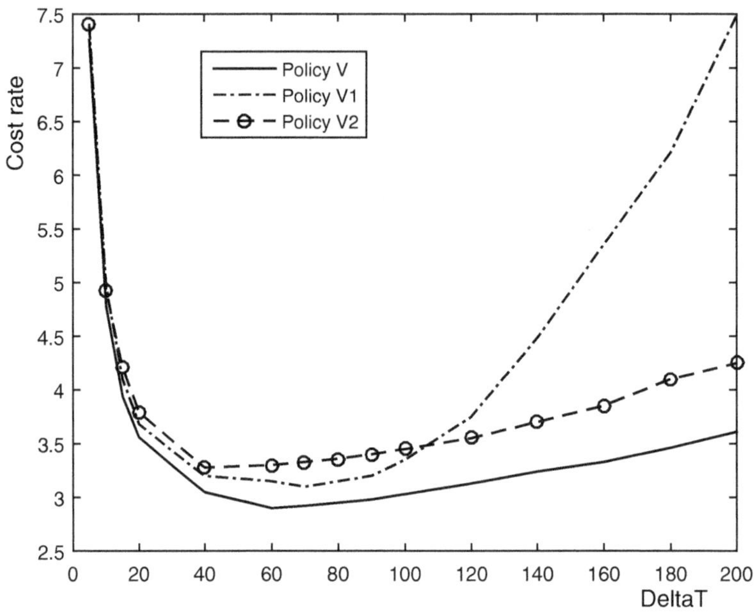

Fig. 4 Cost-rate as a function of inter-inspection interval ΔT

For each optimal maintenance policy (policy P, P1 or P2), the proportion of maintenance actions at maintenance time, i.e. joint replacement of C1 and C2, and individual replacement of C1 or C2 is reported in Table 2.

The results show that the proportion of joint replacement in the proposed opportunistic maintenance policy (policy P) is higher than in the non-opportunistic policy (policy P1). This is because the opportunistic thresholds tend towards a joint replacement of C1 and C2. As for Policy P2 the two components are always replaced together.

Figure 4 shows the relationships between the minimum cost-rate and the inter-inspection interval ΔT for the general opportunistic policy P and its variants (policies P1 and P2). Each point corresponds to an optimal policy with a given value of ΔT. The results show that policy P always provides the lowest cost-rate. We observe also that when $\Delta T < \Delta T^*$ the maintenance cost increases rapidly and the difference between the three policies decreases with decreasing ΔT. However, when $\Delta T > \Delta T^*$, the cost-rate of the non-opportunistic policy (policy P1) increases

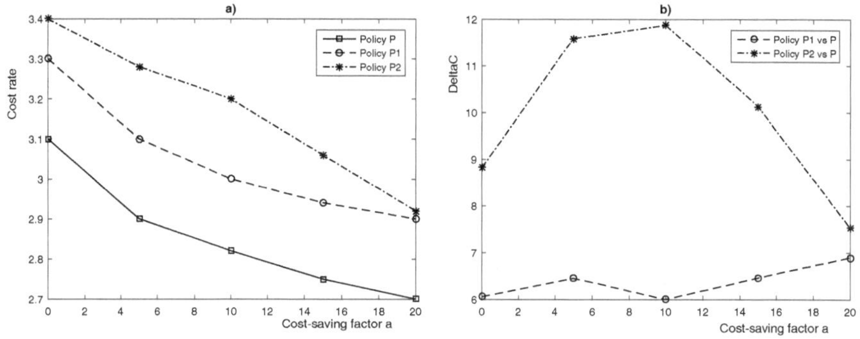

Fig. 5 Cost-rate (**a**) and excess-cost (**b**) as a function of a^G

rapidly with increasing ΔT, while the cost-rate of policies P and P2 increases slowly with increasing ΔT. This implies that the opportunistic replacement and the joint replacement can better compensate for a sub-optimally large ΔT.

4.2 *Impact of Economic Dependence on the Cost*

We investigate the impact of economic dependence on the opportunistic replacement maintenance policy in this section. To do this, we consider the sensitivity of the minimum cost-rate for the general policy (policy P) and its variants (P1 and P2) to the economic dependence degree a^G between the two components of the system.

To assess the performance of these three policies, a relative excess-cost in the minimum cost-rate of the general opportunistic policy P compared to policy Pi ($i = 1, 2$), denoted ΔC_i, is used. It is defined by

$$\Delta C_i = \frac{C_{Pi}^{\infty} - C_P^{\infty}}{C_{Pi}^{\infty}} \cdot 100\%,$$

where C_P^{∞}, C_{Pi}^{∞} are the minimum cost-rate of policy P and Pi ($i = 1, 2$), respectively. Regarding the definition, $\Delta C_i > 0$ means that policy P is more effective than policy Pi and less effective in the opposite case. We vary the economic dependence degree a^G from 0 to 20% while the others parameters remain unchanged. For each value of a^G the minimum cost-rate of each policy is quantified and the excess-cost is then calculated. The obtained results are shown in Fig. 5.

With respect to the cost-saving factor a^G, we see in Fig. 5a that the cost-rate decreases. This is due to the fact that as a^G increases the maintenance costs are reduced. It is not surprising that the proposed opportunistic policy P always provides a lowest cost-rate. This is because policies P1 and P2 are two special cases of policy P.

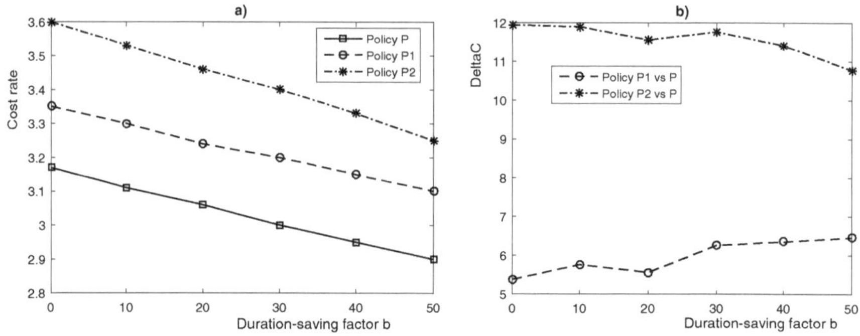

Fig. 6 Cost-rate (**a**) and excess-cost (**b**) as a function of b^G

Figure 5b shows that when $a^G < 10\%$ the excess-cost related to policy P2 increases with an increasing of a^G. This means that the cost-rate of policy P2 decreases more slowly than the cost-rate of policy P as a^G increases. However, when $a^G > 10\%$, the cost-rate of policy P2 decreases more rapidly than the cost-rate of policy P. While the cost-rate of policy P1 decreases more slowly than that of policy P1 with increasing a^G. This is because the two components tend to be replaced together when the cost-saving factor a^G is high.

4.3 Impact of Structural Dependence on the Cost

To study the impact of structural dependence degree on the maintenance cost, let us consider sensitivity with respect to the duration-saving factor b^G. In that way, we vary b^G from 0 to 50%, and keep the others parameters fixed. We determine the minimum cost-rate of each maintenance policy (policy P, P1 or P2) for each value of b^G; we then evaluate the excess-cost. The results obtained are shown in Fig. 6.

Given the results, we see once again that when b^G is increased (or equivalently a reduction on maintenance duration when two components are replaced together) we end up with a decreased cost-rate. We notice that the effect of both the opportunistic policy (P) and non-opportunistic policy (P1) is almost the same, in a similar manner to that for varying a^G. This suggests that for both policies there is a tendency that two components are jointly replaced.

This is to be expected from the opportunistic policy since this is its purpose. However, we might have expected the non-opportunistic policy to show less dependence on a^G and b^G. This can be explained by the fact that when there is no opportunistic replacement, the threshold for preventive replacement compensates (for component C1 in this case). It is lower (than with the opportunistic policy) so that more often than not, the replacement of components is simultaneous (and set-up cost is saved). If it were not the case that replacements are done jointly, then the cost-rate for policy P1 would not depend on a^G and b^G in the way it does. We see this effect as a result of the positive stochastic dependence. If such a dependence did not exist, then the joint replacement of the components when one reaches a preventive replacement threshold would be inefficient. So, when there is no stochastic dependence between components, opportunistic policies become more adequate as the extent of economic dependence increases. This is well known and obvious. However, apparently when a positive stochastic dependence exists, this phenomenon is much less apparent. This is because a non-opportunistic policy will then compensate for the absence of opportunities for replacement by lowering the threshold for preventive replacement of the components. The positive stochastic dependence ensures that replacements are concurrent since components then tend to cross their replacement thresholds together. That said, the phenomenon of components deteriorating together will tend be more apparent when the lifetimes of the components are broadly similar.

4.4 Impacts of Stochastic Dependence on the Cost

To study the impact of stochastic dependence (or state dependence) between two components on the optimum maintenance policy, it is now assumed that the degradation process of each component evolves independently. This can be obtained from the degradation model given by Eq. (12) by setting $\mu^i = \mu^j = 0$. In this way, we reduce the degradation model to two independent gamma process for which the shape and scale parameters can be estimated, using maximum likelihood estimation, from the data simulated with the degradation parameters' values shown in Table 1. The obtained results are reported in Table 3.

The degradation-based opportunistic maintenance policy is then applied. The optimal decision variables are obtained as the following: $\Delta T^* = 120$, $m_p^{1*} = 0.60$, $m_o^{1*} = 0.45$, $m_p^{2*} = 0.55$ and $m_o^{2*} = 0.40$. Note that, when compared with the results given in Sect. 4.1, these optimal values are significantly different. In addition, when applying these optimal decision variables for the case considering the stochastic dependence between components, the cost-rate is obtained as $C^\infty = 3.75$ acu. This result is significantly higher than the one given when the stochastic dependence is considered in degradation modelling ($(3.75-2.90)/2.90) \times 100 = 29.3\%$ higher). This result implies that without considering the stochastic dependence between components can draw in a sub-optimal maintenance policy. Of course, the difference may depend on the both economic and structural dependence degree between the components.

Table 3 Estimated values of degradation parameters without considering stochastic dependence

Component	α^i	β^i
C1	0.1165	0.0100
C2	0.0919	0.0090

5 Summary and Conclusions

The present chapter gives some opportunistic condition-based maintenance policies for multi-dependent component systems. Three kinds of dependencies that may exist between components are investigated. Stochastic dependence or degradation interaction implies that the state (degradation) evolution of a component depends not only on its current state (degradation level) but also on the state (degradation level) of other components. Structural dependence means that replacement of a component leads to a dismantling of other components, and as a consequence, joint replacement of several components can reduce the maintenance duration. Positive economic dependence exists when joint maintenance of a group component is cheaper than performing maintenance on components separately. To select a component to be replaced, two kinds of opportunistic maintenance policies are discussed. The main difference between two kinds of maintenance policies is the decision indicator for maintenance decision-making. In the first type of maintenance policies, namely degradation-based opportunistic policy, the degradation level of each component is used for the maintenance decision-making of both individual and opportunistic preventive maintenance. In that way, the degradation-based opportunistic maintenance policy derives $(2*N + 1)$ decision variables which are needed to be optimized regarding to some optimization criterion such as cost, reliability, etc. From practical point of view, this may lead to some difficulties in the optimization process, especially when the optimal values of these decision variables cannot be obtained by an analytical method. For the second type of maintenance policies, namely predictive reliability-based opportunistic maintenance policy, the predictive reliability estimated at each inspection time is used for triggering both individual preventive maintenance and opportunistic preventive maintenance of a component. Only three decision variables are needed for whatever number of components. From a practical point of view, this highlights the applicability of the maintenance policy. Note however that the predictive reliability may be not easy to obtained especially when the degradation processes of components are dependent. Some efficient prediction methods/approaches may be needed to deal with this problem.

The use and the performance of the degradation-based opportunistic maintenance policy are illustrated through a numerical example of a two-component system considering all three kinds of dependencies between components. The results indicate that (i) dependencies between components have an important impact on the total maintenance cost and should not be ignored; (ii) introducing an opportunistic threshold for replacement makes the maintenance policy more flexible and less sensitive to a sub-optimally large inspection interval. Nonetheless, when a positive stochastic dependence between components exists whereby components tend to deteriorate together, it is less effective to introduce an opportunistic threshold for replacement in order to share maintenance cost. This is a result of having synchronized replacements of components which arise precisely because of degradation dependence. We can therefore claim a general insight that opportunistic maintenance is less opportune when components tend to deteriorate together. It will be very challenging to investigate this claim in a more general context.

Appendix A: Gamma Distribution

A random variable X which is gamma-distributed with shape α^i and rate β^i is denoted

$$X \sim \Gamma(\alpha^i, \beta^i).$$

The corresponding probability density function (PDF) is

$$f_{\alpha^i,\beta^i}(x) = \frac{1}{\Gamma(\alpha^i)} \cdot (\beta^i)^{\alpha^i} \cdot x^{\alpha^i - 1} \cdot e^{-\beta^i \cdot x} \cdot \mathscr{I}_{\{x \geq 0\}},$$

where:

- $\Gamma(\alpha^i) = \int_0^{+\infty} u^{\alpha^i - 1} \cdot e^{-u} du$ denotes a complete gamma function;
- $\mathscr{I}_{\{x \geq 0\}}$ is an indicator function. $\mathscr{I}_{\{x \geq 0\}} = 1$ if $x \geq 0$, $\mathscr{I}_{\{x \geq 0\}} = 0$ and otherwise.

References

Bian L, Gebraeel N (2014) Stochastic framework for partially degradation systems with continuous component degradation-rate-interactions. Nav Res Logist 61:286–303

Bian L, Gebraeel N (2014a) Stochastic modeling and real-time prognostics for multi-component systems with degradation rate interactions. IIE Trans 46:470–482

Castanier B, Grall A, Bérenguer C (2005) A condition-based maintenance policy with non-periodic inspections for a two-unit series system. Reliab Eng Syst Saf 87(1):109–120

Castenier B, Grall A, Berenguer C (2005a) A condition-based maintenance policy with non-periodic inspections for a two-unit series system. Reliab Eng Syst Saf 87:109–120

Cho D-I, Parlar M (1991) A survey of maintenance models for multi-unit systems. Eur J Oper Res 51(1):1–23

Dekker R, Wildeman L, Van Der Duyn Schouten, F (1997) A review of multi-component maintenance models with economic dependence. Math Meth Oper Res 45(3):441–435

Do P, Vu H-C, Barros A, Berenguer C (2015) Maintenance grouping for multi-component systems with availability constraints and limited maintenance teams. Reliab Eng Syst Saf 142:56–67

Do Van P, Barros A, Berenguer C, Bouvard K, Brissaud F (2013) Dynamic grouping maintenance strategy with time limited opportunities. Reliab Eng Syst Saf 120:51–59

Golmakani HR, Moakedi H (2012) Periodic inspection optimization model for a two-component repairable system with failure interaction. Comput Ind Eng 63(3):540–545

Grall A, Dieulle L, Bérenguer C, Roussignol M (2002) Continuous-time predictive-maintenance scheduling for a deteriorating system. IEEE Trans Reliab 51:141–150

Huynh KT, Barros A, Bérenguer C (2014) Multi-level decision-making for the predictive maintenance of k-out-of-n: F deteriorating systems. IEEE Trans Reliab 64(1):94–117
Iung B, Do P, Levrat E, Voisin A (2016) Opportunistic maintenance based on multi-dependent components of manufacturing systems. CIRP Ann Manuf Technol 65(1):401–404
Keizer MCO, Flapper SDP, Teunter RH (2017) Condition-based maintenance policies for systems with multiple dependent components: a review. Eur J Oper Res 261(1):405–420
Liu L, Yu M, Maa Y, Tu Y (2013) Economic and economic-statistical designs of an x control chart for two-unit series systems with condition-based maintenance. Eur J Oper Res 226:491–499
Nguyen K-A, Do P, Grall A (2014) Condition-based maintenance for multi-component systems using importance measure and predictive information. Int J Syst Sci Oper Logistics 1(4):228–45
Nguyen K-A, Do P, Grall A (2015) Multi-level predictive maintenance for multi-component systems. Reliab Eng Syst Saf 44:83–94
Nicolai R, Dekker R (2008) Optimal maintenance of multi-component systems: a review. In: Complex system maintenance handbook. Springer, London, pp 263–286
Ozekici S (1988) Optimal periodic replacement of multicomponent reliability systems. Oper Res 36(4):542–552
Phuc D, Assaf R, Scarf P, Benoit I (2019) Modelling and application of condition-based maintenance for a two-component system with stochastic and economic dependencies. Eng Syst Saf 182:86–97
Rasmekomen N, Parlikad A (2016) Condition-based maintenance of multi-component systems with degradation state-rate interactions. Reliab Eng Syst Saf 148:1–10
Ross S (1996) Stochastic processes. In: Wiley Series in Probability and Statistics. Wiley, New York
Scarf P, Deara M (2003) Block replacement policies for a two-component system with failure dependence. Nav Res Logist 50:70–87
Thomas L (1986) A survey of maintenance and replacement models for maintainability and reliability of multi-item systems. Reliab Eng 16(4):297–309
Tian Z, Liao H (2011) Condition based maintenance optimization for multi-component systems using proportional hazards model. Reliab Eng Syst Saf 96(5):581–589
van der Duyn Schouten F, Vanneste S (1990) Analysis and computation of(n, n)-strategies for maintenance of a two component system. Eur J Oper Res 48:260–274
Van Noortwijk J (2009) A survey of the application of Gamma processes in maintenance. Reliab Eng Syst Saf 94:2–21
Vu H-C, Do P, Barros A, Berenguer C (2014) Maintenance grouping strategy for multi-component systems with dynamic contexts. Reliab Eng Syst Saf 132:233–249
Vu H-C, Do P, Barros A, Berenguer C (2015) Maintenance planning and dynamic grouping for multi-component systems with positive and negative economic dependencies. IMA J. Manag. Math. 23:145–170
Wildeman RE (1996) The art of grouping maintenance, PhD thesis. Derasmus University Rotterdam, Rotterdam
Wildeman R, Dekker R, Smit A (1997) A dynamic policy for grouping maintenance activities. Eur J Oper Res 99:530–551
Zhang N, Fouladirad M, Barros A (2017) Warranty analysis of a two-component system with type i stochastic dependence. Proc Inst Mech Eng O J Risk Reliab 232(3):274–283

Optimal Management of the Flow of Parts for Gas Turbines Maintenance by Reinforcement Learning and Artificial Neural Networks

Luca Bellani, Michele Compare, Piero Baraldi, and Enrico Zio

Symbols and Acronyms

A_k	Action taken at the k-th maintenance event performed
$a_{k,\rho}$	Boolean variable equal to 1 if action ρ is taken at the k-th maintenance event and 0 otherwise
CDF	Cumulative Distribution Function
$C^{rep}(r)$	Repair cost for a part with r maintenance cycles remaining
C^{scrap}	Cost of scrapping a part
$C^{failure}$	Failure penalty
C_k	Cost incurred at the k-th maintenance event
$d_g(t)$	$MNRC$ of the part on the g-th GT at time t
MNRC	Maximum Number of Remaining Cycles

L. Bellani
Aramis s.r.l., Milano, Italy
e-mail: luca.bellani@aramis3d.com

M. Compare (✉)
Aramis s.r.l., Milano, Italy

Department of Energy, Politecnico di Milano, Milano, Italy
e-mail: michele.compare@aramis3d.com; michele.compare@polimi.it

P. Baraldi
Department of Energy, Politecnico di Milano, Milano, Italy
e-mail: piero.baraldi@polimi.it

E. Zio
Dipartimento di Energia, Polytechnic University of Milan, Milano, Italy
e-mail: enrico.zio@polimi.it

A. T. de Almeida et al. (eds.), *Multicriteria and Optimization Models for Risk, Reliability, and Maintenance Decision Analysis*, International Series in Operations Research & Management Science 321, https://doi.org/10.1007/978-3-030-89647-8_20

G	Total number of GTs
g	GT Index
GT	Gas Turbine
MS	Maintenance Shutdown
ANN	Artificial Neural Network
FO	Forced Outage
MRC	Most Residual Cycles
N_{ei}	Total number of RL episodes
n_{ei}	Total number of RL initialization episodes
$Q_\pi(\mathbf{S}_k, A_k)$	State–Action pair value following policy π from the MS or the FO from k-th maintenance on
$Q_\pi(\mathbf{S}_k, A_k)$	State–Action pair value following policy π from the MS or the FO from k-th maintenance on
R	Maximum value of MRNC
r	MRNC index
RL	Reinforcement Learning
$\mathbf{S}_k$	State vector at k-th maintenance event, $\mathbf{S}_k = [S_{k,1}, \ldots, S_{k,R+G+2}]$
SDP	Sequential Decision Problem
H	Number of hours of a GT working cycle
T	Total number of working hours per GT
$\mathcal{T}$	Ordered set including the time instants at which shutdown events occur
$\mathcal{N}_\rho$	Neural network estimating $Q(\mathbf{S}, \rho), \forall \mathbf{S}$
V	Total value of the maintenance expenditures
DM	Decision Maker
$w_r(t)$	Number of parts with MRNC=r available at the warehouse at time t
$F_r(\tau)$	Cumulative Distribution Function of the failure time of a part with r remaining cycles
λ_r	Failure rate of a part with r remaining cycles
λ	Eligibility trace
$\mathbf{1}^{FO}(t)$	FO indicator function
$\mathbf{1}^{MS}(t)$	MS indicator function
K	Index of the last shutdown
δ_k	Index of the GT maintained at the k-th shutdown
α	Learning rate of the network
ϵ	Exploration rate of the algorithm
α_0	Initial learning rate of the ANN
ϵ_0	Initial exploration rate of the RL algorithm
PFM	Part Flow Management
$\hat{q}_\rho(\mathbf{S}_k)$	Estimation of $Q_\pi(\mathbf{S}_k, A_k = \rho)$ provided by ANN $\mathcal{N}_\rho$
$\boldsymbol{\mu}_\rho$	Weights of network $\mathcal{N}_\rho$
P	Number of available actions regarding the part to set on the GT

1 Introduction

Gas Turbines (GTs) are complex systems composed by several expensive capital parts (e.g., buckets, nozzles, shrouds, etc.), which are affected by different degradation processes, such as fracture and fatigue (Yang et al., 2017; Morini et al., 2010; Boyce and Ritchie, 2001), fouling (Tarabrin et al., 1998; Kurz and Brun, 2012; Peters and Ritchie, 2000), corrosion (Eliaz et al., 2002; Goward, 1998), oxidation (Compare et al., 2016). Part degradation can lead the GTs to failure and, thus, to costly Forced Outages (FOs) for performing corrective maintenance actions, in which the failed parts are scrapped and replaced by parts of the same type selected from those available at the warehouse.

Given the criticality of the GT degradation processes, their behaviors have been characterized through attentive engineering analyses, which have also yielded a set of rules determining the maintenance policy for each capital part. In particular, these rules define the interval between scheduled Maintenance Shutdowns (MSs) and impose that every part be scrapped after a prefixed number of working cycles of given duration, provided that it is repaired after each cycle. The reliability of the part decreases at each cycle, but the risk of failure remains acceptable up to the last cycle.

The repaired parts are put back at the warehouse with a reduced number of remaining working cycles, for replacing operating parts in future maintenance. The parts removed from the GTs are replaced by parts taken from the warehouse, either restored or newly purchased. Obviously, once installed on the GTs, the parts are no longer available for replacement of parts of GTs undergoing maintenance in the future (Fig. 1).

The parts repair actions involve both direct workshop costs and indirect costs related to the increased risk of FOs, with consequent penalties for business interruption. On the other hand, repairing the parts gives the possibility of re-using them, with consequent reduction in the number of parts to purchase over the GT operation time horizon. Yet, part repairs close to the end of the GT operation time horizon could lead to the warehouse containing parts ready for installation, but whose value is lost because they cannot be re-used. On the contrary, scrapping parts even if with some remaining cycles reduces the risk of failure and workshop costs, but increases the number of purchases of new parts over the GT operation time horizon.

From the above, it appears that the management of the maintenance (i.e., MSs and FOs) requires decisions on both the removed part (send it to the workshop for repair or scrap it?) and the part to be installed on the GT (new part or part taken from the warehouse?). This leads to the fact that the decisions at every maintenance influence the decisions at the future ones. Given this, the Part Flow Management (PFM) can be framed as a Sequential Decision Problem (SDP, Sutton et al. 1998), wherein a sequence of future maintenance decisions is sought (i.e., the optimal policy), which requires the smallest expected maintenance costs over the GT operation time horizon. This requires considering variables such as the remaining time up to the end of the plant operation, the availability of spares, the costs related to the repair actions, etc.

Despite the relevance of PFM for the safe profitability of GT operation, to the authors' best knowledge, systemic approaches to address it are still lacking.

Currently, PFM is dealt with experience-based rules, such as the Most Residual Cycles (MRC): The removed parts are always repaired and the part with the largest residual life among those available at the warehouse is installed on the GT; a new part is purchased only when the warehouse is empty. MRC ensures the smallest failure probability provided that a part from the warehouse is used. Nonetheless, it has been shown in Compare et al. (2019) that MRC does not necessarily yield optimal policies on a finite time horizon, in which the parts are assumed to not fail and, then, the sequence of MSs is a priori known.

In Compare et al. (2019), the authors have formalized the PFM problem as a SDP and proposed Reinforcement Learning (RL, Sutton et al. 1998; Szepesvári 2010; Kaelbling et al. 1996) for its solution. However, the optimization framework developed in Compare et al. (2019) does not account for the stochastic processes of part failures and the associated FOs, which change the pre-scheduled sequence of MSs and make the optimization problem a time-variant, finite-horizon problem (Werbos et al., 1990; Grondman et al., 2013; Bhatnagar and Abdulla, 2008).

In this work, we extend the framework in Compare et al. (2019) to find optimal PFM policies in the stochastic environment of failure processes. This requires including the real-valued time variable in the state vector: The same warehouse and GT part composition may lead to completely different optimal actions if they present themselves at different times, because of the finite-horizon conditions.

On the other hand, encoding a continuous variable in the state space makes the RL tabular approaches not practicable, as these would require a dense discretization of the time axis leading to the curse of dimensionality (Powell, 2007). To address this issue, we resort to action-value approximation through Artificial Neural Networks (ANNs, Bishop 2006; Haykin et al. 2009), which have been successfully applied in various fields (Tesauro, 1992; Crites and Barto, 1996; Mnih et al., 2013; Silver et al., 2016). The challenge of this approach is that we have to combine two learning processes: ANN learns how to approximate the action value function learned by RL through interactions with the simulated environment. To implement this learning process, we investigate two approaches, whose results are compared with each other and with those of MRC.

The structure of the chapter is as follows. In Sect. 2, we introduce the mathematical formulation of the considered SDP. In Sect. 3, details about the RL algorithms used for optimizing PFM are provided. In Sect. 4, the case study is discussed. Finally, conclusions are drawn in Sect. 5.

2 Problem Setting

Consider an Oil and Gas plant with a number G of GTs, which undergo scheduled maintenance every H units of time. The GTs are operated for T time each.

Every part is assigned a Maximum Number of Remaining Cycles (MNRC), indicated by r, which ranges between $r = 0$, in case of parts that must be scrapped and $r = R$, for new parts. The MNRC is reduced by one upon the installation of the part on a GT: If the GT is stopped, the part will no longer be able to re-perform the entire started cycle.

To formalize the PFM in a stochastic environment, we must consider that the decisions are taken on both scheduled MSs and upon FOs, which occur at random time instants. Namely, if any GT experiences a FO at a time τ after its installation, $\tau \in [0, H]$, then it is immediately repaired and all its future MSs are shifted by τ, as maintenance is always intended to allow the GT working continuously for H hours (Fig. 1). The failure times of the parts obey the exponential distribution with failure rate, λ_r, depending on the MNRC value $r \in \{1, \ldots, R\}$. This entails that the total number of maintenance events, K, is a random variable as well as the time of the events. The failure times of the parts obey the exponential distribution with failure rate, λ_r, depending on the MNRC value $r \in \{1, \ldots, R\}$. The cumulative distribution function (CDF) reads

$$F_r(\tau) = 1 - e^{-\lambda_r \tau} \tag{1}$$

Notice that the choice of describing the part failure behavior by the memory-less exponential distribution with failure rate depending on the MNRC value allows modeling the part degradation mechanism as a Markov process. The resulting step-wise, monotonously increasing behavior of the failure rate can be thought of as a rough approximation of a continuously increasing hazard rate (Zio, 2007).

We assume that the times to perform MSs and FOs, and the times to repair the parts removed from the GT are negligible. Then, the parts repaired are immediately available at the next event.

At any shutdown, the Decision Maker (DM) has to take the following decisions:

- If the maintenance event is a MS, decide whether to repair or scrap the part removed from the maintained GT. $C^{rep}(r)$ is the cost of repairing a part with $r \in \{1, \ldots, R\}$ remaining cycles, whereas C^{scrap} is the cost of scrapping.
- If the maintenance event is a FO, then the part must be scrapped, and a penalty $C^{failure}$ must be paid, which also encodes the extra-costs related to the management of an unplanned event.
- To replace the removed part, decide whether to buy a new part or select one from those available at the warehouse, if any. C^{pur} is the cost of purchasing a new part, whereas the cost of selecting a part from the warehouse is zero, as the repair costs have already been accounted for.

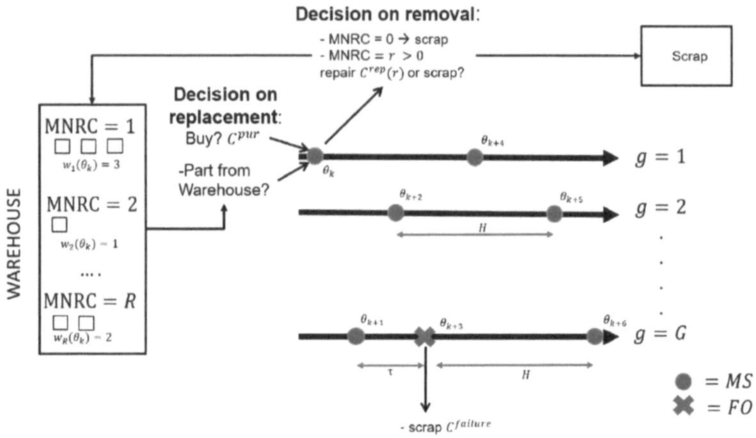

Fig. 1 GT maintenance sequence shift

To keep track of the shutdown temporal sequence, we introduce the ordered set $\mathscr{T} = \{\theta_1, \ldots, \theta_K\}$ including the time instants $\theta_k \in [0, T]$ at which shutdown events, FOs and MSs, occur. K is a random variable indicating the last shutdown within the operational time horizon T: If there are no failures, then K is the total number of scheduled MS on all GTs, which depends on T and H.

To simplify the notation, we define two indicator functions:

$$\mathbf{1}^{FO}(t) = \begin{cases} 1 & \text{if a FO occurs at time } t \\ 0 & \text{otherwise} \end{cases} \tag{2}$$

$$\mathbf{1}^{MS}(t) = \begin{cases} 1 & \text{if a MS occurs at time } t \\ 0 & \text{otherwise} \end{cases} \tag{3}$$

This way, time $t \in \mathscr{T}$ iff $\mathbf{1}^{MS}(t) + \mathbf{1}^{FO}(t) = 1$.

We also introduce the non-negative integer variables $d_g(t)$ and $w_r(t)$ to indicate the MNRC of the capital part on the g-th GT at time t, $g = 1, \ldots, G$, and the number of parts with MNRC equal to r available at the warehouse at time t, respectively, $t \in [0, T]$, $r \in \{1, \ldots, R\}$. Notice that $d_g(t) \in \{0, \ldots, R-1\}$, as the MNRC of a part immediately decreases upon installation. The MNRC of the GT maintained at the k-th shutdown, i.e., at time $t = \theta_k$, is traced by $\delta_k \in \{0, \ldots, R-1\}$.

We assume that there are P available alternative actions regarding the part to set on the GT: The first alternative $p = 1$ refers to the installation of a specifically purchased part, whereas the remaining $P - 1$ alternatives, i.e., $p \in \{2, \ldots, P\}$ refer to the installation of a part currently in the warehouse with MNRC in $\mathscr{P}_p \subseteq \{1, \ldots, R\}$, such that $\bigcup_{p=2}^{P} \mathscr{P}_p = \{1, \ldots, R\}$ and $\mathscr{P}_i \cap \mathscr{P}_j = \emptyset, i \neq j, i \in \{2, \ldots, P\}, j \in \{2, \ldots, P\}$. This entails $2 \leq P \leq R + 1$.

Finally, the Boolean variable $a_{k,\rho} \in \{0, 1\}$ indicates the action $\rho \in \{1, \ldots, 2 \cdot P\}$ taken at the k-th maintenance event at time $t = \theta_k$, $k \in \{1, \ldots, K\}$:

- $a_{k,1} = 1$ when a new part is purchased and installed and the removed part is scrapped.
- $a_{k,\rho} = 1, \rho \in 2, \ldots, P$, when the part with MNRC= $\max\{r \in \mathscr{P}_\rho : w_r(t = \theta_k) > 0\}$ is taken from the warehouse and installed on the GT, whereas the removed part is scrapped.
- $a_{k,P+1} = 1$ when a new part is purchased and installed, and the removed part is repaired.
- $a_{k,\rho} = 1, \rho \in P+2, \ldots, 2 \cdot P$, when the part with MNRC= $\max\{r \in \mathscr{P}_{\rho-P} : w_r(t = \theta_k) > 0\}$ is taken from the warehouse and installed and the removed part is repaired.

The Boolean variable $a_{k,\rho}$ is such that only one action can be taken at the k-th shutdown:

$$\sum_{\rho=1}^{P} a_{k,\rho} = 1 \tag{4}$$

The action taken at the shutdown occurring at time $t = \theta_k$ is

$$A_k = \sum_{\rho=1}^{2 \cdot P} (a_{k,\rho} \cdot \rho) \tag{5}$$

Notice that a part with MNRC $r \in \mathscr{P}_\rho$ can be taken from the warehouse only if it is available at the k-th event, i.e.,

$$a_{k,\rho} \leq \sum_{r \in \mathscr{P}_\rho} w_r(t = \theta_k), \quad \rho \in \{2, \ldots, P\} \tag{6}$$

$$a_{k,\rho} \leq \sum_{r \in \mathscr{P}_{\rho-P}} w_r(t = \theta_k), \quad \rho \in \{P+2, \ldots, 2 \cdot P\} \tag{7}$$

Similarly, a part cannot be repaired if the k-th event is a FO or the removed part has MNRC $r = 0$, i.e.,

$$a_{k,\rho} \leq d_g(t = \theta_k) \cdot \mathbf{1}^{MS}(t = \theta_k), \quad \rho \geq P+1 \tag{8}$$

From the above, the cost incurred at the k-th shutdown reads

$$C_k = (a_{k,1} + a_{k,P+1}) \cdot C^{pur} + \sum_{\rho=1}^{P} a_{k,\rho} \cdot C^{Scrap}$$

$$+\sum_{\rho=P+1}^{2\cdot P} a_{k,\rho}\cdot C^{Rep}(d_g(t=\theta_k))+C^{failure}\cdot \mathbf{1}^{FO}(t=\theta_k) \tag{9}$$

The objective function, to be maximized, reads

$$V=-\mathbb{E}\left[\sum_{k=1}^{K} C_k\right] \tag{10}$$

The value in Eq. (10) is the negative of the expected maintenance expenditures incurred in the entire GT operation time horizon, because RL is usually framed as a maximization.

Notice also that in real industrial applications, the failures of the capital parts mounted on the same GT may be dependent on each other. Nonetheless, here we track a single capital part, only. The extension to applications in which the flows of different capital parts are considered as a whole for a global optimization will be the object of future research work.

3 Algorithm

In this section, we provide some insights about the RL algorithm developed to address PFM, which requires the definition of the environment state, the actions available at each state, and the corresponding rewards (Sutton et al., 1998).

The state at the k-th maintenance event, i.e., at time $t=\theta_k$, is defined by vector $\mathbf{S}_k \in \mathbb{N}^{R+G+1}\times\mathbb{R}$, whose j-th element is

$$\mathbf{S}_{k,j}=\begin{cases} w_j(\theta_k) & \text{if} \quad j\in\{1,\dots,R\} \\ d_{j-R}(\theta_k) & \text{if} \quad j\in\{R+1,\dots,R+G\} \\ \delta_k & \text{if} \quad j=R+G+1 \\ \theta_k & \text{if} \quad j=R+G+2 \end{cases} \tag{11}$$

In words, the first R entries of the state vector define the number of parts available at the warehouse, with different MNRC values; the next G entries, from $R+1$ to $R+G$, indicate the MNRC of the parts installed on the GTs at their respective last MS; the $(R+G+1)$-th entry, δ_k, gives the MNRC of the GT maintained at time $t=\theta_k$; the last entry encodes the time of the shutdown for causing out the maintenance. Notice that:

- The definition of the environment state in Eqs. (11) does not fully satisfy the Markov property (Sutton et al., 1998), as the state vector does not include the time up to the next MS for any GT. This time interval determines the probability of moving from one state to another, as GT parts have higher chances of failing when operated for longer time periods. We have experimentally verified that omitting the information about the remaining time up to the next scheduled event does not significantly impact on the knowledge about the probabilistic behavior of the future evolution of the state, whereas it slightly affects the computational times of the RL.

- Embedding δ_k in the state definition may seem redundant, as the information on the RUL of the parts on the GTs is already encoded in $d_1(t = \theta_k), \ldots, d_G(t = \theta_k)$. However, the occurrence of FOs entails that it is not possible to infer which GT is maintained at θ_k, and the decision to repair the removed part strongly depends on its MNRC (e.g., if the removed part has MNRC$= 0$ it cannot be repaired).

In the tabular RL framework, each state–action pair is described by $Q_\pi(\mathbf{S}_k, A_k)$, which measures the expected return starting from state $\mathbf{S}_k$, taking action A_k and thereafter following policy π (Sutton et al., 1998; Sutton, 1995):

$$Q_\pi(\mathbf{S}_k, A_k) = \mathbb{E}_\pi \left[\sum_{k^*=k}^{K} (-C_{k^*}) | \mathbf{S}_{k^*}, A_{k^*} \right] \tag{12}$$

Given the large dimension of the action-state space including one continuous variable, we resort to action value approximation through ANNs (Bishop, 2006; Haykin et al., 2009). We estimate the value of $Q_\pi(\mathbf{S}_k, A_k)$ using a different network for each action, where each network takes as input the state vector (Eq. (11)) and returns as output the corresponding estimated value.

In detail, there are $2 \cdot P$ different ANNs, $\mathcal{N}_1, \ldots, \mathcal{N}_{2 \cdot P}$, with network weights $\boldsymbol{\mu}_1, \ldots, \boldsymbol{\mu}_{2 \cdot P}$, respectively. The output $\hat{q}_\rho(\mathbf{S}_k | \boldsymbol{\mu}_\rho)$ of network $\mathcal{N}_\rho$ at state $\mathbf{S}_k$ is an estimate of $Q_\pi(\mathbf{S}_k, A_k = \rho)$, $\forall \rho \in \{1, \ldots, 2 \cdot P\}$.

To build the networks with input–output relations in the proper range of values at the beginning of the RL algorithm, the network weights are initialized by a standard supervised training over n_{ei} runs (referred to as episodes) of the PFM, collected by using a pure random policy (i.e., actions are uniformly sampled from the set of applicable actions) and recording states, actions, and maintenance costs, i.e., $\sum_{k=1}^{K} (-C_k)$. This step has proven effective in speeding up convergence (Riedmiller, 2005).

Then, we rely on a procedure derived from the $n-$step gradient–semigradient SARSA algorithm (e.g., van Seijen 2016) to find the best approximation of $Q_\pi(\mathbf{S}_k, A_k)$.

After the first n_{ei} runs, we perform N_{ei} simulations of part flows. At each episode $ei \in \{1, \ldots, N_{ei}\}$, we select the ϵ-greedy action A_k and collect all the transitions $\mathbf{S}_k, A_k \rightarrow \mathbf{S}_{k+1}, A_{k+1}$ and corresponding rewards C_k, $k \in \{1, \ldots, K-1\}$. If $A_k = \rho$, the target value for network $\mathcal{N}_\rho$ at input state $\mathbf{S}_k$ is

$$Y_{k,A_k} = \lambda^{K-k} \cdot G_K + (1-\lambda) \sum_{k^*=k}^{K-1} [\lambda^{k^*-k} \cdot G_{k^*}] \tag{13}$$

where

$$G_{k^*} = \hat{q}_\rho(\mathbf{S}_{k^*+1}|\boldsymbol{\mu}_\rho) + \sum_{i=k}^{k^*}[-C_i], \quad \rho = A_{k^*+1} \tag{14}$$

and

$$G_K = \sum_{i=k}^{K}[-C_i] \tag{15}$$

According to Szepesvári (2010), we have applied a trial-and-error procedure to set the weighting factor $\lambda \in [0, 1]$ as linearly decreasing in $[0, 1]$, i.e., $\lambda = 1 - \frac{ei}{N_{ei}}$ (Hausknecht and Stone, 2016). At the beginning of the optimization, we set $\lambda = 1$, which entails unbiased estimates, as we are using all the collected rewards up to the end of the episode. This, however, results in a large variance of the estimates, because the gathered values of $Q_\pi(S_k, A_k)$ depend on both the stochastic nature of the failure process and the random actions due to the exploration strategy. On the contrary, $\lambda = 0$ entails that we are bootstrapping, i.e., using only the immediate reward and the estimate of the action-value function at the next maintenance event. Bootstrapping results in a smaller variance, due to the randomness in the immediate state and reward transition (i.e., $\mathbf{S}_k, A_k \rightarrow C_k, \mathbf{S}_{k+1}, A_{k+1}$), only, but it is biased, because it depends on the output values of the networks $\hat{q}_{A_{k+1}}(\mathbf{S}_{k+1}|\boldsymbol{\mu}_{A_{k+1}})$, which are only estimates of the true $Q_\pi(\mathbf{S}_{k+1}, A_{k+1})$. However, for values of ei approaching N_{ei}, the output values provided by the networks are quite accurate, especially on the most frequently visited states, whereby $\lambda = 0$ should increase stability.

To train the ANNs, we investigate two different strategies:

- "Incremental": At the end of every RL episode, each input–output value $(\mathbf{S}_k, Y_{k,A_k})$ is passed one-at-a-time to network $\mathcal{N}_\rho$, $\rho = A_k$. Then, we perform a single step of the back-propagation algorithm (Bishop, 2006; Haykin et al., 2009), which updates the network weights according to

$$\boldsymbol{\mu}_{A_k} \leftarrow \boldsymbol{\mu}_{A_k} + \alpha[Y_{k,A_k} - \hat{q}_{A_k}(\mathbf{S}_k|\boldsymbol{\mu}_{A_k})]\nabla\hat{q}_{A_k}(\mathbf{S}_k|\boldsymbol{\mu}_{A_k}) \tag{16}$$

  where $\alpha > 0$ is the learning rate at the $ei - th$ episode. $\alpha = \alpha_0 \cdot \frac{N_\alpha+1}{N_\alpha+ei}$, according to Sutton et al. (1998). Thus, we have to set two parameters: α_0 and N_α.
- "Sliding window": We consider a sliding window of length N_{win} for each network, namely $\mathcal{W}_1, \ldots, \mathcal{W}_{2 \cdot P}$. At each episode, the newly collected input–output values $(\mathbf{S}_k, Y_{k,A_k})$ are appended to the window corresponding to the selected action (i.e., in window $\mathcal{W}_{A_{k^*}}$); the oldest values are removed so that the window length is kept constant. Then, at the end of every RL episode, the network training is performed by providing each neural network $\mathcal{N}_\rho$ with all the data belonging to the corresponding window $\mathcal{W}_\rho$, under the usual techniques of

standard supervised learning. We have used the Levenberg–Marquardt algorithm (Moré, 1978) for weights updating. The "Sliding window" training approach is derived from the "neural fitted Q-iteration (NFQI)" algorithm (Szepesvári, 2010; Riedmiller, 2005)), which instead of considering a window of fixed length trains the networks offline upon the collection of a batch of RL episodes of fixed length. In principle, NFQI is faster than "Sliding window," because it requires to perform fewer training steps. Nonetheless, it suffers from the limitation that it is required to wait a longer number of episodes to change the values of $\hat{q}_\rho(\mathbf{S}_k|\boldsymbol{\mu}_\rho)$ and, thus, the policy. Moreover, we have experimentally verified that in our problem NFQI may lead to catastrophic forgetting (French, 1999): At every training step, the ANNs may completely forget the information learned on previous training batches. This is due to the finite-horizon of the task, which entails that changing the action performed at time θ_k affects all the actions to be taken thereafter. This leads to a large variability in the training batch.

The choice of using SARSA(λ) among the alternative RL algorithms (e.g., Q(λ), Szepesvári 2010) is justified by the fact that being SARSA an on-policy method, it guarantees a robust convergence when used with function approximation (Sutton, 1995, 2015; Tsitsiklis and Van Roy, 1997).

Notice also that the weighting factor λ is different from the failure rate, λ_r (i.e., with subscript), although we indicate them with the same letter. This is due to the large use of this letter in the respective fields.

Finally, notice that the decision of choosing an ANN for every action entails that the ANN updating depends also on the frequency at which the actions are taken. Alternatively, we can consider a single network for all the actions. In this alternative setting, the target value of the network remains Y_{k,A_k} for the action taken, whereas the target value for the other actions is the last available value of $\hat{q}_\rho(\mathbf{S}_k|\boldsymbol{\mu}_\rho)$. The structure of this single network must be sensibly larger, with larger computational times required for updating its weights. This impacts on the RL convergence time, especially for the "Incremental" learning. This setting will be investigated in future research.

4 Case Study

4.1 Case Study Description

In this section, we consider a case study derived from an industrial application. The main characteristics are summarized in Tables 1, 2, and 3.

In the considered Oil and Gas plant there are $G = 2$ GTs (first column in Table 1), each one maintained every $H = 24{,}000$ h (second column) over a time horizon of $T = 239{,}500$ h (third column). The maximum part MNRC, R, is set to 3 (fifth column in Table 1). The time of the first scheduled event on GT $g = 1$ is set to 0, whereas the first MS on GT $g = 2$ (if there are no FOs) is scheduled at

Table 1 General information on the PFM

G	H	T	R	First MS on GT 1	First MS on GT 2	P
2	24,000 h	239,500 h	3	0 h	12,000 h	2

Table 2 Maintenance costs (in arbitrary units) and failure rates of the capital parts

C^{Scrap}	$C^{rep}(r = 1)$	$C^{rep}(r = 2)$	C^{pur}	$C^{failure}$	$\lambda_{r=1}$	$\lambda_{r=2}$	$\lambda_{r=3}$
0	50	50	100	500	$2.5 \cdot 10^{-6}$ h^{-1}	$1.25 \cdot 10^{-6}$ h^{-1}	$4.17 \cdot 10^{-7}$ h^{-1}

Table 3 Initial scenario of the PFM

$w_1(0)$	$w_2(0)$	$w_3(0)$	$d_1(0)$	$d_2(0)$
3	1	0	2	0

Table 4 RL and neural network parameters

N_{ei}	n_{ei}	ϵ_0	N_ϵ	α_0	N_α	N_{win}
12,000	2000	0.3	900	0.01	900	1600

time 12,000 h (sixth and seventh columns in Table 1, respectively). The number of available actions is $2 \cdot P = 4$ (seventh column of Table 1), where $\mathscr{P}_2 = \{1, 2, 3\}$.

The cost values are reported, in arbitrary units, in the first five columns of Table 2, which also shows the failure rates $\lambda_r, r = 1, 2, 3$. These illustrative values entail a failure probability in a maintenance cycle H of 0.06, 0.03, and 0.01, respectively. Notice also that the failures we are referring to do not lead to the complete loss of the entire GT. Rather, we consider as failure the degradation of the functional performance to a level which requires the GT control system to command the stop of the GT for removing the degraded part. The major costs associated with this event are those related to business interruption and to the loss of the part, which is scrapped.

Finally, the first three columns of Table 3 report the number of parts with MNRC equal to 1, 2, and 3 initially available in the warehouse, whereas the MNRC values of the parts installed on GTs $g = 1$ and $g = 2$ are reported in the fourth and fifth column.

The RL parameters for both training approaches are reported in Table 4. The exploration parameter ϵ of the ϵ-greedy policy is $\epsilon = \epsilon_0 \cdot \frac{N_\epsilon+1}{N_\epsilon+ei}$; the learning rate α is updated according to $\alpha = \alpha_0 \cdot \frac{N_\alpha+1}{N_\alpha+ei}$.

Notice that with this setting of parameters, the optimal policy is required to factor in the risk of failure of GT parts, which has low probability and severe consequences. This is a challenging issue for sample-based algorithms (i.e., Monte Carlo (MC) simulation in Sutton et al. (1998)): Few examples are available in the literature of these RL algorithms applied to seek for policies encoding low probable events and they show that convergence may be slow and with large variance in the estimations (Frank et al., 2008).

Notice also that the effect of the actions on the failure events is not simple to learn: While the direct costs of purchase and repair actions are directly linked to the performed actions and immediately encoded into the reward, the costs due to failures of parts are difficult to assign to the actions, as part failure is due not only to the decision on what part is to be set on the GT, but also on repair decisions on the previous MSs: These affect the warehouse composition and, thus, the choice of the parts to set on the GTs at future events.

4.2 Results Discussion

To fairly compare the optimal policies found by RL with that provided by MRC, they have been MC simulated to estimate the average costs. For the RL policies, the episode simulation entails large computational burden, as it requires interaction with the trained networks. Then, the RL policies have been simulated for 20,000 episodes. The MRC policy, instead, has been simulated for 10^6 episodes, given that the computational times are negligible.

Tables 5 and 6 summarize the results of these test simulations. Namely, the first column in Table 5 shows the possible number of FOs occurring over the time horizon; for every number of FO, the three following columns report the mean total maintenance expenditures using MRC, RL "Incremental," and RL "Sliding window" policies, respectively. Table 6 has the same first column as Table 5, whereas the last three columns report the average portion of MC episodes with that number of FOs for MRC, RL "Incremental," and RL "Sliding window" policies, respectively. The last rows of Tables 5 and 6 report the average values and the 68% confidence intervals on the MC estimates, independently on the number of FOs.

From these Tables, we can see that the RL policies outperform MRC with respect to both the total expenditures and expected number of failures. Yet, the policy derived from the "Sliding window" training leads to a number of failures smaller than that of the "Incremental" training (0.614 vs 0.700): Although the policies found by the two training approaches yield similar average maintenance expenditures, however they are quite different, as it can be seen also by comparing the costs of the two policies in case of no FO (first row of Table 5). We can conclude that the policy provided by RL "Sliding window" entails a probability of having part flows without failures larger than those of RL "Incremental" and MRC.

Table 7 gives further insights about the main differences between the policies. We can see that RL "Incremental" is able to find a more profitable part flow policy in the case of no FOs, because it scraps three parts with $MNRC = 1$ (second row, fifth column). Indeed, MRC is not allowed to scrap. The number of purchase actions is the same for the MRC and RL "Incremental" policies, but that found by RL "Incremental" performs one repair less, with a total cost smaller of 50 (in arbitrary units). Notice that scrapping parts with $MNRC = 1$ also decreases the risk of FOs because fewer parts with larger failure rate are set on the GTs. This is the reason why the policy provided by RL "Incremental window" requires to perform 7 purchase actions, i.e., one more than the other two policies, and scrap 6 parts out of 8 with MNRC= 1. In this way, even if the maintenance expenditure is

Table 5 Comparison of maintenance expenditures of MRC, RL "Incremental," and RL "Sliding window"

Number of FOs	MRC	RL "Incremental"	RL "Sliding window"
0	1150	1100	1150
1	1687	1668	1712
2	2223	2230	2272
3	2761	2781	2824
4 or more	3379	3406	3438
Total Maintenance Costs	1549	1498 ± 3.5	1495 ± 2.8

Table 6 Comparison of the average number of failures of MRC, RL "Incremental," and RL "Sliding window"

Number of FOs	MRC	RL "Incremental"	RL "Sliding window"
0	0.47	0.498	0.537
1	0.360	0.342	0.338
2	0.133	0.126	0.103
3	0.031	0.029	0.019
4 or more	0.006	0.005	0.003
Expected Number of FOs	0.743	0.700	0.614

Table 7 Comparison between MRC and RL policies in case of no FO and RL in the deterministic environment

	Number of Purchasing	Repair of Parts with $r = 2$	Repair of Parts with $r = 1$	Scrap of Parts with $r > 0$	Scrap of Parts with $r = 0$
MRC	6	6	5	0	9
RL "Incremental"	6	6	4	3	7
RL "sliding window"	7	7	2	5	6

the same as that of the MRC policy, the probability of failure is smaller because at the MSs events newer parts are set on the GTs with respect to the other two policies.

4.3 Comparison of the Proposed Training Approaches

In this section, we compare the two RL algorithms. To do this, we must bear in mind that the difference in the two resulting policies does not depend exclusively on the different training approaches. Rather, the variability in the results is also due to the aleatory uncertainty in the failure process and the MC error of the sample-based procedure. In this respect, notice that roughly one half of the simulations experiences at least one failure, with consequent change in the MSs sequence.

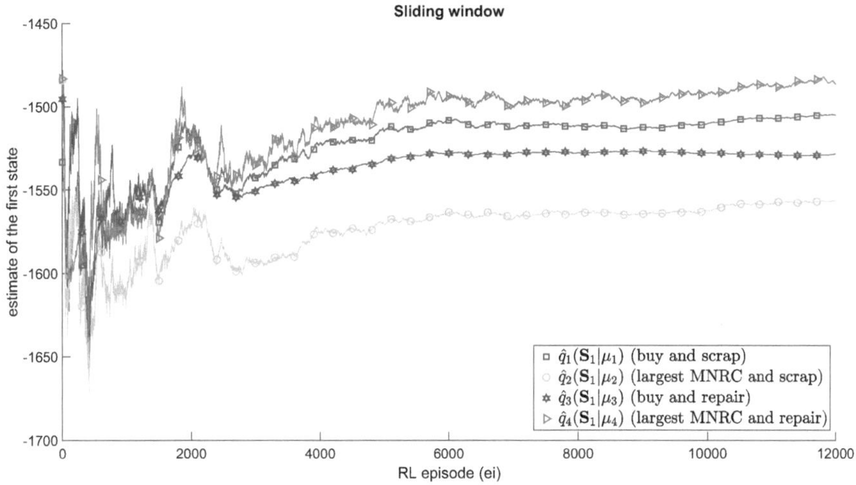

Fig. 2 $\hat{q}_\rho(\mathbf{S}_1)$, $\rho \in \{1, \dots, 4\}$. The learning rate parameters are set to $\alpha_0 = 0.01$ and $N_\alpha = 900$

Moreover, there are several solutions with similar values, which cannot be clearly distinguished by the action-value approximation framework we are dealing with.

To compare the two training strategies described in Sect. 3, the convergence paths of the four ANNs under the "Incremental" and "Sliding window" training approaches are reported in Figs. 2 and 3, respectively.

Both Figures show the values of $\hat{q}_\rho(\mathbf{S}_1)$ $\forall \rho \in \{1, \dots, 4\}$, where $\mathbf{S}_1$ is the state at the first MS (i.e., with warehouse and GT composition as described in Table 3 and maintenance performed at time $\theta_1 = 0$ on GT $g = 1$), versus the number of RL episode $ei \in \{1, \dots, N_{ei}\}$.

From Fig. 2, it appears that "Incremental" training leads all networks to become stable approximately at episode 4000: From that episode on, $\hat{q}_4(\mathbf{S}_1) > \hat{q}_\rho(\mathbf{S}_1)$, $\rho \in \{1, \dots, 3\}$ and action 4 is identified as optimal. Moreover, the value of $\hat{q}_4(\mathbf{S}_1)$ increases almost monotonically along the simulation, and the value of $\hat{q}_4(\mathbf{S}_1)$ at episode $ei = 12\,000$, i.e., at the end of the optimization, is 1486, quite close to the cost value obtained from MC simulations, i.e., 1498 (see Table 5). Notice also that the values $\hat{q}_\rho(\mathbf{S}_1)$ of the other networks are less accurate, because non-optimal actions are explored fewer times under the optimal policy, whereby the corresponding networks are not updated.

From Fig. 3, we can see that the networks trained with the "Sliding window" approach are much less stable than those derived from "Incremental" training. In fact, throughout the training path, all $\hat{q}_\rho(\mathbf{S}_1)$ range in $[-1900, -1450]$, with $\hat{q}_4(\mathbf{S}_1)$ changing its value of about -200 in the last 1000 episodes. Action 3, which here is considered optimal at the end of the algorithm, lies above the other curves only after episode 10,000. Moreover, there are a few episodes around $ei = 11\,500$, in which action 2 has a peak of $\hat{q}_2(\mathbf{S}_1) > \hat{q}_3(\mathbf{S}_1)$. In these episodes, the optimal action would be 2. As already pointed out, this instability of the network weights is mainly due

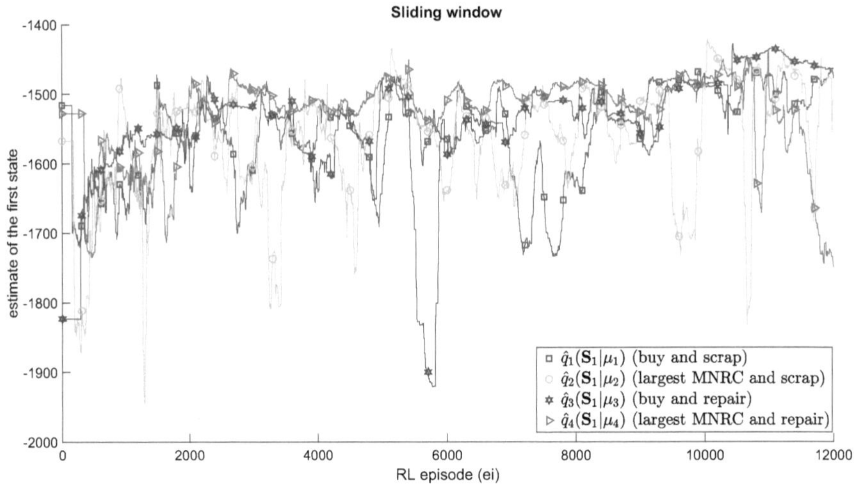

Fig. 3 $\hat{q}_\rho(\mathbf{S}_1)$, $\rho \in \{1, \ldots, 4\}$. The sliding window parameters are set to $N_{win} = 1600$

to the large variability of the episodes and may lead to sub-optimal policies. Notice also that as $\lambda \to 0$ (i.e., when $ei > 10{,}000$), $\hat{q}_3(\mathbf{S}_1)$ tends to get stable. However, the other networks fail to stabilize because they have few samples with input value $\mathbf{S}_1$: When $ei \simeq 12{,}000$, $\epsilon \simeq 2\%$, thus random exploration actions are taken with probability $\frac{0.2}{4} = 0.005$. Each network has a sliding window of length $N_{win} = 1600$ and at every RL episode at least 20 actions are taken: Very roughly, at every RL episode each network $\mathcal{N}_\rho$ collects on average $\frac{20}{4} = 5$ new samples $(\mathbf{S}_k, Y_{k,A_k})$ and, thus, the networks have a memory of about $\frac{1600}{5} = 320$ RL episodes. The first state is always visited, but the average number of samples state $\mathbf{S}_1$ for each non-greedy action is $320 \cdot 0.005 = 1.6$. If a failure occurs upon the non-greedy action ρ, then the maintenance expenditures are over-estimated and, correspondingly, their opposite $\hat{q}_\rho(\mathbf{S}_1)$ underestimates $Q_\pi(\mathbf{S}_1, \rho)$, because for input state $\mathbf{S}_1$ the training set records a value $Y_{1,\rho}$ which is smaller than the real one. The same reasoning applies to all the following states and this increases the instability of the network weights.

To sum up, it may seem that the "Sliding window" approach is worse than the "Incremental" one. However, both approaches have pros and cons, which are summarized as follows:

- The "Incremental" training is much more sensitive to the weights initialization than the "Sliding window" one. We have run several simulations using both approaches and we have experimentally verified that the "Sliding window" method always finds a near-optimal solution (i.e., with average maintenance cost value ranging in [1490, 1510]) independently from the initial weights. For example, Fig. 3 shows that the network corresponding to the optimal action 3 at the beginning of RL is not initialized well, as its estimate of the value of the first state $\hat{q}_3(\mathbf{S}_1)$ is 350 units smaller than the final one. On the other hand, RL with

"Incremental" training is more prone to converge to a sub-optimal policy if the network weights are not properly initialized.

- The "Incremental" approach requires a finer tuning of the training parameters. We have experimentally verified that:
  - The "Incremental" approach converges to the optimal solution in the considered case study for $\alpha_0 \in [0.008, 0.015]$ and $N_\alpha \in [500, 2\,000]$. Larger values of α_0 lead the networks to become unstable, whereas smaller values entail that the networks learn too slowly, with a strong dependence on the initial weights.
  - The "Sliding window" approach converges to a near optimal solution in the considered case study for $N_{win} > 1500$. The larger N_{win} is, the more stable the network weights are and $\hat{q}_\rho(\mathbf{S}_1)$ ranges in a smaller interval of values (Fig. 4 shows an example with $N_{win} = 4000$: $\forall\rho, \hat{q}_\rho(\mathbf{S}_1)$ ranges in $[-1650, -1450]$; there are no "negative peaks" in which $\hat{q}_\rho(\mathbf{S}_1)$ suddenly falls and the curves are generally smoother than those in Fig. 3). However, the computational burden increases with N_{win} and the networks do not become completely stable: In Fig. 4, the optimal action changes at episode $ei \simeq 11\,500$, i.e., nearly at the end of the simulation; $\hat{q}_2(\mathbf{S}_1)$ decreases of about -100 in the last 1000 episodes. To sum up, a trade-off analysis between stability of the network weights and computational effort of the algorithm is required to select the optimal value of N_{win}.
- The "Sliding window" approach is easier and more familiar for the practitioners of the classic supervised offline training approach.

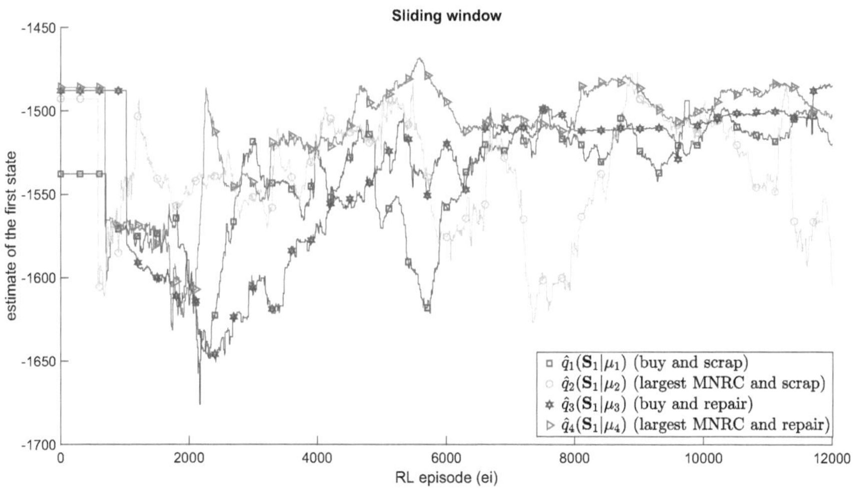

Fig. 4 $\hat{q}_\rho(\mathbf{S}_1)$, $\rho \in \{1, \ldots, 4\}$. The sliding window parameters are set to $N_{win} = 4\,000$

5 Conclusions

This work formalizes the GT part flow management in the Oil and Gas industry as a SDP considering also the stochastic processes of parts failure. RL with function approximation by ANNs is used as solving technique and two different training approaches are investigated, i.e., "Incremental" and "Sliding window." The results of a case study derived from a real industrial application show that RL with both approaches find efficient part flow policies, which increase the GTs reliability, as the expected number of FOs is decreased. The difference between the two training approaches lies in that the "Incremental" training is more stable but requires a finer tuning of the parameters and is more sensitive to the initial settings, whereas the "Sliding window" training is less stable, but converges to a near-optimal solution independently on the initial settings.

References

Bhatnagar S, Abdulla MS (2008) Simulation-based optimization algorithms for finite-horizon Markov decision processes. Simulation 84(12):577–600
Bishop CM (2006) Pattern recognition and machine learning. Springer, Berlin
Boyce B, Ritchie R (2001) Effect of load ratio and maximum stress intensity on the fatigue threshold in Ti–6Al–4V. Eng Fract Mech 68(2):129–147
Compare M, Martini F, Mattafirri S, Carlevaro F, Zio E (2016) Semi-markov model for the oxidation degradation mechanism in gas turbine nozzles. IEEE Trans Reliab 65(2):574–581
Compare M, Bellani L, Cobelli E, Zio E, Annunziata F, Carlevaro F, Sepe M (2019) A reinforcement learning approach to optimal part flow management for gas turbine maintenance. Proc Inst Mech Eng O J Risk Reliab, p 1748006X19869750
Crites RH, Barto AG, (1996) Improving elevator performance using reinforcement learning. In:Advances in neural information processing systems, pp 1017–1023
Eliaz N, Shemesh G, Latanision R (2002) Hot corrosion in gas turbine components. Eng Fail Anal 9(1):31–43
Frank J, Mannor S, Precup D (2008) Reinforcement learning in the presence of rare events. In: Proceedings of the 25th international conference on Machine learning, pp 336–343
French RM, (1999) Catastrophic forgetting in connectionist networks. Trends Cogn Sci 3(4):128–135
Goward G (1998) Progress in coatings for gas turbine airfoils. Surf Coat Technol 108:73–79
Grondman I, Xu H, Jagannathan S, Babuška R (2013) Solutions to finite horizon cost problems using actor-critic reinforcement learning. In: The 2013 international joint conference on neural networks (IJCNN). IEEE, New York, pp 1–7
Hausknecht M, Stone P (2016) On-policy vs. off-policy updates for deep reinforcement learning. In: Deep reinforcement learning: frontiers and challenges, IJCAI 2016 Workshop
Haykin SS et al. (2009) Neural networks and learning machines/Simon Haykin, 3rd ed.
Kaelbling LP, Littman ML, Moore AW (1996) Reinforcement learning: a survey. J Artif Intell Res 4:237–285
Kurz R, Brun K (2012) Fouling mechanisms in axial compressors. J Eng Gas Turbines Power 134(3):1–9
Mnih V, Kavukcuoglu K, Silver D, Graves A, Antonoglou I, Wierstra D, Riedmiller M (2013) Playing Atari with deep reinforcement learning. arXiv preprint arXiv:1312.5602

Moré JJ (1978) The Levenberg-Marquardt algorithm: implementation and theory. In: Numerical analysis. Springer, New York, pp 105–116
Morini M, Pinelli M, Spina PR, Venturini M (2010) Influence of blade deterioration on compressor and turbine performance. J Eng Gas Turbines Power 132(3):1–11
Peters J, Ritchie R (2000) Influence of foreign-object damage on crack initiation and early crack growth during high-cycle fatigue of Ti–6Al–4V. Eng Fract Mech 67(3):193–207
Powell WB (2007) Approximate Dynamic Programming: Solving the curses of dimensionality, vol 703. Wiley, New York
Riedmiller M (2005) Neural fitted q iteration–first experiences with a data efficient neural reinforcement learning method. In: European Conference on Machine Learning. Springer, New York, pp 317–328
Silver D, Huang A, Maddison CJ, Guez A, Sifre L, Van Den Driessche G, Schrittwieser J, Antonoglou I, Panneershelvam V, Lanctot M et al. (2016) Mastering the game of go with deep neural networks and tree search. Nature 529(7587):484
Sutton RS (1995) On the virtues of linear learning and trajectory distributions. In: Proceedings of the workshop on value function approximation, machine learning conference, p 85
Sutton R (2015) Introduction to reinforcement learning with function approximation. In: Tutorial at the conference on neural information processing systems, p 33
Sutton RS, Barto AG et al. (1998) Introduction to reinforcement learning, vol 135. MIT press, Cambridge
Szepesvári C (2010) Algorithms for reinforcement learning. Synth Lect Artif Intell Mach Learn 4(1):1–103
Tarabrin A, Schurovsky V, Bodrov A, Stalder J-P (1998) An analysis of axial compressor fouling and a blade cleaning method, pp 256–261
Tesauro G (1992) Practical issues in temporal difference learning. In: Advances in neural information processing systems, pp 259–266
Tsitsiklis JN, Van Roy B (1997) Analysis of temporal-difference learning with function approximation. In: Advances in neural information processing systems, pp 1075–1081
van Seijen H (2016) Effective multi-step temporal-difference learning for non-linear function approximation. arXiv preprint arXiv:1608.05151
Werbos PJ, Miller W, Sutton R (1990) A menu of designs for reinforcement learning over time. In: Neural Networks for control, vol. 3. MIT Press, Cambridge, pp 67–95
Yang K, He C, Huang Q, Huang ZY, Wang C, Wang Q, Liu YJ, Zhong B (2017) Very high cycle fatigue behaviors of a turbine engine blade alloy at various stress ratios. Int J Fatigue 99:35–43
Zio E (2007) An introduction to the basics of reliability and risk analysis, vol 13. World Scientific, Singapore

Joint Planning of Maintenance and Spare Parts Provision for Industrial Plant

Farhad Zahedi-Hosseini

1 Introduction

In the past few decades, many research papers have been published highlighting the importance of using maintenance analysis for the management of industrial plant (Wang 2012a). Specifically, researchers have concentrated on developing and using analytical models to help reduce equipment downtime and its associated costs including spare parts inventory. However, generally, these models use assumptions which are not easily justified in real-life situations. Equally, to relax some assumptions, we will make the models less suitable to be implemented in industry. Scarf (1997) is an "appeal to maintenance modellers to work with maintenance engineers and managers on real problems" since "too much attention is paid to the invention of new models, with little thought, it seems, as to their applicability". Two decades later, similar observations suggest that still not enough maintenance optimisation research is conducted which may be applicable in real industrial situations (Alrabghi et al. 2017).

Simulation is a useful and flexible modelling environment in which to tackle these important problems (Alrabghi and Tiwari 2015). Our view is that solutions obtained through simulation will bring models and theory closer to practice, not least because simulation tools are accessible to practitioners (Zahedi-Hosseini et al. 2017). Furthermore, since modern manufacturing systems have become more complex due to dependencies and interactions between system components, the use of simulation has grown dramatically over the past few decades (Gupta and Lawsirirat 2006). In maintenance modelling, since many policies are not analytically tractable, simulation is an appropriate tool which offers solutions over

F. Zahedi-Hosseini (✉)
University of Salford, School of Science, Engineering and Environment, Salford, UK
e-mail: f.zahedi@salford.ac.uk

A. T. de Almeida et al. (eds.), *Multicriteria and Optimization Models for Risk, Reliability, and Maintenance Decision Analysis*, International Series in Operations Research & Management Science 321, https://doi.org/10.1007/978-3-030-89647-8_21

analytical approaches (Nicolai and Dekker 2008). Thus, analytical models provide limited solutions to complex maintenance problems. In this chapter, simulation is used to jointly optimise preventive maintenance and spare parts provision for industrial plant under two different industrial configurations. For developing the simulation models (see, Harrell et al. 2011), a discrete-event simulation language known as *ProModel* (ProModel 2016) was used.

In this chapter, two case examples of joint maintenance inventory planning are discussed in detail. The maintenance and spare parts inventory control for a single machine is first considered, where demand for spare parts is driven by plant maintenance requirements. Several inventory policies are used to identify the most cost-effective replenishment policy. The aim of the second case example is to develop simulation models to jointly optimise preventive maintenance and spare parts provisioning for machines working in parallel. The objective is to minimise the occurrence of downtime in production systems where simultaneous machine downtime may halt production and consequently have a significant adverse effect on performance measures. This example highlights contexts for which analytical models cannot be developed due to the underlying difficulty in mathematical analysis and intractability. To compare diverse maintenance and inventory policies for both production configurations, the average cost per unit time, known as cost rate, under steady-state conditions, is used as the optimality criterion.

2 Industrial Context

For the industrial context, applicable to both case examples, imagine a plant with one or more failure modes, which have a maintenance policy of repairing failures as they arise and inspecting the plant critical part every T time units. The aim of the inspection is to identify and timely remove any defects before they cause downtime.

Clearly, the aim would be to minimise the plant operational downtime by reducing the effects of failures and inspection stoppages. Therefore, the decision variable for the maintenance is the optimal inspection interval, T. If a short interval is used for T, the percentage of time that the plant would potentially be operational will be reduced since there would be frequent inspection activities. Alternatively, if a large T is used, then one would not distinguish between this policy and running the plant under a breakdown maintenance regime.

In addition, the availability of spare parts will clearly affect maintenance costs. Keeping a large stock of spare parts will have financial implications for the organisation and the risk of spare parts' obsolescence. Conversely, keeping a small stock of parts might increase the risk of stock-outs, resulting in delays and increased downtime and higher costs for emergency expediting of spare parts.

Although scheduled inspection times are known, the times of demands for spare parts are unknown. Consequently, when, relative to inspection, and in what quantity spares should be ordered is the main question posed in this chapter.

Therefore, to guide the decision-making process for the optimal maintenance of industrial plant, it would be beneficial to use the simulation tool to determine the optimal period for T and establish a cost-effective policy for replenishing the inventory of spare parts.

The main modelling assumptions common in the majority of studies in the literature include the following: (i) perfect maintenance in maintaining identical and independent units; (ii) failures of parts are detected immediately; (iii) maintenance costs are constant, but the cost of preventive maintenance is always lower than corrective maintenance; (iv) duration of maintenance activities are constant or take zero time; and finally, maintenance resources are always immediately available when required. Considering the points listed above highlight the limitations of studies in the literature.

3 Maintenance Strategies

The primary purpose of maintenance optimisation is "to find an effective implementation of maintenance policies" (Zhang and Zeng 2017). The optimality criteria include minimising maintenance costs, reducing machine downtime, or maximising machine availability, to mention only a few examples. Many review papers in the literature give detailed description of the three principal strategies of corrective, preventive, and predictive maintenance (e.g. Van Horenbeek et al. 2013).

Under the corrective maintenance regime, when a component fails, provided spare parts are available, the failed part is replaced by a new one. If no spare is immediately available, downtime will normally occur until parts are replenished in emergency.

Alternatively, systems may be maintained under a preventive inspection maintenance regime where plant is inspected at regular intervals, with a view to identifying and replacing all defective (faulty) parts before they cause failures (e.g. Wang 2008). If the inspection interval is too short, then unnecessary inspection stoppages will add to downtime. Similarly, if the inspection interval is longer than necessary, then random failure of parts in service is increased, resulting in increased downtime. There are several factors that impact upon the determination of the optimum inspection interval: (i) the arrival time of defects; (ii) the rate of arrival; (iii) the delay-time, the time it takes for defects to cause failures; (iv) the frequency of inspections; (v) the cost of inspections; and finally (v) the cost of failures. Many methodologies including the delay-time modelling (DTM) (further described in the subsection below) have been established to determine the optimum inspection interval. DTM describes the failure process in industrial plant in two separate stages: (i) the time it takes, from new or as new, for a defect to arrive, and (ii) the *delay-time*, during which the defect fails. Alternatively, the age-based maintenance strategy, first developed by Barlow and Hunter (1960), may be used. Using this strategy, parts are replaced when they reach their predefined age. In comparison, under the block-based policy, all units are replaced at constant periodic intervals regardless of their age or condition. Under all strategies, whenever units fail, they are replaced provided spare parts are available.

Finally, under the predictive maintenance strategy, the system is monitored, and maintenance is carried out when some signals reach certain limits (e.g. Shafiee et al. 2015). This strategy aims at triggering the preventive maintenance action only when required (see, e.g. Olde Keizer et al. 2017).

In view of different maintenance strategies, downtime and labour costs need to be considered. For example, for the industrial context described in this chapter, bearings used extensively in a production plant can fail unexpectedly and catastrophically (Folger et al. 2014) which will need to be repaired or replaced. Clearly, in this situation, the labour and downtime costs will have a different cost element under failure and preventive replacements.

3.1 Delay-Time Modelling (DTM)

The model of inspection used in this chapter was first introduced by Christer (1976) and applied to an industrial maintenance problem by Christer and Waller (1984a). Since its conception, a few review papers on delay-time modelling and applications have been published (e.g. Wang 2012a).

Delay-time modelling, which appropriately lends itself to be used in industrial plant situation, describes the development of defects in two linked stages. The first stage, as illustrated in Fig. 1, is the time lapse from new, until the arrival of a defect — the *time-to-defect* arrival, u. The second stage, the *delay-time*, h, is the time during which the defect continuously deteriorates until it causes failure. During this latter stage, opportunities arise for inspection, identification of defects, and maintenance intervention before defects cause failures.

Therefore, by definition, the state of the plant is either good or defective before failure (Wang 2012a). The transition from the good to the defective state, which may only be observed by inspection, occurs at a random time and failure occurs some random time later. Baker and Wang (1992) describe the process of estimating the *time-to-defect* and *delay-time* stages, which establish the relationship between the number of failures and the inspection interval. The delay-time concept captures the relationship between failures of items in service, the frequency of inspections, and the identification and replacement of defective parts at inspections, provided spares are available.

The fundamental difference between DTM and other inspection strategies is that under the former, only defective items (if any) are replaced at inspection intervals, rather than block-based replacement, or replacement based purely on the age of the component part.

There are two distinct types of DTM systems which are used to model different industrial situations: (i) single-component or component tracking and (ii) multi-component or complex system. In a single-component system, as shown in Fig. 2, there will be a single dominant failure mode, and the system may be renewed upon

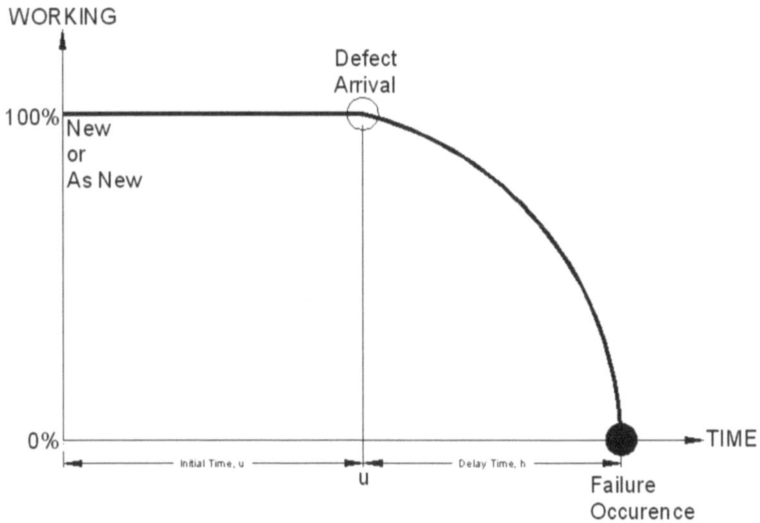

Fig. 1 The delay-time concept

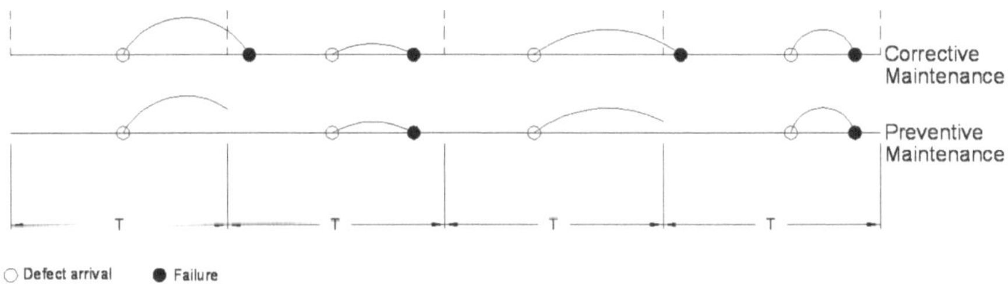

Fig. 2 Defect arrivals and failure occurrences in a single-component system under corrective and preventive maintenance strategies

failure (Wang 2008). For the instance shown in Fig. 2, inspection at the first and third epochs will identify and remove the defects and the system is thus renewed. However, before the second and fourth inspection epochs, component failures occur and the system is renewed again upon replacements. Examples of single-component systems are reported in Yang et al. (2016) and Baker and Wang (1992), for instance.

In contrast, a complex system is one in which many failure modes could arise, and the correction of one failure or the replacement of one defect will have nominal impact upon the overall plant failure characteristics or the steady state of the system. Figure 3(i) depicts an example of a complex system where six defects (1, 2, etc.) arrive over time. If regular inspection takes place, for example, at points A, B and C, then with the assumption of perfect inspection, some defects will be identified and removed before failures occur, as shown in Fig. 3(ii). Considering Fig. 3(ii) further, at inspection point A, two defects have already arrived and are currently in their *delay-times*. Thus, both defects 1 and 2 will be identified and removed at inspection

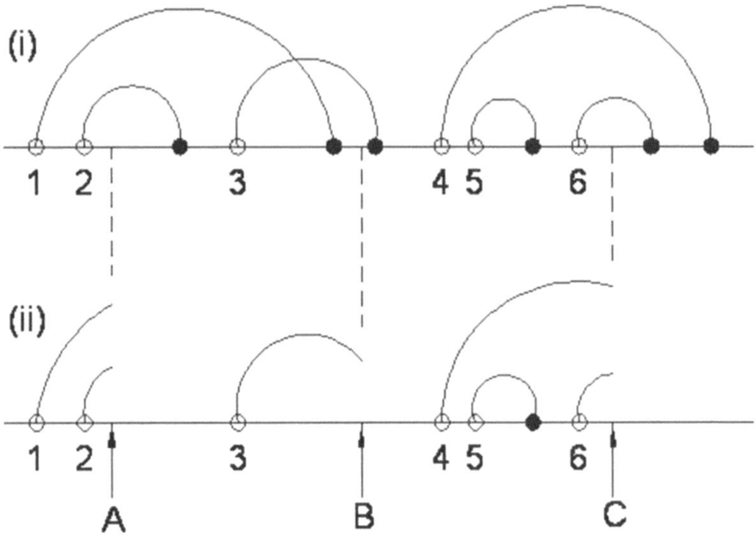

Fig. 3 Defect arrivals and failure occurrences in a complex system of multiple components

point A, either by replacing or repairing before failures occur. Defect 3 arrives in the middle of the period between scheduled inspections A and B and will be identified and removed at inspection point B. Before inspection C, one failure occurs as a result of defect 5. However the inspection at point C identifies and removes both defects 4 and 6 before they cause downtime. Therefore, in this instance, with a suitable length for the inspection interval, 5 out of 6 defects (83%) will be identified and removed. The system may thus be renewed at inspection points A, B, and C if the rate of arrival of defects is constant and the inspections are perfect. Most models in the literature are delay-time-based models of complex systems, and examples include Pietruczuk and Werbinska-Wojciechowska (2017); Lu and Wang (2011); Jones et al. (2010); Akbarov et al. (2008); Christer et al. (1995); and Christer and Waller (1984a, b).

There are also many delay-time-based case study applications reported in the literature. Some examples include Zahedi-Hosseini et al. (2018); Emovon et al. (2016); Liu et al. (2015); Jones et al. (2009, 2010); Akbarov et al. (2008); Arthur (2005); Pillay et al. (2001a, b); Christer et al. (1995); Baker and Wang (1992); Christer (1987); and Christer and Waller (1984a).

4 Inventory Control Strategies

There are two distinct approaches for the replenishment and management of spare parts (Muller 2011). Stock may be reviewed: (i) periodically or (ii) continuously (see, e.g. Santos and Bispo 2016 and Kennedy et al. 2002). Under the periodic review policy, parts may be replenished using the (R, S), (R, s, S), and (R, s, Q) policy (Silver et al. 2016) ~ periodically (R), raising the inventory position to level S, when the stock level drops below s, and by ordering a fixed quantity Q.

In comparison, under the continuous review policy, every time the stock level is depleted, the inventory levels are checked. Then, either a sufficient quantity, up-to-level S, is ordered if the inventory position reaches or drops below s ~ the (s, S) policy, or a fixed quantity of parts is ordered when the inventory position reaches or drops below s ~ the (s, Q) policy. When there is a per unit demand, both the (s, S) and (s, Q) policies give the same result when $Q = S - s$.

Under all policies, there are three costs which are traded off, namely, ordering, holding, and shortage costs. The ordering cost is fixed for the unit purchase of spares under normal circumstances or for ordering spares in emergencies. The holding cost will fluctuate due to capital and space cost implications. Finally, if there is insufficient number of spares available at the right time, shortage costs will be incurred which will add to downtime. The number of parts to keep in the store will depend on the frequency of component failure and the replenishment lead time. However, the cost of keeping inventory and the risk of spare part obsolescence must be considered too. All these costs will be balanced under different policies to produce an overall optimum cost. In joint optimisation problems, it is important to select maintenance and inventory policies which are suitable to the industrial situation.

5 Joint Optimisation for a Single Machine

The industrial context that was described in Sect. 2 is a typical industrial problem which is reported in the literature including Wang (2012b). To ensure that the developed simulation models and experiments in this chapter were realistic, 15 companies and 6 academics were surveyed to provide the information required for the costs and parameter values used in the models.

Therefore, simulation models were developed for jointly optimising the maintenance and inventory control provision for a paper mill where bearings are critical components, which deteriorate and fail randomly.

We suppose that the paper mill has many (typically over 100) identical bearings, where simultaneous defects may arise, and bearings fail randomly based on the two-stage delay-time concept described previously, in Sect. 3.1. The arrival of defects is exponentially distributed, which is consistent with the delay-time model of a complex system discussed in Wang (2012b), for example. Based on our survey, $\lambda = 0.05$ per week, and the *delay-time* has a Weibull distribution with $\alpha = 10$ and $\beta = 3$. When failures occur, provided spare parts are available, bearings are replaced, which takes 9 h for each replacement. While an individual machine is down due to bearing failure, the cost rate is £1,000 per hour.

Bearing condition is continuously monitored but periodically reported by external specialists at the beginning of each inspection epoch, by processing of the raw condition data. This means the inspection process incurs cost but no plant downtime. Therefore, all bearings are inspected in parallel, and defective bearings are replaced preventively, which takes 4 h for each bearing. It is logical to assume that during replacement, downtime defects do not arise or grow, and the bearings do not age or fail. The system is assumed to be operating under steady-state conditions.

The demand for the critical component is generated through failures of bearings in service and preventively replacing all defective bearings at scheduled inspections every T time units. Demand is satisfied from the inventory in the store provided there are enough spares or by ordering spares in an emergency. Based on the survey, lead time for ordering spares =3 weeks, and the lead time for ordering shortages in an emergency =1 day. Orders are placed and received at the beginning of each order and receipt day, respectively. However, orders arrive prior to reviewing the current inventory if it coincides with an order placing day. The order cost =£100, and the cost rate of holding inventory =1% of item cost per week. The purchase cost of one bearing is =£2,000, and the shipment cost of each bearing in an emergency is =£1,000.

In this chapter, several periodic review replenishment policies, and their variants, are compared as explained and discussed in detail in Sect. 4. T (the inspection interval) $= kR$ (the review priod) is set for $k > 0$ for policy variants. In all cases, the simulation models seek values of the decision variables, T and R, that minimise the long run overall expected cost per unit time. However, other decision variables depend on the exact inventory policy considered.

Three principal joint policies and one variant in each case were considered, namely, $(R, S, T = R)$; $(R, S, T = 2R)$; $(R, s, S, T = R)$; $(R, s, S, T = 2R)$; $(R, s, Q, T = R)$; and $(R, s, Q, T = 2R)$. While the analysis and discussion in this section is entirely focused on the cases $T = kR$ with $k = 1$ *and* 2, other values such as $k = 0.5$, 3, *and* 4 were also investigated. However, for the range of parameter values in our studies, other values of k were not found to be cost-optimal. For optimising our system, *ProModel*'s own optimisation tool, *SimRunner* (ProModel 2010), was integrated with the simulation models, which ensured that the optimal cost was achieved by running multiple combinations of certain variables. Although in practice several decision variables may be used as a focus in an optimisation study, in our study we have used the minimisation of the overall cost, which is most common in the optimisation of maintenance inventory problems (Van Horenbeek et al. 2010).

Table 1 illustrate that among all joint policies considered in this chapter, the $(R, S, T = 2R)$ policy has the lowest total cost per unit time ~ inspecting the bearings in the paper mill plant every 10 weeks and ordering spares every 5 weeks. The results also indicate that the $(R, s, S, T = 2R)$ policy is also cost-minimal with $T = 10$ and $R = 5$ since $S^* - s^* = 1$ (S^* and s^* are the optimum values of S and s, respectively).

Among all policies considered and their variants, the second and third lowest cost rates are also associated with the $(R, S, T = 2R)$ policy, inspecting every 11 and

Table 1 Cost rate and decision variables value for various joint policies

	$(R,S,T=R)$		$(R,s,S,T=R)$			$(R,s,Q,T=R)$			$(R,S,T=2R)$		$(R,s,S,T=2R)$			$(R,s,Q,T=2R)$		
T	£ cost rate	S	£ *cost* rate	s	S	£ cost rate	s	Q	£ cost rate	S	£ cost rate	s	S	£ cost rate	s	Q
5	698.07	3	698.07	2	3	704.70	2	2	704.72	3	704.72	2	3	709.52	2	2
6	660.35	3	660.35	2	3	664.08	2	2	651.64	2	653.44	2	3	654.22	1	2
7	640.27	3	640.27	2	3	644.57	2	2	629.64	3	629.64	2	3	636.09	1	2
8	624.79	3	624.79	2	3	624.12	2	2	612.04	3	612.04	2	3	618.88	1	2
9	611.81	3	611.81	2	3	612.00	2	2	600.53	3	600.53	2	3	601.08	1	2
10	612.64	4	612.64	2	3	612.48	2	2	***595.69***	**3**	***595.69***	**2**	**3**	601.45	2	2
11	614.21	4	614.21	2	3	616.04	2	2	596.38	3	596.38	2	3	603.15	2	2
12	616.87	4	616.87	2	3	620.06	2	2	603.55	3	603.55	2	3	611.00	2	2
13	622.11	4	622.11	2	3	627.82	2	2	606.88	3	606.88	2	3	613.53	2	2
14	627.84	4	627.84	2	3	635.90	2	2	610.97	3	610.97	2	3	615.81	2	2
15	633.85	4	633.85	2	3	638.71	2	2	618.62	3	618.62	2	3	622.85	2	2

Lowest cost rate for each specific policy. **Policy(s) with lowest cost rate overall**

9 weeks and ordering every 5.5 and 4.5 weeks, respectively. Considering the two lowest cost rate policies, the cost-optimal policy performs more frequent inspections at the cost of keeping more spares, resulting in the reduction of the overall cost. From the maintenance point of view, there seems to be a balance between the increased cost of inspections and the decreased cost of failures. Similarly, from an inventory control view point, the increased cost of keeping inventory is balanced against the possible reduction of stock-outs and its associated costs. In the maintenance inventory optimisation problem, less frequent ordering of higher spare quantities seems to potentially reduce the possibility of stock-outs under some joint policies.

For this kind of studies, sensitivity analysis is used in several publications in the literature, which "test the robustness of a suggested model by varying inputs and investigating if the results are in line with the expected outcome", Boulet et al. (2009), for example. Therefore, in our studies, for the cost-optimal policy $(R, S, T = 2R)$, sensitivity to various parameters such as defect arrival intensity, machine downtime, inspection frequency, and various costs was investigated. The effect of the defect arrival intensity on the overall cost rate is as expected since the cost rates for 0.5λ and 2λ are nearly halved and doubled, respectively. When the machine downtime cost rate is halved and doubled, the overall cost rate is reduced by 22% and increased by 41% with respect to the baseline, respectively. Furthermore, the optimal T shows the greatest impact when the inspection becomes infrequent and the cost is halved. Finally, varying the cost of inspection is as expected since the optimal T moves in the expected direction and the greatest effect is displayed when inspection is frequent. The sensitivity analysis suggests that the effect of the ordering cost, holding spares, or replenishment lead time on the cost rate is minimal.

In summary, in the context of the paper mill plant that was studied in this chapter, it is cost-optimal to order twice as frequently as to inspect, which increases planned costs and decreases unplanned costs due to a reduction in bearing failure and hence downtime.

6 Joint Optimisation for Parallel Machines

As in the previous case study (described in Sect. 5), the industrial context used here is also a paper mill with many bearings as the *critical* components in the plant. However, for this second case, we are considering a parallel production system comprising two identical paper rolling machines. It is assumed that bearings are the common inventory in the plant.

In this new system, we assume a complex system delay-time model for each machine. Defects arise based on a Poisson process, and bearings fail in service based on the delay-time model (Christer and Waller 1984b). At the beginning of each inspection epoch, both machines are inspected simultaneously since the monitoring of raw inspection data is analysed and reported by the external specialists every T time units. Defects that are identified are preventively replaced, which take place consecutively — Machine 1 and then Machine 2. This ensures there is no simultaneous stoppage of both machines, preventing simultaneous downtime. Furthermore, if a failure replacement coincides with a scheduled inspection, then bearing replacement, if required, would not commence until downtime due to failure replacement is ended. If downtime occurs due to two concurrent failures on both machines, or a failure on one machine while preventive replacement is taking place on the other machine, then a simultaneous downtime cost at a rate of £10,000 per hour is incurred. Figure 4 illustrates a flowchart depicting the process of capturing and recording simultaneous machine downtime. Preventive replacement of a bearing does not cause simultaneous downtime since a preventive replacement will be postponed until a failure replacement is complete. It is this kind of complexity that simulation modelling can tackle, and mathematical modelling is intractable. Apart from specific assumptions discussed in this section, all other general assumptions considered in Sect. 5 would also apply here.

As before, the survey provided the relevant data, which ensured the *input data* to the simulation models were based on realistic information. Therefore, the data used in this model is the same as the data used for the single-machine scenario in Sect. 5.

Since in the first case study, the (R, S) periodic review replenishment policy proved cost-optimal, the same policy and its variants are used so that a direct comparison may be made between the two industrial situations discussed in Sects. 5 and 6. Therefore, the joint maintenance inventory $(R, S, T = kR)$ policy was used for some positive, rational number k. As before, under such a policy, the inventory position is reviewed every R time unit, and, if required, an order is placed to bring the stock level up to level S. Two policy variants are considered here when T and R coincide, with $k = 1$ *and* 2, respectively. In both cases, the *inspection* process precedes the stock *review* activity, so that the number of spares used at inspection are considered before an order is placed, if required. Therefore, specific values of the three decision variables, R, T, and S, each with two degrees of freedom, that minimise the long-run total cost per unit time, or cost rate, are sought as the optimisation criteria.

The results displayed in Table 2 demonstrate that the $(R, S, T = 2R)$ policy with $T = 10$, $R = 5$, and $S = 3$ produces the lowest cost rate for inspecting, reviewing, and ordering bearings in the plant. Apart from k=1 and 2, other values were also investigated but were not cost-effective. The next best four policies are still the $(R, S, T = 2R)$ policy, inspecting bearings every 11, 9, 12, and 13 weeks, respectively. The simulation results for this model are consistent with the results obtained for the single-machine industrial situation discussed in Sect. 5, which imply that performing multiple stock reviews of critical components in the plant, between inspections, is cost-effective.

Again, sensitivity analysis of the optimal policy to various parameter variation was performed. It was found that the cost-optimal policy is most sensitive to the defect arrival intensity and α, the scale parameter of the Weibull *delay-time* distribution. Clearly, the former parameter affects the maintenance costs, and the latter regulates the failure intensity and the number of defects identified at inspections. The cost-optimal policy is less sensitive to the parameter values of

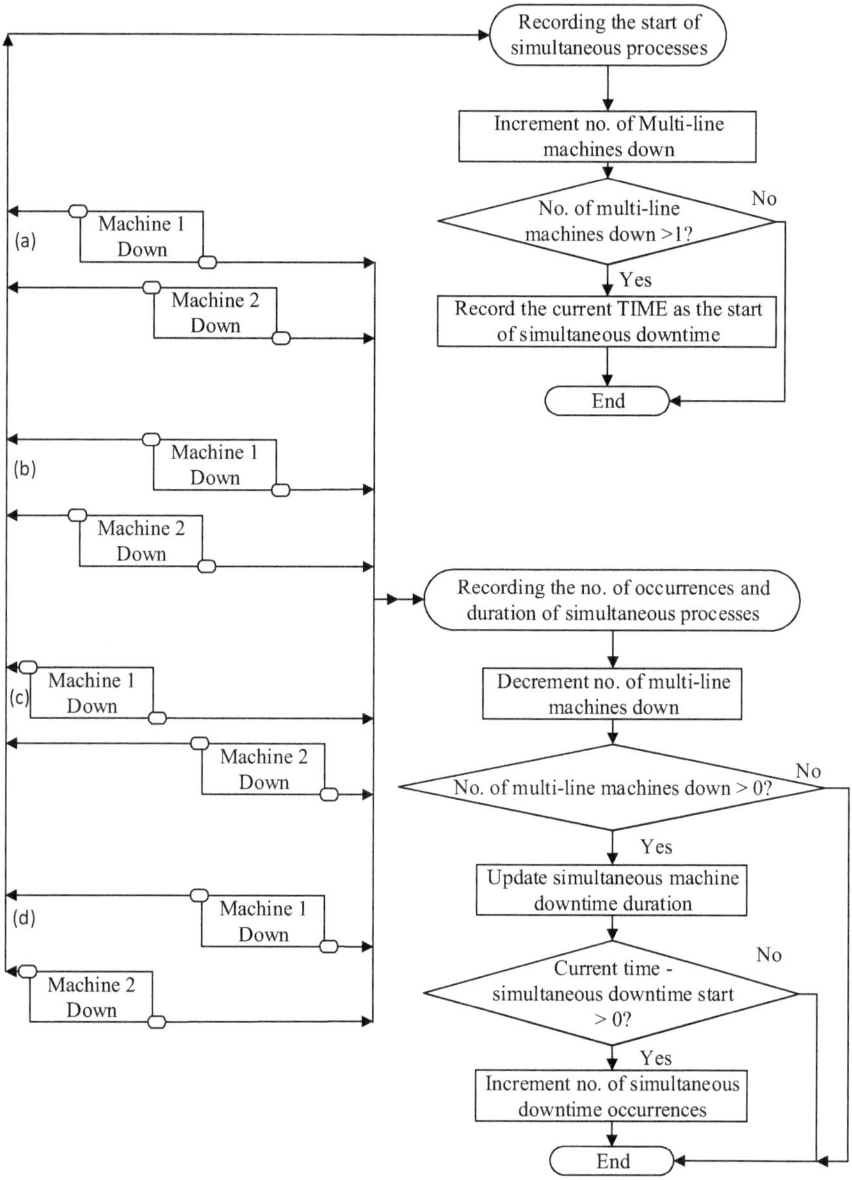

Fig. 4 Flowchart depicting the logic used in capturing and recording simultaneous machine downtime occurrences

spare unit cost, normal lead time, shortage lead time, and simultaneous machine downtime cost rate, since the latter two are rare under maintenance inventory joint optimisation.

Table 2 Cost rate and decision variable value for two policies

	(R,S,T = R)		*(R,S,T = 2R)*	
T	*£ cost rate*	*S*	*£ cost rate*	*S*
6	1269.58	3	1256.21	2
7	1220.76	3	1206.23	3
8	1192.68	4	1176.37	3
9	1175.31	4	1158.81	3
10	1174.59	4	***1149.23***	***3***
11	1176.59	4	1151.84	3
12	1183.63	4	1161.36	3
13	1190.48	4	1164.50	3
14	1203.50	4	1181.91	3

Lowest cost rate for each specific policy.

Policy with lowest cost rate overall

7 Conclusions

Several simulation models were developed for jointly optimising the maintenance and spare parts inventory for a single-machine scenario compared to a parallel production facility. In the latter case, it is assumed that simultaneous downtime of parallel machines incurs significant cost to the organisation. Simulation models were specifically developed for an industrial situation comprising paper rolling machines in a paper mill. The aim was to jointly optimise the planned maintenance inspection interval *T*, and the review period *R*, for maintaining a single stock keeping unit – bearings. *SimRunner*, *ProModel*'s optimisation tool, was used to find the optimal policy in each case. Without a modelling tool, it will be unclear when inspections should be performed, and when spares should be ordered – and in what quantity, in a given context.

The results show that under both industrial situations, it is cost-efficient to place multiple orders between inspection cycles. Among various parameters, the emergency shipment cost and the defect arrival rate have the least and most impact on the cost-optimal policies, respectively. The sensitivity analysis shows that results are broadly in line with expectations, which, in part, validates the output results from the simulation models.

The joint optimisation work discussed in this chapter may be extended in several directions. The assumption of "perfect inspection", which assumes the identification and removal of all defective components present in the system at the time of inspection, may be relaxed. However, to model a more realistic industrial scenario will need access to reliable data to make the experimentation meaningful. Models are also based on the assumption of immediate replacement of all defective components, identified through inspection, provided spares are available. In real industrial situations, the replacement of some or all of the components may be delayed until the next replacement cycle, which may be more cost-effective, especially if spare parts are not immediately available. If the item is in a minor defective state and the spare part is not available, one can wait and postpone replacement rather than rushing into an emergency replenishment.

The simulation models in this chapter are developed for specific industrial situations and production configurations. Unlike analytical models, simulation-based models require computation time for experimentation, which will inevitably take some time to produce results. Moreover, discrete-event simulation (by its nature), together with an optimisation tool, will not necessarily produce an exact optimum solution because the search space is not continuous.

References

Akbarov A, Wang W, Christer AH (2008) Problem identification in maintenance modelling: a case study. Int J Product Res 46:1031–1046

Alrabghi A, Tiwari A (2015) State of the art in simulation-based optimisation for maintenance systems. Comput Ind Eng 82:167–182

Alrabghi A, Tiwari A, Savill M (2017) Simulation-based optimisation of maintenance systems: industrial case studies. J Manufact Syst 44:191–206

Arthur N (2005) Optimization of vibration analysis inspection intervals for an offshore oil and gas water injection pumping system. Proc Instit Mech Eng Part E J Process Mech Eng 219:251–259

Baker RD, Wang W (1992) Estimating the delay-time distribution of faults in repairable machinery from failure data. IMA J Mathemat Appl Business Ind 3:259–281

Barlow R, Hunter L (1960) Optimum preventive maintenance policies. Operat Res 8:90–100

Boulet JF, Gharbi A, Kenn JP (2009) Multiobjective optimization in an unreliable failure-prone manufacturing system. J Quality Maint Eng 15:397–411

Christer AH (1976) Innovative Decision Making. In: Bowen KC, White DJ (eds) Proceedings of the NATO conference on the role and effectiveness of theories of decision in practice. Hodder and Stoughton, pp 368–377

Christer AH (1987) Delay-time model of reliability of equipment subject to inspection monitoring. J Operat Res Soc 38:329–334

Christer AH, Waller WM (1984a) Delay time models of industrial inspection maintenance problems. J Operat Res Soc 35:401–406

Christer AH, Waller WM (1984b) Reducing production downtime using delay-time analysis. J Operat Res Soc 35:499–512

Christer AH, Wang W, Baker RD, Sharp JM (1995) Modelling maintenance practice of production plant using the delay-time concept. IMA J Manag Mathemat 6:67–83

Emovon I, Norman RA, Murphy AJ (2016) An integration of multi-criteria decision making techniques with a delay time model for determination of inspection intervals for marine machinery systems. Appl Ocean Res 59:65–82

Folger R, Rodes J, Novak D (2014) Bearing killer: preventing common causes of bearing system damage – Part 1. Maint Eng 14:12–15

Gupta A, Lawsirirat C (2006) Strategically optimum maintenance of monitoring-enabled multi-component systems using continuous-time jump deterioration models. J Quality Maint Eng 12:306–329

Harrell C, Ghosh BK, Bowden RO (2011) Simulation using ProModel, 3rd edn. McGraw Hill

Jones B, Jenkinson I, Wang J (2009) Methodology of using delay-time analysis for a manufacturing industry. Reliability Eng Syst Safety 94:111–124

Jones B, Jenkinson I, Yang Z, Wang J (2010) The use of Bayesian network modelling for maintenance planning in a manufacturing industry. Reliability Eng Syst Safety 95:267–277

Kennedy WJ, Patterson JW, Fredendall LD (2002) An overview of recent literature on spare parts inventories. Int J Product Econ 76:201–215
Liu X, Wang W, Peng R (2015) An integrated production and delay-time based preventive maintenance planning model for a multi-product production system. J Maint Reliability 17:215–221
Lu WY, Wang W (2011) Modelling preventive maintenance based on the delay time concept in the context of a case study. J Maint Reliability 3:4–10
Muller M (2011) Essentials of inventory management. 2nd, AMACOM
Nicolai RP, Dekker R (2008) Optimal maintenance of multi-component systems: a review. In: Kobbacy KAH, Murthy DNP (eds) Complex system maintenance handbook. Springer, London, pp 263–286
Olde Keizer MCA, Flapper SDP, Teunter RH (2017) Condition-based maintenance policies for systems with multiple dependent components: a review. Eur J Operat Res 261:405–420
Pietruczuk AJ, Werbińska-Wojciechowska S (2017) Block inspection policy model with imperfect maintenance for single-unit systems. Proc Eng 187:570–581
Pillay A, Wang J, Wall AD, Ruxton T (2001a) A maintenance study of fishing vessel equipment using delay-time analysis. J Quality Maint Eng 7:118–128
Pillay A, Wang J, Wall AD (2001b) Optimal inspection period for fishing vessel equipment: a cost and downtime model using delay time analysis. Marine Technol 38:122–129
ProModel (2010) SimRunner user guide. ProModel Corporation
ProModel (2016) ProModel user guide. ProModel Corporation
Santos AF, Bispo CF (2016) Simulation based optimisation package for periodic review inventory control. In: Roeder TMK, Szechtman R, Zhou E, Huschka T, Chick SE (eds) Proceedings of the winter simulation conference
Scarf PA (1997) On the application of mathematical models in maintenance. Eur J Operat Res 99:493–506
Shafiee M, Finkelstein M, Bérenguer C (2015) An opportunistic condition-based maintenance policy for offshore wind turbine blades subjected to degradation and environmental shocks. Reliability Eng Syst Safety 142:463–471
Silver EA, Pyke DF, Thomas DJ (2016) Inventory management and production planning and scheduling, 3rd edn. Wiley
Van Horenbeek A, Pintelon L, Muchiri P (2010) Maintenance optimization models and criteria. Int J Syst Assurance Eng Manag 1:189–200
Van Horenbeek A, Buré J, Cattrysse D, Pintelon L, Van Steenwegen P (2013) Joint maintenance and inventory optimization systems: a review. Int J Product Econ 143:499–508
Wang W (2008) Delay time modelling. In: Kobbacy KAH, Murthy DNP (eds) Complex system maintenance handbook. Springer, London, pp 345–370
Wang W (2012a) An overview of the recent advances in delay-time-based maintenance modelling. Reliability Eng Syst Safety 106:165–178
Wang W (2012b) A stochastic model for joint spare parts inventory and planned maintenance optimisation. Eur J Operat Res 216:127–139
Yang L, Ma X, Zhai Q, Zhao Y (2016) A delay time model for a mission-based system subject to periodic and random inspection and postponed replacement. Reliability Eng Syst Safety 150:96–104
Zahedi-Hosseini F, Scarf PA, Syntetos AA (2017) Joint optimisation of inspection maintenance and spare parts provisioning: a comparative study of inventory policies using simulation and survey data. Reliability Eng Syst Safety 168:306–316. https://doi.org/10.1016/j.ress.2017.03.007
Zahedi-Hosseini F, Syntetos AA, Scarf PA (2018) Optimisation of inspection policy for multi-line production systems. Eur J Ind Eng 12:233–251
Zhang X, Zeng J (2017) Joint optimization of condition-based opportunistic maintenance and spare parts provisioning policy in multiunit systems. Eur J Operat Res 262:479–498

Some New Advances in Modeling for Performance-Based Maintenance Services

Tongdan Jin, Yisha Xiang, Jin Qin, and Vinod Subramanyam

1 Introduction

Product support and maintenance post the installation are becoming more important as the global economy continues to shift to a service-oriented business paradigm. For example, with the expansion of the commercial aviation industry, the maintenance, repair, and overhaul (MRO) market that supports it is expected to grow. Total MRO spending worldwide is expected to rise to \$116 billion by 2029, up from \$81.9 billion in 2019 (Wyman 2019). In defense industry, the annual operation and sustainment cost for US military equipment alone is \$63 billion, and the MRO activities are supported by 678,000 DoD personnel along with hundreds of private contractors (Smith 2007). According to a study by Accenture (Dennis and Kambil 2003) after-sales services including part supply contribute only 25% of revenues across all manufacturing companies but are responsible for 40–50% of profit stream. Many firms have begun to recognize the importance of the after-sales services and start to offer extend warranty and repair contracts. In particular, some commercial companies have a heavy reliance on MRO that is treated as a key enabler for their business success (e.g., Southwest Airlines, JetBlue, Jaguar, Rolls-Royce, and Pratt & Whitney). This is due to the fact that the availability of their products and services directly depend on the success of MRO program. When new products are expected

T. Jin (✉) · V. Subramanyam
Ingram School of Engineering, Texas State University, San Marcos, TX, USA
e-mail: tj17@txstate.edu

Y. Xiang
Industrial, Manufacturing and Systems Engineering, Texas Tech University, Lubbock, TX, USA

J. Qin
School of Management, The University of Science and Technology of China, Anhui, Hefei, China

A. T. de Almeida et al. (eds.), *Multicriteria and Optimization Models for Risk, Reliability, and Maintenance Decision Analysis*, International Series in Operations Research & Management Science 321, https://doi.org/10.1007/978-3-030-89647-8_22

to be released with higher reliability and longer service time, it is anticipated that MRO services continue its growth, generate more profits, and become more relevant. Therefore, rigorous planning models and implementation strategies are increasingly important as the complexity and integration of new products continue to increase.

Traditional MRO services are performed under the so-called material-based contracts (MBC). Namely, service providers are paid each time a maintenance task is accomplished. While some industries are more sensitive to operational costs, achieving performance outcomes are the top priorities in many other industries. For example, in capital-intensive industries, such as aerospace and defense sectors, key performance outcomes including system readiness, equipment reliability, and mission success rate are of great importance (Nowicki et al. 2008). Though cost is always important, these customers are relatively less sensitive to the maintenance expense than the system uptime. In these performance-oriented industries, a new support contracting mechanism, performance-based maintenance (PBM), has emerged and received much attention in recent years. In the US military, PBM is also called "performance-based logistics (PBL)" (DoD 2005); and in commercial airlines, it is referred to as "power by the hour." One of the earliest PBL programs dates back to 1998 when Lockheed Martin offered a maintenance solution to the US DoD for supporting F-117 Nighthawk, which tied its compensation to the fighter's performance outcome (Mirzahosseinian and Piplani 2011). During the past decade, PBL has been popularized in the after-sales market for servicing capital-intensive goods in both private and public sections, such as Lockheed Martin, Rolls-Royce, General Electric, and Boeing. A more general concept that encompasses both PBM and PBL is called performance-based contracting (PBC). It is a product support strategy used to achieve certain measurable performance goals. The primary means of accomplishing this are through incentivized, long-term service contracts with mutually agreed and measurable levels of performance outcomes defined by the customer and agreed on by contractors. Besides the defense and airline sectors, applications of PBC have been also reported in healthcare delivery, power industry, and transportation industry. For instance, PBC is adopted in highway pavement maintenance and railroad track maintenance (Anastasopoulos et al. 2010; D'Angelo et al. 2018).

Pertaining to PBM in system support and operation, Jing and Tang (2017) suggest that expected backorders, average parts availability, and average waiting time are the preferred key performance measures perceived by the service provider. Under these criteria, a probabilistic constrained inventory minimization model is further formulated in a multi-echelon, multi-system spare parts setting. Glas and Kleemann (2017) analyze 21 business cases and conclude that the success of PBC relies on clear responsibilities, quantifiable performance indicators, transparent measurement, cooperative culture, and a precise utilization profile of core assets. Hur et al. (2018) propose an inventory optimization algorithm to find the spares requirement of aircraft components during the end-of-life period with aircraft availability as a performance metric under the PBL contracting. Comprehensive reviews on this research stream have been made by Selviaridis and Wynstra (2015) and Glas et al. (2018).

The primary goal of PBM is to reduce the system's ownership cost while ensuring the reliability and availability during its useful lifetime. Therefore, PBM is a holistic maintenance solution, and differs from the traditional MBC in that the customer buys the system performance instead of paying the spare parts and repair labors transacted in a maintenance process. Specifically, the users of the equipment define the performance goal and sign the maintenance contract with the service provider who is committed to the attainment of the performance goal. Meanwhile, the service provider is incentivized to meet or exceed the performance goal in order to reap a higher service revenue. PBM potentially motivates the service provider to design reliability into product's early development and manufacturing stages, because the efforts will be paid off by reduced maintenance and repair costs in field use. PBM is a key enabler to the implementation and operation of PBL contract that gains growing popularity in the US defense sector. Studies show that aircraft reliability and operational availability have increased by 15–20% in the US military, showing the effectiveness of PBM programs (Garvey 2005; Kratz 2005).

Despite the success of PBM in defense sector, its theoretical studies and potential applications in other industries still remain in an early stage. Recently, there is a growing stream of literature exploring the design and implementation of PBM program in a broad application scope (Nowicki et al. 2008; Kang and McDonald 2010; Randall et al. 2010; Jin and Wang 2012; Xiang et al. 2017; Qiu et al. 2017). For instance, Nowicki et al. (2008) define two incentive payment models based on linear and exponential functions. Kang and McDonald (2010) use design of experiments to identify critical logistics factors that impact the readiness and life cycle cost of light armored vehicles. In a recent work by Qiu et al. (2017), their model considers maintenance error, and the system is replaced upon a hard failure or preventively replaced if the cumulative soft failure exceeds a predefined threshold. The objective of their model is to maximize the expected net revenue of the supplier operating under a PBC.

In general, these studies attempt to address two fundamental questions: (1) what are the key drivers behind the system reliability and availability under the PBM contracting? and (2) how do they interact with each other and jointly influence the decisions made by the customer and the supplier? This chapter reviews the recent findings and further present two PBM contracting models in terms of maximizing the service profit and minimizing the life cycle cost, respectively.

The remainder of the chapter is organized as follows. Section 2 presents a four-step procedure of designing and managing a PBM contract. Section 3 defines five key performance measures and further explores the interactions between operational availability and its underlying drivers. In Sect. 4, we investigate a PBM contract model for maximizing the service profit. Section 5 compares the PBM contracts when the goal is to maximize the service profit, minimize the cost rate, and maximize the availability. In Sect. 6, some discussions about the future research are provided. Section 7 concludes the chapter.

Abbreviation	Comment
AC	Alternating current
CBM	Condition-based maintenance
CM	Corrective maintenance
CPAF	Cost plus award fee
CPFF	Cost plus fixed fee
CPIF	Cost plus incentive fee
CUU	Cost per unit usage
DC	Direct current
FMECA	Failure mode, effects, and criticality analysis
IG	Inverse Gaussian
LCC	Life cycle cost
LF	Logistics footprint
LRT	Logistics response time
LRU	Line replaceable unit
MBC	Material-based contract
MLDT	Mean logistics delay time
MR	Mission reliability
MRO	Maintenance, repair, and overhaul
MTBF	Mean time between failures
MTTR	Mean time to repair
OA	Operational availability
OEM	Original equipment manufacturer
PBC	Performance-based contract
PBL	Performance-based logistics
PBM	Performance-based maintenance
PM	Preventive maintenance
RFRW	Free replacement warranty
TAT	Turnaround time
WT	Wind turbine

2 Overview of Performance-Based Maintenance

2.1 The Evolution of Maintenance Strategy

Maintenance policy can be classified into three categories depending on whether the action is triggered by the time/usage, the health condition, or the incentives. Corrective maintenance (CM), also known as "run-to-failure maintenance," is triggered by the time as it is executed only if the system fails randomly. Preventive maintenance (PM) is a schedule-based process that is performed based on a predefined time, cumulated usage, or degradation level. For instance, block replacement is a type of PM in which all items are replaced at a predefined time regardless of the actual age. Condition-based maintenance (CBM) also belongs to the PM category, yet the

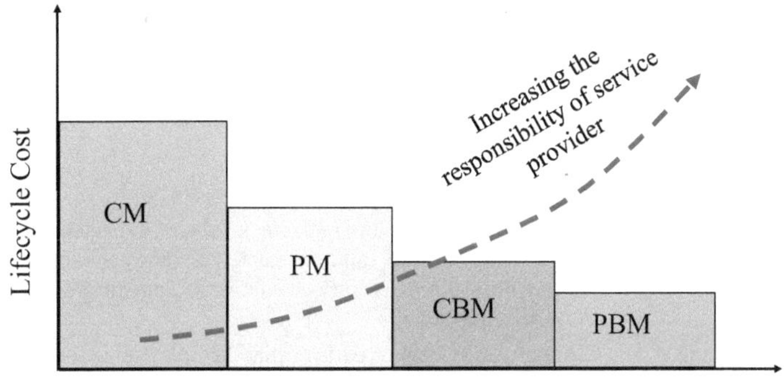

Fig. 1 The evolution of asset management strategy

repair or replacement is triggered by the degradation level or the health condition which is often derived from in situ sensory data. PBM belongs to incentive-driven, instead of transaction-based, maintenance strategy. It differs from CM and PM in that PBM takes a holistic approach to plan the MRO service considering the interests of multiple stakeholders whereas CM and PM often seek the optimal replacement decision from a single stakeholder, either the service provider (e.g., cost minimization) or the end user (e.g., availability maximization). Therefore, multi-criteria optimization, e.g., cost minimization and availability maximization simultaneously, is considered to be viable approach to implementing the PBM service.

Figure 1 shows how the asset management strategy evolved in the past several decades, which is driven by continuous reduction of the ownership cost and the transferring of more responsibilities to the service provider.

Though the goal of different maintenance strategies might be similar, i.e., minimizing the cost rate or ensuring the equipment uptime, CM, PM, and CBM usually emphasize the technical aspect by focusing on when to execute the inspection, replacement, or repair action. PBM is designed with business or profitability aspect. The attention is often paid to implementing an appropriate maintenance and support logistics to meet the predefined contractual goal. Under the PBM framework, the service provider can choose a specific replacement policy, such as time-, usage-, or condition-based criterion, to manage the maintenance contract as long as it is mutually agreed between the supplier and the customer. Perhaps, the key difference between PBM and other maintenance policies is that in PBM design for reliability, replacement time, and spares provisioning can be jointly coordinated across the product life cycle.

Table 1 A four-step procedure of PBM implementation

Step	Objectives	Detailed tasks
1	Identifying performance outcomes	System readiness, mission completion, assurance of spare parts supply
2	Defining performance measures	Operational availability, spare parts availability, stocking-out probability, part failure rate, stock fill rate, expected backorders, mean time between failures, mean time between replacements, mean time to repair, cost per unit usage, logistics response time, logistics footprint
3	Determining contract goals	Maximize system or parts availability, maximize service profit, maximize fill rate, minimize backorders, minimize failure rate, minimize repair waiting time, minimize the life cycle cost, minimize the maintenance cost rate
4	Choosing incentive mechanisms	Fixed cost payment, cost plus payment, cost plus incentive fee, cost sharing or cost reimbursement.

2.2 *Implementation of Performance-Based Maintenance*

The planning and implementation of a PBM contract can be divided into four steps: (1) identifying performance outcomes; (2) determining performance measures; (3) defining performance criteria; and (4) designing incentive payment. The objectives and tasks of each step are summarized in Table 1. To effectively implement a PBM program, reliability performance must be appropriately translated into measurable value so that it can be assessed quantitatively over the contract period. Meanwhile, performance measures can also be used as service criteria, decision variables, or constraints to guide the actions of the service supplier.

3 Identifying and Defining Performance Measures

3.1 *Five Key Performance Measures*

Though various performance measures have been proposed to assess the outcome of a PBM contract, the following quantitative metrics are commonly adopted: operational availability, mission reliability, cost per unit usage, logistics response time, and logistic footprint. Below, these metrics are elaborated in terms of their interactions.

According to the US DoD (DoD 2005), operational availability (OA) is defined as "a measure of a degree to which an item is in an operable state and can be committed at the start of a mission when the mission is called for at a random point in time." Let A_o be the steady-state value of OA, it can be estimated as.

$$A_o = \frac{MTBF}{MTBF + MTTR + MLDT}, \tag{1}$$

where MTBF is the mean time between failures of the system. MTTR is the mean time to repair, and MLDT is the mean logistics delay time for repair. In particular, MTTR is the hands-on time (i.e., touch labor) to recover the system given the spare part is available. MLDT represents the waiting time associated with the arrival of spare parts, maintenance crew, and necessary tools.

Mission reliability (MR) measures the capability of an item or system to accomplish the required task for the duration of a specified working condition. Essentially MR defines the probability of failure-free operation during the time period to complete the task. Let t_m be the time duration of the mission, then.

$$MR = P_r\{T \geq t_m\} \geq \gamma, \tag{2}$$

where γ for $0 \leq \gamma \leq 1$ represents the desired probability of failure-free operation during t_m. An alternative approach would be to measure the cumulative operating time of a system prior to its failure or scheduled maintenance. Other metrics such as failure rate or failure intensity rate can also be used to characterize the system MR. For instance, assume the item's lifetime is exponentially distributed with failure rate λ. Based on Eq. (2), MTBF and λ must meet the following criterion in order to satisfy the required mission reliability γ. That is

$$MTBF = \frac{1}{\lambda} \geq -\frac{t_m}{\ln \gamma},$$

Similar relation between MTBF and MR can be derived if the lifetime follows other distributions such as Weibull and log-normal.

Logistics response time (LRT) is the duration of calendar time from when a failure occurred to the time when it is fixed. Since LRT is highly correlated with MLDT and MTTR, it can be approximated as follows:

$$LRT = MLDT + MTTR \tag{3}$$

For instance, if a failure occurs at 6 am in the morning, and it takes 2 hours for the staff to report the failure to the service provider, and it takes another 5 hours for the provider to restore the system. Then LRT $= 2 + 5 = 8$ hours in the case.

Cost per unit usage (CUU) is defined as the total operation and support cost of a system divided by the usage factor. Typical usage factors include hours, miles, rounds, or launches depending on the nature of the system. For instance, the usage of a rocket is measured by the number of launches while aircraft engine usage is often measured by flight hours, take-off, and landing frequency. In addition, overhead costs associated with MRO activities are also part of the operation and support cost. As such, CUU can be estimated as

$$CUU = \frac{\text{total operation and support cost}}{\text{usage factor}} \tag{4}$$

Logistics footprint (LF) quantifies the size of logistics supply chain needed to sustain the operation of a fleet of systems. Measurable indices include but not limit to spare parts inventory, personnel, facilities, truck/transport fleet, and number of sub-contractors. Since LF encompasses a variety of elements, each element should be characterized, measured, and monitored individually and independently. If the unit cost of individual elements is available, it is appropriate to envelop the entire embodiment of logistics support based on the following formula:

$$LF = c_1 I + c_2 P + c_3 F + c_4 V + c_5 S \tag{5}$$

where c_1, c_2, c_3, c_4, and c_5 represent the unit cost of spare parts inventory, maintenance crew, facility, transportation, and contractor supply. Note that, I, P, F, V, and S represent the size or capacity of spare parts inventory, personnel, facilities, transportation, and contractor supply, respectively.

3.2 *Operational Availability Under Corrective Replacement*

OA is treated as the primal performance measure that ultimately governs the other four performance measures, i.e., MR, LRT, LF, and CUU. Originally from Espiritu et al. (2012), Fig. 2 depicts the interaction of five performance measures. To meet the OA target, the service provider may choose to improve MR through reliability growth and redundancy allocation scheme or compress the LRT by deploying a responsive logistics network. Higher MR implies a larger MTBF or MTBR, which leads to a smaller amount of failures. Hence it saves repair cost and spare parts investment. This ultimately reduces the size of LF such as repair facility and crew members. On the other hand, to reduce the LRT, the supplier needs to escalate the safety stock level of the spares inventory, implement expedited transportation mode, or deploy a larger maintenance crew. These decisions further influence the outcomes of, or more likely increase, the cost of LF and CUU.

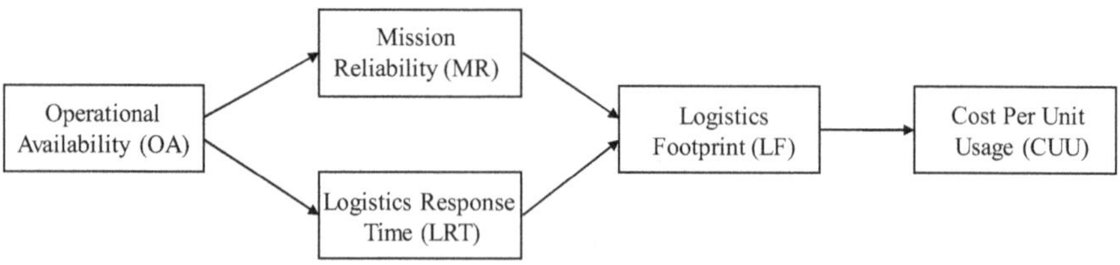

Fig. 2 Interactions of the core performance measure in PBM

It has been shown that OA is primarily governed by six performance drivers, namely, intrinsic failure rate, system usage, spare parts stock level, maintenance policy, repair turnaround time, and hands-on replacement time (Jin et al. 2013). For single-item systems subject to corrective maintenance, a unified availability model synthesizing six drivers is presented as follows (Jin and Wang 2012):

$$A(\lambda, s, \rho, n, t_r, t_s) = \frac{1}{1 + \rho\lambda t_s + \rho\lambda t_r \left(1 - \sum_{k=0}^{s} \frac{(n\rho\lambda t_r)^k e^{-n\rho\lambda t_r}}{k!}\right)}, \tag{6}$$

where

λ = failure rate.
ρ = system usage rate, and $\rho \in [0, 1]$.
s = spare parts stocking level.
n = number of operating units in the field or fleet size.
t_r = repair turnaround time between the spares inventory and the repair center.
t_s = hands-on time of replacing the failed item.

Equation (6) is derived under the assumption that the repair center has ample capacity which is modeled as $M/G/\infty$ queue. For a system comprised multiple types of components each having different operational availability, the system availability is given by

$$A_s = \prod_{i=1}^{K} A_i(\lambda_i, s_i, \rho_i, n_i, t_{ir}, t_{is}) \tag{7}$$

where K is the number of components or items in a system, and A_i (λ_i, s_i, ρ_i, n_i, t_{ir}, t_{is}) is the availability of component i for $i = 1, 2, \ldots, K$.

3.3 Operational Availability Under Age-Based Replacement

In this section, we extend Eq. (6) to age-based PM. Let τ be the predefined replacement time. An item may survive through τ or fail prior to τ. The former is callled planned or scheduled replacement and the latter is called failure replacement. Technically an item survived at τ is still a functioning unit, though its reliability has deteriorated. In general, less time and effort are required to recondition a degraded unit than repairing a failed unit. We use t_p to denote the turnaround time (TAT) for reconditioning a deteriorated unit. In other words, TAT stands for the time lapse from when the item is removed from the system to when it is reconditioned in the repair center and put back in the inventory. Similarly t_r denotes the TAT for repairing a failed unit, and $t_r > t_p$.

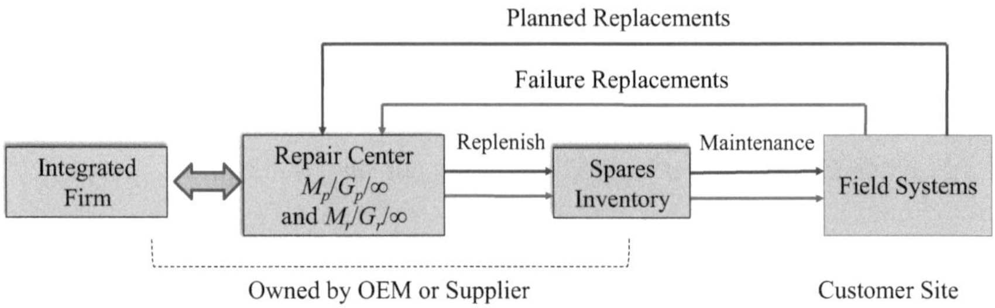

Fig. 3 PBM service with age-based replacement policy

Figure 3 shows an integrated product-service system where the supplier sells new product as well as offering maintenance services in the after-sales market. It is assumed that the repair center has ample capacity to carry out reconditioning and repairing jobs, modeled as $M_p/G_p/\infty$ and $M_r/G_r/\infty$ queues, respectively. Then the operational availability for a single-item system under age-based PM policy is obtained as

$$A = \frac{\int_0^\tau R(t;\rho)\,dt}{\int_0^\tau R(t;\rho)\,dt + t_s + \left(t_p R(\tau;\rho) + t_r F\left(\tau;\rho\right)\right)\Pr\{O > s\}}, \tag{8}$$

Note that O is the random variable representing the spare parts demand to the inventory. For detailed derivation of Eq. (8), readers are referred to Jin et al. (2013). Equation (8) comprehends eight performance drivers, namely, inherent reliability $R(t)$, usage rate ρ, replacement time τ, spare parts stock level s, fleet size n, hands-on repair time t_s, parts reconditioning turn-around time t_p, and parts repair turn-around time t_r. Note that $R(t;\rho)$ is the reliability function considering the usage factor. For instance, Weibull reliability function considering the usage rate is expressed as.

$$R(t;\rho) = \exp\left(-\left(\frac{\rho t}{\eta}\right)^\beta\right), \quad \text{for } t \geq 0 \tag{9}$$

where η and β are the scale and shape parameters, respectively. Figure 4 is a four-layer diagram that decomposes the OA into other performance measures under various maintenance or replacement policies. An important observation is that OA involves multi-stakeholders that are manifested by the service provider and the customer. Therefore, the outcome of OA is jointed determined by the actions of the service provider and the customer.

Recently, Xiang et al. (2017) derived a system availability model based on CBM policy; they further propose a performance-based contracting scheme to attain three objectives: minimizing the cost, maximizing the availability, or maximizing service profit. The advantage of CBM over age- or time-based PM is that it allows for accomendating unit-to-unit degradation heterogeneity.

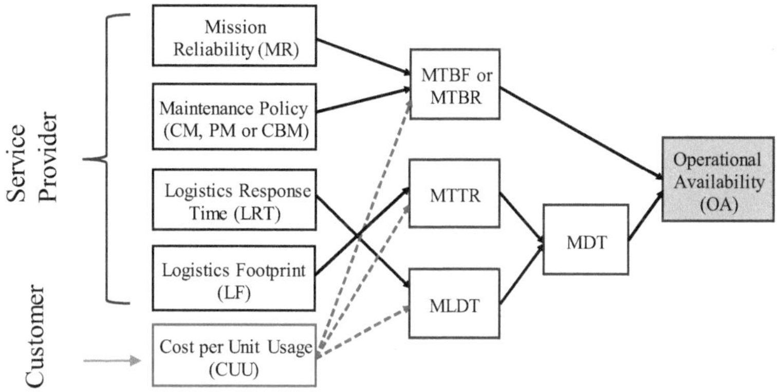

Fig. 4 The correlations between OA and other performance measures

4 Contracting for Profit Maximization

4.1 System Life Cycle Cost

The life cycle cost (LCC) of capital equipment typically consists of four cost elements: design and development, manufacturing, operation and support, and retirement or decommission. Though the actual cost may vary, operation and support cost usually accounts for 50–70% of the LCC, manufacturing accounts for 20–30%, design and development is about 10–20%, and decommission is 5% (DoD 2016). A major force driving the transition from MBC to PBM is to reduce the operation and maintenance cost, hence lowering product life cycle cost.

Without loss of generality, the cost model below is derived upon single-item systems, but it can be extended to a multi-item system fleet. Let $C(\lambda, s)$ denote the life cycle cost for n single-item systems deployed in a customer site. The following cost models adopted from Öner et al. (2010) and Jin and Wang (2012) are used to estimate the fleet life cycle cost. The notation of parameters and variables are listed in Table 2.

$$C(\lambda, s) = D(\lambda) + nc(\lambda) + I(\lambda, s) \tag{10}$$

where

$$D(\lambda) = B_1 \exp\left(\varphi \frac{\lambda_{\max} - \lambda}{\lambda - \lambda_{\min}}\right), \quad \text{for } \lambda_{min} < \lambda \leq \lambda_{max} \tag{11}$$

Table 2 Notation of the product life cycle model

B_1	Baseline system design cost
B_2	Baseline system manufacturing cost
B_3	Incremental manufacturing cost
T	Duration of maintenance service contract period
ϕ	Degree of difficulty of reliability growth in design
ν	Degree of difficulty of reliability growth in manufacturing
c_r	Unit repair cost of field returned item or line replaceable unit
γ	Annual interest rate
$\lambda_{\min}$	Minimum achievable failure rate
$\lambda_{\max}$	Maximum acceptable failure rate
$C(\lambda, s)$	Life cycle cost of a system fleet
$D(\lambda)$	Design cost of the system
$c(\lambda)$	Manufacturing cost of a single system or item
$I(\lambda, s)$	Spare parts inventory cost

$$c(\lambda) = B_2 + B_3\left(\frac{1}{\lambda^{\nu}} - \frac{1}{\lambda_{\max}^{\nu}}\right), \quad \text{for } \lambda_{min} < \lambda \le \lambda_{max} \tag{12}$$

$$I(\lambda, s) = sc(\lambda) + c_r n\rho\lambda\frac{(1+\gamma)^T - 1}{\gamma(1+\gamma)^T} \tag{13}$$

Here $D(\lambda)$, $c(\lambda)$, and $I(\lambda, s)$ represent the costs associated with design, manufacturing, and spare parts inventory during the contractual period, respectively. Note that $\lambda_{\max}$ represents the largest acceptable failure rate by the customer. Similarly $\lambda_{\min}$ is the lowest failure rate that could be possibly achievable by the service provider. In Eq. (11), ϕ is a positive parameter capturing the difficulty in increasing the reliability in product design phase, and in Eq. (12), ν is a positive parameter capturing the difficulty in increasing the reliability in manufacturing phase.

4.2 Incentive Payment Model

Several payment schemes are available to incentivize the service suppler to attain the contract goal, including cost plus fixed fee (CPFF), cost plus award fee (CPAF), and cost plus incentive fee (CPIF). In CPFF, the contractor receives a predetermined fee that was agreed upon at the time of contract formation. Essentially it transfers a substantial amount of risks to the supplier. However, if product reliability is well established and the majority of operational uncertainties and failure mechanisms are known to the supplier, the supplier can charge the customer a reasonable amount of cost that mitigates any unexpected reliability risks.

In CPAF, customer pays a fee to its contractor based on the contractor's work performance. Sometimes the fee is determined subjectively by an award board or committee, whereas in others the fee is based upon the observed performance metrics such as operational availability and mission reliability.

In CPIF, the customer pays a larger amount of fee as incentives to the contractor who is able to meet or exceed the performance goals that are predefined and mutually agreed. For a linear CPIF function, it consists of a fixed fee and a reward fee, and the latter is often proportional to the achieved system performance. For instance, if the operational availability (A) is designated as the performance measure, then the linear CPIF function is given as

$$G(A)=\begin{cases} a+b_1\left(A-A_{\min}\right) & A\geq A_{\min} \\ a+b_2\left(A-A_{\min}\right) & A<A_{\min} \end{cases} \tag{14}$$

where $A_{\min}$ is the target or contractual operational availability. Here a is the fixed payment regardless the performance outcome. Parameters b_1 and b_2 are the reward or penalty rate, respectively. A larger b_1 (or b_2) implies that the supplier receives more compensation (or penalty) given the same A.

Intuitively, further increasing the system availability becomes more difficult if A is already high. The following exponential CPIF model proposed by Nowicki et al. (2008) aims to provide more compensation as A approaches unity. That is

$$G(A)=\begin{cases} \exp\left(c+d_1\left(A-A_{\min}\right)\right) & A\geq A_{\min} \\ \exp\left(c+d_2\left(A-A_{\min}\right)\right) & A<A_{\min} \end{cases} \tag{15}$$

Note that c, d_1, and d_2 are model parameters. Assume the minimum required availability by the customer is $A_{\min}=0.8$. Figure 5 plots the exponential CPIF payment curve in two cases. In Case 1, $c=7$, $d_1=5$, and $d_2=2$; in Case 2, $c=7$, $d_1=3$, and $d_2=1$. It shows that the total compensation $G(A)$ increases exponentially with d_1, yet it also decreases exponentially with d_2 given the same A. Compared with the linear CPIF in Eq. (14), the exponential model offers a larger amount of incentives to the supplier because of the higher reward rate for $A\geq A_{\min}$.

4.3 Service Profit Maximization

We first present a PBM contract model by assuming that the supplier or original equipment manufacturer (OEM) is capable of achieving the operational availability target. Assume a system consists of K components for $i=1, 2, \ldots, K$. In addition, components of the same type may also be repeatedly used in a system with m_i being the number of component i. For instance, a modern wind turbine system is often configured with three identical blades, and failure of one blade brings the system to down state. The following profit maximization model originally proposed by Espiritu et al. (2012) is used for illustration:

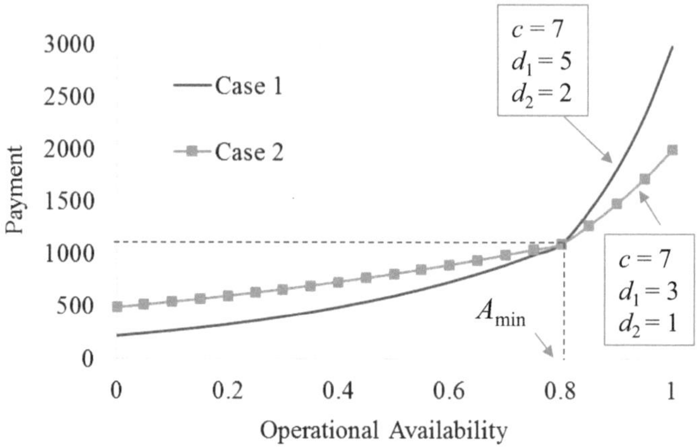

Fig. 5 Cost plus incentive fee with exponential compensation

Model P1:
Max:

$$P\left(\lambda, \mathrm{s}; \rho\right) = G\left(A_s\left(\lambda, \mathrm{s}; \rho\right)\right) - \sum_{i=1}^{K}\left(D_i\left(\lambda_i\right) - B_{1,i}\right) - n\sum_{i=1}^{K} m_i\left(c_i\left(\lambda_i\right) - B_{2,i}\right) - \sum_{i=1}^{K} I_i\left(\lambda_i, s_i; \rho\right) \tag{16}$$

Subject to:

$$A_s\left(\lambda, \mathrm{s}; \rho\right) = \prod_{i=1}^{K}\left(A_i\left(\lambda_i, s_i; \rho\right)^{m_i} \geq A_{\min}\right. \tag{17}$$

$$\lambda_{\min,i} < \lambda_i \leq \lambda_{\max,i} \qquad \text{for } i = 1, 2, \ldots, K \tag{18}$$

The objective function (16) is formulated to maximize the expected service profit for a fleet of n systems during the contractual period. Note that λ_i and s_i are the failure rate and the spare parts stock level for the ith component type. They are the decision variables. When systems are not fully utilized, the system usage rate ρ is used in Eqs. (16) and (17). $A_s(\lambda, \mathrm{s}; \rho)$ and $A_i(\lambda_i, s_i; \rho)$ represent the system operational availability and the component operational availability of type i, respectively.

Model P1 belongs to the mixed integer nonlinear programming problem. This type of problem is difficult to solve because of the mix of nonlinearity and combinatorial natures of integer programs. Genetic algorithm combined with heuristic search (Coit et al. 2004; Marseguerra et al. 2005) has shown to be effective in exploring the solution space to find the optimal or near optimal solution at a reasonable computation cost. Next we use a genetic algorithm to search for the optimal values of λ_i and s_i for $i = 1,2, \ldots, K$ when Model P1 is applied in wind generation industry.

4.4 Application to Wind Power Industry

A wind turbine (WT) is a complex electro-mechanical system comprising multiple components (also known as line replicable unit or LRU), including blades, main shaft/bearing, gearbox, generator, AC/DC converters, and control mechanisms. High system availability is desirable as wind farmers are able to maximize the energy yield given the intermittent wind speed. Due to the complexity of wind turbines, the maintenance and repair services are undertaken by the OEM or third-party logistics suppliers. A PBM contract is designed with the focus on three components: blades, main shaft/bearing, and the gearbox, as they dominate the failures of wind turbines. Reliability and cost related to component design, manufacturing, spare parts, and repairs are listed in Tables 3. These parameters are estimated based on the reports in Tavner et al. (2007) and NREL (2012).

The profit maximization for the wind turbine fleet is solved based on Model P1. Genetic algorithm is used to find the optimal or near optimal values for λ_i and s_i, and the results are summarized in Table 4. Under the linear reward with $a = \$3 \times 10^7$ and $b_1 = \$3 \times 10^8$ in Eq. (14), the supplier can reap nearly \$25.06 million profit during 5 years by keeping the system availability at 0.9889. The availability is higher

Table 3 Reliability and cost parameters of wind turbine components

Index	$i = 1$	$i = 2$	$i = 3$
Component type	Blade	Main shaft/bearing	Gearbox
m_i	3	1	1
λ_{max} (failure/year)	0.2898	0.0312	0.1306
λ_{min} (failure/year)	0.1560	0.0168	0.0703
B_1 (\$)	3,330,000	675,000	1,936,500
B_2 (\$)	333,000	67,500	193,650
B_3 (\$)	20,000	7000	12,000
ϕ	0.02	0.02	0.02
ν	0.6	0.6	0.6
c_r (\$/repair)	40,000	50,000	60,000
t_r (days)	45	90	120
t_s (days)	3	4	6

Table 4 Availability and service profit under a 5-year contract

	$\tau = 5$ years, $A_{\min} = 0.97$, $\rho = 1$, and $n = 50$ systems					
Reward function	Linear			Exponential		
Index	$i = 1$	$i = 2$	$i = 3$	$i = 1$	$i = 2$	$i = 3$
Component type	Blade	Main shaft/bearing	Gearbox	Blade	Main shaft/bearing	Gearbox
λ (failure/year)	0.1799	0.0311	0.1203	0.1792	0.0312	0.1187
s	7	0	5	7	0	5
A (single component)	0.99807	0.99722	0.99741	0.99808	0.99721	0.99748
A (component cluster)	0.99422	0.99722	0.99741	0.99427	0.99721	0.99748
A (system)	0.9889			0.9890		
Total profit ($\$ \times 10^6$)	25.06			24.78		

than the customer's requirement $A_{\min} = 0.97$. This is the result of OEM's motivation to maximize the service profit.

Model P1 is also solved under the exponential CPIF payment in Eq. (15). To make a fair comparison, we set $c = 17.217$ and $d_1 = d_2 = 8.745$ such that the base and the maximum revenues are the same as the linear CPIF. The results show that under the exponential CPIF, the profit is reduced to \$24.78 million. However, the system availability still reaches 0.989, slightly higher than the linear model. This observation shows that the OEM must spend more efforts to gain the same amount of the profit under the exponential reward function. In other words, exponential CPIF enables the customer to incentivize the supplier's performance while reducing the cost of ownership. The inventory decision in both cases happened to be the same. In particular, seven spare blades and five spare gearboxes are allocated. Spare parts for the main shaft/bearing are not needed because of its high reliability with short repair turnaround time. This result is quite useful. It shows that if reliability and repair logistics are appropriately coordinated, we are able to achieve high system availability with zero spares inventory.

5 Contracting Under Condition-Based Maintenance

5.1 Maintenance Optimization Model

In this section, we propose a PBM planning model under imperfect CBM policy. We assume that system failure can only be detected through inspection. The system is periodically inspected every δ time interval. Upon inspection, if the cumulative deterioration level $X(t)$ is between the maintenance threshold ξ and the failure threshold s, a preventive maintenance action is performed; if $X(t)$ exceeds the failure threshold s, corrective maintenance is carried out; otherwise, do nothing. The preventive repair is assumed to be imperfect with the so-called (p, q) rule. Namely, the system upon PM is restored to an as-good-as-new state with probability p and remains in the same state just prior to the PM action with probability $q = 1 - p$. The times required for performing inspection, PM, and CM actions are random variables. The expected PBM service profit rate is denoted with $\pi(\cdot)$, and the reward rate with $g(\cdot)$. Let A denote the achieved system availability and $C(\cdot)$ denote the expected cost rate. Originally proposed by Xiang et al. (2017), the optimization model with the objective of maximizing profit is given by.

Model P2:

$$\begin{aligned} &\max \quad \pi(\delta, \xi) = g(A(\delta, \xi)) - C(\delta, \xi) \\ &(\delta^*, \xi^*) = \arg \ \max \{\pi(\delta, \xi)\} \\ &\delta > 0, \xi > 0 \end{aligned} \tag{19}$$

For comparison purposes, we also propose two benchmark models with the objectives of minimizing cost and maximizing availability, respectively.

Model P3:

$$\begin{aligned} &\min \quad C(\delta, \xi) \\ &(\delta^*, \xi^*) = \arg \quad \min\{C(\delta, \xi)\} \\ &\delta > 0, \xi > 0 \end{aligned} \tag{20}$$

Model P4:

$$\begin{aligned} &\max \quad A(\delta, \xi) \\ &(\delta^*, \xi^*) = \arg \quad \min\{A(\delta, \xi)\} \\ &\delta > 0, \xi > 0 \end{aligned} \tag{21}$$

5.2 *System Availability and Cost Rate*

Next we present the average availability and the associated costs derived from the proposed CBM policy. We define the "renewal cycle" as the period of time between two consecutive perfect maintenance actions (e.g., perfect preventive repair or corrective repair). A cycle can be the interval between two corrective repairs, two perfect preventive repairs, or a perfect preventive repair and a corrective repair. If a preventive repair does not bring the system to an as-good-as-new state, it is not considered as a renewal point. Downtime may include the time of inspection and preventive or corrective repair. Table 5 lists the key notation used in the model development.

According to Xiang et al. (2017), the expected uptime per maintenance cycle can be estimated as

Table 5 Notation for PBM contracting with CBM

C_{CM}	Cost of corrective maintenance
C_{PM}	Cost of preventive maintenance
C_{insp}	Cost of inspection
μ_{CM}	Expected corrective maintenance time
μ_{PM}	Expected preventive maintenance time
μ_{insp}	Expected inspection time

$$E\left(T_{\text{up}}\right)=\sum_{j=1}^{\infty} j\delta\times\Pr\left(\text{device preventively repaired perfectly at the } j^{\text{th}} \text{ inspection}\right)$$
$$+\sum_{j=1}^{\infty} j\delta\times\Pr\left(\text{device correctively repaired at the } j^{\text{th}} \text{ inspections}\right) \tag{22}$$

Next we derive the expected downtime. The expected total downtime is as follows:

$$\begin{aligned}E\left(T_{\text{down}}\right)&=E\left(\text{inspection time}\right)+E\left(\text{preventive repair time}\right)\\&\quad+E\left(\text{corrective repair time}\right)\\&=\sum_{i=1}^{2}\sum_{j=1}^{\infty}E_{i,j}\left(T_{\text{down}}\right)\end{aligned} \tag{23}$$

Therefore, we have the expected cost rate as follows:

$$C=\frac{\sum_{i=1}^{2}\sum_{j=1}^{\infty}E_{i,j}\left(\text{Cost}\right)}{E\left(T_{\text{down}}\right)+E\left(T_{\text{up}}\right)} \tag{24}$$

Note that the policy that minimizes the cost rate or maximizes the average availability is also the optimal solution that maximizes the profit rate. Define

$$\overline{A}\left(\delta,\xi\right)=1-A\left(\delta,\xi\right)=E\left(T_{\text{down}}\right)/E\left(T_{\text{up}}\right)+E\left(T_{\text{down}}\right) \tag{25}$$

Then, maximization of the availability A is equivalent to the minimization of $\overline{A}$ in Eq. (25). Let ω be the ratio of $\overline{A}$ to the cost rate given the same maintenance policy, defined as

$$\omega=\overline{A}\left(\delta,\xi\right)/C\left(\delta,\xi\right)=\sum_{i=1}^{2}\sum_{j=1}^{\infty}E_{i,j}\left(T_{\text{down}}\right)/\sum_{i=1}^{2}\sum_{j=1}^{\infty}E_{i,j}\left(\text{Cost}\right) \tag{26}$$

If $C_{\text{CM}}\,/\,C_{\text{nsp}}=u_{\text{CM}}\,/\,u_{\text{insp}}$ and $C_{\text{PM}}\,/\,C_{\text{insp}}=u_{\text{PM}}\,/\,u_{\text{insp}}$, the ratio is a constant, which implies that Models P3 and P4 are equivalent and the problem of maximizing the profit rate has the same optimal solution as that of minimizing the cost rate or maximizing the average availability. We expect the optimal solutions to be different when these constant parameters are not linearly related. In the next section, we provide numerical examples to illustrate the applications of these models.

5.3 Numerical Experiments

In this section, we investigate how the profit-centric approach affects the optimal policies. We compare cost rate, availability, and profit rate between the profit-centric approach and the two benchmarks.

Consider a system that deteriorates over time, and the system state can be described by a gamma process $\{X(t),\ t \geq 0\}$ with random effect z that controls heterogeneity across units. Given z, the process $\{X(t)\}$ has independent increments: for $0 \leq t_1 < t_2$, $X(t_2)$-$X(t_1)$ is independent of $X(t_1)$ and has a gamma distribution $Ga(\alpha(t_2\text{-}t_1), z)$ with $\alpha_0 = 0$. The conditional density function of $X(t_2)$ - $X(t_1)$ for given z is $f(x|z)$, where the gamma density function is given by Lawless and Crowder (2004)

$$f\ (x\,|z) = \Gamma(\alpha\ (t_2 - t_1))^{-1} z^{-\alpha(t_2-t_1)} x^{\alpha(t_2-t_1)-1} e^{-x/z},\ \text{for } x > 0 \tag{27}$$

It is mathematically convenient to assume $Ga(\theta, \gamma^{-1})$ as the distribution of z^{-1}; hence, the degradation process has a closed form for probability density function and cumulative distribution function. The marginal density of $X(t)$ is as follows:

$$f(x) = B(\alpha t, \theta)^{-1} \frac{\gamma^{\theta} x^{\alpha t - 1}}{(x+\gamma)^{\alpha t + \theta}}, \tag{28}$$

where $B(\alpha t,\ \theta) = \Gamma(\alpha t)\Gamma(\theta)/\ \Gamma(\alpha t + \theta)$. We note that $\theta X(t)/\gamma\alpha t$ has an F distribution. Therefore, the distribution function of the degradation $X(t)$ is

$$F\ (x;\, x_0, t) = \mathcal{F}_{2\alpha t, 2\theta}\left(\frac{\theta x}{\gamma \alpha t}\right) \tag{29}$$

Without loss of generality, let $\gamma = \theta$ so that z^{-1} has a mean 1 and variance θ^{-1}.

Suppose $s = 50$, $x_0 = 0$, $\alpha = 1$, and the random effect variable z controls heterogeneity across units. z^{-1} has a gamma distribution $Ga(5,\ 1/5)$. We fix the inspection time and inspection cost ($\mu_{\text{inspt}} = 0.2$ and $c_{\text{inspt}} = 5$), and different levels of PM and CM repair times (cost) are selected for the purpose of sensitivity analysis. We assume that $c_{\text{PM}} < c_{\text{CM}}$ and $\mu_{\text{PM}} < \mu_{\text{CM}}$; otherwise it would not be necessary to perform preventive maintenance. We assume that the relationship between rewards and availability is stepwise linear, and the reward function is given by

$$g(A) = \begin{cases} 0, & \text{if } A \leq 0.3 \\ 5 + \kappa\ (A\text{-}0.3)\,, & \text{otherwise} \end{cases} \tag{30}$$

If the availability is below 0.3, the reward is zero and the service provider is penalized for poor performance. Two levels of rewards are considered, $\kappa = 30$, and $\kappa = 50$. We are interested in exploring how optimal policies change with different κ. Rosenbrock's method is used to find the optimal inspection intervals and maintenance threshold (Bazaraa et al. 2006). This method does not employ line searches but rather takes discrete steps along the search directions. At each iteration, the new directions established by the Rosenbrock procedure are linearly independent and orthogonal. Results of the numerical examples are summarized in Tables 6 and 7. Note that C denotes the cost rate, A denotes the availability, and P denotes the profit rate.

From Tables 6 and 7, we can see that the profit rates from the profit-centric approach are better than those from the two benchmark approaches even when κ is small ($\kappa = 30$), which illustrates that simply minimizing cost rates or maximizing availabilities would not lead to best profit rates. Also note that the improvement in profit rates are per time unit, and the total gains period would become larger over time. We also observe that when incentives for better performance outcomes are low, maintenance policies under cost minimization are close to policies under the profit-centric approach. As this incentive increases, policies under the availability maximization approach are closer to the ones under the profit-centric approach.

Both tables also indicate that the performance outcome, i.e., availability, has a dominant role in determining the profit rate when there is more incentive for improvement in the average availability. Since cost parameters are in general much harder to obtain, and the repair times are often more available and accurate, we can implement the optimal policies from the max availability model and still get satisfactory profits when the reward is performance-based and the incentive to improve the performance is high.

Based on the above observations, we can conclude that the traditional cost minimization approach in MRO services might not lead to optimal profits in many cases when reward is determined by the performance. Another important finding is that policies that maximize availability provide reasonably good profits in most numerical examples examined. If cost parameters are not available, the optimal policies under the availability maximization can be used as an appropriate substitute for profit maximization.

6 Future Research Direction

6.1 Maintenance Planning Under a Variable System Fleet

During the new product introduction, the installed base or the fleet size continue to grow as more systems are shipped and installed at customer sites. Given the same product reliability, it is anticipated that more failure returns are generated due to the expanding installed base. To optimize the spare parts stock level, the stream of the aggregate fleet failures over the maintenance horizon is of our interest. In fact, there are two stochastic processes involved when one manages the repair and spare parts supply for a growing installed base. One is the number of installed systems

Table 6 Min cost versus max availability versus max profit ($\kappa = 30$)

				Min cost					Max availability					Max profit				
c_{PM}	c_{CM}	μ_{PM}	μ_{CM}	δ	ξ	C	A	P	δ	ξ	C	A	P	δ	ξ	C	A	P
50	1000	2	24	3.01	32.38	3.73	0.85	18.23	3.74	36.00	3.94	0.87	18.05	3.31	34.46	3.81	0.87	18.40
		2	12	3.23	32.83	3.77	0.86	17.90	6.68	36.97	5.39	0.88	17.11	3.35	35.05	3.89	0.88	18.40
		2	6	3.23	32.83	3.79	0.86	17.99	11.12	39.27	8.61	0.91	14.74	3.74	36.00	4.03	0.89	18.58
50	500	2	24	3.98	37.24	3.24	0.85	18.31	3.74	36.00	3.94	0.87	18.71	3.98	36.07	3.26	0.87	18.71
		2	12	4.28	38.14	3.35	0.87	18.75	6.68	36.97	3.64	0.88	18.86	4.48	37.61	3.35	0.88	19.18
		2	6	3.89	35.32	3.35	0.89	19.25	11.12	39.27	4.84	0.91	18.51	5.36	38.93	3.57	0.90	19.51
50	250	2	24	9.08	39.37	2.55	0.82	18.06	3.74	36.00	3.94	0.87	19.06	4.17	37.44	2.80	0.86	19.13
		2	12	7.02	39.53	2.67	0.87	19.56	6.68	36.97	2.76	0.88	19.74	5.63	40.82	2.70	0.88	19.83
		2	6	7.46	38.91	2.76	0.90	20.20	11.12	39.27	2.96	0.91	20.39	9.07	39.40	2.84	0.91	20.50

Table 7 Min cost versus max availability versus max profit ($\kappa = 50$)

				Min cost					Max availability					Max profit				
c_{PM}	c_{CM}	μ_{PM}	μ_{CM}	δ	ξ	C	A	P	δ	ξ	C	A	P	δ	ξ	C	A	P
50	1000	2	24	3.01	32.38	3.73	0.85	29.32	3.74	36.00	3.94	0.87	29.38	3.43	35.28	3.83	0.87	29.49
		2	12	3.23	32.83	3.77	0.86	29.00	6.68	36.97	5.39	0.88	28.78	3.74	36.00	4.00	0.88	30.01
		2	6	3.23	32.83	3.79	0.86	29.17	11.12	39.27	8.61	0.91	26.97	4.63	33.32	4.25	0.88	29.97
50	500	2	24	3.98	37.24	3.24	0.85	29.34	3.74	36.00	3.94	0.87	30.04	3.74	36.00	3.28	0.87	30.04
		2	12	4.28	38.14	3.35	0.87	30.15	6.68	36.97	3.64	0.88	30.52	5.30	36.73	3.44	0.88	30.75
		2	6	3.89	35.32	3.35	0.89	30.99	11.12	39.27	4.84	0.91	30.74	5.66	38.85	3.62	0.90	31.56
50	250	2	24	9.08	39.37	2.55	0.82	28.46	3.74	36.00	3.94	0.87	30.38	4.21	37.72	2.78	0.86	30.40
		2	12	7.02	39.53	2.67	0.87	31.04	6.68	36.97	2.76	0.88	31.41	5.66	38.85	2.71	0.89	31.58
		2	6	7.46	38.91	2.76	0.90	32.16	11.12	39.27	2.96	0.91	32.62	9.07	39.40	2.84	0.91	32.73

by time t, denoted as $N(t)$. The other is the random failures from individual systems between [0, t]. Given a system installed, the failure behavior can be modeled as a renewal process. Let $S(t)$ be the aggregate fleet failures between [0, t] and $F(t)$ be the time to failure distribution of individual systems, then $S(t)$ can be estimated by.

$$S(t) = H(t) + \sum_{i=1}^{N(t)} H(t - W_i), \quad \text{for } t > 0 \tag{31}$$

where

$$H(t) = F(t) + \int_0^t H(t - x) f(x) dx, \tag{32}$$

Here $H(t)$ is the number of renewals for the initial system installed at time zero, and $H(t\text{-}W_i)$ represents the renewals for the ith system installed at W_i with $0 < W_1 < W_2 < \ldots < W_{N(t)}$. It is quite difficult to directly compute $S(t)$ in Eq. (31) due to the stochastic nature of W_i and $N(t)$. The situation become even worse as most empirical time to failure distributions do not have a closed form solution of $H(t)$.

However, the mean and variance of $S(t)$ can be explicitly characterized if the following conditions regarding the system installation and failure processes are satisfied. These are

- The size of the installed base during [0, t], denoted as $N(t)$, increases following the Poisson counting process with installation rate of λ (i.e., systems per unit time), that is.

$$\Pr\{N(t) = n\} = \frac{(\lambda t)^n \exp(-\lambda t)}{n!}, \quad \text{for } n = 0, 1, 2, \ldots \tag{33}$$

- In case the time to failure is exponentially distribution with failure rate α, then the renewal function becomes.

$$H(t) = \alpha t, \quad \text{for} \quad t \geq 0 \tag{34}$$

Now a closed form estimate for the mean and the variance of $S(t)$ can be obtained as follows (Jin and Tian 2012).

$$E[S(t)] = \alpha t + \frac{1}{2} \alpha \lambda t^2 \tag{35}$$

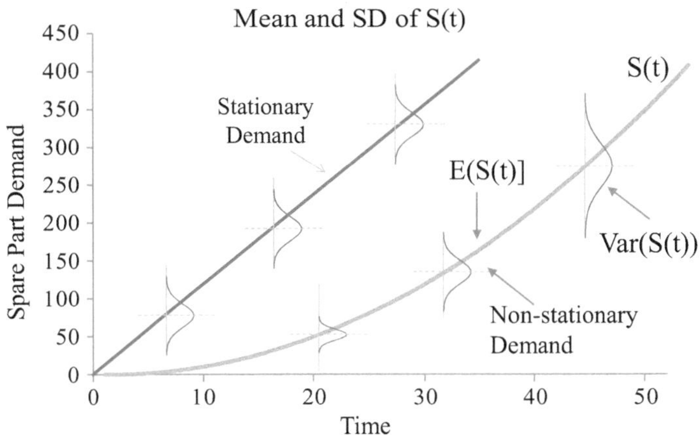

Fig. 6 Stationary vs. nonstationary demands (SD = standard deviation)

$$Var\ (S(t)) = \alpha t + \frac{1}{2}\alpha\lambda t^2 + \frac{1}{3}\alpha^2\lambda t^3 \tag{36}$$

Notice that $E[S(t)]$ and $Var(S(t))$ are functions of t in quadratic and cubic formats, respectively. This clearly indicates that $S(t)$ is a nonstationary process with time-varying mean and variance. Figure 6 shows that $S(t)$ differs from a stationary failure steam because the former has a non-constant or time-varying variance.

6.2 *Performance-Based Maintenance in Post-Warranty Service*

Jointly optimization models of maintenance and warranty repair have been investigated in literature. Recently, Chien (2019) uses generalized Polya process to model repairable systems in which the reliability becomes worse with the amount of cumulative failures. Optimal replacement periods are derived to minimize the long-run expected cost rate during the warranty period. After the warranty expires, the service provider can continue to sustain the system reliability by tracking and monitoring the degradation level. But few warranty policies and post-warranty maintenance models are developed under the condition-based monitoring framework. Since CBM is able to ensure the just-in-time replacement, under-maintenance or excessive replacements can be mitigated. From the manufacturer's perspective, maintenance expense can be saved; from the customer's perspective, unexpected downtime costs are avoided. Therefore, integrating CBM into warranty and post-warranty maintenance contract is a win-win solution to the manufacturer and the consumer.

Recently, Shang et al. (2018) adopt inverse Gaussian (IG) process to design a renewable free replacement warranty (RFRW) policy in which the replacement decision depends on the degradation threshold. IG process has been shown an effective model to characterize the random effects and covariates and is frequently used to model the monotone degradation process (Ye and Chen 2014; Peng et al. 2017). Under the IG-based RFRW policy, the replacement is triggered by the degradation threshold, not the self-announcing failure. Shang et al. (2018) further show that in the monopoly market, the manufacturer acts as a dual role (i.e., both leader and monopolist) and determines the optimal warranty period and sale price by maximizing the profit. In the competitive market, the manufacturer acts as a single role (i.e., leader) and the IG-based RFRW is used as the only competitive strategy. As such, the manufacturer needs to determine optimal warranty period and optimal replacement threshold to maximize the after-sales service profit.

In summary, during the post-warranty period, the customer aims to minimize the maintenance cost rate by considering a hybrid PM effect: the reduction of degradation level and the age. Two scenarios need to be considered for computing the expected cost rate. In the first scenario, the historical degradation level is treated as a random value that is not observable to the customer. In the second scenario, the degradation level is observable to the customer, but the value may not be accurate.

7 Conclusions

This study has presented several analytical models for planning performance-based maintenance services from the system life cycle cost perspective. We begin with reviewing five performance measures and their interdependency. Operational availability is treated as the core performance measure as it ultimately influences and governs other four performance measures. We also present two unified availability models under corrective maintenance and preventative maintenance, respectively. Four different optimization models are formulated with the goal of maximizing service profit or system availability or minimizing the cost rate subject to uncertain usage and operational conditions. The following managerial insights are derived. First, if reliability allocation and repair logistics are appropriately coordinated, zero spare parts inventory is technically achievable and financially attractive. Second, when cost parameters are not available or uncertain, the optimal policies under the availability maximization can be used as a good substitute for maximizing the service profit. In the future, we will generalize the proposed decision models to incorporate more realistic conditions such as variable installed base, reliability growth, and extended warranty. In that way, the cost savings the equipment owner is able to achieve under performance-based contracting can be compared with other programs in a broader scope.

References

Anastasopoulos PC, McCullouch BG, Gkritza K, Mannering FL, Sinha K (2010) Cost savings analysis of performance-based contracts for highway maintenance operations. J Infrastruct Syst 16(4):251–263

Bazaraa MS, Sherali HD, Shetty CM (2006) Nonlinear programming: theory and algorithms. Wiley, Hoboken

Chien, YH (2019) Optimal periodic replacement policy for a GPP repairable product under the free-repair warranty. Quality Technology and Quantitative Management, 16(3):347–354. https://doi.org/10.1080/16843703.2017.1422218

Coit DW, Jin T, Wattanapongsakorn N (2004) System optimization considering component reliability estimation uncertainty: multi-criteria approach. IEEE Trans Reliability 53(3):369–380

D'Angelo G, Bressi S, Giunta M, Presti DL, Thom N (2018) Novel performance-based technique for predicting maintenance strategy of bitumen stabilised ballast. Construction and Building Materials 161:1–8

Dennis MJ, Kambil A (2003) Service management: building profits after the sale. Supply Chain Manag Rev 7(3):42–48

DoD (2005) Guide for achieving reliability, availability, and maintainability. The U.S. Department of Defense, published by Defense Systems Information and Analysis Center, https://www.dsiac.org/resources/reference_documents. Accessed on 10 Sept 2018

DoD (2009) Department of defense reliability, availability, maintainability, and cost rationale report manual, Prepared by the Office of the Secretary of Defense in Collaboration with The Joint Staff, June 1, 2009. Available at http://www.acq.osd.mil/se/docs/DoD-RAM-C-Manual.pdf. Accessed on 4 Oct 2017

DoD (2016) Operating and support cost management guidebook. US Department of Defense Report, available at https://www.dau.mil/guidebooks/Shared%20Documents%20HTML/OS%20Cost%20Guide.aspx. Access on 12 Feb 2019

Espiritu J, Sung CH, Jin T, Huang HZ (2012) Contracting for performance-based maintenance service under profit maximization. In Proceedings of the 18th ISSAT international conference, Boston, MA, USA, pp 335–339

Garvey L (2005) Navy success with PBL. In Proceedings of DAU/PMI conference, Transforming DoD Logistics

Glas AH, Kleemann FC (2017) Performance-based contracting: contextual factors and the degree of buyer supplier integration. J Business Ind Marketing 32(5):677–692

Glas AH, Henne FU, Essig M (2018) Missing performance management and measurement aspects in performance-based contracting: a systematic process-based literature analysis of an astonishing research gap. Int J Operat Product Manag 38(11):2062–2095

Hur M, Keskin BB, Schmidt CP (2018) End-of-life inventory control of aircraft spare parts under performance based logistics. Int J Product Econ 204:186–203

Jin T, Tian Y (2012) Optimizing reliability and service parts logistics for a time-varying installed base. Eur J Operat Res 218(1):152–162

Jin T, Wang P (2012) Planning performance based contracts considering reliability and uncertain system usage. J Operat Res Soc 63(10):1467–1478

Jin T, Xiang Y, Cassady R (2013) Understanding operational availability in performance-based logistics and maintenance services. In Proceedings of reliability and maintainability symposium, pp. 1–6, in IEEE Xplore Database

Jing H, Tang LC (2017) A risk-based approach to managing performance-based maintenance contracts. Quality Reliability Eng Int 33(4):853–865

Kang K, McDonald M (2010) Impact of logistics on readiness and life cycle cost: a design of experiments approach. In Proceedings of Winter Simulation Conference, pp. 1336–1346

Kratz L (2005) Logistics transformation. In proceedings of Mid-Atlantic Logistics Conference, International Society of Logistics

Lawless JF, Crowder MJ (2004) Covariates and random effects in a gamma process model with application to degradation and failure. Lifetime Data Analy 10(3):213–227
Marseguerra M, Zio E, Podofillini L (2005) Multi-objective spare parts allocation by means of genetic algorithms and Monte Carlo simulations. Reliability Eng Syst Safety 87(3):325–335
Mirzahosseinian H, Piplani R (2011) A study of repairable parts inventory system operating under performance-based contract. Eur J Operat Res 214(2):256–261
Nowicki D, Kumar U, Steudel H, Verma D (2008) Spares provisioning under performance-based logistics contract: profit-centric approach. J Operat Res Soc 59(3):342–352
NREL (2012) Distributed generation energy technology operations and maintenance costs, published by National Renewable Energy Laboratory, available at https://www.nrel.gov/analysis/tech-cost-om-dg.html. Accessed on 30 Aug 2018
Öner KB, Kiesmüller GP, van Houtum GJ (2010) Optimization of component reliability in the design phase of capital goods. Eur J Operat Res 205(3):615–624
Peng W, Li Y-F, Yang Y-J, Mi J, Huang H-Z (2017) Bayesian degradation analysis with inverse Gaussian process models under time-varying degradation rates. IEEE Trans Reliability 66(1):84–96
Qiu Q, Cui L, Shen J, Yang L (2017) Optimal maintenance policy considering maintenance errors for systems operating under performance-based contracts. Comput Ind Eng 112:147–155
Randall WS, Pohlen TL, Hanna JB (2010) Evolving a theory of performance based logistics using insights from service dominant logic. J Business Logist 31(2):35–62
Selviaridis K, Wynstra F (2015) Performance-based contracting: a literature review and future research directions. Int J Product 53(12):3505–3540
Shang L, Si S, Sun S, Jin T (2018) Optimal warranty design and post-warranty maintenance for products under inverse Gaussian degradation. IISE Trans 100(10):913–927
Smith SJ (2007) Equipping the American forces with a reliable and robust condition based maintenance plus capability. USAF Report, available at: https://www.slideserve.com/. Accessed on 20 Feb 2019
Tavner PJ, Xiang J, Spinato F (2007) Reliability analysis for wind turbines. Wind Energy 10(1):1–18
Wyman O (2019) Global fleet & MRO market forecast commentary 2019–2029. Available at https://www.oliverwyman.com/our-expertise/insights/2019/jan/global-fleet-mro-market-forecast-commentary-2019-2029.html. Accessed on 26 May 2019
Xiang Y, Zhu Z, Coit DW, Feng Q (2017) Condition-based maintenance under performance-based contracting. Comput Ind Eng 111(9):391–402
Ye ZS, Chen N (2014) The inverse Gaussian process as a degradation model. Technometrics 56(3):302–311

Selective Maintenance Optimization Under Uncertainties

Yu Liu, Tangfan Xiahou, and Tao Jiang

1 Introduction

All the engineered systems are subject to deterioration with the increase of their usage. In most cases, a failure of engineering systems will incur unexpected production delay and/or economic loss, and even cause a severe threat to personal safety. Various maintenance actions have been extensively conducted to recover degraded systems to better states and prolong the residual lifetime of aging systems (Wang 2002). In many industrial and military environments, systems, such as maritime vessels, military aircrafts, and nuclear power plants, are intended to complete a sequence of missions with a finite break between two adjacent missions. Due to limited maintenance resources, such as budget, time, manpower, in the break, it is, however, impossible to perform all the desirable maintenance action during the break. Alternatively, only a subset of maintenance actions is selected from all the available maintenance actions to be performed during the break, so as to ensure the success of the next mission to the maximum extent. The specific maintenance optimization problem is termed as selective maintenance (Cassady et al. 2001a).

The earliest reported work on selective maintenance can be tracked to Rice et al. (1998) in which the selective maintenance problem for parallel-series systems composed of s-independent and identical components with a constant failure rate was converted into a mathematical programming model. A more general model for selective maintenance was formulated by Cassady et al. (2001b) in which the lifetime distribution of each component followed the Weibull distribution, and one of three maintenance actions, i.e., minimal repair, corrective replacement,

Y. Liu (✉) · T. Xiahou · T. Jiang
School of Mechanical and Electrical Engineering, University of Electronic Science and Technology of China, Chengdu, China
e-mail: yuliu@uestc.edu.cn

A. T. de Almeida et al. (eds.), *Multicriteria and Optimization Models for Risk, Reliability, and Maintenance Decision Analysis*, International Series in Operations Research & Management Science 321, https://doi.org/10.1007/978-3-030-89647-8_23

and preventive replacement, can be selected for each component. Following these pioneering works, selective maintenance has been intensively explored in the past two decades from various angles, and these works can be categorized into the four classes:

1. *Imperfect maintenance:* Most deteriorating system/components after repairs may not be "as good as new" or "as bad as old," but in the condition somewhere between these two extremes. Many imperfect maintenance models have been developed to characterize failure pattern of aging systems/components after repairs (Pham and Wang 1996). Imperfect maintenance was firstly introduced into selective maintenance by Liu and Huang (2010a), and an exponential function was put forth to link the age reduction coefficient with the consumed maintenance cost. A hybrid model was used by Pandey et al. (2013a) to characterize the failure intensity of repaired components from the perspective of both a hazard adjustment and an age reduction. In the context of multistate systems, Pandey et al. (2013b) assumed that a multistate component can be recovered to a better state after an imperfect maintenance. The associated maintenance cost and time are functions of the two states before and after maintenance. A comprehensive review on selective maintenance under imperfect maintenance was given in Pandey et al. (2013c).
2. *Miscellaneous system configurations:* By taking account of the multistate nature of engineering systems, Liu and Huang (2010a) and Pandey et al. (2013c) investigated the selective maintenance problem for multistate systems composed of binary-state components and multistate components, respectively. The stochastic dependency among the failures of components were investigated by Maaroufi et al. (2013) and Dao and Zuo (2016) for binary-state systems and multistate systems, respectively. The propagated failures with global effect and failure isolation phenomena was taken into account by Maaroufi et al. (2013) for binary-state systems. The economic dependence associated with sharing of setting up and repairing of multiple identical components was considered by Dao et al. (2014). The selective maintenance for a large-scale k-out-of-n: G system and a fleet of systems were studied by Diallo et al. (2018) and Schneider and Richard Cassady (2015)), respectively.
3. *Mission profiles:* The long-term operation issue for selective maintenance was studied by Iyoob et al. (2006) with the assumption that a system is to perform a sequence of equally spaced, identical missions with breaks between missions. Maillart et al. (2009) explored the selective maintenance for the case of multiple consecutive missions, and due to limited budget, only a subset of failed components can be correctively replaced by new ones. With the assumption that all the failed components can be immediately recovered to functioning state by minimal repairs, Pandey et al. (2016) determined the optimal number of breaks for a finite mission planning horizon. As loading condition in a mission may vary with respect to time, Dao and Zuo (2017) developed a Monte Carlo simulation method to evaluate the probability of mission success and enumerated the optimal selective maintenance strategy.

4. *Optimization algorithms:* With the increase of the system size, the full enumeration method becomes computationally inefficient to identify the optimal selective maintenance strategy. Some advanced optimization algorithms, such as genetic algorithm (GA) (Liu and Huang 2010a), differential evolution algorithm (DEA) (Pandey et al. 2013b), and ant colony optimization (ACO) (Liu et al. 2018), have been utilized to resolve the resulting optimization problems.

A review on selective maintenance of multi-component systems can be found in Cao et al. (2018a). It is noteworthy that many reported works have not sufficiently taken account of various inevitable uncertainties that may exist in selective maintenance optimization. These uncertainties, oftentimes, can produce non-ignorable impact on maintenance decision-making. This chapter aims at providing a comprehensive review of the recent studies on selective maintenance under uncertainties. Additionally, two selective maintenance models under either stochastic maintenance durations or imperfect observations are elaborated and exemplified, respectively.

The reminder of this chapter is organized as follows. In Sect. 2, the basic selective maintenance model and the possible uncertainties in selective maintenance are introduced. In Sect. 3, four existing selective maintenance models under uncertainties are reviewed, and two new selective maintenance models under uncertainties are proposed. A brief closure is given in Sect. 4.

2 Basic Selective Maintenance Model and Uncertainties

2.1 Basic Selective Maintenance Model

A system composed of multiple components is intended to complete a sequence of consecutive missions as shown in Fig. 1. Maintenance activities can only be executed during the break between two adjacent missions. At the end of the last mission, some components have failed or degraded. Due to limited maintenance resources (e.g., budget and break duration), it is unaffordable to repair/replace all the failed or aging components. A maintenance decision has to be made at the beginning of the break to determine a subset of components to be repaired, so as to ensure the success of the next mission.

The assumptions in the basic selective maintenance model are as follows (Cassady et al. 2001b; Liu and Huang 2010a):

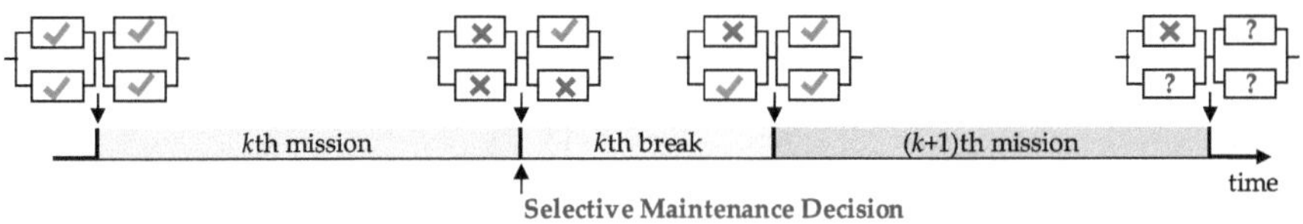

Fig. 1 An illustration of a system executing consecutive missions

1. A system consists of M repairable and s-independent components connected with an arbitrary configuration, such as series-parallel, bridge, and network. Both the system and its components are binary-state, i.e., working perfectly and failed completely. The system configuration remains unchanged throughout all the missions.
2. The state of component l at the beginning of the kth ($k \in \{1, 2, \ldots, N\}$) mission is represented by a binary variable, denoted as $X_{l,k}$, and

$$X_{l,k} = \begin{cases} 1 & \text{if component } l \text{ is functioning} \\ 0 & \text{if component } l \text{ is failed} \end{cases}.$$

At the end of the kth mission, the state of component l is denoted by a binary variable, denoted as $Y_{l,k}$, and

$$Y_{l,k} = \begin{cases} 1 & \text{if component } l \text{ is functioning} \\ 0 & \text{if component } l \text{ is failed} \end{cases}.$$

Because maintenance activities are not allowed to be carried out in the course of a mission, the relationship $X_{l,k} \geq Y_{l,k}$ always holds.

3. Maintenance actions can only be executed during the breaks between two adjacent missions. A set of optional maintenance actions, including do nothing, minimal repair, corrective/preventive replacement, and imperfect corrective/preventive maintenance, can be selected to be performed on either failed or functioning components.
4. The durations of the kth mission and the kth break are denoted as L_k and Z_k, respectively. The states and effective ages of all the components at the end of the kth mission are exactly known.

On the basis of these assumptions, the survival probability of component l at the end of the $(k+1)$th mission is given by:

$$r_l(k+1) = \exp\left(-\int_0^{L_{k+1}} \lambda_{l,k+1}\left(U_{l,k+1}+s\right) ds\right) \cdot X_{l,k+1} = \frac{R_l\left(U_{l,k+1}+L_{k+1}\right)}{R_l\left(U_{l,k+1}\right)} \cdot X_{l,k+1}, \tag{1}$$

where $\lambda_{l,k+1}(t)$ is the failure intensity function of component l in the $(k+1)$ th mission; $U_{l,k+1}$ is the effective age of component l at the beginning of the $(k+1)$ th mission; $R_l(t)$ is the reliability function of component l. Consequently, selective maintenance can be formulated as an optimization problem as follows:

$$\begin{aligned} \max_{\mathbf{C}(k)} \quad & R(k+1) \\ \text{s.t.} \quad & C(k) \leq C_k \; , \\ & Z(k) \leq Z_k \end{aligned} \tag{2}$$

where $\mathbf{C}(k) = [C_1(k), C_2(k), \ldots, C_M(k)]$ is the vector of all the decision variables; $R(k+1)$ is the probability of the entire system successfully completing the $(k+1)$th mission, and it is a function of the survival probabilities of all the components. $C(k)$ and $Z(k)$ are the total maintenance cost and time of a particular maintenance strategy for the k th break, respectively. C_k is the limited maintenance budget for the k th break. A subset of components together with the corresponding maintenance actions are to be determined by the resulting optimization problem.

2.2 Uncertainties in Selective Maintenance

The basic selective maintenance model in Eq. (2) is the simplest one without taking account of various uncertainties. The decision-making of the selective maintenance optimization is affected by a set of factors, such as maintenance efficiency, maintenance cost, durations of missions and breaks, durations of maintenance actions, degradation parameters of components, and condition of components at the end of the last mission. In engineering practices, these factors inevitably suffer from uncertainties, producing non-ignorable impact on selective maintenance optimization. Prior to introducing extended selective maintenance models under uncertainties, we itemize the possible uncertainties in selective maintenance.

1. Uncertainty associated with maintenance efficiency. As maintenance actions could be imperfect, several imperfect maintenance models, such as the Kijima types I and II models (Liu and Huang 2010a), and the hybrid model (Pandey et al. 2013a) have been utilized to characterize the efficiency of maintenance actions (Pandey et al. 2013c). However, the age reduction coefficient in the Kijima models and/or the hazard adjustment coefficient in the hybrid model cannot be precisely evaluated due to the variations of the degree of the expertise of repairmen. Alternatively, in lieu of a constant value, the age reduction coefficient, taking any value in the range of [0, 1], is represented by a random variable with a probability density function (PDF) of $f_{B_{l,k}}\left(b_{l,k}\right)$ in Khatab and Aghezzaf (2016) and Duan et al. (2018)).
2. Uncertainty associated with maintenance cost. In the basic selective maintenance model, the constraint pertaining to the total maintenance cost and maintenance budget are deterministic. However, in real-world applications, the maintenance cost associated with each maintenance action may be uncertain, and thus, the total maintenance cost being not greater than the maintenance budget is a probabilistic constraint. In this regard, the probability of the total maintenance cost being not greater than C_k must be not less than a pre-specified probability p_0 (Ali et al. 2013).
3. Uncertainty associated with the durations of missions and breaks. The basic selective maintenance model assumes that the durations of both missions and breaks are deterministic and known. Such an assumption may not be valid as predicting the duration of a mission and break is difficult in many real-world

situations. For example, due to the randomness of weather condition, the arrival and departure of aircrafts cannot be exactly known in advance. The assumption of determinacy is released in Khatab et al. (2017a, b), and the durations of both missions and breaks are treated stochastically. The durations of the kth break and the $(k+1)$th mission are governed by PDFs $f_{Z_k}(t)$ $(Z_k \in [z_k^{\min}, z_k^{\max}])$ and $f_{L_{k+1}}(t)$ $(L_{k+1} \in [l_{k+1}^{\min}, l_{k+1}^{\max}])$, respectively.

4. Uncertainty associated with the performance capacities and state transition intensities of components. Due to insufficient data and unpredictable external working conditions, both the performance capacity and state transition intensities of components cannot be precisely known. In such a case, these quantities can be represented by fuzzy numbers rather than crisp values.
5. Uncertainty associated with the duration of each maintenance action. Similar to the uncertainty associated with maintenance efficiency, the duration of a maintenance action may also be uncertain due to the variations of the degree of the expertise of repairmen. Let a_l denote a maintenance action for component l and T_l represent the duration of the maintenance action. In this study, the duration of each maintenance action, i.e., T_l, is a random variable with PDF $f_{T_l}(t)$ $(T_l \in [t_l^{\min}, t_l^{\max}])$.
6. Uncertainty associated with observed condition of components. The condition of components can be indirectly reflected by the observations collected from various ways, such as vibration analysis and ultrasonic analysis. However, measurement errors, limited accuracy/precision of sensors or inspection instruments, poor diagnostic tools or algorithms, and non-rigorous interpretations varying from person to person are inevitable in engineering practices. Observations associated with the conditions of components oftentimes contain noise and uncertainty.

3 Selective Maintenance Models Under Uncertainties

In this section, six extended selective maintenance models under uncertainties are introduced. We will first review Models 1–4 reported in the literature and then propose two new selective maintenance models, i.e., Models 5 and 6. A brief summary of the features of the six models is presented in Table 1.

3.1 Model 1: Uncertainty Associated with Maintenance Efficiency

The uncertainty associated with maintenance efficiency was incorporated into selective maintenance in Khatab and Aghezzaf (2016) and Duan et al. (2018). If the Kijima type II model is used to characterize the efficiency of an imperfect

Table 1 Summary of the features of the six models

Model	Uncertainty type	Decision variables	Optimization objective(s)
1	Maintenance efficiency	Maintenance cost to each component	Probability of a system successfully completing the next mission
2	Maintenance cost	Maintenance cost to each component	Total maintenance cost
3	Durations of missions and breaks	Maintenance cost to each component	Probability of a system successfully completing the next mission
4	Performance and transition parameters of components	Maintenance cost to each component	Probability of a system successfully completing the next mission
5	Durations of breaks and maintenance actions	Maintenance cost to each component and the order of each selected maintenance action	Probability of a system successfully completing the next mission
6	Imperfect observations	Maintenance cost to each component	Expectation and variance of the probability of a system successfully completing the next mission

maintenance action, the survival probability of component l at the end of the $(k + 1)$th mission is formulated as (Khatab and Aghezzaf 2016; Duan et al. 2018):

$$r_l(k+1) = \left(\frac{\int_0^1 R_l\left(b_{l,k} V_{l,k} + L_{k+1}\right) f_{B_{l,k}}\left(b_{l,k}\right) db_{l,k}}{\int_0^1 R_l\left(b_{l,k} V_{l,k}\right) f_{B_{l,k}}\left(b_{l,k}\right) db_{l,k}} \right) \cdot X_{l,k+1}, \tag{3}$$

where $V_{l,\,k}$ is the effective age of component l at the end of the kth mission. The associated cost for each imperfect maintenance action has an inverse relation with the expected value of the age reduction coefficient.

The selective maintenance optimization model aims at minimizing the total maintenance cost to successfully complete the $(k + 1)$th mission with respect to a required reliability level $R_{k\,+\,1}$ along with a limited break duration Z_k, and it is given by (Khatab and Aghezzaf 2016; Duan et al. 2018):

$$\begin{aligned} \min_{\mathbf{C}(k)} \quad & C(k) \\ \text{s.t.} \quad & R(k+1) \geq R_{k+1} \cdot \\ & Z(k) \leq Z_k \end{aligned} \tag{4}$$

3.2 Model 2: Uncertainty Associated with Maintenance Cost

The uncertainty associated with maintenance cost was incorporated into selective maintenance in Ali et al. (2013) in which a stochastic programming method was developed when the maintenance cost of each maintenance action is treated as a random variable. Suppose that each subsystem can be maintained simultaneously by the respective maintenance team, in the extended selective maintenance model, the maximum total maintenance time of all the subsystems must be less than or equal to the duration of a break. Therefore, the extended selective maintenance model can be formulated as (Ali et al. 2013):

$$\begin{array}{ll} \max\limits_{\mathbf{C}(k)} & R(k+1) \\ \text{s.t.} & \Pr\{C(k) \leq C_k\} \geq p_0 \\ & \max\{Z_i(k)\} \leq Z_k \end{array}, \tag{5}$$

where the probability of a system successfully completing the $(k + 1)$th mission is the objective to be maximized. The first constraint in Eq. (5) indicates that the probability of the total maintenance cost $C(k)$ being not less than maintenance budget C_k must be not less than a pre-specified probability p_0. $Z_i(k)$ is the total maintenance time to complete all the selected maintenance action for subsystem i. Alternatively, the selective maintenance model can be written as (Ali et al. 2013):

$$\begin{array}{ll} \min\limits_{\mathbf{C}(k)} & k_1 E[C(k)] + k_2\sqrt{Var[C(k)]} \\ \text{s.t.} & R(k+1) \geq R_{k+1} \\ & \max\{Z_i(k)\} \leq Z_k \end{array}, \tag{6}$$

where the total maintenance cost is to be minimized and the probability of the system successfully completing the $(k + 1)$th mission should reach a required reliability level R_{k+1}. $E[\cdot]$ and $Var[\cdot]$ are the expectation and variance of the total maintenance cost, respectively. k_1 and k_2, taking values from $[0, 1]$ and $k_1 + k_2 = 1$, indicate the relative importance between the expectation and variance of the total maintenance cost.

3.3 Model 3: Uncertainty Associated with the Durations of Missions and Breaks

The uncertainty associated with the durations of missions and breaks was incorporated into selective maintenance in Khatab et al. (2017a, b). If the Kijima type II model is used to characterize the efficiency of an imperfect maintenance action, the survival probability of component l at the end of the $(k + 1)$th mission is formulated as (Khatab et al. 2017a, b):

$$r_l(k+1) = \left(\frac{\int_{l_{k+1}^{\min}}^{l_{k+1}^{\max}} R_l\left(b_{l,k} V_{l,k} + l_{k+1}\right) f_{L_{k+1}}\left(l_{k+1}\right) dl_{k+1}}{R_l\left(b_{l,k} V_{l,k}\right)} \right) \cdot X_{l,k+1}. \tag{7}$$

The extended selective maintenance model, aiming at minimizing the total maintenance cost while satisfying a required reliability level and a probabilistic constraint of the total maintenance time, is formulated as (Khatab et al. 2017a, b):

$$\begin{array}{ll} \min\limits_{\mathbf{C}(k)} & C(k) \\ \text{s.t.} & R(k+1) \geq R_{k+1} \\ & \Pr\{Z(k) \leq Z_k\} \geq \tau_s \end{array}, \tag{8}$$

where τ_s can be interpreted as a required service ratio, that is, the probability of all the scheduled maintenance actions being completed within Z_k should be not less than the ratio. If the duration of each maintenance action is also stochastic and all the selected maintenance actions are executed sequentially, $Z(k)$ is equal to the summation of all the selected maintenance actions for a particular maintenance strategy (Liu et al. 2018; Khatab et al. 2017b).

3.4 Model 4: Uncertainty Associated with Performance Capacities and State Transition Intensities of Components

The uncertainty associated with performance capacities and state transition intensities of components was incorporated into selective maintenance in Cao et al. (2018b). In the context of multistate systems, triangular fuzzy numbers have been used to represent the imprecise performance capacity and state transition intensities of multistate components (Cao et al. 2018b). By using the fuzzy Markov model and the reliability evaluation approach proposed by Liu and Huang (2010b), the fuzzy probability of a system completing the next mission can be assessed. The corresponding selective maintenance model is formulated as (Cao et al. 2018b):

$$\begin{array}{l} \max\limits_{\mathbf{C}(k)} R_\alpha(k+1) \\ \text{s.t. } C(k) \leq C_k \end{array}, \tag{9}$$

where $R_\alpha(k+1)$ is the α-cut level set of the probability of a system successfully completing the $(k+1)$ th mission, and $\alpha \in [0, 1]$. If $\alpha = 1$, the optimization objective in Eq. (9) degenerates to the case of crisp values, whereas $\alpha = 0$ provides the maximum possible range of the probability of a system successfully completing the $(k+1)$th mission under fuzzy uncertainty.

3.5 Model 5: Uncertainty Associated with the Durations of Breaks and Maintenance Actions

In this subsection, an extended selective maintenance model with considering the uncertainty associated with the durations of breaks and maintenance actions is developed.

3.5.1 Stochastic Durations of Maintenance Actions and Breaks

Let a_l denote a maintenance action for component l and T_l represent the duration of the maintenance action. In this study, both the duration of each maintenance action, i.e., T_l, and the duration of the break between the kth and $(k+1)$th missions, i.e., Z_k, are random variables. The PDFs of T_l and Z_k are denoted by $f_{T_l}(t)$ ($T_l \in \left[t_l^{\min}, t_l^{\max}\right]$) and $f_{Z_k}(t)$ ($Z_k \in \left[z_k^{\min}, z_k^{\max}\right]$), respectively.

As both the durations of maintenance actions and breaks are uncertain, the sequence of selected maintenance actions is crucial to the completion of the next mission. The purpose of Model 5 is to identify an optimal maintenance sequence and explore the impact of the maintenance sequence on the success of the next mission. For simplicity, only replacement for each component is considered herein, and the replacement cost of component l is denoted as c_l. It is no doubt that the proposed method can be extended to a generalized case in which a variety of maintenance options are available to each component.

3.5.2 Sequence of Maintenance Actions

In this study, a selective maintenance strategy is a planned maintenance sequence in a break. To construct a sequence planning for all the maintenance actions, a binary decision variable, denoted as $H_{v,l}(k)$, is used to represent the order of a maintenance action a_l for component l in the kth break. Thus, one has:

$$H_{v,l}(k) = \begin{cases} 1 \text{ if the } v\text{th maintenace action is performed on component } l \\ 0 \text{ otherwise} \end{cases},$$

where $v \in \{1, 2, \ldots, M\}$ denotes the index of the selected maintenance action a_l in a maintenance sequence. It is worth noting that, in a break, at most one maintenance action can be selected for each component due to the limited maintenance resources, i.e., $\sum_{v=1}^{M} H_{v,l}(k) \le 1$ for any $l \in \{1, 2, \ldots, M\}$. On the other hand, each selected maintenance action can only be placed on a particular position of a maintenance sequence, i.e., $\sum_{l=1}^{M} H_{v,l}(k) \le 1$ for any $v \in \{1, 2, \ldots, M\}$.

As a result, the total maintenance cost $C(k)$ of a particular maintenance sequence for the kth break is equal to:

$$C(k) = \sum_{v=1}^{M} \sum_{l=1}^{M} H_{v,l}(k) \cdot c_l. \tag{10}$$

3.5.3 The Probability Distribution of the Number of Completed Maintenance Actions

As all the selected maintenance actions are conducted sequentially, the number of maintenance actions that can be completed in a break is random inherently. Let T_v^{s} denote the stochastic duration of the vth ($v \in \{1, 2, \ldots, M\}$) maintenance action in a particular maintenance sequence, and one has $T_v^{\mathrm{s}} = \sum_{l=1}^{M} H_{v,l}(k) \cdot T_l$. The PDF of T_v^{s}, denoted as $g_v(t)$, can be written as:

$$g_v(t) = \sum_{l=1}^{M} H_{v,l}(k) \cdot f_{T_l}(t). \tag{11}$$

The corresponding cumulative duration of completing the first m ($m \in \{1, 2, \ldots, M\}$) maintenance actions in a particular maintenance sequence, denoted as T_m^{c}, can be formulated as:

$$T_m^{\mathrm{c}} = \sum_{i=1}^{m} T_i^{\mathrm{s}} = \sum_{v=1}^{m} \sum_{l=1}^{M} H_{v,l}(k) \cdot T_l. \tag{12}$$

Given the PDF of T_v^{s}, i.e., $g_v(t)$, the cumulative distribution function (CDF) of T_m^{c}, denoted as $F_{T_m^{\mathrm{c}}}(t)$ ($m \in \{1, 2, \ldots, M\}$), can be derived via a multidimensional convolution as follows:

$$F_{T_m^{\mathrm{c}}}(t) = \int_0^t \left[\int_0^{t_m^{\mathrm{c}}} g_m\left(t_m^{\mathrm{c}} - t_{m-1}^{\mathrm{c}}\right) \cdots \int_0^{t_2^{\mathrm{c}}} g_2\left(t_2^{\mathrm{c}} - t_1^{\mathrm{c}}\right) g_1\left(t_1^{\mathrm{c}}\right) dt_1^{\mathrm{c}} \ldots dt_{m-1}^{\mathrm{c}} \right] dt_m^{\mathrm{c}}. \tag{13}$$

Let $N_k(t)$ denote the number of completed maintenance actions at time t in the kth break. The probability of the first $m(m \in \{1, 2, \ldots, M\})$ maintenance actions being completed in the kth break, denoted as $P_{k,m}$, is given by:

$$\begin{aligned} P_{k,m} = \Pr\{N_k(Z_k) = m\} &= \int_{z_k^{\min}}^{z_k^{\max}} \Pr\{N_k(Z_k) = m | Z_k = t\} f_{Z_k}(t) dt \\ &= \int_{z_k^{\min}}^{z_k^{\max}} \left(F_{T_m^{\mathrm{c}}}(t) - F_{T_{m+1}^{\mathrm{c}}}(t) \right) f_{Z_k}(t) dt. \end{aligned} \tag{14}$$

3.5.4 The Proposed Selective Maintenance Model

Based on Eq. (14), by taking account of all the possible m maintenance actions, the probability of the system successfully completing the $(k+1)$th mission can be formulated as:

$$\begin{aligned} R(k+1) &= \sum_{m=0}^{M} R(k+1|N_k(Z_k)=m)\cdot \Pr\{N_k(Z_k)=m\} \\ &= \sum_{m=0}^{M} R(k+1|N_k(Z_k)=m)\cdot P_{k,m}. \end{aligned} \tag{15}$$

where $R(k+1|\, N_k(Z_k)=m)$ is the conditional reliability of the system successfully completing the $(k+1)$th mission under the condition that the first m components in the maintenance sequence can be completely repaired. Given the states of all the components at the end of the kth mission and the maintenance action for each component, our specific selection maintenance problem can be formulated as:

$$\begin{aligned} \max_{\mathbf{H}(k)} \quad & R(k+1) \\ \text{s.t.} \quad & C(k) \le C_k \\ & \textstyle\sum_{l=1}^{M} H_{v,l}(k) \le 1 \;, \\ & \textstyle\sum_{v=1}^{M} H_{v,l}(k) \le 1 \\ & H_{v,l}(k) = 1 \;\text{ or }\; 0 \end{aligned} \tag{16}$$

where $\mathbf{H}(k)=[H_{1,1}(k), H_{1,2}(k), \ldots, H_{1,M}(k); H_{2,1}(k), H_{2,2}(k), \ldots, H_{2,M}(k); \ldots; H_{M,1}(k), H_{M,2}(k), \ldots, H_{M,M}(k)]$ is the matrix of all the decision variables. The first constraint is the budget C_k of the kth break. The basic selective maintenance model in Eq. (2) only identifies the maintenance actions for all the components. However, as both the durations of maintenance actions and breaks are uncertain, the sequence of selected maintenance actions is crucial to the next mission completion. The proposed selective maintenance model in Eq. (16) not only identifies the maintenance actions for all the components but also determines the sequence of selected maintenance actions.

3.5.5 Illustrative Example 1

The illustrative example is a simple four-component system, as shown in Fig. 2. Suppose that at the end of the last mission, i.e., the kth mission, all the components are failed. The duration of the $(k+1)$th mission is set to be 0.2 months. The distributions of the failure time of all the components can be characterized by exponential distribution, and the failure rate of each component during the $(k+1)$th mission is given in Table 2. The maintenance action of each component and the associated costs and the stochastic durations are tabulated in Table 2. The stochasticity of the durations of maintenance actions is quantified by gamma

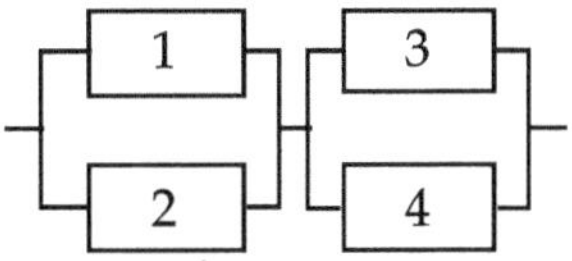

Fig. 2 A four-component system

Table 2 Parameters of components (unit of failure rate: month^{-1}, unit of cost: US $1000)

ID (l)	$\lambda_{l,\,k+1}(t)$	a_l	c_l	Gamma distribution	
				Shape	Scale
1	1.0	a_1	5.0	20	0.2
2	3.0	a_2	3.0	10	
3	1.0	a_3	10.0	30	
4	2.0	a_4	7.0	25	

Table 3 Results of Strategy 1

ID (m)	$H_{m,\,l}(k)$	$P_{k,\,m}$	$R(k+1 \mid N_k = m)$
1	$H_{1,\,1}(k)$	0.0000	0
2	$H_{2,\,3}(k)$	0.2056	0
3	$H_{3,\,3}(k)$	0.4891	0.7518
4	$H_{4,\,4}(k)$	0.3052	0.8633
$R(k+1)$	–	0.6419	–

Table 4 Results of Strategy 2

ID (m)	$H_{m,\,l}(k)$	$P_{k,\,m}$	$R(k+1 \mid N_k = m)$
1	$H_{1,\,4}(k)$	0.1206	0
2	$H_{2,\,3}(k)$	0.1805	0
3	$H_{3,\,2}(k)$	0.3937	0.5160
4	$H_{4,\,1}(k)$	0.3052	0.8633
$R(k+1)$	–	0.4666	–

distributions with the shape and scale parameters listed in Table 2. The break between the kth and the $(k + 1)$th missions is stochastic and characterized by a uniform distribution in the range of [10,20] hours. The maintenance budget is 25×1000 US dollars.

To demonstrate the impact of the sequence of maintenance actions on the probability of the system successfully completing the $(k + 1)$th mission, we examined the case where all the components are to be replaced sequentially in the break. Table 3 gives the results for Strategy 1 of $[H_{1,\,1}(k) = 1,\ H_{2,\,2}(k) = 1,\ H_{3,\,3}(k) = 1,\ H_{4,\,4}(k) = 1]$, that is, Components 1, 2, 3, and 4 are replaced in order. Likewise, the results of Strategy 2, that is, $[H_{1,\,4}(k) = 1,\ H_{1,\,3}(k) = 1,\ H_{3,\,2}(k) = 1,\ H_{4,\,1}(k) = 1]$, are given in Table 4. As observed from Table 4, it is found that exchanging the order of replacing components can yield a different value of $R(k + 1)$.

Moreover, we enumerated all the feasible solutions of the maintenance actions and found that the optimal sequence of the maintenance actions is $[H_{1,\,3}(k) = 1,\ H_{2,\,1}(k) = 1,\ H_{3,\,2}(k) = 1,\ H_{4,\,4}(k) = 1]$, that is, to replace Component 3 first and then replace Components 1, 2, and 4 in order. With this strategy, the maximum probability of the system successfully completing the $(k + 1)$th mission, i.e., $R(k + 1)$, is 0.7313.

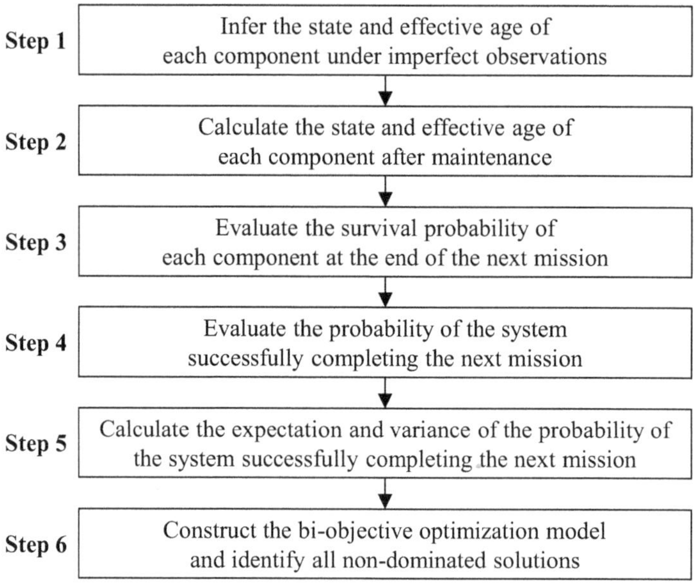

Fig. 3 Flowchart of the proposed robust selective maintenance method

3.6 Model 6: Uncertainty Associated with Imperfect Observations

As a selective maintenance strategy is determined based on the condition of all the components, the imperfection of observations can affect the selective maintenance decision-making. In this subsection, an extended selective maintenance model with considering the uncertainty associated with imperfect observations is developed. The proposed robust selective maintenance method contains six steps, as shown in Fig. 3. In Step 1, the state and effective age of a particular component under imperfect observations are inferred. The state and effective age of a component after executing the maintenance are computed in Step 2. After evaluating the survival probability of a component (Step 3) and the probability of a repaired system successfully completing the next mission (Step 4), the expectation and variance of the probability of the system successfully completing the next mission are obtained in Step 5. As a result, in Step 6, a bi-objective optimization model is formulated aiming at identifying a robust selective maintenance strategy with a high expectation and a small uncertainty. The procedures of formulizing the optimization model are detailed in the ensuing section.

3.6.1 State and Effective Age Under Imperfect Observations

At the end of the kth mission, the observed state of component l is represented by a binary variable, denoted as $Y_{i,k}^{\mathrm{o}}$, and

$$Y_{l,k}^{\mathrm{o}} = \begin{cases} 1 & \text{if component } l \text{ is observed in the functioning state} \\ 0 & \text{if component } l \text{ is observed in the failure state} \end{cases}.$$

In this study, the observation matrix of component l is adopted to quantify the stochastic relation between the observed values and true values of states, and it is defined as:

$$\mathbf{O}_l = \begin{bmatrix} 1-\delta_l^{\mathrm{I}} & \delta_l^{\mathrm{I}} \\ \delta_l^{\mathrm{II}} & 1-\delta_l^{\mathrm{II}} \end{bmatrix},$$

where δ_l^{I} represents the probability of component l being observed in the failure state when the component is still functioning and δ_l^{II} vice versa.

Based on Eq. (1), the survival probability of component l at the end of the kth mission can be obtained. The prior state probability of component l at the end of the kth mission, denoted as $\Pr\{Y_{l,\,k}\}$, is derived as:

$$\Pr\left\{Y_{l,k}\right\} = \begin{cases} r_l(k) & Y_{l,k} = 1 \\ 1 - r_l(k) & Y_{l,k} = 0 \end{cases}. \tag{17}$$

Therefore, by the Bayes rule, the posterior state distribution of component l under the imperfect observation $Y_{l,k}^{\mathrm{o}}$ can be formulated as:

$$\Pr\left\{Y_{l,k}|Y_{l,k}^{\mathrm{o}}\right\} = \frac{\Pr\left\{Y_{l,k}, Y_{l,k}^{\mathrm{o}}\right\}}{\Pr\left\{Y_{l,k}^{\mathrm{o}}\right\}} = \frac{\Pr\left\{Y_{l,k}^{\mathrm{o}}|Y_{l,k}\right\}\Pr\left\{Y_{l,k}\right\}}{\sum_{Y_{l,k}}\Pr\left\{Y_{l,k}^{\mathrm{o}}|Y_{l,k}\right\}\Pr\left\{Y_{l,k}\right\}}. \tag{18}$$

where $\Pr\left\{Y_{l,k}^{\mathrm{o}}|Y_{l,k}\right\}$ is the conditional state probability that can be obtained by the observation matrix.

The true value and observed value of the operating time of component l in the kth mission are denoted as $T_{l,k}^{\mathrm{w}}$ ($T_{l,k}^{\mathrm{w}} \in [0, L_k]$) and $T_{l,k}^{\mathrm{w,o}}$, respectively. The observed value of the effective age of component l at the end of the kth mission is denoted as $V_{l,k}^{\mathrm{o}}$. Based on the observed and true values of the state, the relations between the observed value $V_{l,k}^{\mathrm{o}}$ and true value $V_{l,\,k}$ of the effective age of component l at the end of kth mission can be categorized into four cases as tabulated in Table 5. The probabilities of component l being in the four cases can be obtained by Eq. (18).

In Cases 1 and 2 of Table 5, component l remains functioning at the end of the kth mission; therefore, the true value of the effective age of at the end of the kth mission is a certain value, i.e., $V_{l,\,k} = U_{l,\,k} + L_k$.

Table 5 Relations between $V_{l,k}$ and $V_{l,k}^{\mathrm{o}}$

Case	$Y_{l,k}^{\mathrm{o}}$	$T_{l,k}^{\mathrm{w,o}}$	$V_{l,k}^{\mathrm{o}}$	$Y_{l,k}$	$V_{l,k}$
1	1	L_k	$U_{l,k}+L_k$	1	$U_{l,k}+L_k$
2	0	$t_{l,k}^{\mathrm{w,o}}$	$U_{l,k}+t_{l,k}^{\mathrm{w,o}}$	1	$U_{l,k}+L_k$
3	1	L_k	$U_{l,k}+L_k$	0	$U_{l,k}+T_{l,k}^{\mathrm{w}}$
4	0	$t_{l,k}^{\mathrm{w,o}}$	$U_{l,k}+t_{l,k}^{\mathrm{w,o}}$	0	$f_{V_{i,k}\mid V_{i,k}^{\mathrm{o}}}\left(v_{i,k}\mid v_{i,k}^{\mathrm{o}}\right)$

In Case 3 of Table 5, component l is in the failure state at the end of the kth mission, whereas the component is observed in the functioning state. In this case, there is a conflict between the observed state and true state, and the true value of the effective age $V_{l,k}$ is the sum of $U_{l,k}$ and $T_{l,k}^{\mathrm{w}}$. The (truncated) PDF of $T_{l,k}^{\mathrm{w}}$ can be derived as:

$$f_{T_{l,k}^{\mathrm{w}}}(t)=-\frac{1}{1-r_l(k)}\frac{1}{R_l\left(U_{l,k}\right)}\frac{dR_l\left(U_{l,k}+t\right)}{dt}. \tag{19}$$

Consequently, the PDF of $V_{l,k}$ is given by $f_{V_{l,k}}(t)=f_{T_{l,k}^{\mathrm{w}}}\left(t-U_{l,k}\right)$.

In Case 4 of Table 5, component l is in the failure state at the end of the kth mission, meanwhile the component is observed in the failure state. In this case, a conditional PDF, denoted by $f_{T_{l,k}^{\mathrm{w,o}}\mid T_{l,k}^{\mathrm{w}}}\left(t_{l,k}^{\mathrm{w,o}}\mid t_{l,k}^{\mathrm{w}}\right)$, is introduced to quantify the difference between the observed and true values of the operating time. Specifically, the truncated normal distribution with mean $t_{l,k}^{\mathrm{w}}$ and variance σ_l^2 within the interval $[0,L_k]$ is used to characterize the stochastic relation between the observed and true values of the operating time. The variance σ_l^2 is used to quantify the uncertainty of the operating time. Therefore, the conditional PDF of $V_{l,k}^{\mathrm{o}}$, under the condition that $V_{l,k}=v_{l,k}$, is given by $f_{V_{l,k}^{\mathrm{o}}\mid,V_{l,k}}\left(v_{l,k}^{\mathrm{o}}\mid v_{l,k}\right)=f_{T_{l,k}^{\mathrm{w,o}}\mid T_{l,k}^{\mathrm{w}}}\left(v_{l,k}^{\mathrm{o}}-U_{l,k}\mid t_{l,k}^{\mathrm{w}}\right)$. Consequently, the posterior PDF of the effective age, under the condition that $V_{l,k}^{\mathrm{o}}=v_{l,k}^{\mathrm{o}}$, can be calculated by using the Bayes formula as follows:

$$f_{V_{l,k}\mid V_{l,k}^{\mathrm{o}}}\left(v_{l,k}\mid v_{l,k}^{\mathrm{o}}\right)=\frac{f_{V_{l,k}^{\mathrm{o}}\mid V_{l,k}}\left(v_{l,k}^{\mathrm{o}}\mid v_{l,k}\right)f_{V_{l,k}}\left(v_{l,k}\right)}{\int_{V_{l,k}}f_{V_{l,k}^{\mathrm{o}}\mid V_{l,k}}\left(v_{l,k}^{\mathrm{o}}\mid v_{l,k}\right)f_{V_{l,k}}\left(v_{l,k}\right)dv_{l,k}}. \tag{20}$$

3.6.2 State and Effective Age of a Component After Maintenance

The state and effective age of a component at the beginning of the $(k+1)$th mission are only affected by the state and effective age of the component at the end of the k th mission as well as the maintenance action performed on the component. Therefore, the state and effective age of component l after the kth break fall into one of the following two cases.

Case 1: $Y_{l,k} = 1$

In this case, if no maintenance cost is apportioned to component l in the kth break, the effective age of the component at the beginning of the $(k + 1)$th mission is then left unchanged and one has $U_{l,k+1} = U_{l,k} + L_k$. If component l is replaced by a new one, then $U_{l,k+1}$ is set to be zero. Moreover, if an imperfect maintenance is performed on component l, based on the Kijima type II model, one has $U_{l,k+1} = b_{l,k}(U_{l,k} + L_k)$.

Case 2: $Y_{l,k} = 0$

If component l fails before the end of the kth mission and will not be repaired in the kth break, the component remains in the failure state and the effective age is left unchanged. Likewise, the effective age is set to be zero with a perfect maintenance. If the component is imperfectly maintained, based on the Kijima type II model, the effective age changes to be $U_{l,k+1} = b_{l,k}V_{l,k}$.

3.6.3 Probability of a System Successfully Completing the Next Mission

Based on Eq. (1), the conditional survival probability of component l at the end of the $(k + 1)$th mission, under the condition that $U_{l,k+1} = u_{l,k+1}$, is given by:

$$r_l\left(k+1|U_{l,k+1}=u_{l,k+1}\right)=R_l\left(u_{l,k+1}+L_{k+1}\right)/R_l\left(u_{l,k+1}\right)\cdot X_{l,k+1}. \quad (21)$$

Consequently, based on the system structure function and the conditional survival probability of each component, the conditional probability of the entire system successfully completing the $(k + 1)$th mission, under the condition that $\mathbf{U}_{k+1} = \mathbf{u}_{k+1} = \{u_{1,k+1}, u_{2,k+1}, \ldots, u_{M,k+1}\}$, denoted as $R_s(k + 1| \mathbf{U}_{k+1} = \mathbf{u}_{k+1})$, can be evaluated readily. Therefore, the expectation and variance of the probability of the entire system successfully completing the $(k + 1)$th mission are, respectively, formulated as:

$$\begin{aligned} m_s(k+1) &= E\left[R_s\left(k+1|\mathbf{U}_{k+1}=\mathbf{u}_{k+1}\right)\right] \\ &= \int_{U_{1,k+1}}\int_{U_{2,k+1}}\cdots\int_{U_{M,k+1}} R_s\left(k+1|\mathbf{U}_{k+1}=\mathbf{u}_{k+1}\right)du_{1,k+1}du_{2,k+1}\ldots du_{M,k+1}, \end{aligned} \quad (22)$$

$$D_s(k+1)=E\left[R_s(k+1|\mathbf{U}_{k+1}=\mathbf{u}_{k+1})^2\right]-m_s(k+1)^2. \quad (23)$$

3.6.4 Robust Selective Maintenance Model

Based on Eqs. (22) and (23), a robust selective maintenance model, aiming at maximizing the expectation and minimizing the variance of the probability of a repairable system successfully completing the $(k + 1)$th mission, can be formulated as:

$$\begin{array}{ll} \max\limits_{\mathbf{C}(k)} & [m_s(k+1), \ -D_s(k+1)] \\ \text{s.t.} & \sum_{l=1}^{M} C_l(k) \le C_k \\ & C_l(k) = c_l^0 + c_l(k) \\ & c_l(k) \le c_l^{\text{rp}} \quad l \in \{1, 2, \ldots, M\} \\ & c_l(k) \ge 0 \quad l \in \{1, 2, \ldots, M\} \end{array} \tag{24}$$

where $\mathbf{C}(k) = [C_1(k), C_2(k), \ldots, C_M(k)]$ is the vector of all the decision variables; c_l^{rp} is the replacement cost of component l; c_l^0 is the fixed maintenance cost; and $c_l(k)$ is the variable corrective/preventive maintenance cost for component l in the kth break. The basic selective maintenance model is essentially a single-objective optimization problem. However, due to the uncertainty associated with imperfect observations, the probability of a system successfully completing the next mission becomes uncertain. In this case, a maintenance strategy with maximum expectation may not be a credible decision if the result contains a huge uncertainty. Therefore, unlike the aforementioned works that treated the expectation of the probability of a system successfully completing the next mission as the objective to be maximized, the new robust selective maintenance model herein is formulated as a bi-objective optimization problem.

3.6.5 Illustrative Example 2

The four-component system, shown in Fig. 2, is exemplified here to demonstrate the effectiveness of the proposed robust selective maintenance method. The failure time of each component complies with the Weibull distribution, and the parameters of each component, e.g., scale parameter θ_l and shape parameter β_l of the Weibull distribution, maintenance costs, the effective age at the beginning of the kth mission, and the imperfectly observed values of the state and operating time, are tabulated in Table 6. The observed value of the effective age of each component at the end of the $(k + 1)$th mission is also tabulated in Table 6.

Suppose that the durations of the kth and $(k + 1)$th missions both take value of $Z_k = Z_{k+1} = 10$ days. The observation errors are assumed to be $\delta_l^{\text{I}} = \delta_l^{\text{II}} = 0.2$ and $\sigma_l = 2$ $(l \in \{1, 2, 3, 4\})$. The enumeration method is used to determine the Pareto optimal set. Given the maintenance budget $C_k = 70 \times 1000$ US dollars, four non-dominated solutions can be found in the feasible domain, and the Pareto optimal

Table 6 Parameters of components (unit of cost: US $1000, unit of time: day)

ID	θ_l	β_l	c_l^0	$c_l^{\min}$	c_l^{rp}	$Y_{l,k}^{\text{o}}$	$U_{l,k}$	$T_{l,k}^{\text{w,o}}$	$V_{l,k}^{\text{o}}$
1	22	2.4	1.8	1.8	35	0	3	6.0	9.0
2	24	2.2	1.8	2.0	38	1	3	10.0	13.0
3	25	1.9	2.2	2.4	44	1	4	10.0	14.0
4	28	1.6	2.0	2.0	42	0	6	7.5	13.5

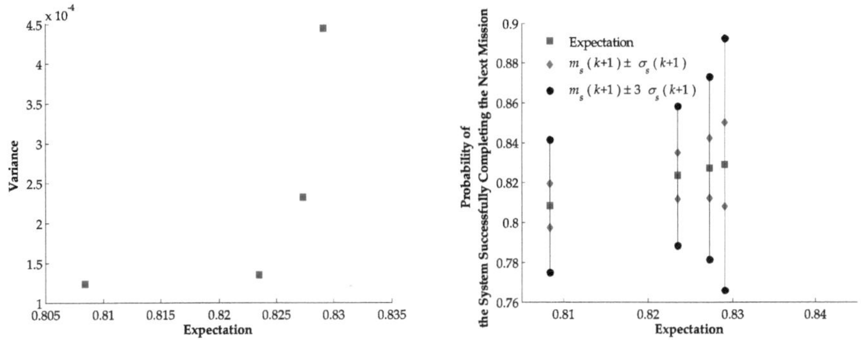

(a) Pareto optimal set

(b) Confidence intervals

Fig. 4 Four non-dominated solutions of the illustrative case

set of all non-dominated solutions is depicted in Fig. 4a. The best-compromised strategy can be chosen based on the following four alternative criteria:

Criterion 1: The non-dominated solution with the maximum value of $m_s(k+1)$
Criterion 2: The non-dominated solution with the minimum value of $D_s(k+1)$ or $\sigma_s(k+1)$
Criterion 3: The non-dominated solution with the maximum value of $m_s(k+1) - \sigma_s(k+1)$
Criterion 4: The non-dominated solution with the maximum value of $m_s(k+1) - 3\sigma_s(k+1)$

$\sigma_s(k+1)$ is the square root of $D_s(k+1)$. For each non-dominated solution, $m_s(k+1)$, $D_s(k+1)$, the confidence intervals of $m_s(k+1) \pm \sigma_s(k+1)$ and $m_s(k+1) \pm 3\sigma_s(k+1)$, and the actual total maintenance cost $C(k)$ are listed in Table 7. Meanwhile, the confidence intervals of each non-dominated solutions are depicted in Fig. 4b. As a result, the best-compromised strategies under each criterion (highlighted in Table 7) can be determined. For instance, because the second non-dominated solution takes the maximum value of $m_s(k+1) - 3\sigma_s(k+1)$ in Table 7, the best-compromised maintenance costs for Components 1, 2, 3, and 4 under Criterion 4 are 36.8, 3.8, 23.1, and 4.0 × 1000 US dollars, respectively.

In general, the criterion for choosing the best-compromised strategy is essentially application-dependent. The best-compromised strategy may vary from case to case. For concreteness, a power supply system in a public hospital is extremely risk-averse; therefore, decision-makers intend to select a non-dominated maintenance strategy under Criterion 3 or 4 to maximize the lower bound of the probability of the system successfully completing the next mission as an unexpected black out can incur a huge risk to patients. However, the same power supply system in a school can be risk-neutral to some extent, and thus decision-makers may choose the best-compromised solution based on Criterion 1 to achieve a maximum expected value. It is noteworthy that the best-compromised solutions under Criteria 1 and 2

Table 7 Four non-dominated solutions (unit of cost: US $1000)

ID	Maintenance costs $\mathbf{C}(k)$	Criterion 1	2	3	4	$C(k)$
1	(36.8, 23.8, 4.6, 4.0)	0.8084	**[1.237 × 10^{-4}]**	[0.7973, 0.8195]	[0.7751, 0.8418]	69.2
2	(36.8, 3.8, 23.1, 4.0)	0.8235	1.352 × 10^{-4}	[0.8119, 0.8351]	[**0.7886**, 0.8584]	67.7
3	(36.8, 3.8, 18.5, 8.4)	0.8273	2.329 × 10^{-4}	[**0.8121**, 0.8426]	[0.7815, 0.8731]	67.5
4	(36.8, 0.0, 23.1, 8.4)	**0.8291**	4.445 × 10^{-4}	[0.8080, 0.8502]	[0.7658, 0.8923]	68.3

are exactly equivalent to the optimal solutions by maximizing the expectation only (single objective) and minimizing the variance only (single objective), respectively.

4 Conclusions and Discussions

In this chapter, we have reviewed four existing selective maintenance models which take account of several potential uncertainties in selective maintenance. In addition, two new selective maintenance models under uncertainties were proposed. The first proposed model, i.e., Model 5, is able to address the uncertainty associated with the durations of breaks and maintenance actions. The second proposed model, i.e., Model 6, can cope with the uncertainty associated with imperfect observations. As demonstrated in two illustrative examples, the two proposed models can manage these uncertainties appropriately in decision-making.

It is worth noting that, in selective maintenance optimization, if there is one (or more than one) type of stochastic uncertainty that can affect the survival functions of components in the next mission, such as the uncertainties associated with maintenance efficiency, durations of missions and breaks, failure law, and observations, the probability of a system successfully completing the next mission is a random quantity rather than a constant value. Such variation can be quantified either by mean and variance as Model 6 for simplification or by a full probability distribution/confidence interval. The expectation of the mission success as used in most of the existing studies may not be a wise choice for selective maintenance optimization under uncertainty.

Acknowledgments The authors greatly acknowledge grant support from the National Natural Science Foundation of China under contract number 71771039.

References

Ali I, Raghav YS, Khan MF, Bari A (2013) Selective maintenance in system reliability with random costs of repairing and replacing the components. Commun Stat Simul Comput 42:2026–2039
Cao W, Jia X, Hu Q, Zhao J, Wu Y (2018a) A literature review on selective maintenance for multi-unit systems. Qual Reliab Eng Int 34:824–845
Cao W, Jia X, Liu Y, Hu Q (2018b) Selective maintenance optimization for fuzzy multi-state systems. J Intell Fuzzy Syst 34:105–121
Cassady CR, Pohl EA, Murdock WP (2001a) Selective maintenance modeling for industrial systems. J Qual Maint Eng 7:104–117
Cassady CR, Murdock WP, Pohl EA (2001b) Selective maintenance for support equipment involving multiple maintenance actions. Eur J Oper Res 129:252–258
Dao CD, Zuo MJ (2016) Selective maintenance for multistate series systems with S-dependent components. IEEE Trans Reliab 65:525–539
Dao CD, Zuo MJ (2017) Optimal selective maintenance for multi-state systems in variable loading conditions. Reliab Eng Syst Saf 166:171–180
Dao CD, Zuo MJ, Pandey M (2014) Selective maintenance for multi-state series–parallel systems under economic dependence. Reliab Eng Syst Saf 121:240–249
Diallo C, Venkatadri U, Khatab A, Liu Z (2018) Optimal selective maintenance decisions for large serial k-out-of-n: G systems under imperfect maintenance. Reliab Eng Syst Saf 175:234–245
Duan C, Deng C, Gharaei A, Wu J, Wang B (2018) Selective maintenance scheduling under stochastic maintenance quality with multiple maintenance actions. Int J Prod Res 56:1–19
Iyoob IM, Cassady CR, Pohl EA (2006) Establishing maintenance resource levels using selective maintenance. Eng Econ 51:99–114
Khatab A, Aghezzaf EH (2016) Selective maintenance optimization when quality of imperfect maintenance actions are stochastic. Reliab Eng Syst Saf 150:182–189
Khatab A, Aghezzaf EH, Djelloul I, Sari Z (2017a) Selective maintenance optimization for systems operating missions and scheduled breaks with stochastic durations. J Manuf Syst 43:168–177
Khatab A, Aghezzaf ELH, Diallo C, Djelloul I (2017b) Selective maintenance optimisation for series-parallel systems alternating missions and scheduled breaks with stochastic durations. Int J Prod Res 55:3008–3024
Liu Y, Huang H (2010a) Optimal selective maintenance strategy for multi-state systems under imperfect maintenance. IEEE Trans Reliab 59:356–367
Liu Y, Huang HZ (2010b) Reliability assessment for fuzzy multi-state systems. Int J Syst Sci 41:365–379
Liu Y, Chen Y, Jiang T (2018) On sequence planning for selective maintenance of multi-state systems under stochastic maintenance durations. Eur J Oper Res 268:113–127
Maaroufi G, Chelbi A, Rezg N (2013) Optimal selective renewal policy for systems subject to propagated failures with global effect and failure isolation phenomena. Reliab Eng Syst Saf 114:61–70
Maillart LM, Cassady CR, Rainwater C, Schneider K (2009) Selective maintenance decision-making over extended planning horizons. IEEE Trans Reliab 58:462–469
Pandey M, Zuo MJ, Moghaddass R, Tiwari MK (2013a) Selective maintenance for binary systems under imperfect repair. Reliab Eng Syst Saf 113:42–51
Pandey M, Zuo MJ, Moghaddass R (2013b) Selective maintenance modeling for a multistate system with multistate components under imperfect maintenance. IIE Trans 45:1221–1234
Pandey M, Liu Y, Zuo MJ (2013c) Selective maintenance for complex systems considering imperfect maintenance efficiency, reliability modeling with applications: essays in honor of Professor Toshio Nakagawa on his 70th birthday, p 17
Pandey M, Zuo MJ, Moghaddass R (2016) Selective maintenance scheduling over a finite planning horizon. Proc Inst Mech Eng O J Risk Reliab 230:162–177
Pham H, Wang H (1996) Imperfect maintenance. Eur J Oper Res 94:425–438

Rice W, Cassady CR, Nachlas J. (1998) Optimal maintenance plans under limited maintenance time. Proceedings of the seventh industrial engineering research conference, pp 1–3

Schneider K, Richard Cassady C (2015) Evaluation and comparison of alternative fleet-level selective maintenance models. Reliab Eng Syst Saf 134:178–187

Wang H (2002) A survey of maintenance policies of deteriorating systems. Eur J Oper Res 139:469–489